Genome-Wide Association Studies
From Polymorphism to Personalized Medicine

Over the last 20 years, Genome-Wide Association Studies (GWAS) have revealed a great deal about the genetic basis of a wide range of complex diseases, and they will undoubtedly continue to have a broad impact as we move to an era of personalized medicine. This authoritative text, written by leaders and innovators from both academia and industry, covers the basic science as well as the clinical, biotechnological, and pharmaceutical potential of these methods.

With special emphasis given to highlighting pharmaco-genomics and population genomics studies using next-generation technology approaches, this is the first book devoted to combining association studies with single nucleotide polymorphisms, copy number variants, haplotypes and expressed quantitative trait loci. A reliable guide for newcomers to the field, as well as for experienced scientists, this is a unique resource for anyone interested in how the revolutionary power of genomics can be applied to solve problems in complex disease.

KRISHNARAO APPASANI is the Founder and Chief Executive Officer of GeneExpression Systems, a global conference-producing organization focusing on biomedical and physical sciences. He is an award-winning scientist and also the editor of *Epigenomics: From Chromatin Biology to Therapeutics* (2012), *MicroRNAs: From Basic Science to Disease Biology* (2007), and *RNA Interference: From Basic Science to Drug Development* (2005), all published by Cambridge University Press.

Genome-Wide Association Studies

FROM POLYMORPHISM TO PERSONALIZED MEDICINE

Edited by

Krishnarao Appasani
GeneExpression Systems, Inc.

Forewords by

Stephen W. Scherer
University of Toronto, Canada

and

Peter M. Visscher
University of Queensland, Australia

CAMBRIDGE
UNIVERSITY PRESS

CAMBRIDGE
UNIVERSITY PRESS

University Printing House, Cambridge CB2 8BS, United Kingdom

Cambridge University Press is part of the University of Cambridge.

It furthers the University's mission by disseminating knowledge in the pursuit of education, learning and research at the highest international levels of excellence.

www.cambridge.org
Information on this title: www.cambridge.org/9781107042766

First published 2016

Printed in the United States of America by Sheridan Books, Inc.

A catalogue record for this publication is available from the British Library

Library of Congress Cataloguing in Publication data
Genome-wide association studies : from polymorphism to personalized medicine / edited by Krishnarao Appasani ; forewords by Stephen W. Scherer and Peter M. Visscher.
p. ; cm.
Includes bibliographical references and index.
ISBN 978-1-107-04276-6 (hardback)
I. Appasani, Krishnarao, 1959–, editor.
[DNLM: 1. Genome – Wide Association Study. 2. Genetic Testing. 3. Genomics. QU 550]
RB155
616′.042–dc23
2015017921

ISBN 978-1-107-04276-6 Hardback

Dedicated to

My friend, ***Arthur Beck Pardee***
Emeritus Professor
Harvard University Medical School, USA

Contents

Contributors

Krishnarao Appasani
GeneExpression Systems, Inc.
Waltham, Massachusetts, USA

Raghu Appasani
University of Massachusetts Medical School
Worcester, Massachusetts, USA &
MINDS Foundation USA & India
Boston, Massachusetts, USA

Neha S. Bhise
Department of Pharmacotherapy and Translational Research
Center for Pharmacogenomics
College of Pharmacy, University of Florida
Gainesville, FL, USA &
Department of Experimental and Clinical Research
University of Minnesota
St. Paul, Minnesota, USA

Elena Bosch
Institut de Biologia Evolutiva (UPF-CSIC)
Departament de Ciències Experimentals i de la Salut
Universitat Pompeu Fabra
Parc de Recerca Biomèdica de Barcelona
Barcelona, Catalonia, Spain

Erwin Brosens
PhD student Pediatric Surgery & Clinical Genetics
Erasmus MC–Sophia Children's Hospital
GE Rotterdam, The Netherlands

Margit Burmeister
Molecular & Behavioral Neuroscience Institute
University of Michigan
Ann Arbor, Michigan, USA

Ferran Casals
Genomics Core Facility
Departament de Ciències Experimentals i de la Salut
Universitat Pompeu Fabra
Barcelona, Catalonia, Spain

Lata Chauhan
Department of Pharmacotherapy and Translational Research
Center for Pharmacogenomics
College of Pharmacy
University of Florida
Gainesville, Florida, USA

Mengjie Chen
Department of Biostatistics
School of Public Health
Yale University
New Haven, Connecticut, USA &
Presently at Department of Biostatistics and Genetics
University of North Carolina Chapel Hill, NC, USA

Dongjun Chung
Department of Biostatistics
School of Public Health
Yale University
New Haven, Connecticut, USA

Eske M. Derks
Department of Psychiatry
Amsterdam Medical Center
University of Amsterdam
Amsterdam, The Netherlands

Nevenka Dimitrova
Philips Research
Briarcliff Manor, New York, USA

Martin Eden
Manchester Centre for Health Economics
The University of Manchester
Manchester, UK

H. Eussen
Department of Clinical Genetics
Erasmus Medical Centre
GE Rotterdam, The Netherlands

Joel Gelernter
Department of Genetics
Yale University School of Medicine
New Haven, Connecticut, USA

James Hicks
Cold Spring Harbor Laboratory
Cold Spring Harbor, New York, USA

Takae Hirasawa
Dept. of Epigenetics Medicine
Graduate School of Medicine and Engineering
University of Yamanashi
Yamanashi, Japan

Randall C. Johnson
BSP CCR Genetics Core
Leidos Biomedical Research, Inc
Frederick National Lab
Frederick, Maryland, USA

René S. Kahn
Department of Psychiatry
Rudolf Magnus Institute of Neuroscience
University Medical Center
Utrecht, The Netherlands

Sitharthan Kamalakaran
Philips Research North America
Briarcliff Manor, New York, USA

Tuomas O. Kilpeläinen
The Novo Nordisk Foundation Center for Basic Metabolic Research
Section of Metabolic Genetics
Faculty of Health and Medical Sciences
University of Copenhagen
Copenhagen, Denmark

A. de Klein
Department of Clinical Genetics

Erasmus Medical Centre
GE Rotterdam, The Netherlands

Takeo Kubota
Dept. of Epigenetics Medicine
Graduate School of Medicine and Engineering
University of Yamanashi
Yamanashi, Japan

Chao-Qiang Lai
Jean Mayer USDA Human Nutrition Research Center on Aging
Tufts University
Boston, Massachusetts, USA

Jatinder K. Lamba
Department of Pharmacotherapy and Translational Research
Center for Pharmacogenomics
College of Pharmacy, University of Florida
Gainesville, Florida, USA

Yu-Chi Lee
Jean Mayer USDA Human Nutrition Research Center on Aging
Tufts University
Boston, Massachusetts, USA

Cong Li
Department of Biostatistics
School of Public Health
Yale University
New Haven, Connecticut, USA

Yiyi Ma
Nutrition and Genomics Laboratory
Jean Mayer USDA Human Nutrition Research Center on Aging
Tufts University
Boston, Massachusetts, USA &
Presently at Boston University School of Medicine
Boston, Massachusetts, USA

Urko M. Marigorta
Institut de Biología Evolutiva
Universitat Pompeu Fabra
Biomedical Research Park (PRBB)
Barcelona, Spain &
School of Biology

Georgia Institute of Technology
Atlanta, GA, USA

Sadeep Medhasi
Division of Pharmacogenomics and Personalized Medicine
Mahidol University Faculty of Medicine
Bangkok, Thailand &
Somdech Phra Debaratana Medical Center
Ramathibodi Hospital
Bangkok, Thailand

Barry Merriman
Thermo Fisher Scientific, Inc.
Carlsbad, California, USA &
Presently at Human Longevity, Inc. San Diego, California, USA

Kunio Miyake
Department of Epigenetics Medicine
Graduate School of Medicine and Engineering
University of Yamanashi
Yamanashi, Japan

Arcadi Navarro
Institute of Evolutionary Biology
Universitat Pompeu Fabra, Biomedical Research Park
Barcelona, Spain &
National Institute for Bioinformatics, Barcelona, Catalonia, Spain &
Centre for Genomic Regulation, Barcelona, Catalonia, Spain &
Institució Catalana de Recerca i Estudis Avançats (ICREA)
Barcelona, Catalonia, Spain

Dorien H. Nieman
Department of Psychiatry
Amsterdam Medical Center
University of Amsterdam
Amsterdam, The Netherlands

Sheikh Nizamuddin
CSIR-Centre for Cellular and Molecular Biology
Hyderabad, India

José M. Ordovás
Department of Epidemiology
Centro Nacional Investigaciones Cardiovasculares &
Instituto Madrileño de Estudios Avanzados en Alimentación
Madrid, Spain

Kouichi Ozaki
Laboratory for Cardiovascular Diseases
RIKEN Center for Integrative Medical Sciences
Yokohoma City, Kanagawa, Japan

Laurence D. Parnell
Jean Mayer USDA Human Nutrition Research Center on Aging
Tufts University
Boston, Massachusetts, USA

Katherine Payne
Manchester Center for Health Economics
The University of Manchester
Manchester, UK

Apichaya Puangpetch
Division of Pharmacogenomics and Personalized Medicine
Mahidol University Faculty of Medicine
Bangkok, Thailand &
Somdech Phra Debaratana Medical Center
Ramathibodi Hospital
Bangkok, Thailand

Juan Antonio Rodriguez
Institut de Biología Evolutiva
Universitat Pompeu Fabra
Biomedical Research Park (PRBB)
Barcelona, Spain &
National Institute for Bioinformatics (INB), Spain

Pål Sætrom
Department of Computer and Information Science
Department of Cancer Research and Molecular Medicine
Norwegian University of Science and Technology
Trondheim, Norway

Stephen W. Scherer
The Centre for Applied Genomics
The Hospital for Sick Children
University of Toronto
Toronto, Ontario, Canada

Caren E. Smith
Jean Mayer USDA Human Nutrition Research Center on Aging
Tufts University
Boston, Massachusetts, USA

K.G. Snoek
Department of Pediatric Surgery
Erasmus Medical Centre–Sophia Children's Hospital
GE Rotterdam, The Netherlands

Sven Stringer
Department of Psychiatry
Amsterdam Medical Center
University of Amsterdam &
Department of Psychiatry
Rudolf Magnus Institute of Neuroscience
University Medical Center
Utrecht, The Netherlands

Chonlaphat Sukasem
Division of Pharmacogenomics and Personalized Medicine
Mahidol University Faculty of Medicine
Bangkok, Thailand &
Somdech Phra Debaratana Medical Center
Ramathibodi Hospital
Bangkok, Thailand

Rakesh Tamang
CSIR-Centre for Cellular and Molecular Biology &
Osmania University, Hyderabad &
University of Calcutta
Kolkata, India

Toshihiro Tanaka
Laboratory of Cardiovascular Diseases
RIKEN, Yokohoma City, Kanagawa, Japan &
Department of Human Genetics and Disease Diversity
Graduate School of Medical and Dental Sciences
Tokyo Medical and Dental University
Tokyo, Japan

Man-Hung Eric Tang
Cold Spring Harbor Laboratory
Cold Spring Harbor, New York, USA &
Department of Oncology
Clinical Sciences
Lund University, Lund, Sweden

Kumarasamy Thangaraj
Evolutionary and Medical Genetics Laboratory

CSIR-Centre for Cellular and Molecular Biology
Hyderabad, India

Laurent F. Thomas
Department of Cancer Research and Molecular Medicine
Norwegian University of Science and Technology
Trondheim, Norway

D. Tibboel
Department of Pediatric Surgery
Erasmus Medical Center–Sophia Children's Hospital
GE Rotterdam, The Netherlands

Elisa M. Trucco
Department of Psychiatry
Molecular & Behavioral Neuroscience Institute
University of Michigan
Ann Arbor, Michigan, USA

Vinay Varadan
Case Comprehensive Cancer Center
Case Western Reserve University
Cleveland, USA

D. Veenma
Department of Pediatric Surgery & Clinical Genetics
Erasmus Medical Centre–Sophia Children's Hospital
GE Rotterdam, The Netherlands

Sandra Villafuerte
Department of Psychiatry
Molecular & Behavioral Neuroscience Institute
University of Michigan
Ann Arbor, Michigan, USA

Peter M. Visscher
Queensland Brain Institute
University of Queensland
St. Lucia, Brisbane, Queensland, Australia

Edwin Wang
Bioinformatics and Systems Biology
National Research Council Canada &
McGill University Center for Bioinformatics
Montreal, Canada

Cheryl A. Winkler
Senior Investigator
SAIC-National Cancer Institute, NIH
Frederick, Maryland, USA

Can Yang
Department of Mathematics
Hong Kong Baptist University
Kowloon Tong, Hong Kong

Meredith Yeager
Core Genotyping Facility
SAIC-National Cancer Institute, NIH
Gaithersburg, Maryland, USA

Naif Zaman
Bioinformatics and Systems Biology
National Research Council Canada &
McGill University Center for Bioinformatics
Montreal, Canada

Michael Q. Zhang
Cold Spring Harbor Laboratory
Cold Spring Harbor, New York, USA &
The University of Texas at Dallas
Richardson, Texas, USA &
Tsinghua University
Beijing, China

Hongyu Zhao
Department of Biostatistics
School of Public Health
Yale University
New Haven, Connecticut, USA

Foreword

Stephen W. Scherer

I am grateful to Krishnarao Appasani for making the Herculean effort to prepare this volume on the rapidly expanding fields of Genome-Wide Association Studies (GWAS) and personalized medicine, and for inviting me to offer a few of my own introductory statements.

The complexity of the human disease state has always been an area of human curiosity. Over the last decade, GWASs have enabled us to expand our understanding of complex diseases using genetic-based approaches. We now see GWAS as a technology platform that promises to help move us into the era of personalized medicine.

Genome-Wide Association Studies: From Polymorphism to Personalized Medicine edited by Dr. Appasani has assembled the contributing chapters into five main areas encompassing: an introduction to GWAS in medicine, GWAS in pharmacogenomics, different classes of genetic variants for GWAS, new technologies including next-generation sequencing, and population genetics. The component chapters will be highly valuable, not only to those who are experimentally active in these aspects of research, but also to those interested in potential drug discovery applications. The historical perspectives offered also bring forward a unique vantage point into ongoing and future research in this field.

I believe that this book will become a reliable guide for anyone attempting to understand the successes with GWAS, planning new experiments, as well as its potential for the advancement of medicine. We approach a time where advances in genome sequencing technologies will deliver the long-awaited $1,000 genome, which promises to enable capture of all classes of genetic variants in a single experiment empowering new and innovative future studies. As such, studying the content of this book will allow us to pause and reflect, of where the field has come from, and where it needs to now go.

August 05, 2015

Professor Stephen W. Scherer, PhD. FRSC
Director of The Centre for Applied Genomics and
Director, McLaughlin Centre
University of Toronto
Toronto, Ontario, Canada

Foreword

Peter M. Visscher

Almost 10 years of GWAS: a feast of discoveries

Genome-wide association studies (GWASs) are less than 10 years old but have revolutionized discoveries in human genetics. GWAS is an experimental design based upon association because it exploits the fact that genetic variants that are close together tend to be statistically correlated. Driven by advances in array technologies these genome surveys of genetic variation have led to the discovery of thousands of DNA variants that are associated with complex traits, including many diseases. They have also led to new insights in human evolution and population differences.

Despite its undeniable success in finding many replicable associations between genetic variants and complex traits, GWAS has been criticized for a number of supposed flaws and with ever-shifting goalposts. Initial criticism was that it wouldn't work at all. This was followed by critique that not enough variants were detected per trait. When tens to hundreds of variants were detected for a disease or trait, the criticism became that not enough of the genetic variation was explained by those variants. Then, when a significant chunk of variation was accounted for, the perceived problem was that there was no biological insight. And finally, when that critique was shown by empirical data to be unjustified, the supposed fault with GWAS has become that its findings have not yet been translated into the clinic. And that for an experimental design less than 10 years old! I have never fully understood the criticism and attribute it to a mixture of ignorance about what association studies are designed for and prejudice with respect to the nature of genetic variation. As emphasized in this book, GWAS has shown that there are many genes with small effects that contribute to risk of common disease. That is not a shortcoming of the design, but revealing the true state of nature.

The facts speak for themselves, as highlighted and summarized nicely in this book. For many diseases and quantitative traits, tens to hundreds of robust associations have been discovered, including for diseases such as schizophrenia for which no genes were identified before the advent of GWAS. But GWAS has gone well beyond the discovery of robust associations. It has informed us about the nature of trait variation – genetic variation underlying disease and quantitative traits is highly polygenic and a surprisingly large proportion of genetic variation is tagged by common SNPs. It had led to the discoveries of new biological pathways involved in disease and has led to new biological knowledge about how DNA variation can affect gene regulation. Although it is early days, there are also signs that GWAS findings will lead to new drug development and may contribute to the adoption of early intervention strategies by stratifying people according to their genetic risk.

This book contains chapters from many of the players who have contributed to GWAS discovery in the last decade. It provides a helpful overview of the many facets of genome-wide data, their analysis and the interpretation of discoveries. We are living in an era of high-throughput hypothesis generating science, and, as demonstrated in this book, GWAS is a testament to the variety of discoveries that can be made when taking a whole genome approach to genetics, combined with large experimental sample sizes.

July 20, 2015

Peter M. Visscher, PhD.
Professor and Chair of Quantitative Genetics
Queensland Brain Institute
University of Queensland
St. Lucia, Brisbane, Queensland, Australia

Preface

DNA neither cares nor knows. DNA just is. And we dance to its music
Richard Dawkins, English evolutionary biologist and science writer (1941–)

In 1994, more than two decades ago, Francis Collins (the present Director of National Institutes of Health) expressed his view that "finding genes is like trying to find a needle in a haystack." In the pre-genomic era (prior to the Human Genome Project), geneticists were primarily focused on family-based linkage studies that examined simple Mendelian disorders, hereditary diseases caused by the malfunction of a single gene. In those days, it was difficult to study complex diseases like cancer or diabetes, which are influenced by multiple genes and multiple environmental factors. After completion of the sequencing of the "human genome" in early 2001, the big buzz in biology switched to understanding the complex diseases. From that time onwards, scientific efforts became increasingly globalized and increasingly intimate collaboration between academia and industry developed. Advances in genomic technologies have contributed to the development and use of new tools such as Genome-Wide Association Studies. Genome-Wide Association Studies (GWAS) were designed to survey the role of common genetic variations in complex human diseases. It was thought that GWAS would have the advantage of not relying on prior knowledge of biological pathways compared with "candidate gene-based studies," and it was also expected that GWAS would have higher power and finer resolution to identify genetic variants in the genome.

Another global collaborative effort was initiated in 2003, with the "HapMap Project" being established to characterize the haplotype patterns in the human genome and subsequently to identify single nucleotide polymorphisms. In the past decade GWAS has been adopted to study the genomics of complex traits of several diseases ranging from leukemia, obesity, diabetes, and coronary heart diseases to neurological diseases. GWAS research has aided in the discovery of hundreds of common variants whose allele frequencies are statistically correlated with various illnesses and traits. However, these studies did not establish any significant biological or clinical relevance in terms of prognosis and/or treatment.

Genome-wide Association Studies: From Polymorphism to Personalized Medicine is intended for those in the biotechnology, genetics, genomics, pharmaco-genomics, and molecular medicine fields. There are a few books already available covering genomic structural variants or copy number variations (Kehrer-Sawatzki, H. and Cooper, D.N. (2009) *Copy Number Variation and Disease*, Basel, Switzerland; Karger A.G., Zeggini, E. and Morris, A. (2010) *Analysis of Complex Disease Association Studies*. New York, USA, Academic Press; Feuk, L. (2012) *Genomic Structural Variants*, New Jersey, USA, Springer Press; Gondro, C., van der Werf, J. and Hayes, B. (2013) *Genome-Wide Association Studies and Genomic Prediction*, New Jersey, USA, Springer). For example, Kehrer-Sawatzki and Cooper (2009) highlighted the copy number variations, whereas Zeggini and Morris (2010) nicely covered the importance of disease association studies. The book by Feuk (2012) emphasized the protocols involved with identifying the structural variants, whereas a recent book by Gondro *et al.* (2013) on GWAS focused mainly on statistical approaches. This present book differs, in that it is the first text completely devoted to combining association studies with single nucleotide polymorphisms (SNPs), copy number variations (CNVs), haplotypes, and expressed quantitative trait loci (eQTL). Special emphasis is placed on pharmaco-genomic and population genomics studies using next-generation technology approaches. This book also focuses on the use of association studies in the context of disease biology and personalized medicine. The goal is for this book to serve as a reference for graduate students, post-doctoral researchers, and teachers and as an explanatory analysis for executives and scientists in biotechnology and pharmaceutical companies. Our hope is that this volume will serve as a prologue to the field for both newcomers and those already active in the field. We have carefully chosen the chapters, written by experts in the field from both academia and industry, and have divided the chapters into appropriate sections to support the theme expressed in the subtitle of this book: *From Polymorphism to Personalized Medicine.*

Many people have contributed to making our involvement in this project possible. We thank our teachers for their excellent teaching, guidance, and mentorship, which has helped us to bring about this educational enterprise. We are extremely grateful to all of the contributors to this book, without whose commitment this book would not have been possible. Many people have had a hand in the preparation of this book. Each chapter has been passed back and forth between the authors for criticism and revision; hence, each chapter represents a joint contribution. We thank our readers, who have made the hours spent putting together this volume worthwhile. We are indebted to the staff of Cambridge University Press, and in particular Katrina Halliday for her generosity and efficiency throughout the editing of this book; she truly understands the urgency and need of this volume. We also extend our appreciations to Megan Waddington and Victoria Parrin for their excellent cooperation during the development of this volume. We want to thank Professor Stephen Scherer, a Canadian Geneticist, and one of the pioneers in the field of copy number variations, for his kindness in writing the Foreword to this book. We also want to thank Professor Peter Visscher, an Australian Quantitative Geneticist, and one of the pioneers in the field of genome association studies, for his thoughtfulness in writing the Foreword to this book. Last, but not least, we thank Shyamala Appasani for her understanding and support during the development of this interesting volume.

This book is the third joint project of father and son. A portion of the royalties will be contributed to the Dr. Appasani Foundation (a non-profit organization devoted to bringing social change through the education of youth in developing nations) and The MINDS Foundation (Mental Illness and Neurological Diseases), which is committed to taking a grassroots approach to providing high-quality mental healthcare in rural India.

Krishnarao Appasani
Raghu K. Appasani

Abbreviations

ABPA	approximate Bayesian polygenic analysis
ACCE	**a**nalytic validity, **c**linical validity, **c**linical utility and associated **e**thical, legal and social implications
ACTG	Adult AIDS Clinical Trials Group
AD	alcohol dependence
ADHD	attention deficit/hyperactivity disorder
ADME	absorption, distribution, metabolism, and elimination
ADR	adverse drug reaction
AFP	alphafetoprotein
AI	allelic imbalance
ALD	admixture linkage disequilibrium
ALPS	autoimmune lymphoproliferative syndrome
AMD	age-related macular degeneration
AML	acute myeloid leukemia
ANI	ancestral North Indian
ANTAC	asymptotically normal estimation with thresholding after adjusting covariates
APA	alternative polyadenylation
AR	allelic ratios
ART	assisted reproductive technologies
ARV	antiretroviral
ASD	autism spectrum disorder
ASI	ancestral South Indian
AUD	alcohol use disorder
BCH	Boston Children's Hospital
BDNF	brain-derived neurotrophic factor
BEDMR	breakpoint-enriched differentially methylated region
BER	breakpoint-enriched region
BGA	biogeographical ancestry
BGI	Beijing Genome Institute

BLUP	best linear unbiased prediction
BPD	bipolar disorder
CAD	coronary artery diseases
cART	combined antiretroviral therapy
CAST	Cohort Allelic Sums Test
CBA	cost–benefit analysis
CBS	circular binary segmentation
CBT	cognitive behavioral therapy
CCFA	Crohns and Colitis Foundation of America
CCMB	Centre for Cellular and Molecular Biology
CCR5	chemokine co-receptor
CD/CV	Common Disease/Common Variant
CDA	cytidine deaminase
CDRI	Central Drug Research Institute
CEA	cost-effectiveness analysis
CFH	complement factor H
cGGM	conditional Gaussian graphical models
CGH	comparative genomic hybridization
CGS	CpG-related SNP
CGS-C	CpG-related SNP with the allele to create CpG dinucleotides
CGS-Cp	proxy CGS-C
CGS-D	CpG-related SNP with the allele to disrupt CpG dinucleotides
CGS-Di	index CGS-D
cM	centimorgan
CMA	cost-minimization analysis
CMC	Combined Multivariate and Collapsing Method
CNP	copy number polymorphism
CNS	central nervous systems
CNV	copy number variations
COGS	cost of goods and services
CS	cleavage site
CSIR	Council of Scientific and Industrial Research
CTPS	cytidine-5′-triphosphate synthetase
CUA	cost–utility analysis
DCDA	dichorionic diamnionic
DCK	deoxycytidine kinase
DCTD	deoxycytidylate deaminase
dCTP	deoxycytidine triphosphate
DHS	DNase I hypersensitivity site
DMP	DNA methylation position
DMR	differentially methylated region
DNMT	DNA methyltransferase
DOHaD	Developmental Origin of Health and Diseases
DSM	Diagnostic and Statistical Manual of Mental Disorders
DZ	dizygotic

EEG	electroencephalography
eGFR	estimated glomerular filtration rate
EH	expression heterogeneity
EM	expectation–maximization
EMMPAT	Evolutionary Mixed Model for Pooled Associated Testing
ENCODE	Encyclopedia of DNA Elements
eQED	eQTL electrical diagrams
eQTL	expressed quantitative trait loci
ES	exome sequencing
ETV-RAM	etravirine-resistance-associated mutation
EWAS	epigenome-wide association studies
FAB	French–American–British
FDR	false discover rate
FISH	fluorescence *in situ* hybridization
FMRP	Fragile X mental retardation protein
FPKM	fragments per kilobase of exon per million fragments mapped
GABA	gamma-aminobutyric acid
GCTA	genome-wide complex trait analysis
GIANT	Genomic Investigation of ANthropometric Traits
GO	gene ontology
GST	glutathione-*S*-transferases
GWA	genome-wide association
GWAS	Genome-Wide Association Studies
HAART	highly active antiretroviral therapy
HCAEC	human coronary artery endothelial cells
HCASMC	human coronary artery smooth muscle cell
HCG	human chorionic gonadotropin
HDL	high-density lipoprotein
HDL-C	high-density lipoprotein cholesterol
HIV	human immunodeficiency virus
HWE	Hardy–Weinberg Equilibrium
IBD	identical-by-descent
IBS	identity-by-state
ICE	inter-sample correlation emended
ICF	immunodeficiency, centromere instability and facial anomalies (syndrome)
ICGC	International Cancer Genome Consortium
IGIB	Institute of Genomics and Integrative Biology
IGV	Indian Genome Variation
IICB	Indian Institute of Chemical Biology
IMTECH	Institute of Microbial Technology
InDel	Insertion/Deletions
InI	integrase inhibitor
ISC	International Schizophrenia Consortium
ITD	internal tandem duplication

ITRC	Industrial Toxicological Research Centre
KSR	kinase suppressor of ras
LAPS	Lifetime Alcohol Problems Score
LCL	lymphoblast cell line
LCR	low-copy number region
LD	linkage disequilibrium
LINE	long interspersed elements
LMM	linear mixed model
LoF	loss of function
LORS	low-rank representation and sparse regression
LR	likelihood ratio
MAF	minor allele frequencies
MALD	mapping disease genes by admixture linkage disequilibrium
MALDI/TOF	matrix-assisted laser desorption/ionization–time of flight
MAP	mitogen-activated protein
MAPK	mitogen-activated protein kinase
MBD	methyl-CpG-binding domain (protein)
MCDA	monochorionic diamnionic
MCL	gene ontology-guided Markov cluster (algorithm)
MET	motivational enhancement therapy
MFE	minimum free energy
MGS	Molecular Genetics of Schizophrenia
MHC	major histocompatibility complex
MI	myocardial infarction
miRSNP	SNPs in microRNAs target site
MLS	Michigan Longitudinal Study
MOMA	methylation oligonucleotide microarray analysis
MRD	minimal residual disease
MSA	multiple system atrophy
MSS	maternal serum screening
VP	Million Veteran Program
MZ	monozygotic
NAHR	non-allelic homologous recombination
NAPHA	National Access to Antiretroviral Programs for People who have AIDS
NCATS	National Center for Advancing Translational Science
NDK	nucleoside diphosphate kinase
NFκB	nuclear factor κB
NGS	next-generation sequencing
NHGRI	National Human Genome Research Institute
NIPT	non-invasive prenatal testing
NNIBP	NNRTI binding pocket
NNRTS	non-nucleoside reverse transcriptase inhibitor
NRTI	nucleoside (or nucleotide) reverse transcriptase inhibitor
OATP	organic anion transporting polypeptide

OMIM	Online Mendelian Inheritance in Man
OR	odds ratio
ORF	opening reading frame
$P(x>0)$	percentage of CpG sites with methylation level greater than 0
PA	protease inhibitors (PI),
PAGE	Population Architecture using Genomics and Epidemiology
PAS	polyadenylation signal
PC	principal components
PCA	principal components analysis
PCR	polymerase chain reaction
PD	pharmacodynamics
PGC	Psychiatric Genomics Consortium
PGx	pharmacogenomics
PK	pharmacokinetics
POMC	pro-opiomelanocortin
PROCARDIS	Precocious Coronary Artery Disease
QALY	Quality Adjusted Life Year
QC	quality control
QQ	quantile–quantile
QTL	quantitative trait loci
RA	rheumatoid arthritis
RefSeq	Reference Sequence
REML	restricted maximum likelihood
RISC	RNA-induced silencing complex
ROMA	representational oligonucleotide microarray analysis
RPKM	reads per kilobase per million reads mapped
RR	relative risk
RR	ribonucleotide reductase
RRBS	reduced representative bisulfite sequencing
RVM	relevance vector machine
SAGE	Study of Addiction: Genetics and Environment
SCID	Structured Clinical Interview for DSM-IV
SCZ	schizophrenia
SFS	site frequency spectrum
SINE	short interspersed elements
SMC	smooth muscle cell
SNP	single nucleotide polymorphisms
STR	short tandem repeat
SV	structural variation
SVA	surrogate variable analysis
SVM	support vector machines
T1D	type 1 diabetes
T2D	type 2 diabetes
TCGA	The Cancer Genome Atlas
TDT	transmission disequilibrium test

TF	transcription factor
TSF	twelve-step facilitation
TSS	transcription start site
TTTS	twin-to-twin transfusion syndrome
UC	ulcerative colitis
VA	Veterans Administration
VHA	Veterans Healthcare Association
VTA	ventral tegmental area
WGS	whole-genome sequencing
WT	wild-type
WTCCC	Wellcome Trust Case Control Consortium
WTP	willingness-to-pay
XCI	X chromosome inactivation
YBP	years before present

Part I

Genome-wide association studies

1 Introduction to genome-wide association studies and personalized medicine

Krishnarao Appasani and Raghu K. Appasani

My scientific studies have afforded me great gratification; and I am convinced that it will not be long before the whole world acknowledges the results of my work.

Gregor J. Mendel, Austrian botanist/geneticist (1822–1884)

Ronald A. Fisher, an English statistical geneticist, showed for the first time that a complex quantitative trait can be explained by Mendelian inheritance if multiple genes affect the trait (Fisher, 1918). Thus, one can infer accurate statistical predictions of a complex trait requiring the identification of many small-effect variants, which, in combination, can explain a large fraction of variance in the phenotype. Before 1990, a number of examples of pharmacogenetic traits, usually binary, were published and reviewed (Nebert *et al.*, 2008). Most of them adhere to simple Mendelian inheritance and are controlled by one or a very small number of large-effecter genes. These breakthroughs in genotype–phenotype associations helped to establish expectations of individualized genetic risk prediction. In the pre-genomic era, the genetic dissection of complex diseases was done through classical linkage studies (Lander and Botstein, 1989), and candidate gene-based association studies (Cousin *et al.*, 2003; Patnala *et al.*, 2013). The classical linkage study is a powerful approach to identify rare and high penetrant disease variants or genes, whereas the candidate gene approach was limited to a few genetic markers that are involved in the pathogenesis of complex diseases.

The genome-wide association studies (GWAS) approach was first proposed by Risch and Merikangas in 1996 as a statistical method to detect common variants with modest genetic effects compared to linkage studies. A global collaborative effort called the "HapMap project" was initiated in 2003 to characterize the haplotype patterns in the human genome and subsequently to identify single nucleotide polymorphism (SNPs or snips; The International HapMap

Genome-Wide Association Studies: From Polymorphism to Personalized Medicine, ed. K. Appasani. Published by Cambridge University Press.

Consortium, 2005). HapMap resources guided the design and analysis of genetic association studies that shed light on structural variation and recombination, and identified loci that are involved in natural selection during human evolution. Since the discovery of GWAS and SNPs, association studies have been increasingly employed to reveal the relationships between SNPs and particular disease conditions. The first large set of results from a GWAS were reported in 2005 and 2006 (Klein *et al.*, 2005; Dewan *et al.*, 2006). In 2007, the Wellcome Trust Case Control Consortium published the first large, well-designed GWAS for complex diseases to employ a SNP chip that had good coverage of the genome (Wellcome Trust Case Control Consortium, 2007). The GWAS technique has already supplied useful information about the genetic basis of various diseases including: asthma, cancer, diabetes, heart disease, and mental disorders. The GWAS technique seeks out statistical differences among sets of polymorphic genetic markers between sufferers of a given disease and a control group. The use of GWAS also helped to extend the field of human genetics further using expression quantitative trait loci studies, individualized drug therapy, and personalized medicine strategies.

Scope of this book

This text consists of 23 chapters, grouped into 5 parts, and many of the aforementioned applications are described within various sections of the book, which are summarized as follows.

Part I: Genome-wide association studies

This section consists of four chapters. Phenotypes are composites of the observable traits of organisms and living individuals that originate from the expression of the instructions recorded in the organism's DNA under the influence of environmental factors. Heritability is a global measure that quantifies the overall contribution of genetic factors to a phenotype. Researchers working on such disparate fields such as livestock selection, medical genetics, behavioral economics, or evolutionary biology need to understand the genetic basis of phenotypes. Arcadi Navarro's colleagues describe the study of the genetics of polygenic traits and its importance to include GWAS in Chapter 2. Hongyu Zhao and his colleagues in Chapter 3 provide recent progress in GWAS with respect to statistics, including heritability estimation, association mapping and risk prediction. The limited availability of GWAS performed on populations of non-European ancestry has not precluded a widespread replication of risk variants across populations being observed. Indeed, large inter-population replicability rates allow researchers to reject a scenario of susceptibility variants being continent-specific, and rather favor a scenario of causal variants being common and shared across human populations. These results are shadowed by a poor and unbalanced availability of cross-population data, but the extension of GWAS across ancestries remains one of the best tactics to finally unveil the location of causal variants in susceptibility loci. In Chapter 4, Martinez-Marigorta and colleagues detail that GWAS replicability varies with time and space across human populations.

Part II: Genome-wide studies in disease biology

This section consists of five chapters. Applications of GWAS in diseases such as myocardial infarction, cancer, and neurodevelopmental disorders are described in this section. Body mass index is the ratio of body weight to height squared, and is the most commonly used index of adiposity and the diagnostic criterion for obesity, a leading risk factor for type 2 diabetes, cardiovascular disease, cancer, and premature death. Approximately half of the factors contributing to body mass index are heritable, and unraveling the specific genetic variation that contributes to this heritability is vital for understanding the biological mechanisms that regulate adiposity. The implementation of the GWAS approach has dramatically increased the speed of gene discovery for body mass index, which is detailed in Chapter 5 by Kilpelainen. Myocardial infarction is a common disease and among the leading causes of death in the world. Pathogenesis of myocardial infarction depends on complex interactions of environmental and genetic factors. Kouichi Ozaki and Toshihiro Tanaka's group adopted GWAS for the first time in human disease biology especially for myocardial infarction and identified a disease susceptibility gene in the Japanese population, and those studies are detailed in Chapter 6. Genome-wide studies have been carried out for decades in familial studies of Mendelian diseases with high penetrance. As genotyping technology continued to advance, genotyping chips became available with increasing numbers of SNPs. The first SNP chip with genome-wide coverage was made available by the mid-2000s, with the first studies using this technology published in 2005 (Klein *et al.*, 2005). Yeager's group applied GWAS studies in the study of cancer, which is described in Chapter 7.

Schizophrenia is a severe mental disorder with a typical onset in adolescence. Symptoms include: delusions, hallucinations, anhedonia, blunted affect, and disorganized speech. In addition, the majority of patients with schizophrenia show some level of cognitive dysfunction. In general, patients with schizophrenia have deficits in most cognitive domains (e.g., attention, memory, and executive functioning). In 2009, the first GWAS study on schizophrenia was published. In Chapter 8, Derks and his colleagues describe the details of such genome-wide association analysis in this psychiatric disease. Advances in methodologies for epigenetic analysis, such as Beadchip microarrays and next-generation sequencing, enable the investigation of the epigenetic status at individual loci, multiple loci, or the whole genome (Appasani, 2012). These new approaches also enable epigenome-wide association studies (EWAS). Several lines of evidence suggest that epigenetic abnormalities can be induced by environmental factors. Thus, clinical epigenetic research not only needs to target congenital disorders, but also needs to investigate acquired chronic diseases including common mental and neurodevelopmental disorders, in which epigenetic abnormalities may reside at multiple genomic loci. In recent years, the numbers of patients suffering from such chronic diseases have been reported to be increasing in Asian countries (for example, Japan). EWAS studies have mainly been performed for cancer and diabetes mellitus. However, for the first time, Kubota and his colleagues performed and compared EWAS in both brain tissue and blood samples for neurodevelopmental disorders, and the results are described in Chapter 9.

Part III: Single nucleotide polymorphisms, copy number variants, haplotypes and eQTLs

This section consists of five chapters describing the details of the inter-relationship of microRNAs and SNPs, copy number variants (CNVs), and expression quantitative trait loci (eQTLs). MicroRNAs are small, single-stranded RNAs of about 22 nucleotides involved in gene regulation by binding to 3′ untranslated regions of messenger RNAs (Bartel, 2004). Gene silencing by microRNAs is an important mechanism in physiological processes, and its deregulation can lead to complex diseases such as cancer (Garzon *et al.*, 2006). SNPs in the coding sequence of messenger RNAs have been well studied, but not much work has been done in the non-coding regions of the genome, such as the 3′ untranslated regions of messenger RNAs, which harbor many functional sequence elements involved in gene regulation. One type of functional element that can be disrupted by SNPs is the microRNAs target site. SNPs in microRNAs target sites (miRSNPs) can change the affinity between the miRNA seed sequence and its target messenger RNA, resulting in deregulation of gene expression, and possibly in phenotype differences and diseases (Sethupathy and Collins, 2008). In Chapter 10, Thomas and Sætrom summarize the details of the SNPs in microRNAs target sites and map them on the breast cancer susceptibility genes.

Alcohol use disorder is under partial genetic control as a result of common variants in several genes, each with a small effect. Additionally, environmental exposures also play an important role in the development of alcohol use disorder. In recent years, it has been increasingly recognized that genetic and environmental factors are inter-dependent, suggesting that the expression of genetic liability depends on environmental factors. Furthermore, risk factors known as intermediate phenotypes are under the same gene–environment influence and may provide important clues to further understand this heterogeneous disorder. Therefore, Burmeister and colleagues studied the linkage complex associations in alcohol use disorder, and these are described in Chapter 11 of this volume. As we described earlier, GWAS is a method that can link human diseases and traits to specific haplotypes using SNPs. Using GWAS and genotyping approaches, traits and diseases can be associated with large DNA gains or losses. These are referred to as copy number polymorphisms or copy number variations, and can either be inherited as an Mendelian trait or arise as a *de novo* event. In Chapter 12, Brosens *et al.* describe the copy number variations observed in monozygotic twins, their timing and effect, and their impact on human traits and diseases.

Recent findings have suggested that links between the epigenetic status and genetic variants may underline the functionality of SNPs. Most importantly, DNA methylation has been most frequently linked to several human diseases such as cardiovascular diseases, diabetes, obesity, dyslipidemia, and cancer. CpG-related SNPs (CGSs) constitute a group of SNPs with a particular relationship to DNA methylation. By definition, CGSs refer to those SNPs which can change the formation of CpG dinucleotides that have been established as the primary target site of DNA methylation. CGSs have been found to contribute a significant fraction of allele-specific methylation regions in the human genome (Shoemaker *et al.*, 2010; Gertz *et al.*, 2011), and more than 80% of CGSs were shown to play a regulatory role in DNA methylation (Zhi *et al.*, 2013). In Chapter 13, Ma *et al.* describe their group's first contributions to studying CGSs and their relationship with DNA methylation through genome-wide scale and integrated bioinformatics analysis. These studies help not only in providing a candidate functional mechanism to link SNPs and DNA methylation, but also in its potential contributions to personalized medicine.

Expression quantitative trait loci (eQTL) mapping is a powerful approach to detecting transcriptional regulatory relationships at the scale of the genome. In eQTL studies, gene expression levels measured by high-throughput technologies, such as microarrays and RNA-Seq, are treated as quantitative traits. By simultaneously capturing many regulatory interactions, expression quantitative trait loci offer valuable insights into the genetic architecture of expression regulation (Rockman and Kruglyak, 2006). The ultimate goal of eQTL studies is to elucidate how genetic variations affect phenotypes by using gene expression levels as intermediate molecular phenotypes (Nica and Dermitzakis, 2008). Integrative analysis of variations in transcriptome and next-generation sequencing will result in unprecedented accuracy in eQTL detection and interpretation. These studies will hopefully bring insights into the molecular pathogenesis of complex traits. Thus, in Chapter 14, Chen *et al.* summarize the details of the eQTL analysis and mapping studies.

Part IV: Next-generation sequencing technology and pharmaco-genomics

This section consists of six chapters. Most rare diseases have a genetic basis, and are inherited in a Mendelian fashion. They are usually monogenic disorders segregating in families or are sporadic, being autosomal or sex-linked, and dominant or recessive (McKusick, 1994). Before the advent of the new next-generation sequencing technologies, positional cloning was the most commonly used technique for the analysis of the genetic basis of Mendelian diseases. Linkage and homozygosity mapping have benefitted from the increasing densities of genetic and physical mapping. The new method of exome sequencing helps to screen for known or new functional variants in previously discovered candidate genes and genes functionally related to them. In several cases, exome sequencing has resolved or improved misdiagnoses (Ku *et al.*, 2012). A combination of next-generation sequencing and *in silico* approaches have helped to prioritize and catalog a number of rare variants in both complex traits and disease phenotypes. In Chapters 15 and 16, Casals and Bosch describe the use of next-generation sequencing as a method to discover and understand the biology of several human rare and complex diseases, respectively.

Breast cancer is the most common type of malignancy among women in many countries around the world. It is well established that multiple genetic and epigenetic factors play an important role in breast cancer. Epigenetic characterization of cancer using DNA methylation profiling of tumors and their corresponding normal profiles has shown that the methylation landscapes are quite disrupted in cancer. Therefore, it is important to identify the mechanistic cross-talk between epigenetic modifications, genome instability, and transcriptional programs within breast cancers. Existing methodologies provide little insight into the mechanisms that drive these epigenetic and genetic changes on a genome-wide scale. In order to investigate these mechanisms Dimitrova and her colleagues and collaborators have performed systematic identification of differentially methylated regions and Alu-enriched loci in relation to major genome rearrangements and breakpoint enriched regions, and they are summarized in Chapter 17. A network analysis approach helps to elucidate networks underlying signaling mechanisms that govern cancer cell survival and proliferation and imply selective pressures for the evolutionary convergence of cancer genomic alterations. To study the genomic alteration in breast cancer, in Chapter 18, Zaman and Wang describe the details of a signaling network analysis approach used in breast cancer and implications for personalized medicine.

Acute myeloid leukemia is a clonal disorder and a very heterogeneous disease with various subtypes classified based on morphology, immunophenotype, and cytogenetics. The nucleoside analog, *cytarabine*, has been the mainstay of acute myeloid leukemia chemotherapy for more than 40 years. However, extensive inter-patient variation in treatment response, development of resistance, and inadequate response to first-line therapy remain the major hurdles to effective chemotherapy. The integration of pharmaco-genetic markers with prognostic markers in larger clinical cohorts would advance our ability to design personalized therapy in patients to achieve the greatest therapeutic benefit. In Chapter 19, Lamba and colleagues summarize the recent advances in pediatric acute myeloid leukemia pharmaco-genomics. Over the past decade, a vast amount of literature has been published about the significant pharmaco-genetic associations on the use of antiretroviral drugs, growing our knowledge base. Genotype–phenotype associations will require the use of advanced technologies for genome-wide screening and statistical analysis in order to draw a definitive comprehensive association. In Chapter 20, Sukasem *et al.* describe the combination of genome technology, computational, and bioinformatics used to study multi-ethnic pharmaco-genomics.

Part V: Population genetics and personalized medicine

This section consists of three chapters. It is a well-established fact that the anatomically modern human originated in Africa about 160,000 years ago and migrated out-of-Africa. In this process, several populations arose and each one of them has their own evolutionary history. Genetic drift, endogamy practices, admixture, and natural selection are a few examples of evolutionary phenomena leading to genetic diversity among populations around the world, including susceptibility and resistance to genetic diseases, infectious diseases, therapeutic response to drugs, etc. Understanding these phenomena is much more relevant and important in a country like India, which has the richest ethnic, cultural, linguistic, and social diversity in the world. In Chapter 21, Thangaraj and colleagues clearly describe the diverse population genomics and specific variations in India and draw conclusions on the development of "ethnicity-based genomic medicine" strategies.

Personalized medicine is an emerging field that holds promise for major advances in prevention and care at the patient level, and major reductions in the cost of healthcare at the societal level. In personalized medicine, genomics generally plays the role of the ultimate molecular diagnostic. Solving disease at a population scale involves aligning very powerful stakeholders, which often involves obtaining the support of kings, princes, presidents, prime ministers or billionaires. In Chapter 22, Merriman details the "genomic medicine" approach by highlighting the various "ethnicity-based genome sequencing" efforts. According to Merriman, this genomic medicine strategy will allow countries to de-risk and accelerate its execution, removing barriers of competency and capacity, and focusing on integrating the benefits of genomics into their national healthcare systems for the benefit of their people.

Use of the term "personalized medicine" is now ubiquitous (Hamburg and Collins, 2010). However, there is little consistency in how personalized medicine is either defined or used in clinical practice. Various authors have attempted to bring together the different definitions of personalized medicine. Redekop and Mladsi have defined personalized medicine as: "the use of combined knowledge (genetic or otherwise) about a person to predict disease susceptibility, disease prognosis, or treatment response and thereby improve that person's health" (Redekop and Mladsi, 2013). Producing economic evidence within the emerging field of personalized medicine is vital to support the timely and evidence-based introduction of technologies which have a potentially beneficial role in healthcare systems worldwide. Therefore, in Chapter 23, health economists Payne and Eden give an overview of the issues to be considered when evaluating the economic impact of the technologies used to personalize medicine.

Future perspectives

According to McClellan and King, the main advantage of GWAS research results has helped us to discover hundreds of common variants whose allele frequencies are statistically correlated with various illnesses and traits, but these studies did not establish any significant biological or clinical relevance in terms of prognosis and/or treatment (McClellan and King, 2010). We mention throughout this volume that patients' genomes were compared (with those of healthy genomes) using SNP chips in the ($100 million) "HapMap" project. However, these studies did not provide the molecular etiology of a given disease (Wade, 2011; Visscher *et al.*, 2012). The current generation of GWAS have contributed to identifying novel genes associated with common complex diseases. On the other hand, another approach called "Whole-Genome Sequencing" provides the ability to identify rare alleles with larger effects that are not detected by GWAS. Ideally, a combination of both these approaches will be utilized to provide a more complete view of human genetic variation (Burnham and Hayden, 2012). It is important to mention that although we can obtain GWAS and whole-genome sequencing data from a large number of patients, effect sizes for the majority are of small-effect variants which are simply too miniscule to be detected, even with a practicably attainable sample size. Future GWAS will have to explore structural variations, gene–gene interactions, epigenetic and gene–environment interactions. In a nutshell, a holistic approach is needed to contribute to the future of "personalized medicine."

In conclusion, we agree with the comments of Nebert and Zhang that the idealistic goal of personalized medicine and individualized drug therapy, which needs a holistic understanding of each individual patient's unique -omics readout, is most likely unattainable for the vast majority of complex traits (Nebert and Zhang, 2012). In the "human genome era," we have achieved several breakthroughs in our understanding of the biology, genetics and pathogenesis of human diseases. In the coming "human circuitry era" (Nybo, 2014; understanding the gene–protein networks of various tissues/organs, and all the neural circuits in brain) using "genomic medicine" approaches, we hope to unravel more mysteries of disease biology which will open up new research avenues and lay a strong foundation for the development of the new field of "personalized medicine."

Due to space constraints very few relevant papers were cited here. We apologize for not citing all other contributions.

REFERENCES

Appasani, K. (2012). Epigenomics: From chromatin biology to therapeutics. Cambridge University Press, Cambridge.

Bartel, D. P. (2004). MicroRNAs: genomics, biogenesis, mechanism and function. *Cell*, **116**, 281–297.

Burnham, L.R. and Hayden, M.R. (2012). Whole-genome sequencing: the new standard of care? *Science*, **336**, 1112–1113.

Cousin, E., Genin, M., Mace, S., *et al.* (2003). Association studies in candidate genes: strategies to select SNPs to be tested. *Hum. Hered.*, **56**, 151–159.

Dewan, A., Liu, M., Hartman, S., *et al.* (2006). HTRA1 promoter polymorphism in wet age-related macular degeneration. *Science*, **314**, 989–992.

Fisher, R.A. (1918). The correlation between relatives on the supposition of Mendelian inheritance. *Trans. R. Soc. Edinb.*, **52**, 399–433.

Garzon, R., Fabbri, M., Simmino, A., *et al.* (2006). MicroRNA expression and function in cancer. *Trends Mol. Med.*, **12**, 580–587.

Gertz, J., Varley, K.E., Reddy, T.E., *et al.* (2011). Analysis of DNA methylation in a three-generation family reveals widespread genetic influence on epigenetic regulation. *Publ. Libr. Sci. Genet.*, **7**, e1002228.

Hamburg, M.A. and Collins, F.S. (2010). The path to personalized medicine. *New Engl. J. Med.*, **363**, 301–304.

Klein, R.J., Zeiss, C., Chew, E.Y., *et al.* (2005). Complement factor H polymorphism in age-related macular degeneration. *Science*, **308**, 385–389.

Ku, C.S., Cooper, D.N., Polychronakos, C., *et al.* (2012). Exome sequencing: dual role as a discovery and diagnostic tool. *Ann. Neurol.*, **71**, 5–14.

Lander, E.S. and Botstein, D. (1989). Mapping Mendelian factors underlying quantitative traits using RFLP linkage maps. *Genetics*, **121**, 185–199.

McClellan, J. and King, M.C. (2010). Genetic heterogeneity in human diseases. *Cell*, **141**, 210–217.

McKusick, V.A. (1994). Mendelian inheritance in man: A catalogue of human genes and genetic disorders. Johns Hopkins University Press, Baltimore, MD. 11th Edition. XLV.

Nebert, D.W. and Zhang, G. (2012). Personalized medicine: temper expectations. *Science*, **337**, 910.

Nebert, D.W., Zhang, G. and Vesell, E.S. (2008). From human genetics and genomics to pharmacogenetics and pharmacogenomics: past lessons, future directions. *Drug Metab. Rev.*, 40, 187–224.

Nica, A.C. and Dermitzakis, E. T. (2008). Using gene expression to investigate the genetic basis of complex disorders. *Hum. Mol. Genet.*, **17**, R129–R134.

Nybo, K. (2014). Profile of Eric Lander. *BioTechniques*, **56**, 107.

Patnala, R., Clements, J. and Batra, J. (2013). Candidate gene association studies: a comprehensive guide to useful in silico tools. *BioMed Central. Genet.*, **14**, 39.

Redekop, W.K. and Mladsi, D. (2013). The faces of personalized medicine: a framework for understanding its meaning and scope. *Value Health*, **16**, S4–9.

Risch, N. and Merikangas, K. (1996). The future of genetic studies of complex human diseases. *Science*, **273**, 1516–1517.

Rockman, M.V. and Kruglyak, L. (2006). Genetics of global gene expression. *Nature Rev. Genet.*, **7**, 862–872.

Sethupathy, P. and Collins, F.S. (2008). MicroRNA target site polymorphisms and human disease. *Trends Genet.*, **24**, 489–497.

Shoemaker, R., Deng, J., Wang, W., *et al.* (2010). Allele-specific methylation is prevalent and is contributed by CpG-SNPs in the human genome. *Gene Res.*, **20**, 883–889.

The International HapMap Consortium. (2005). A haplotype map of the human genome. *Nature*, **437**, 1299–1320.

Visscher, P.M., Brown, M.A., McCarthy, M.I. *et al.* (2012). Five years of GWAS discovery. *Am. J. Hum. Genet.*, **90**, 7–24.

Wade, N. (2011). *New York Times*, March 20.

Wellcome Trust Case Control Consortium (2007). Genome-wide association study of 14,000 cases of seven common diseases and 3,000 shared controls. *Nature*, **447**, 661–678.

Zhi, D., Aslibekyan, S., Irvin, M.R., *et al.* (2013). SNPs located at CpG sites modulate genome–epigenome interaction. *Epigenetics*, **8**, 802–806.

2 GWAS: a milestone in the road from genotypes to phenotypes

Urko M. Marigorta, Juan Antonio Rodriguez and Arcadi Navarro

2.1. Introduction: phenotypes and genetic variation

Phenotypes are composites of the observable traits of organisms and living individuals that originate from the expression of the instructions recorded in the organism's DNA under the influence of environmental factors. Researchers working in such disparate fields as livestock selection, medical genetics, behavioral economics, or evolutionary biology need to understand the genetic basis of phenotypes. For instance, plant breeders aim to predict traits such as crop response to fertilizers (Hospital, 2009); clinical geneticists intend to trace genetic mutations that result in diseases – abnormal phenotypes characterized by pathology (Sullivan *et al.*, 2012); behavioral economists try to understand the genetic underpinnings of human behavioral variation (Navarro, 2009); and evolutionary biologists try to detect the molecular signature of natural selection in genes related to adaptive traits, such as lactase persistence (Hurst, 2009).

Despite its outstanding scientific and economic interest, studying the genetics of phenotypes is not devoid of complexities. Most traits, such as human height, tend to present continuous variation across individuals. This is because they are controlled by large numbers of genes and each causal variant explains but a tiny fraction of the overall phenotypic variation. In this regard, genome-wide association studies (GWAS) have arisen as one of the most powerful tools to unravel the alleles that underlie individual phenotypic variation. This chapter reviews the bases of the study of the genetics of polygenic traits and provides a brief historical account of the developments in the field until the current wave of GWAS.

2.2. The study of the genetic architecture of phenotypes

2.2.1. Forces shaping human genetic variation

Many different tools of statistical genomics, including GWAS, have been designed with the aim of mapping phenotype diversity to the underlying causal genetic

Genome-Wide Association Studies: From Polymorphism to Personalized Medicine, ed. K. Appasani. Published by Cambridge University Press.

factors that vary across individuals. The two main forces increasing genetic diversity in human genomes are mutation and recombination. Additional forces, such as genetic drift and natural selection, govern the fate of extant genetic variation in populations. Together, all of them shape the degree of phenotypic variability present across humans.

Two major classes of genetic variation can be distinguished in our genomes according to their size: point mutations, and structural variation (Frazer *et al.*, 2009). Point mutations are substitutions of a single base and are known as single nucleotide polymorphisms (SNPs). SNPs are the smallest class of polymorphism in terms of the number of affected nucleotides and constitute the most prevalent change in human genomes (e.g., over 62 million SNPs were deposited in the public repository dbSNP, release 141, as of May 2014). Given the short span of genes in the human genome, most SNPs are "intergenic" rather than "genic." That feature, however, does not preclude a functional effect, as intergenic SNPs may have regulatory roles. Structural variation (SV) is the other main source of genetic diversity. It includes variability that ranges from small "indels" (insertions/deletions) to large chromosomal rearrangements (Frazer *et al.*, 2009). Each type of SV has distinctive rates of mutation and evolution, and their role in human phenotypic variation has been confirmed abundantly.

Recombination constitutes the second major evolutionary force increasing genetic variability (Slatkin, 2008). Instead of creating new variants, variability is increased by placing different variants in the same chromosome, thus creating new combinations of alleles. Rates of recombination vary across the genome, and "hotspots" of 1–2 kb length with >100-fold increased rates can be distinguished (Myers *et al.*, 2005). Recombination has a strong impact on linkage disequilibrium (LD), which refers to the correlation between alleles located at different genomic positions. LD is directly related to the notion of haplotype: combinations of variants along the same chromosome that are inherited together through gametes. In a population, LD occurs when some combinations of alleles are found as haplotypes more often than expected by chance (Balding *et al.*, 2007). The main cause of LD is the appearance of new alleles through mutation. In contrast, the reshuffling of haplotypes by recombination reduces it. One of the most used statistics to measure LD between pairs of loci is r^2; that is, the statistical correlation of alleles at pairs of SNPs. This measure is very relevant to ascertain the tagSNPs that best cover genetic variation within LD blocks (de Bakker *et al.*, 2005).

Genetic drift, natural selection, and migratory processes shape genetic diversity. For humans, our recent origins in Africa and our subsequent worldwide dispersal are the major determinants of the distribution of variability across populations and of the patterns of LD between genomic variants (Rosenberg *et al.*, 2002). Over the last 30 years, the availability of DNA sequence data has permitted the exhaustive description of human genetic variation, which is currently being refined with the latest developments in genotyping and sequencing technologies. Overall, African populations harbor greater amounts of genetic diversity than populations that underwent the bottleneck effects of out-of-Africa migrations (Goldstein and Chikhi, 2002). Nevertheless, the explosive expansion in human census size has resulted in the vast majority of human polymorphism being rare (minor allele frequency <0.05; Keinan and Clark, 2012). Due to their late appearance in recent times, rare variants tend to be confined to continental populations (Gravel *et al.*, 2011). On the contrary, most common variants predate human expansion across major landmasses and are shared across populations (The International HapMap 3 Consortium, 2009).

2.2.2. Epidemiology and complex traits

Epidemiology is usually defined as studying the occurrence and determinants of phenotypes in populations (Rothman, 2002). Beyond its traditional focus on disease manifestation, epidemiological techniques can be used to study any anthropometric trait, such as height or hair color. Usually, causation is multifactorial and variable across individuals. A key step in any epidemiological study is the definition of the target population, the group of individuals from which conclusions on phenotypes – usually disease – can be extracted.

Two basic study designs are used in epidemiology: cohort and case-control studies. Individuals from a cohort are initially disease-free and ascertained because of a shared characteristic or risk exposure. These individuals are followed over a period of time, and the point estimate of relative risk (RR) can be drawn. Specifically, relative risk measures the ratio of disease probability in exposed persons over that in people who are unexposed. Although the gold standard in the field, cohort studies are costly and take long times because only a fraction of individuals end up developing the disease under study. The case-control study is an immediate alternative in which individuals with a binary phenotype of interest – again, usually disease status – are identified and their suspected exposures are compared with those of controls. Ideally, control individuals must be unbiased representatives of the same population. From a contingency table, the association of exposures and case-control status is measured by means of the odds ratio (OR), the ratio of the odds of exposure in cases to the odds of exposure in the controls (Rothman, 2002).

Genetic epidemiology focuses on the study of the role of genetic as well as of environmental factors (Thomas, 2004; Ziegler *et al.*, 2010). Its primary focus is inherited variation and comparisons among relatives. Genetic epidemiology aims to study if any given disease "runs in families" and to understand the role of genetics in its inheritance patterns. Together with cohort and case-control studies, family-based designs are the cornerstones of the field. Overall, the process of genetic epidemiology can be thought of as a step-by-step task whose final aim is the unravelling of the genetic architecture of complex traits. Eventually, the full description of the genetic architecture of any given phenotype consists of the mapping of all genetic variants associated with it and the determination of the causal variants that the associated markers were tagging, together with their effect sizes and interactions with other genetic and environmental factors.

2.2.3. Heritability: sizing the causes of phenotypic variation

The study of patterns of familial aggregation constitutes the first step to ensure that genetic variants shape any phenotype of interest. Familial resemblance refers to the increase in phenotypic correlation among relatives compared to unrelated individuals (Ziegler *et al.*, 2010). A comparison of resemblances between individuals with different degrees of relatedness (e.g., siblings, cousins, second cousins) allows for an initial exploration of the range of possible genetic architectures of a phenotype (i.e., number of risk alleles, inheritance pattern, and so on). Nevertheless, recurrence ratios among relatives can arise due to either genetic or environmental factors, and sophisticated techniques have been developed to assign a role of genetic variation in the trait of interest (Scurrah *et al.*, 2005).

Heritability is a population parameter that weights the extent to which phenotypic variation is due to genetic factors (Visscher *et al.*, 2008). Specifically, the variation observed across individuals is partitioned into unobserved genetic and environmental categories. Each of these partitions captures part of the variance, and the ratio of variances weights the influence of genetic variability in the phenotypic variance of a trait. Further partition of the genetic component of variance leads to measures of h^2 (narrow-sense heritability), which is the fraction of variance attributable to additive genetic factors, that is, due to the additive effects of the specific alleles inherited by an individual. Figure 2.1 reports heritability estimates for a range of human complex traits. It is clear that heritability levels are highly heterogeneous across conditions, populations, and time. For

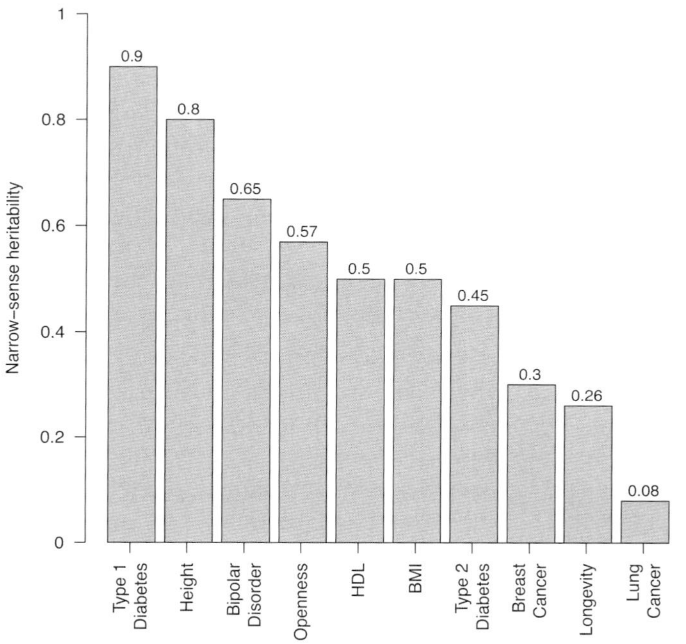

Figure 2.1. Heritability estimates for 10 human diseases and complex traits. The estimates have been gathered from Bouchard and McGue (2003) and Visscher *et al.* (2012). The average is shown if a range in heritability estimates was available.

instance, the generalization of good nutrition and medical care has resulted in large heritability for height in developed countries. In contrast, the low estimate for lung cancer indicates that non-genetic risk factors such as smoking habits play the largest role.

To estimate heritability, the correlation among relatives can be used to compare observed versus expected resemblance in families (Visscher *et al.*, 2008). Classically used methods are the regression of the offspring on the parental phenotypes, or the difference in correlation between monozygotic (MZ) and dizygotic (DZ) twins (study designs vary depending on whether twins have or have not been raised together). However, the incorporation of complex phenomena such as genome–environment interactions can be extremely difficult and may result in biased heritability estimates (Boomsma *et al.*, 2002). Notably, the increasing availability of array-based sets of genetic markers and of sequencing data allows for estimates of heritability taken from sets of unrelated individuals (Yang *et al.*, 2010).

Several misconceptions plague the interpretation of heritability (Visscher *et al.*, 2008). It is important to keep in mind that heritability measures are population-specific estimates and can vary with time, sex, and age. Traits with large heritabilities can be strongly affected by changes in the environment (e.g., the recent global increase in human height). Hence, differences in environment must be ruled out to conclude that differences among populations are genetic in origin (Feldman and Lewontin, 1975).

2.3. Historical overview of GWAS: antecedents and current state of the art

2.3.1. Inferring the architecture of human complex traits: the "Common Disease/Common Variant" hypothesis

Genetic architecture refers to the underlying genetic basis of a phenotypic trait, composed of the distribution of causal variants, allele frequencies, effect sizes, and the patterns of pleiotropy, dominance, and epistasis they maintain (Visscher *et al.*, 2010). Disease phenotypes have classically been the main focus of studies investigating the genetic architecture of complex traits. For instance, the first experimental studies in the early 1950s and 1960s found strong associations between common variations in ABO and HLA and several types of cancer and autoimmune diseases (Bodmer and Bonilla, 2008). Interestingly, these alleles present big effect sizes along with large allele frequencies. For instance, 8% of individuals of European ancestry carry HLA-B27 alleles associated with ankylosing spondylitis, with the original OR estimates being over 80 (Schlosstein *et al.*, 1973). In the 1980s, the study of complex disease genetics with positional cloning methods was accelerated thanks to the availability of genomic maps through restriction fragment length polymorphism (Botstein *et al.*, 1980). Yet, the most successful findings were unravelled as false positives by later studies (Botstein and Risch, 2003), and the main successes were shown to be restricted to familial forms of disease (e.g., *BRCA1* and risk for young-onset breast cancer). The failure of linkage methods when applied to complex phenotypes prompted a fierce debate on the methodology necessary to unravel disease causal variants.

In the late 1990s, several landmark studies explored the methodological challenge of unravelling disease-associated variants of low risk (Lander, 1996; Risch and Merikangas, 1996; Chakravarti, 1999; Reich and Lander, 2001). For instance, Risch and Merikangas (1996) compared linkage to association methods and showed the latter would be more empowered to find low-risk alleles in the absence of allelic heterogeneity. The authors also acknowledged that strict significance thresholds would be necessary to avoid an inflation of false positives (because large numbers of polymorphisms from across the whole genome ought to be tested).

A variety of papers also explored the issue of disease architecture from the perspective of disease evolution and using simulations (Pritchard, 2001; Reich and Lander, 2001; Rosenberg *et al.*, 2002). These studies explored wide ranges of the effects of risk variants on the fitness of carriers (i.e., the effect a risk variant will have upon the probabilities of survival and reproduction of an individual) to answer questions on the frequency and number of alleles expected in loci playing a major role in the architecture of complex traits. These studies showed that a regime of weak purifying selection would prevent the fixation of disease variants, leaving them at intermediate frequencies. Given its correlation with heterozygosity (Visscher *et al.*, 2012), these alleles would contribute most of the genetic variance of disease. In other words, simulation studies suggested that a relatively restricted set of common variants (frequencies >0.05) could account for a large proportion of the heritability.

All these ideas helped in the establishment of the "Common Disease/Common Variant" (CD/CV) hypothesis. This paradigm states that common variants in susceptibility genes account for most of the genetic risk of complex disease and other polygenic traits. The slight deleterious effects of these variants in the fitness of affected individuals explain their relatively high allele frequencies. Their frequency would counterweight the low relative risks of these variants, thus explaining the large prevalence of complex disease in modern populations. Along with strong criticisms from some authors (Weiss and Terwilliger, 2000), the CD/CV hypothesis was established as the paradigm in human genetics and paved the way for the wave of LD-based association studies.

2.3.2. Linkage disequilibrium mapping and the candidate gene era

At the turn of the new millennium and parallel to the ongoing Human Genome Project, LD-based association mapping was established as the choice method to unravel the genetic bases of complex traits. This method aims at establishing statistically supported associations between genetic markers and the phenotype of interest. In contrast to linkage, association mapping looks for the transmission of the phenotype with alleles instead of with loci. In the context of disease studies, genetic associations unravel when specific alleles are more frequent in groups of affected than of non-affected individuals (Ziegler *et al.*, 2010).

Associations between alleles and disease arise either when the tested marker is causal (*direct association*) or in the more likely scenario that the variant under study presents strong LD with the marker that in turn causes disease (*indirect association*). In consequence, association studies present two requirements regarding genetic variation. First, prior biological knowledge is necessary to select loci that "make biological sense" (Thomas, 2004), or contain clues of participation in the aetiology of the phenotype of interest. Second, dense spacing of markers is necessary to cover extensively the ascertained candidate loci. SNPs are the marker of choice due to their abundance, but an understanding of the strength of LD in human genomes is necessary for a proper design of population-based association studies. A very initial study based on simulations estimated that LD would not extend, on average, beyond 3 kb (Kruglyak, 1999). Posterior studies based on real data established the existence of discrete haplotype blocks that extend for several tens of kilobases before being punctuated by recombination hotspots (Daly *et al.*, 2001). The possibility of characterizing haplotype blocks by the genotyping of a low number of SNPs (named tagSNPs) that may be shared across populations (Gabriel *et al.*, 2002) prompted the establishment of the HapMap project in 2002. The widespread availability of SNP markers expedited the publication of candidate gene association studies using case-control designs (Hirschhorn and Altshuler, 2002). In this design, frequencies of variants at the ascertained SNPs are compared in populations of cases and controls. Their penetrance and relative risks can be approximated in association studies by calculating the strength of association using, for instance ORs in the case of binary traits. Conditions that are better approximated as quantitative traits (e.g., height or blood pressure) can be assessed through linear regression methods.

The identification and replication of several associations through candidate gene approaches – such as *CTLA4* and type 1 diabetes, or *NOS2* and Crohn's disease – created a huge wave of hype in the community. However, the lack of replication of significant findings has constituted a major challenge of candidate gene studies. An early comprehensive review of >600 studies found that only 6 of 166 putative associations had been consistently replicated by 2002 (Hirschhorn and Altshuler, 2002). Other meta-analysis described large heterogeneity in the reported effect sizes (Ioannidis *et al.*, 2001). A wealth of statistical, biological and sociological reasons has been put forward to explain replication failures (Hirschhorn and Altshuler, 2002). Among these, population stratification (Pritchard *et al.*, 2000), multiple testing (Devlin and Roeder, 1999), and outcome publication bias are often cited. Nonetheless, true findings can also fail to replicate due to lack of statistical power in replication studies. Typical sample sizes used in candidate–gene association studies (in the hundreds of samples) may not have the power to statistically distinguish the low effect sizes of variants associated with complex disease (Chanock *et al.*, 2007). Finally, true biological heterogeneity may account for the lack of replication of the findings of association studies across populations, hinting at a role of genetic differences in the varying prevalence of disease across human continents. For more details about replication see (Marigorta *et al.*, 2011) and Chapter 4 in this book.

2.3.3. The current gold standard: genome-wide association studies

The plethora of questionable associations that infested the field of candidate–gene studies could have ruined the prospects of ever discovering the genetic bases of disease and other complex traits. However, the development of genotyping techniques improved the availability of polymorphism to a density of up to several SNPs per kilobase. Additionally, the unravelling of LD patterns informed the possibility of capturing ~80% of the predicted >10 million common SNPs segregating in human populations with a scattered selection of 0.5–1 million SNPs (Visscher *et al.*, 2012). Technological development of commercial chips for high-throughput genotyping made it feasible to look for common variants associated with complex disease by means of genome-wide association studies.

The first GWAS were published in 2005 and 2006 (Klein *et al.*, 2005; Dewan *et al.*, 2006). Even if using few markers and samples (<100,000 SNPs genotyped upon <200 individuals), both studies managed to find common variants associated with age-related macular degeneration thanks to the large effect size of causal variants (OR>2). In 2007, the Wellcome Trust Case Control Consortium published a GWAS for seven different diseases using a set of 3000 shared controls (The Wellcome Trust Case Control Consortium[WTCCC], 2007). The WTCCC paper became a landmark due to the large number of samples used (2000 per disease), the confirmation of the low risk of population stratification in European GWAS, the setting of genome-wide significance thresholds and the replication of previous signals. Dropping costs of commercial arrays and the development of computational tools helped in the exponential increase of published GWAS (Hindorff *et al.*, 2009; Clarke *et al.*, 2011).

GWAS present two key differences with respect to candidate gene studies. First, there is an inherent issue related to numbers. Previously, a few tens of markers and, at most, a few hundreds of individuals were analyzed. In contrast, GWAS test millions of markers (many of which are just imputed, that is, *inferred* using statistical techniques) in thousands of individuals gathered from diverse cohorts (Figure 2.2). Second, GWAS are said to be "hypothesis-free": they do look for common risk effects, but they do so by looking at SNPs scattered across the genome and without any *a priori* list of candidate loci (Ziegler *et al.*, 2010). Thus, the adjustment for multiple testing constitutes an important decision in GWAS. The two most used cut-offs are 5×10^{-7} (WTCCC) and 5×10^{-8} (strict Bonferroni), but there are several available methods available to select proper significance thresholds (McCarthy *et al.*, 2008).

At the time of writing, more than 1900 published studies were recorded in the public catalogue of GWAS maintained by the *NIH Office of Population Genomics* (Hindorff *et al.*). However, GWAS present a heavy bias towards using exclusively individuals of European genetic ancestry. In 2010, an early survey of GWAS publication patterns found that >80% of studies did not use any cohort of non-European ancestry (Rosenberg *et al.*, 2010), although this bias has decreased in recent years (Fu *et al.*, 2011). Interestingly, non-European GWAS present challenges related to imputation ability, genomic coverage, and statistical power (due to the ascertainment of SNPs in commercial arrays). However, several reasons fuel

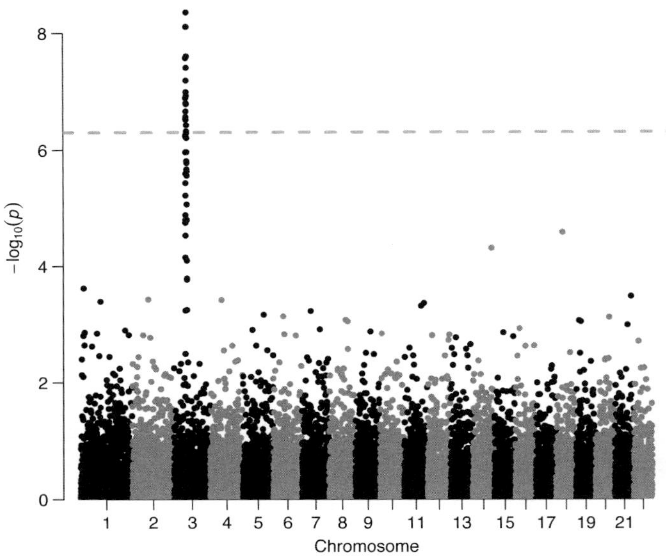

Figure 2.2. Example of a Manhattan Plot depicting the association of markers tested in a GWAS. SNPs are displayed along the chromosomes from left to right. Height in the *y*-axis shows the strength of association (–log10 of the *p*-value) with the trait under study. Each dot corresponds to the association status of a single SNP. In the example above, several SNPs located in chromosome 3 form a neat peak of association because they present high LD among themselves. This is the only signal surviving the strict significance threshold to avoid multiple testing (horizontal dashed line at $P = 5 \times 10^{-7}$). This simulated example has been created using the R package *qqman*.

the case for GWAS generalization across populations to better achieve the objectives of complex disease mapping (Rosenberg *et al.*, 2010). The ability to detect disease variants can vary if they have different effect size or present disparate allele frequencies across populations (Adeyemo and Rotimi, 2009). Additionally, use of diverse populations allows us to take advantage of the variation in LD across populations and thus help in the fine-mapping to narrow the location of causal variants (Visscher *et al.*, 2012). In any case, preliminary comparisons of GWAS replicability have reported high rates of concordance across populations (Marigorta and Navarro, 2013).

2.3.4. Knowledge gained from GWAS and the missing heritability

There are over 2000 loci that have been robustly associated with disease (Visscher *et al.*, 2012). The number of loci identified for each disease has increased exponentially when compared to associations discovered and replicated through the candidate gene approach. This observation emphasizes the limitations of an approach based on biological candidates compared to the "hypothesis-free" design of GWAS. Pathway analysis of the discovered loci has provided unsuspected insights into biological mechanisms of human complex phenotypes. The range of new understanding applies to specific traits (e.g., *IL23R* role in ankylosing spondylitis) as well as shared etiology across phenotypes (e.g., loci associated to disparate autoimmune disorders).

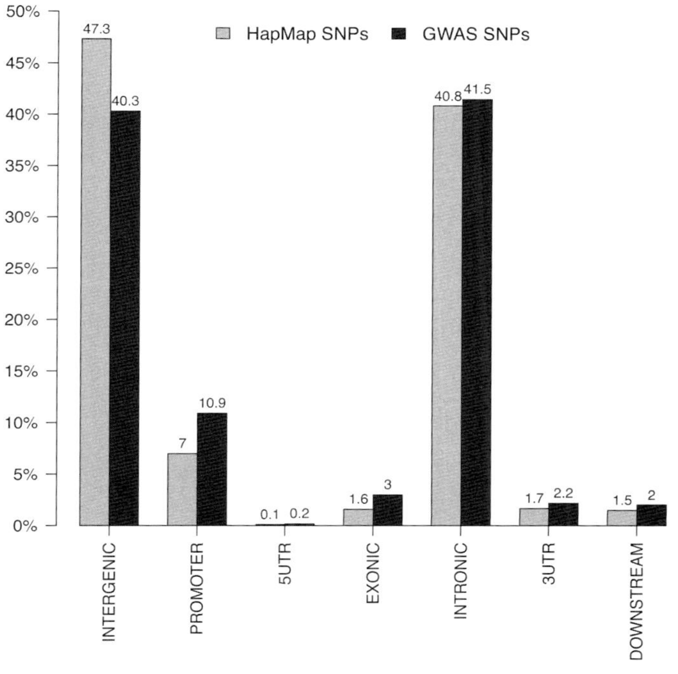

Figure 2.3. Enrichment/depletion of GWAS signals according to the annotation category of associated variants. For each functional category, light gray bars indicate the percentage of variants annotated in HapMap, a repository of common SNPs. Dark gray bars indicate the corresponding figure for SNPs associated with complex traits that were available at the GWAS catalog as of June 2014. Intergenic SNPs is the only annotation category that shows depletion in GWAS.

An interesting aspect of GWAS lies in the translational application of several "druggable" hits (Lander, 2011). The analysis of SNPs discovered by GWAS also shows three interesting patterns related to functionality and the heritability explained. First, even if genic regions present the largest enrichment in associated variants (Figure 2.3), the vast majority of GWAS findings locate outside transcriptional units (Hindorff *et al.*, 2009). In fact, oftentimes association signals map at gene-poor regions (e.g. 8q24 associations for several cancers). Second, associated SNPs are enriched for expressed quantitative trait loci (eQTLs) and abundant clues of pleiotropy have been described (Nicolae *et al.*, 2010). Finally, the distribution of allele frequencies of GWAS variants contains an excess of common variants (minor allele frequencies (MAF)>20%) that collectively explain substantial proportions of the genetic architecture of traits (Park *et al.*, 2011).

Despite these successes, the bulk of genetic susceptibility remains unexplained. For instance, the 180 variants associated with height in a large GWAS meta-analysis account for less than 20% of the heritability of the trait (Lango Allen *et al.*, 2010). This phenomenon has been coined the "case of the missing heritability" (Maher, 2008). Several hypotheses have been put forward to explain the missing heritability problem of human genetics (Maher, 2008; Goldstein, 2009;

Manolio *et al.*, 2009; Eichler *et al.*, 2010). For instance, stringent correction for multiple testing to avoid false positive findings can swamp the signal of alleles with very small effect sizes. This latter model of genetic architecture, known as the "infinitesimal" model, posits that a myriad (hundreds or thousands) of genome-wide scattered genetic variants account for the variation of phenotypes across individuals (Gibson, 2012). A 2008 study fitted an exponential distribution using the effect size of the alleles associated with height and predicted that 93,000 causal SNPs would be required to explain the heritability of the trait (Goldstein, 2009). Further evidence pointing at the presence of thousands of very low effect variants comes from the joint consideration of the effects of all genotyped SNPs instead of looking at individual variants. Linear models built using all genotyped SNPs are able to explain 45% of height heritability and the percentage of variance explained by each chromosome correlates perfectly with chromosome length (Yang *et al.*, 2010). Heritability would not be "missing" but "hidden" by the strict significance cut-offs used in GWAS.

Other authors champion alternative explanations, such as heritability being explained by rare variants of large effect that are not captured in commercial genotyping arrays, an unexplored role of epistatic interactions (GWAS usually assume additive effects), the hidden effects of gene-by-environment interactions, the presence of parent-of-origin effects (Eichler *et al.*, 2010), or the inherent phenotypic heterogeneity among patients that could be a recipe for the failure of GWAS (e.g., the huge heterogeneity across patients of psychiatric disorders).

The prospects for personalized genetic medicine depend on the ability to translate GWAS findings into the clinical setting. One immediate possibility lies in the development of genetic profiles based on GWAS risk markers available through consumer testing (Lee *et al.*, 2008 ; Jakobsdottir *et al.*, 2009, Kraft *et al.*, 2009). Potentially, genetic profiles based on risk markers may distinguish between high-risk and low-risk groups of individuals. Additive models constructed from few GWAS variants have been shown to lack the power to pinpoint individuals that will develop a given complex disease (Jakobsdottir *et al.*, 2009), and although the calculation of scores from training sets using liberal thresholds somehow improves the picture (Evans *et al.*, 2009), discriminative accuracy presents great variation across diseases. Current protocols include genetic risk through family history, but a large proportion of complex cases subjects do not have close diseased relatives (Wray and Visscher, 2010). Indeed, 50% of genetic variance occurs within families and thus genetic risk of disease changes across relatives with the same family history. Current estimations show that genome-wide SNP data from unrelated people have enough precision to predict phenotypes when risk estimates are calculated with >100,000 individuals (Goddard, 2009). The current performance of predictions for individual genomes remains unclear (Jelier *et al.*, 2011; Burga and Lehner, 2012) and only the combination of larger sample sizes and improved genomic coverage will enhance the hopes for personalized genetic testing.

REFERENCES

Adeyemo, A. and Rotimi, C. (2009). Genetic variants associated with complex human diseases show wide variation across multiple populations. *Publ. Health Genom.*, **13**, 72–79.

Balding, D.J., Bishop, M. and Cannings, C. (2007). *Handbook of Statistical Genetics*. John Wiley & Sons, Ltd, Chichester.

Bodmer, W. and Bonilla, C. (2008). Common and rare variants in multifactorial susceptibility to common diseases. *Nature Genet.*, **40**, 695–701.

Boomsma, D., Busjahn, A. and Peltonen, L. (2002). Classical twin studies and beyond. *Nature Rev. Genet.*, **3**, 872–882.

Botstein, D. and Risch, N. (2003). Discovering genotypes underlying human phenotypes: past successes for Mendelian disease, future approaches for complex disease. *Nature Genet.*, **33**(Suppl), 228–237.

Botstein, D., White, R.L., Skolnick, M. and Davis, R. W. (1980). Construction of a genetic linkage map in man using restriction fragment length polymorphisms. *Am. J. Hum. Genet.*, **32**, 314–331.

Bouchard, T.J., Jr. and McGue, M. (2003). Genetic and environmental influences on human psychological differences. *J. Neurobiol.*, **54**, 4–45.

Burga, A. and Lehner, B. (2012). Beyond genotype to phenotype: why the phenotype of an individual cannot always be predicted from their genome sequence and the environment that they experience. *FEBS J.*, **279**, 3765–3775.

Chakravarti, A. (1999). Population genetics – making sense out of sequence. *Nature Genet.*, **21**, 56–60.

Chanock, S.J., Manolio, T., Boehnke, M., *et al.* (2007). Replicating genotype–phenotype associations. *Nature*, **447**, 655–660.

Clarke, G.M., Anderson, C.A., Pettersson, F.H., *et al.* (2011). Basic statistical analysis in genetic case-control studies. *Nature Protoc.*, **6**, 121–133.

Daly, M.J., Rioux, J.D., Schaffner, S.F., Hudson, T.J. and Lander, E.S. (2001). High-resolution haplotype structure in the human genome. *Nature Genet.*, **29**, 229–232.

De Bakker, P.I., Yelensky, R., Pe'er, I., *et al.* (2005). Efficiency and power in genetic association studies. *Nature Genet.*, **37**, 1217–1223.

Devlin, B. and Roeder, K. (1999). Genomic control for association studies. *Biometrics*, 55, 997–1004.

Dewan, A., Liu, M., Hartman, S., *et al.* (2006). *HTRA1* promoter polymorphism in wet age-related macular degeneration. *Science*, **314**, 989–992.

Eichler, E.E., Flint, J., Gibson, G., *et al.* (2010). Missing heritability and strategies for finding the underlying causes of complex disease. *Nature Rev. Genet.*, **11**, 446–450.

Evans, D.M., Visscher, P.M. and Wray, N.R. (2009). Harnessing the information contained within genome-wide association studies to improve individual prediction of complex disease risk. *Hum. Mol. Genet.*, **18**, 3525–3531.

Feldman, M.W. and Lewontin, R.C. (1975). The heritability hang-up. *Science*, **190**, 1163–1168.

Frazer, K.A., Murray, S.S., Schork, N.J. and Topol, E.J. (2009). Human genetic variation and its contribution to complex traits. *Nature Rev. Genet.*, **10**, 241–251.

Fu, J., Festen, E.A. and Wijmenga, C. (2011). Multi-ethnic studies in complex traits. *Hum. Mol. Genet.*, **20**, R206–213.

Gabriel, S.B., Schaffner, S.F., Nguyen, H., *et al.* (2002). The structure of haplotype blocks in the human genome. *Science*, **296**, 2225–2229.

Gibson, G. (2012). Rare and common variants: twenty arguments. *Nature Rev. Genet.*, **13**, 135–145.

Goddard, M. (2009). Genomic selection: prediction of accuracy and maximisation of long term response. *Genetica*, **136**, 245–257.

Goldstein, D.B. (2009). Common genetic variation and human traits. *New Engl. J. Med.*, **360**, 1696–1698.

Goldstein, D.B. and Chikhi, L. (2002). Human migrations and population structure: what we know and why it matters. *Annu. Rev. Genom. Hum. Genet.*, **3**, 129–152.

Gravel, S., Henn, B.M., Gutenkunst, R.N., *et al.* (2011). Demographic history and rare allele sharing among human populations. *Proc. Natl Acad. Sci. USA*, **108**, 11983–11988.

Hindorff, L.A., MacArthur, J., (European Bioinformatics Institute), Wise, A., *et al.* (2009) A Catalog of Published Genome-Wide Association Studies. www.genome.gov/gwastudies.

Hindorff, L.A., Sethupathy, P., Junkins, H.A., *et al.* (2009). Potential etiologic and functional implications of genome-wide association loci for human diseases and traits. *Proc. Natl Acad. Sci. USA*, **106**, 9362–9367.

Hirschhorn, J.N. and Altshuler, D. (2002). Once and again-issues surrounding replication in genetic association studies. *J. Clin. Endocrinol. Metab.*, **87**, 4438–4441.

Hospital, F. (2009). Challenges for effective marker-assisted selection in plants. *Genetica*, **136**, 303–310.

Hurst, L.D. (2009). Fundamental concepts in genetics: genetics and the understanding of selection. *Nature Rev. Genet.*, **10**, 83–93.

Ioannidis, J.P., Ntzani, E.E., Trikalinos, T.A. and Contopoulos-Ioannidis, D.G. (2001). Replication validity of genetic association studies. *Nature Genet.*, **29**, 306–309.

Jakobsdottir, J., Gorin, M.B., Conley, Y.P., Ferrell, R.E. and Weeks, D.E. (2009). Interpretation of genetic association studies: markers with replicated highly significant odds ratios may be poor classifiers. *PLoS Genet.*, **5**, e1000337.

Jelier, R., Semple, J.I., Garcia-Verdugo, R. and Lehner, B. (2011). Predicting phenotypic variation in yeast from individual genome sequences. *Nature Genet.*, **43**, 1270–1274.

Keinan, A. and Clark, A.G. (2012). Recent explosive human population growth has resulted in an excess of rare genetic variants. *Science*, **336**, 740–743.

Klein, R.J., Zeiss, C., Chew, E.Y., *et al.* (2005). Complement factor H polymorphism in age-related macular degeneration. *Science*, **308**, 385–389.

Kraft, P., Wacholder, S., Cornelis, M.C., *et al.* (2009). Beyond odds ratios – communicating disease risk based on genetic profiles. *Nature Rev. Genet.*, 10, 264–269.

Kruglyak, L. 1999. Prospects for whole-genome linkage disequilibrium mapping of common disease genes. *Nature Genet.*, **22**, 139–144.

Lander, E.S. (1996). The new genomics: global views of biology. *Science*, **274**, 536–539.

Lander, E.S. (2011). Initial impact of the sequencing of the human genome. *Nature*, **470**, 187–197.

Lango Allen, H., Estrada, K., Lettre, G., *et al.* (2010). Hundreds of variants clustered in genomic loci and biological pathways affect human height. *Nature*, **467**, 832–838.

Lee, S.H., Van Der Werf, J.H., Hayes, B.J., Goddard, M.E. and Visscher, P.M. (2008). Predicting unobserved phenotypes for complex traits from whole-genome SNP data. *PLoS Genet.*, **4**, e1000231.

Maher, B. (2008). Personal genomes: the case of the missing heritability. *Nature*, **456**, 18–21.

Manolio, T.A., Collins, F.S., Cox, N.J., *et al.* (2009). Finding the missing heritability of complex diseases. *Nature*, **461**, 747–753.

Marigorta, U.M. and Navarro, A. (2013). High trans-ethnic replicability of GWAS results implies common causal variants. *PLoS Genet.*, **9**, e1003566.

Marigorta, U.M., Lao, O., Casals, F., *et al.* (2011). Recent human evolution has shaped geographical differences in susceptibility to disease. *BMC Genom.*, **12**, 55.

McCarthy, M.I., Abecasis, G.R., Cardon, L.R., *et al.* (2008). Genome-wide association studies for complex traits: consensus, uncertainty and challenges. *Nature Rev. Genet.*, **9**, 356–369.

Myers, S., Bottolo, L., Freeman, C., McVean, G. and Donnelly, P. (2005). A fine-scale map of recombination rates and hotspots across the human genome. *Science*, **310**, 321–324.

Navarro, A. (2009). Genoeconomics: promises and caveats for a new field. *Ann. NY Acad. Sci.*, **1167**, 57–65.

Nicolae, D.L., Gamazon, E., Zhang, W., *et al.* (2010). Trait-associated SNPs are more likely to be eQTLs: annotation to enhance discovery from GWAS. *PLoS Genet.*, **6**, e1000888.

Park, J.H., Gail, M.H., Weinberg, C.R., *et al.* (2011). Distribution of allele frequencies and effect sizes and their interrelationships for common genetic susceptibility variants. *Proc. Natl Acad. Sci. USA*, **108**, 18026–18031.

Pritchard, J.K. (2001). Are rare variants responsible for susceptibility to complex diseases? *Am. J. Hum. Genet.*, **69**, 124–137.

Pritchard, J.K., Stephens, M., Rosenberg, N.A. and Donnelly, P. (2000). Association mapping in structured populations. *Am. J. Hum. Genet.*, 67, 170–181.

Reich, D.E. and Lander, E.S. (2001). On the allelic spectrum of human disease. *Trends Genet.*, **17**, 502–510.

Risch, N. and Merikangas, K. (1996). The future of genetic studies of complex human diseases. *Science*, **273**, 1516–1517.

Rosenberg, N.A., Pritchard, J.K., Weber, J.L., *et al.* (2002). Genetic structure of human populations. *Science*, **298**, 2381–2385.

Rosenberg, N.A., Huang, L., Jewett, E.M., *et al.* (2010). Genome-wide association studies in diverse populations. *Nature Rev. Genet.*, **11**, 356–366.

Rothman, K.J. (2002). *Epidemiology: An Introduction*, Oxford University Press, Oxford.

Schlosstein, L., Terasaki, P.I., Bluestone, R. and Pearson, C.M. (1973). High association of an HL-A antigen, W27, with ankylosing spondylitis. *New Engl. J. Med.*, **288**, 704–706.

Scurrah, K., Gurrin, L., Palmer, L. and Burton, P. (2005). Estimation of genetic and environmental factors for binary traits using family data. *Stat. Med.*, **24**, 1613–1617; author reply 1617–1618.

Slatkin, M. (2008). Linkage disequilibrium – understanding the evolutionary past and mapping the medical future. *Nature Rev. Genet.*, **9**, 477–485.

Sullivan, P.F., Daly, M.J. and O'Donovan, M. (2012). Genetic architectures of psychiatric disorders: the emerging picture and its implications. *Nature Rev. Genet.*, **13**, 537–551.

The International HapMap 3 Consortium (2009). Integrating common and rare genetic variation in diverse human populations. *Nature*, **467**, 52–58.

The Wellcome Trust Case Control Consortium (2007). Genome-wide association study of 14,000 cases of seven common diseases and 3,000 shared controls. *Nature*, **447**, 661–678.

Thomas, D.C. (2004). *Statistical Methods in Genetic Epidemiology*, Oxford University Press, Oxford.

Visscher, P.M., Hill, W.G. and Wray, N.R. (2008). Heritability in the genomics era – concepts and misconceptions. *Nature Rev. Genet.*, **9**, 255–266.

Visscher, P.M., McEvoy, B. and Yang, J. (2010). From Galton to GWAS: quantitative genetics of human height. *Genet. Res. (Camb.)*, **92**, 371–379.

Visscher, P.M., Brown, M.A., McCarthy, M.I. and Yang, J. (2012). Five years of GWAS discovery. *Am. J. Hum. Genet.*, **90**, 7–24.

Weiss, K.M. and Terwilliger, J.D. (2000). How many diseases does it take to map a gene with SNPs? *Nature Genet.*, **26**, 151–157.

Welter, D., MacArthur, J., Morales, J., *et al.* (2014). The NHGRI GWAS Catalog, a curated resource of SNP-trait associations. *Nucl. Acids Res.*, Jan; **42**, (Database issue); D1001–D1006.

Wray, N.R. and Visscher, P.M. (2010). Narrowing the boundaries of the genetic architecture of schizophrenia. *Schizophr. Bull.*, **36**, 14–23.

Yang, J., Benyamin, B., McEvoy, B.P., *et al.* (2010). Common SNPs explain a large proportion of the heritability for human height. *Nature Genet.*, **42**, 565–569.

Ziegler, A., Konig, I.K. and Pahlke, F. (2010). *A Statistical Approach to Genetic Epidemiology*, Wiley VCH, Weinheim.

3 Introduction to statistical methods in genome-wide association studies

Can Yang, Cong Li, Dongjun Chung, Mengjie Chen, Joel Gelernter and Hongyu Zhao

3.1. Introduction

After the completion of the Human Genome Project (Lander *et al.*, 2001; Venter *et al.*, 2001) and initiation of the International HapMap Project (Sachidanandam *et al.*, 2001), genome-wide association studies (GWAS) were designed to survey the role of common genetic variations in complex human diseases. It was expected that GWAS would have the advantage of not relying on prior knowledge of biological pathways compared with "candidate gene" studies (Tabor *et al.*, 2002; Wang *et al.*, 2005), because it assays a dense set of single-nucleotide polymorphisms (SNPs) across the whole genome. This advantage allows GWAS to overcome the bias of "candidate gene" studies due to incomplete prior knowledge. It was also expected that GWAS would have higher power and finer resolution to identify genetic variants of modest effects compared to family-based linkage studies (Risch & Merikangas, 1996).

The success of identifying genes for age-related macular degeneration (AMD) under the GWAS paradigm (Klein *et al.*, 2005) convinced the genetics community on the efficiency and feasibility of the GWAS approach to identify unknown disease-associated variants. This study used a commercial genotyping array and assayed about 100,000 SNPs throughout the human genome. It identified the association of complement factor H (CFH) with AMD. The success of finding a common risk allele with an odds ratio (OR) of 4.6 in a small sample set of 96 cases and 50 controls has generated considerable excitement in the genetics community. The p-value of the strongest SNP association surpassed the genome-wide significance threshold after the Bonferroni correction. More importantly, this finding was replicated in the following-up studies (Donoso *et al.*, 2010). Undoubtedly, this encouraging finding raised the confidence among researchers to detect genetic variants that underlie various complex diseases through GWAS.

Genome-Wide Association Studies: From Polymorphism to Personalized Medicine, ed. K. Appasani. Published by Cambridge University Press.

In 2007, the Wellcome Trust Case Control Consortium (WTCCC) published the results of seven GWAS, including Bipolar Disorder, Coronary Artery Disease, Crohn's Disease, Hypertension, Rheumatoid Arthritis, Type 1 Diabetes, and Type 2 Diabetes (The Wellcome Trust Case Control Consortium, 2007). The WTCCC study is considered the starting point of large-scale GWAS (Visscher *et al.*, 2012). Since then, an increasing number of GWAS have been conducted and over 10,000 loci have been reported to be significantly associated with at least one complex trait (see the web resource of GWAS catalog (Hindorff et al., 2009), http://www.genome.gov/gwastudies/).

Despite the success of GWAS, these significantly associated SNPs can only explain a small portion of genetic contributions to complex diseases (T.A. Manolio *et al.*, 2009). For example, human height is a highly heritable trait whose heritability is estimated to be around 80%; that is, 80% of variation in height within the same population can be attributed to genetic effects (Visscher *et al.*, 2008). Based on large-scale GWAS, about 180 loci have been reported to be significantly associated with human height (Allen *et al.*, 2010). However, these loci together can only explain about 5–10% of variation in height (Visscher, 2008; T.A. Manolio *et al.*, 2009; Allen *et al.*, 2010). This phenomenon is referred to as "the missing heritability" (Maher, 2008; T.A. Manolio *et al.*, 2009; T. Manolio, 2010). Since then, finding the missing heritability has drawn much attention from worldwide researchers. In 2010, Yang *et al.* showed that 45% of the heritability for human height can be explained by using all genotyped common SNPs (Yang *et al.*, 2010). This result implies that most of the heritability is not missing: given the limited sample size, many individual effects of genetic markers are too weak to pass the genome-wide significance, and thus those risk genetic variants remain undiscovered. So far, people have found similar genetic architectures for many other complex traits (Visscher *et al.*, 2012), such as metabolic syndrome traits (Vattikuti *et al.*, 2012), psychiatric disorders (S.H. Lee *et al.*, 2012; Cross-Disorder Group of the Psychiatric Genomics Consortium, 2013a), that is, the phenotype is affected by many genetic variants with small or modest effects, which is usually referred to as "polygenicity." The polygenicity of complex traits is further supported by recent GWAS with larger sample sizes, in which more associated common SNPs with moderate effects have been identified (e.g. Morris *et al.*, 2012).

In this chapter, the concept of heritability is first introduced in Section 3.2, because it characterizes the global contribution of genetic factors to a phenotype. Statistical methods for association mapping which aim at identifying local parts of genome related to the phenotype are presented in Section 3.3, and statistical methods for genetic risk prediction are provided in Section 3.4. Simulation is conducted in Section 3.5 to illustrate how these concepts are connected. Future directions are discussed in Section 3.6 and our conclusion is given in Section 3.7.

3.2. Statistical methods for heritability estimation

3.2.1. Overview of heritability estimation

We present an overview of heritability estimation in this subsection, and provide more details in the following subsections. Let us first introduce the concepts about "heritability" (Visscher *et al.*, 2008). Broad-sense heritability (H^2) is defined as the proportion of the phenotypic variance that can be attributed to the genetic factors. It can be further partitioned into three components corresponding to additive genetic effects, dominant effects, and epistasis (Cordell, 2009; Wan *et al.*, 2010), respectively. The first component that corresponds to the additive effects is defined as "narrow-sense" heritability (h^2). Based on family data, heritability can be either estimated from balanced designs using linear regression (Falconer, 1996), or unbalanced designs using linear mixed models (LMM; Hopper, 1993).

Recently, an LMM-based approach was proposed to estimate heritability based on the genotype data of thousands of samples (Yang *et al.*, 2010). Because information from ungenotyped markers can only be partially captured in this estimation due to the imperfect linkage disequilibrium (LD) between them and genotyped markers, the variance explained by genotyped SNPs should be less than narrow-sense heritability. For convenience, the estimated variance explained by the genotyped markers is referred to as chip-heritability (h_g^2). In GWAS, many associated risk variants have been identified at the genome-wide significance level (e.g., p-value $< 5 \times 10^{-8}$), and heritability explained by them is denoted as h^2_{gwas}.

For the missing heritability problem, it is originally referred to as the gap between h^2 and h^2_{gwas}. In fact, this gap can be further decomposed into two parts: $h^2 - h_g^2$ and $h_g^2 - h^2_{\text{gwas}}$. The former exists due to the fact that the available genotyping platform can only partially capture genetic information, such as imperfect LD between genotyped markers and ungenotyped causal markers. The latter is due to the limited sample size which means that many SNPs with small or moderate effects remain undiscovered.

In the following subsections, we will provide a brief introduction of heritability estimation based on both pedigree data and GWAS data. We will emphasize the connections among the above concepts and present a unified framework based on LMM.

3.2.2. Heritability estimation based on pedigree data

R.A. Fisher developed the theoretical basis for heritability estimation (Fisher, 1918), in which the observed phenotypic similarity between relatives was related to their degree of genetic relationship. In statistical genetics, genetic covariance between relatives is used to describe their genetic resemblance. Before providing the details of the models used for heritability estimation based on pedigree data, we would like to introduce the connections among these important concepts: genetic covariance between relatives and variance of genetic effects (additive, dominance effects, and epistasis).

Let σ_P^2 be the variance of the phenotype and it can be decomposed as

$$\sigma_P^2 = \sigma_A^2 + \sigma_D^2 + \sigma_I^2 + \sigma_E^2, \tag{1}$$

where $\sigma_A^2, \sigma_D^2, \sigma_I^2$, and σ_E^2 correspond to the variance of additive effects, dominance effects, epistasis and environmental effects, respectively. A general formula[1] for the genetic covariance between two relatives has been derived based on these variance components (Falconer, 1996). To obtain a simple interpretation, epistasis is not considered at first and the genetic covariance between a pair of individuals (indexed by j_1 and j_2) is given by

$$\mathrm{Cov}(j_1, j_2) = K_{\mathrm{IBD}}(j_1, j_2)\sigma_A^2 + K_{\mathrm{IBD}_2}(j_1, j_2)\sigma_D^2, \tag{2}$$

where $K_{\mathrm{IBD}}(j_1, j_2)$ is the expected proportion of their genomes sharing one chromosome identical-by-descent (IBD); $K_{\mathrm{IBD}_2}(j_1, j_2)$ is the expected proportion of their genomes with both chromosomes IBD. Taking first-degree relatives as examples, $K_{\mathrm{IBD}}(j_1, j_2) = 1/2$ and $K_{\mathrm{IBD}_2}(j_1, j_2) = 0$ for parent–offspring pairs; $K_{\mathrm{IBD}}(j_1, j_2) = 1/2$ and $K_{\mathrm{IBD}_2}(j_1, j_2) = 1/4$ for full sibling pairs. The dominance effects do not contribute to the genetic covariance of parent–offspring pairs, because they require sharing two copies of genes while parents can only transmit one copy of each gene to their offspring (Falconer, 1996; Visscher *et al.*, 2008). Thus, the estimated heritability based on parent–offspring pairs is indeed narrow-sense heritability (h^2). When full sibling pairs are available, it becomes possible to estimate both broad-sense heritability and narrow-sense heritability.

Now let us consider an illustrative example for heritability estimation. Suppose there are pairs of full siblings available from n distinct pedigrees. Let $i = 1, \ldots, n$ be the index of pedigrees and $j = 1, 2$ be the index of the full siblings. Their phenotypic values (e.g., height) are collected in $\mathbf{Y} = [y_{ij}] \in \mathbb{R}^{n \times 2}$ and the i-th row of $\mathbf{Y}$ is denoted as $\mathbf{y}_i$. Consider the following model: N

$$\mathbf{y}_i = \boldsymbol{\mu} + \mathbf{g}_i + \mathbf{e}_i, \tag{3}$$

where $\boldsymbol{\mu}$ is the population mean, $\mathbf{g}_i$ denotes the genetic component that affects the phenotype, and $\mathbf{e}_i \sim N(0, \sigma_E^2\mathbf{I})$ corresponds to environmental effects.[2] Assuming $\mathbf{g}_i$ and $\mathbf{e}_i$ are independent, then we have

$$\mathrm{Cov}(\mathbf{y}_i) = \mathrm{Cov}(\mathbf{g}_i) + \mathrm{Cov}(\mathbf{e}_i). \tag{4}$$

According to Equation (2),

$$\Pr(\mathbf{y}_i | \mu, \sigma_A^2, \sigma_D^2, \sigma_E^2) = \mathcal{N}(\boldsymbol{\mu}, \boldsymbol{\Sigma}), \tag{5}$$

where

$$\sum = \mathrm{Cov}(\mathbf{y}_i) = \begin{pmatrix} 1 & \frac{1}{2} \\ \frac{1}{2} & 1 \end{pmatrix}\sigma_A^2 + \begin{pmatrix} 1 & \frac{1}{4} \\ \frac{1}{4} & 1 \end{pmatrix}\sigma_D^2 + \begin{pmatrix} 1 & 0 \\ 0 & 1 \end{pmatrix}\sigma_E^2.$$

At this point, we may view $\mathbf{y}_i (i = 1, \ldots, n)$ as data points drawn from the distribution $\mathcal{N}(\boldsymbol{\mu}, \boldsymbol{\Sigma})$. Therefore, as long as n is much larger than the number of

[1] This formula is derived under the following assumptions: random mating, no inbreeding, Hardy–Weinberg equilibrium, and linkage equilibrium. Detailed derivation can be found in Falconer (1996) and Lynch and Walsh (1998).

[2] Here $\mathbf{I}$ is a 2×2 identity matrix. To keep the notation uncluttered, we do not explicitly state the dimension of the identity matrix because it can be clearly determined in the context.

parameters in model (5), we can obtain accurate estimates of these variance components $(\sigma_A^2, \sigma_D^2, \sigma_E^2)$, and then obtain narrow-sense heritability and broad-sense heritability as follows:

$$h^2 = \frac{\sigma_A^2}{\sigma_A^2 + \sigma_D^2 + \sigma_E^2}, \quad H^2 = \frac{\sigma_A^2 + \sigma_D^2}{\sigma_A^2 + \sigma_D^2 + \sigma_E^2}. \tag{6}$$

Although the above model might be too simple to capture complexity in real data analysis, it can be extended to handle more general and complex pedigrees (Hopper, 1993). Let us first consider the case that all the data come from regular pedigrees, that is, relatives from n pedigrees have the same genetic structure and the number of relatives is the same across all n pedigrees, say q. For data from regular pedigrees, the phenotype data $\mathbf{y}_i \in \mathbb{R}^q$ can still be considered as a data point sampled from a multivariate normal distribution:

$$\mathbf{y}_i \sim N(\boldsymbol{\mu}, \mathbf{K}_{\text{IBD}}\, \sigma_A^2 + \mathbf{K}_{\text{IBD}_2}\, \sigma_D^2 + \sigma_E^2 \mathbf{I}), \quad i = 1, \ldots, n, \tag{7}$$

where $\mathbf{K}_{\text{IBD}}$ and $\mathbf{K}_{\text{IBD}_2}$ are the matrix corresponding to $K_{\text{IBD}}(j_1, j_2)$ and $K_{\text{IBD}_2}(j_1, j_2)$ in Equation (2). Assuming the independence of different pedigrees, the log-likelihood of model (7) can be written as

$$LL(\boldsymbol{\mu}, \sigma_A^2, \sigma_D^2, \sigma_E^2) = -\frac{n}{2} logdet(\boldsymbol{\Sigma}) - \sum_{i=1}^{n} \frac{1}{2} (\mathbf{y}_i - \boldsymbol{\mu})^T \boldsymbol{\Sigma}^{-1} (\mathbf{y}_i - \boldsymbol{\mu}) + \text{constant}, \tag{8}$$

where $\boldsymbol{\Sigma} = \mathbf{K}_{\text{IBD}}\, \sigma_A^2 + \mathbf{K}_{\text{IBD}_2}\, \sigma_D^2 + \sigma_E^2 \mathbf{I}$. Parameters can be estimated by maximizing this log-likelihood.

Because pedigree data collected from human families are often not regular, model (7) needs to be further extended to handle data from general pedigrees (Lange *et al.*, 1976). Let q_i be the size of the i-th pedigree. Likewise, the phenotype vector $\mathbf{y}_i \in \mathbb{R}^{q_i}$ comes from $N(\boldsymbol{\mu}_i, \boldsymbol{\Sigma}_i)$, where $\boldsymbol{\Sigma}_i = \mathbf{K}_{\text{IBD}}^{(i)}\, \sigma_A^2 + \mathbf{K}_{\text{IBD}_2}^{(i)}\, \sigma_D^2 + \sigma_E^2 \mathbf{I}$. The log-likelihood of $\mathbf{y}_i$ can be written as

$$LL_i(\boldsymbol{\mu}_i, \sigma_A^2, \sigma_D^2, \sigma_E^2) = -\frac{1}{2} logdet(\boldsymbol{\Sigma}_i) - \frac{1}{2} (\mathbf{y}_i - \boldsymbol{\mu}_i)^T \boldsymbol{\Sigma}_i^{-1} (\mathbf{y}_i - \boldsymbol{\mu}_i) + \text{constant}. \tag{9}$$

Although Equation (9) looks very similar to Equation (8), it corresponds to an unbalanced design (because each pedigree allows to have different sizes). Maximum likelihood methods were introduced by Lange *et al.* (1976) using the expectation–maximization (EM) algorithm. Now people often use restricted maximum likelihood (REML) methods (Thompson, 1977b, 1977a; Searle *et al.*, 2006) for parameter estimation.

Other extensions of model (7) have also been investigated. To account for common environmental factors shared by relatives, another variance component can be added into the model. Then the covariance matrix becomes

$$\boldsymbol{\Sigma} = \mathbf{K}_{\text{IBD}}\, \sigma_A^2 + \mathbf{K}_{\text{IBD}_2}\, \sigma_D^2 + \mathbf{K}_C\, \sigma_C^2 + \sigma_E^2 \mathbf{I}, \tag{10}$$

where σ_C^2 is the variance component corresponding to shared environmental factors, and $\mathbf{K}_C$ is the matrix specifying relatives who share common

Table 3.1. Coefficients of additive effects, dominance effects and epistasis for different pairs of relatives.

		K_{IBD}	K_{IBD_2}	K^2_{IBD}	$K^2_{\text{IBD}_2}$	$K_{\text{IBD}}K_{\text{IBD}_2}$
MZ twin		1	1	1	1	1
First-degree	Full sibling	1/2	1/4	1/4	1/16	1/8
	Parent–offspring	1/2	0	1/4	0	0
Second-degree	Grandparent–offspring	1/4	0	1/16	0	0
Third-degree	First cousins	1/8	0	1/64	0	0

environmental factors. For example, $\mathbf{K}_C(j_1, j_2) = 1$ when j_1 and j_2 are monozygotic twins and live together. The value in $\mathbf{K}_C$ can also be chosen according to the time that the pair of relatives have lived together (Hopper & Mathews, 1982, 1983), while inappropriate specification of $\mathbf{K}_C$ may make some variance componenents unidentifiable. To allow epistasis, the covariance matrix can be modified as follows (Hopper, 1993; Falconer, 1996):

$$\Sigma = \mathbf{K}_{\text{IBD}}\,\sigma_A^2 + \mathbf{K}_{\text{IBD}_2}\sigma_D^2 + \mathbf{K}^2_{\text{IBD}}\sigma^2_{AA} + \mathbf{K}^2_{\text{IBD}_2}\,\sigma^2_{DD} + \mathbf{K}_{\text{IBD}}\mathbf{K}_{\text{IBD}_2}\,\sigma^2_{AD} + \sigma^2_E\mathbf{I} \tag{11}$$

where $\mathbf{K}^2_{\text{IBD}}\,\sigma^2_{AA} + \mathbf{K}^2_{\text{IBD}_2}\,\sigma^2_{DD} + \mathbf{K}_{\text{IBD}}\mathbf{K}_{\text{IBD}_2}\,\sigma^2_{AD}$ represents the genetic covariance contributed by pairwise interactions of all loci. The values in these matrices for different pairs of relatives are summarized in Table 3.1.

3.2.3. Heritability estimation based on GWAS data

So far, we have discussed how to estimate heritability from pedigree data. In the genomic era, the availability of genome-wide SNP markers can provide new insight in understanding the relationship between the genotype and phenotype. For instance, many complex traits have been shown to be highly polygenic, that is, there exist many genetic variants with small or moderate effects jointly affecting those complex traits (Visscher *et al.*, 2012). In this subsection, we introduce these recently developed statistical methods (e.g., Yang et al., 2010; Yang *et al.*, 2011a), which can be used to explore genetic architectures based on genome-wide markers.

3.2.3.1. A basic LMM

Suppose there are n subjects with their genotype matrix $\mathbf{G} = [g_{im}] \in \mathbb{R}^{n\times M}$ and their phenotype vector $\mathbf{y} \in \mathbb{R}^{n\times 1}$, where M is the number of SNP markers and $g_{im} \in \{0, 1, 2\}$ is the number of copies of the reference allele for the m-th SNP of the i-th individual. Consider the following LMM:

$$\begin{aligned} \mathbf{y} &= \mathbf{X}\boldsymbol{\beta} + \mathbf{W}\mathbf{u} + \mathbf{e}, \\ \mathbf{u} &\sim N(0, \sigma_u^2\mathbf{I}), \\ \mathbf{e} &\sim N(0, \sigma_e^2\mathbf{I}), \end{aligned} \tag{12}$$

where $\mathbf{X} \in \mathbb{R}^{n\times c}$ is the design matrix which collects the intercept of the regression model and all covariates, such as age, sex, and a few principal components (PCs) of the genotype data (PCs are used for adjustment of the population structure, as shown in Section 3.3); $\boldsymbol{\beta}$ represents fixed effects; c is the number of fixed effects; $\mathbf{W} = [w_{im}] \in \mathbb{R}^{n \times M}$ is the standardized genotype matrix given by

$$w_{im} = \frac{(g_{im} - p_m)}{\sqrt{2p_m(1 - p_m)M}}, \tag{13}$$

where p_m is the frequency of the reference allele; $\mathbf{u}$ is the random effect from $\mathcal{N}(0, \sigma_u^2 \mathbf{I})$, and $\mathbf{e}$ is the residual error from $N(0, \sigma_e^2 \mathbf{I})$. The fixed effect $\boldsymbol{\beta}$ and two variance components (σ_u^2, σ_e^2) need to be estimated from data. Clearly, LMM (12) assumes the individual effects of genetic variants additively affect the phenotype.

The likelihood of model (12) can be written as

$$\Pr(\mathbf{y}|\,\boldsymbol{\beta}, \mathbf{u}, \sigma_u^2, \sigma_e^2) = N(\mathbf{y}|\mathbf{X}\boldsymbol{\beta} + \mathbf{W}\mathbf{u}, \sigma_e^2 \mathbf{I}) \ , \ \textit{with} \ \Pr(\mathbf{u}) = N(0, \sigma_u^2 \mathbf{I}). \tag{14}$$

To obtain the likelihood without random effects, we integrate out $\mathbf{u}$ as

$$\Pr(\mathbf{y}|\,\boldsymbol{\beta}, \sigma_u^2, \sigma_e^2) = \int \Pr(\mathbf{y}|\,\boldsymbol{\beta}, \mathbf{u}, \sigma_u^2, \sigma_e^2)\Pr(\mathbf{u})d\mathbf{u}. \tag{15}$$

Because both $\Pr(\mathbf{y}|\,\boldsymbol{\beta}, \mathbf{u}, \sigma_u^2, \sigma_e^2)$ and $\Pr(\mathbf{u})$ are Gaussian, this integration has an analytically close form (a detailed calculation of integration (15) can be found, for example, in chapter 2 of Bishop and Nasrabadi (2006):

$$\Pr(\mathbf{y}|\,\boldsymbol{\beta}, \sigma_u^2, \sigma_e^2) = N(\mathbf{X}\boldsymbol{\beta}, \mathbf{W}\mathbf{W}^T \sigma_u^2 + \sigma_e^2 \mathbf{I}), \tag{16}$$

that is, the phenotype $\mathbf{y}$ is modeled based on a multivariate normal distribution with mean $\mathbf{X}\boldsymbol{\beta}$ and variance $\mathbf{W}\mathbf{W}^T \sigma_u^2 + \sigma_e^2 \mathbf{I}$. Then the identity-by-state (IBS)-based genetic relationship matrix is defined as

$$\mathbf{K}_{\text{IBS}} = \mathbf{W}\mathbf{W}^T, \tag{17}$$

and model (16) can be written as

$$\mathbf{y} \sim N(\mathbf{X}\boldsymbol{\beta}, \mathbf{K}_{\text{IBS}} \sigma_u^2 + \sigma_e^2 \mathbf{I}). \tag{18}$$

Now we can see the basic idea of the LMM-based approach for heritability estimation is that the relationship between the phenotype and genotype is modeled as multivariate normal distribution (18). The phenotypic similarity is captured by the covariance $\boldsymbol{\Sigma} = \mathbf{K}_{\text{IBS}} \sigma_u^2 + \sigma_e^2 \mathbf{I}$, where $\mathbf{K}_{\text{IBS}}$ could be considered as a good approximation of $\mathbf{K}_{\text{IBD}}$ due to the availability of genome-wide dense markers used for its construction. Based on this interpretation, the estimated heritability $\left(\frac{\sigma_u^2}{\sigma_u^2 + \sigma_e^2}\right)$ seems to correspond to narrow-sense heritability h^2. However, this estimate could have a downward bias in the GWAS setting (Vattikuti *et al.*, 2012; Zaitlen *et al.*, 2013). This could be understood by noticing that $\mathbf{K}_{\text{IBS}}$ actually captures two sources of information: the larger elements in $\mathbf{K}_{\text{IBS}}$ estimate genetic similarities of related individuals, while the smaller elements estimate genetic similarities of unrelated individuals based on genotyped SNP markers. The obtained ratio $\left(\frac{\sigma_u^2}{\sigma_u^2 + \sigma_e^2}\right)$ will lie between narrow-sense heritability h^2 and chip-heritability h_g^2. To estimate h_g^2 accurately, the Yang–Visscher method (Yang *et al.*, 2010) advocated the removal of closely related individuals.

Next, we continue our discussion about this LMM-based approach for chip-heritability estimation. Checking Equations (12–16), we find that it has several assumptions:

- First, to model the relationship between the phenotype and genotypes, Equation (12) makes use of all genotyped SNPs rather than the causal SNPs of the phenotype. The phenotype of interest is believed to be affected by a fraction of the genome, which suggests that a large number of noise SNPs have been included in the model. If this is true, how much will it affect the chip-heritability estimation?
- Second, in Equation (12), the effect size $\mathbf{u}$ and residual $\mathbf{e}$ are assumed to be normally distributed. Is this normality assumption critical?
- Third, the genotype matrix $\mathbf{G}$ is standardized using Equation (13). Essentially, this standardization implicitly assumes that the variance of the per-allele effect size at the m-th SNP is proportional to $\frac{1}{p_m(1-p_m)}$. As a result, variants with lower MAFs tend to have larger effect sizes. How much does this assumption influence the result?
- Fourth, LD of the genotyped SNPs widely exists, and the LD pattern across the genome is not even. Without modeling this effect, will contributions of causal SNPs be overestimated in high-LD regions and underestimated in low-LD regions?

Recently, simulation has been done to investigate the impacts of these assumptions (Speed *et al.*, 2012). Violations of the first three assumptions were found to have only minor effects on chip-heritability estimation, but the LD effects can be substantial. An empirical approach was proposed to adjust for the LD effects in chip-heritability estimation (Speed *et al.*, 2012).

3.2.3.2. Extensions of LMM

The exclusion of related individuals results in a reduction of the sample size, which leads to a larger standard error of the estimated chip-heritability. Recently, a modified version of the LMM-based approach was proposed to handle related individuals (Zaitlen *et al.*, 2013). When related individuals are collected in GWAS, $\mathbf{K}_{\text{IBS}}$ (constructed from genome-wide markers) can be decomposed into two parts: one for related individuals and another for unrelated individuals. Specifically, a threshold τ is applied to $\mathbf{K}_{\text{IBS}}$ such that the elements in $\mathbf{K}_{\text{IBS}}$ below τ are set to be 0, denoted as $\mathbf{K}_{\text{IBS}>\tau}$ ($\tau = 0.025$ or 0.05 was suggested in Zaitlen *et al.*, 2013). When $\mathbf{K}_{\text{IBS}>\tau}$ is used to replace $\mathbf{K}_{\text{IBS}}$, model (16) becomes

$$\mathbf{y} \sim N(\mathbf{X}\boldsymbol{\beta}, \mathbf{K}_{\text{IBS}>\tau}\,\sigma_g^2 + \sigma_e^2\mathbf{I}). \tag{19}$$

The logic here is that the large entries in $\mathbf{K}_{\text{IBS}}$ are good estimates of the pairwise IBD of related individuals, and the estimated heritability corresponds to narrow-sense heritability $h^2 = \frac{\sigma_g^2}{\sigma_g^2+\sigma_e^2}$.

In order to estimate chip-heritability, the following model has been proposed

$$\mathbf{y} \sim N(\mathbf{X}\boldsymbol{\beta}, \mathbf{K}_{\mathrm{IBS}>\tau}\,\sigma_\tau^2 + \mathbf{K}_{\mathrm{IBS}}\,\sigma_u^2 + \sigma_e^2\mathbf{I}), \tag{20}$$

where σ_u^2 corresponds to the phenotypic variance captured by the genotyping platform, and $\sigma_\tau^2 = \sigma_g^2 - \sigma_u^2$ corresponds to the variance contributed by other genetic factors captured by the pedigree structure. Thus, narrow-sense heritability and chip-heritability can be jointly obtained from model (20):

$$h^2 = \frac{\sigma_\tau^2 + \sigma_u^2}{\sigma_\tau^2 + \sigma_u^2 + \sigma_e^2}, \quad h_g^2 = \frac{\sigma_u^2}{\sigma_\tau^2 + \sigma_u^2 + \sigma_e^2}. \tag{21}$$

Model (20) can be further extended (Zaitlen *et al.*, 2013) to estimate broad-sense heritability (epistasis is ignored here) as follows:

$$\mathbf{y} \sim N(\mathbf{X}\boldsymbol{\beta}, \mathbf{K}_{\mathrm{IBS}>\tau}\,\sigma_\tau^2 + \mathbf{K}_{\mathrm{IBD}_2}\,\sigma_d^2 + \sigma_e^2\mathbf{I}), \tag{22}$$

where $\mathbf{K}_{\mathrm{IBD}_2}$ can be estimated from genome-wide dense markers (e.g., Browning and Browning, 2013). Then broad-sense heritability[3] can be obtained as

$$H^2 = \frac{\sigma_\tau^2 + \sigma_d^2}{\sigma_\tau^2 + \sigma_d^2 + \sigma_e^2}. \tag{23}$$

Besides heritability estimation, the availability of genome-wide markers enables us to explore the genetic architecture of complex traits by genome partitioning (Yang *et al.*, 2011a). The following genome-partition model can be considered as a straightforward extension of model (12):

$$\begin{aligned} \mathbf{y} &= \mathbf{X}\boldsymbol{\beta} + \sum_k \mathbf{W}_k\mathbf{u}_k + \mathbf{e}, \\ \mathbf{u}_k &\sim N(0, \sigma_{u_k}^2\mathbf{I}), \\ \mathbf{e} &\sim N(0, \sigma_e^2\mathbf{I}), \end{aligned} \tag{24}$$

where $\mathbf{W}_k$ is the k-th part of the genome, and $\sigma_{u_k}^2$ is the corresponding variance component. For example, this model can be applied to estimate the phenotypic variance explained by each chromosome (Yang *et al.*, 2011a), where $\mathbf{W}_k$ corresponds to the standardized genotype matrix of the k-th chromosome.

Although LMM-based approaches considered here focus on quantitative traits, they can be extended to case-control studies (S.H. Lee *et al.*, 2011), where ascertainment bias needs to be considered carefully (Golan and Rosset, 2013).

3.3. Association mapping

Heritability estimation is a valuable tool to evaluate the overall genetic contribution, but it cannot identify specific genetic variants that underlie human diseases. Association mapping is such a tool that aims at pinpointing the disease-associated genetic variants.

[3] Epistasis is ignored here.

Many statistical methods have been proposed to identify associated genetic variants for both quantitative traits and case-control studies. Excellent review of these methods can be found in a series of review papers (Balding, 2006; Asimit and Zeggini, 2010; Bansal *et al.*, 2010). A major challenge of association mapping for GWAS is the abundance of spurious associations caused by population structures and cryptic relatedness (Price *et al.*, 2010). Recently, LMM arises as an effective tool in association mapping based on GWAS data (Kang *et al.*, 2010; Lippert *et al.*, 2011; Korte *et al.*, 2012; Svishcheva *et al.*, 2012; Zhou and Stephens, 2012). Not only are efficient algorithms of LMM available to perform analysis of large-scale GWAS data (e.g., FastLMM: Lippert *et al.*, 2011; and GEMMA: Zhou and Stephens, 2012), but also it is very effective for correcting confounding by population structures, family structures, and cryptic relatedness (Price *et al.*, 2010; Sul and Eskin, 2013). We begin our introduction of association mapping methods with principal components analysis (PCA)-based approaches for correcting population structures, and then relate them to LMM.

In the sample collection stage of GWAS, it is not an easy task to control the homogeneity of the population background. When the samples are collected within genetically heterogeneous populations, allele frequencies at many loci can be different among subpopulations. If disease prevalence also differs across these subpopulations, the proportions of cases and controls sampled from these subpopulations will be different. Consequently, the observed difference of allele or genotype frequencies between cases and controls at many loci could be due to the population structure rather than disease association (Marchini *et al.*, 2004). Failing to adjust for population stratification will lead to spurious associations. PCA arises as an effective tool to deal with this problem (Price *et al.*, 2006; Zhu *et al.*, 2002). More specifically, PCA is applied to extract a small number of dimensions that can capture the population structure as much as possible. Let $\mathbf{W} = \mathbf{U}\mathbf{S}\mathbf{V}^T$ be the singular value decomposition of the standardized genotype matrix $\mathbf{W}$, as given in Equation (13). As we know, the first column $\mathbf{V}_1$ of $\mathbf{V}$ corresponds to the direction that captures the largest variance of genotypes, and the second column $\mathbf{V}_2$ represents the direction that captures the largest variance in the space orthogonal to $\mathbf{V}_1$, etc. Then, $\mathbf{U}$ can be considered as the coordinates after projecting $\mathbf{W}$ into the space spanned by $\mathbf{V}$, where the i-th row of $\mathbf{U}$ is the coordinate of the i-th individual. If we only consider the first few directions in $\mathbf{V}$, we can visualize all the samples in a low-dimensional space. Here we take HapMap data as an extreme example.[4] These samples are collected from 11 sites around the world (please refer to the caption of Figure 3.1 for information on these 11 sites). The first two columns of $\mathbf{U}$ are shown in Figure 3.1. As we clearly see, population structures can be well represented by the first few columns of $\mathbf{U}$. For example, individuals from CHB, CHD, and JPT form a cluster in the southwest of Figure 3.1 because Chinese and Japanese are very similar to each other, and individuals in CEU and TSI form another cluster in the northwest.

[4] In a real GWAS, collected samples will not be as diverse as the HapMap samples.

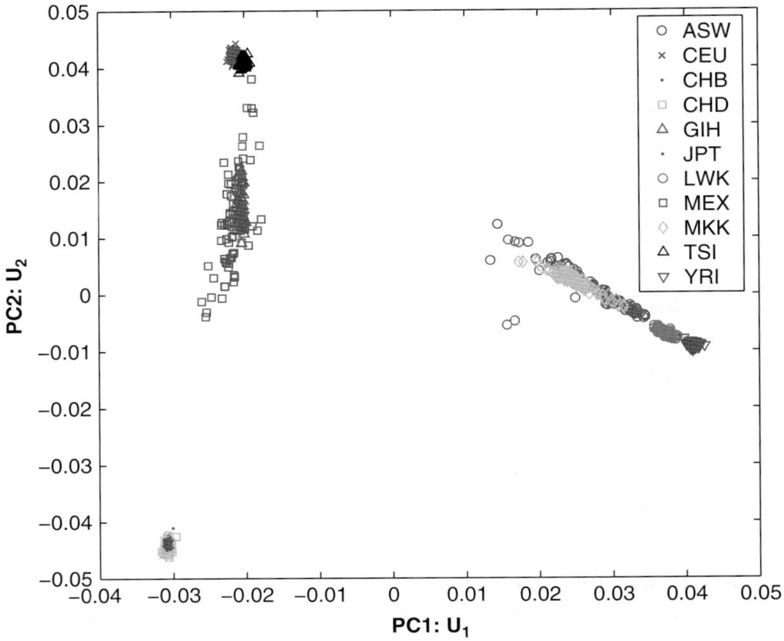

Figure 3.1. PCA analysis of HapMap data. Genotypes of 1184 individuals from 11 sites around the world are available at the HapMap website. Non-founders are removed from PCA analysis, and 988 individuals remain. ASW, African ancestry in Southwest USA; CEU, Utah residents with Northern and Western European ancestry from the CEPH collection; CHB, Han Chinese in Beijing, China; CHD, Chinese in Metropolitan Denver, Colorado; GIH, Gujarati Indians in Houston, Texas; JPT, Japanese in Tokyo, Japan; LWK, Luhya in Webuye, Kenya; MXL, Mexican ancestry in Los Angeles, California; MKK, Maasai in Kinyawa, Kenya; TSI, Toscani in Italia; YRI, Yoruba in Ibadan, Nigeria. A black and white version of this figure will appear in some formats. For the color version, please refer to the plate section.

To adjust for population stratification, the first k columns of $\mathbf{U}$, denoted as $\mathbf{U}^{(k)}$, can be included as covariates in the standard linear regression model:

$$\mathbf{y} = \mathbf{X}\boldsymbol{\beta} + \mathbf{U}^{(k)}\boldsymbol{\gamma} + \mathbf{G}_j\boldsymbol{\alpha} + \mathbf{e}, \tag{25}$$

where $\mathbf{X} \in \mathbb{R}^{n \times c}$ is the design matrix including the intercept (a column vector of 1) and covariates (e.g., sex and age); $\boldsymbol{\gamma} \in \mathbb{R}^k$ is the coefficient vector for $\mathbf{U}^{(k)}$; $\mathbf{G}_j \in \mathbb{R}^{n \times 1}$ is the genotype vector of the j-th SNP marker, and α is its effect size. By including $\mathbf{U}^{(k)}\gamma$, the effect of population structures on the phenotype $\mathbf{y}$ has been adjusted such that a large number of spurious associations can be avoid when performing hypothesis testing (26)

$$H_0 : \alpha = 0 \text{ vs. } H_A : \alpha \neq 0. \tag{26}$$

For case-control studies, linear regression can be replaced by logistic regression. Statistical analysis based on different models (e.g., linear regression and logistic regression) has been well implemented in the toolkit named PLINK (Purcell *et al.*, 2007).

Although PCA-based approaches can correct population stratification, they are not adequate when some related individuals are included in GWAS. In the presence of relatedness among samples, standard linear regression may not be valid. To account for the relatedness among samples, we consider the following LMM (Kang *et al.*, 2010):

$$\begin{aligned} \mathbf{y} &= \mathbf{X}\boldsymbol{\beta} + \mathbf{G}\boldsymbol{\alpha} + \mathbf{u} + \mathbf{e}, \\ \mathbf{u} &\sim N(0, \sigma_u^2\mathbf{K}), \\ \mathbf{e} &\sim N(0, \sigma_e^2\mathbf{I}), \end{aligned} \tag{27}$$

where $\mathbf{u}$ and $\mathbf{e}$ are random effects from $\mathcal{N}(0, \sigma_u^2\mathbf{K})$ and $\mathcal{N}(0, \sigma_u^2\mathbf{I})$, respectively; $\mathbf{K}$ is an $n \times n$ genetic relationship matrix; σ_u^2 and σ_e^2 are the two variance components. Typically, $\mathbf{K}$ is pre-computed from genome-wide markers using Equation (17). Although the estimated variance components (σ_u^2 and σ_e^2) may not provide a meaningful heritability estimation, as we discussed in Section 2.3.1, interpretation of these two variance components is not that important, as long as they successfully account for the relatedness among samples. Hypothesis testing (26) is of the main interest here.

Now we discuss the connection between PCA-based approaches (25) and LMM (27). Noticing that $\mathbf{K}$ in model (27) has eigen-decomposition as $\mathbf{K} = \mathbf{W}\mathbf{W}^T = (\mathbf{USV}^T)(\mathbf{USV}^T)^T = \mathbf{US}^2\mathbf{U}^T$, $\mathbf{U}^{(k)}$ in model (25) collects the eigenvectors belonging to the largest k eigenvalues. In the PCA-based model, $\mathbf{U}^{(k)}$ is considered as the design matrix of fixed effects, where k is chosen by users. While in LMM, all information in $\mathbf{K}$ is used and the two variance components are obtained automatically during model fitting. In this sense, the PCA-based model can be considered as a low-rank approximation of LMM. For illustration purpose, we simulated a phenotype vector based on the HapMap data set, in which 100 causal SNPs jointly contributed 50% phenotypic variance, i.e., $h^2 = 50\%$. The QQ-plots of association mapping is shown in Figure 3.2. The genomic inflation factors[5] were 8.9313, 1.0242, and 0.9996 for association mapping without adjustment, with PC adjustment, and with LMM adjustment, respectively. This result illustrates that both PCA-based approaches and LMM can effectively correct population stratification.

3.4. Risk prediction

3.4.1. Overview

Accurately predicting disease risk in individuals is an important task in human genetics studies. Risk prediction models provide a scientific basis to identify individuals with higher disease risk, and facilitate their disease prevention.

[5] The genomic inflation factor λ (Devlin and Roeder, 1999) is calculated as the median of all observed χ^2 test values divided by 0.456, where 0.456 is the median of the χ^2 distribution with degree of freedom $df = 1$, which is the null distribution. The inflation factor λ measures the overall deviation of the observed statistics from the null. Based on the assumption that only a small fraction of SNPs associated with the phenotype, λ should be very close to 1. In the presence of polygenic inheritance, λ can be slightly greater than 1 (Yang *et al.*, 2011b).

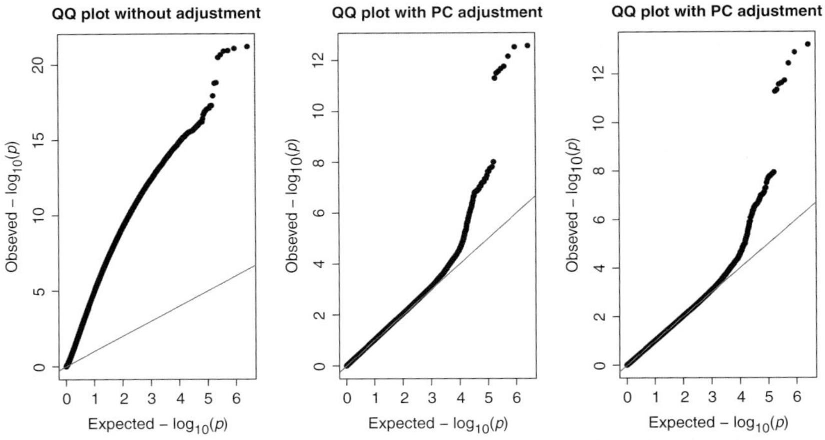

Figure 3.2. The QQ-plots of association mapping. Without adjusting population structure, many false positives will be produced based on the inflated *p*-values. After correction of population stratification, the genomic inflation factors are close to 1 for both PC-based adjustment and LMM-based adjustment.

Although hundreds of GWAS have been conducted on many complex human traits in recent years, there has been only limited success in translating these GWAS data into clinically useful risk prediction models. Most genetic risk prediction models, including those used by direct-to-customer genetic testing companies (e.g., 23andme), are constructed based on the SNPs with genome-wide significance. However, such SNPs typically account for only a small fraction of heritability and thus cannot provide satisfactory prediction accuracy (T.A. Manolio *et al.*, 2009). Recently, common SNPs have been shown to be able to explain a substantial fraction of heritability (Yang *et al.*, 2010). Researchers have started to build prediction models using whole-genome SNPs (de los Campos *et al.*, 2010). In the following, we first introduce some widely used methods in risk prediction, and then discuss their connection.

3.4.2. Models

Suppose we have the phenotype vector $\mathbf{y} \in \mathbb{R}^{n\times 1}$, the matrix $\mathbf{X} \in \mathbb{R}^{n\times c}$ collecting all covariates, and the genotype matrix $\mathbf{G} \in \mathbb{R}^{n\times M}$, where n is the number of samples, c is the number of covariates, and M is the number of SNPs. Let $\mathbf{X}_0 \in \mathbb{R}^{n_0\times c}$ and $\mathbf{G}_0 \in \mathbb{R}^{n_0\times M}$ be the covariate matrix and genotype matrix of new samples with unknown phenotype values, where n_0 is the number of new samples. The goal of genetic risk prediction is to build up a model based on available sample information ($\mathbf{y}$, $\mathbf{X}$ and $\mathbf{G}$), and then apply this model to predict phenotype values of the new samples (based on $\mathbf{X}_0$ and $\mathbf{G}_0$). Because genetic risk prediction is mainly interested in how to use genetic information, we ignore the covariates ($\mathbf{X}$) at first and focus on the genotype matrix $\mathbf{G}$.

In high-dimensional prediction problems, linear models are often competitive with nonlinear models. We consider two popular linear models for risk prediction here: ridge regression and lasso (e.g., see chapter 3 in Hastie *et al.*, 2009), because these two models are built up based on quite different assumptions. We will explain their model assumptions after introducing their model structures. Consider ridge regression (28)

$$\hat{\boldsymbol{\beta}}^{\text{Ridge}}(\lambda) = \min_{\boldsymbol{\beta}} \frac{1}{2} \| \mathbf{y} - \mathbf{W}\boldsymbol{\beta} \|^2 + \lambda \| \boldsymbol{\beta} \|_2^2, \tag{28}$$

where $\mathbf{W}$ is the standardized version (13) of $\mathbf{G}$, $\boldsymbol{\beta}$ is the regression coefficient vector, $\| \boldsymbol{\beta} \|_2^2 = \sum_{m=1}^{M} \boldsymbol{\beta}_m^2$ is the squared ℓ_2 norm, and λ is the regularization parameter controlling model complexity. A closed-form solution to (28) is given by

$$\hat{\boldsymbol{\beta}}^{\text{Ridge}}(\lambda) = (\mathbf{W}^T\mathbf{W} + \lambda\mathbf{I})^{-1}\mathbf{W}^T\mathbf{y}. \tag{29}$$

Direct application of Equation (29) may not be possible because it involves the inverse of an $M \times M$ matrix. By using the Woodbury matrix identity, Equation (29) can be rewritten as

$$\hat{\boldsymbol{\beta}}^{\text{Ridge}}(\lambda) = \mathbf{W}^T(\mathbf{W}\mathbf{W}^T + \lambda\mathbf{I})^{-1}\mathbf{y}. \tag{30}$$

Equation (30) only involves the inverse of an $n \times n$ matrix, which makes the computation possible. In fact, efficient computation algorithms are available for ℓ_2 regularized regression and classification, the singular value decomposition of $\mathbf{W}$ is the critical step to reduce computational costs. The details can be found in Hastie and Tibshirani (2004). Now consider the lasso problem (31)

$$\hat{\boldsymbol{\beta}}^{\text{Lasso}}(\lambda) = \min_{\boldsymbol{\beta}} \frac{1}{2} \| \mathbf{y} - \mathbf{W}\boldsymbol{\beta} \|^2 + \lambda \| \boldsymbol{\beta} \|_1, \tag{31}$$

where $\| \boldsymbol{\beta} \|_1 = \sum_{m=1}^{M} |\boldsymbol{\beta}_m|$ is the ℓ_1 norm of $\boldsymbol{\beta}$. Comparing lasso (31) with ridge regression (28), the only difference is that the ℓ_1 norm is used instead of the squared ℓ_2 norm. However, this change will make $\hat{\boldsymbol{\beta}}^{\text{Lasso}}(\lambda)$ quite different from $\hat{\boldsymbol{\beta}}^{\text{Ridge}}(\lambda)$, that is, many elements in $\hat{\boldsymbol{\beta}}^{\text{Lasso}}(\lambda)$ will be exactly zero for large λ, while $\hat{\boldsymbol{\beta}}^{\text{Ridge}}(\lambda)$ will not. Therefore, the underlying assumption of lasso is that the true coefficient $\boldsymbol{\beta}$ should be sparse, while ridge regression assumes $\boldsymbol{\beta}$ to be dense. For simplicity, we shall use the term "a sparse model" for a regression model with many zeros in its coefficient vector $\boldsymbol{\beta}$, and "a dense model" for a model with many nonzeros in $\boldsymbol{\beta}$.

Next, we investigate the risk prediction problem from the perspective of random-effects models (Goddard *et al.*, 2009; de los Campos *et al.*, 2013), and we shall see their connections with ridge regression and lasso. Consider a random-effects model (32)

$$\begin{aligned} \mathbf{y} &= \mathbf{W}\mathbf{u} + \mathbf{e}, \\ \mathbf{u} &\sim N(0, \sigma_u^2\mathbf{I}), \\ \mathbf{e} &\sim N(0, \sigma_e^2\mathbf{I}), \end{aligned} \tag{32}$$

where the effects of SNPs are viewed as random and drawn from $\mathcal{N}(0, \sigma_u^2\mathbf{I})$. This model can be written in the form of Gaussian distributions as

$$\begin{aligned}\Pr(\mathbf{u}) &= \mathcal{N}(0, \sigma_u^2 \mathbf{I});\\ \Pr(\mathbf{y}|\mathbf{u}, \sigma_u^2, \sigma_e^2) &= \mathcal{N}(\mathbf{W}\mathbf{u}, \sigma_e^2 \mathbf{I}).\end{aligned} \tag{33}$$

Because the marginal distribution of $\mathbf{u}$ is Gaussian and the conditional distribution of $\mathbf{y}$, given $\mathbf{u}$, is also Gaussian, we can analytically write the distribution of $\mathbf{y}$ and the conditional distribution of $\mathbf{u}$, given $\mathbf{y}$ as[6]

$$\Pr(\mathbf{y}|\sigma_u^2, \sigma_e^2) = N(0, \mathbf{W}\mathbf{W}^T \sigma_u^2 + \sigma_e^2 \mathbf{I}); \tag{34}$$

$$\Pr(\mathbf{u}|\mathbf{y}, \sigma_u^2, \sigma_e^2) = N\left(\left(\mathbf{W}^T\mathbf{W} + \frac{\sigma_e^2}{\sigma_u^2}\mathbf{I}\right)^{-1} \mathbf{W}^T\mathbf{y}, \left(\mathbf{W}^T\mathbf{W} + \frac{\sigma_e^2}{\sigma_u^2}\mathbf{I}\right)^{-1}\right). \tag{35}$$

Clearly, the posterior mean of $\mathbf{u}$ is given by

$$E(\mathbf{u}|\mathbf{y}, \sigma_u^2, \sigma_e^2) = \left(\mathbf{W}^T\mathbf{W} + \frac{\sigma_e^2}{\sigma_u^2}\mathbf{I}\right)^{-1} \mathbf{W}^T\mathbf{y}, \tag{36}$$

which is exactly the same with Equation (29) when $\lambda = \frac{\sigma_e^2}{\sigma_u^2}$. For the random-effects model, parameters σ_u^2 and σ_e^2 are typically estimated from data, by maximizing the likelihood of $\mathbf{y}$ (34), while the regularization parameter λ in (28) are often determined by cross-validation. Equation (36) is also known as the best linear unbiased prediction (BLUP) of the random effects $\mathbf{u}$.

So far we have seen the connection between ridge regression and random-effects models. Now we consider a well known sparse model in the machine learning literature, named the "relevance vector machine" (RVM; Tipping, 2001), and we shall see that RVM is a natural extension of random-effect model (32). The RVM is given as follows:

$$\begin{aligned}\mathbf{y} &= \sum_{j=1}^{M} \mathbf{W}_m u_m + \mathbf{e},\\ u_m &\sim N(0, \sigma_m^2),\\ \mathbf{e} &\sim N(0, \sigma_e^2 \mathbf{I}),\end{aligned} \tag{37}$$

where $\mathbf{W}_m$ is the m-th column of $\mathbf{W}$ and u_m is the m-th element of $\mathbf{u}$. Unlike model (32) in which all u_ms share a common variance σ_u^2, this model is more flexible by allowing each u_m a different σ_m^2. An empirical Bayes method (also known as evidence approximation or type II maximum likelihood in machine learning) has been developed to estimate model parameters (Tipping, 2001; Tipping and Faul, 2003). It can be shown that, during the optimization process, many σ_m^2 will become zero, which implies the posterior mean of corresponding u_m will be zero. This suggests that a sparse regression model can also be achieved in the framework of random-effects models.

To model the effect of covariates in risk prediction, it is straightforward to add them into model (28) as follows:

[6] Details can be found, for example, in chapter 2 of Bishop and Nasrabadi (2006).

$$(\hat{\beta}_0, \hat{\beta}) = \min_{\beta} \frac{1}{2} \| \mathbf{y} - \mathbf{X}\beta_0 - \mathbf{W}\beta \|^2 + \lambda \| \beta \|_2^2, \tag{38}$$

which can be shown to be equivalent to LMM (12).

3.5. Simulation study

We have introduced heritability estimation, association mapping, and risk prediction in GWAS. Comprehensive simulation studies have been conducted to compare the performance of different statistical methods. For example, a recent paper (Lippert *et al.*, 2013) studied the performance of LMM under various settings. In this section, we will mainly focus on the following interesting questions to explore the connection among heritability estimation, association mapping, and risk prediction:

- As we discussed in Section 4.2, LMM (ridge regression) is a dense model. However, there are only a fraction of SNPs associated with the phenotype of interest. Does LMM work for heritability estimation in such a case?
- If a substantial fraction of heritability can be explained by genotyped common SNPs, why can the identified SNPs at genome-wide significance only explain a small proportion of heritability? What is the reason for the gap between h_g^2 and h_{gwas}^2?
- If chip-heritability is 50%, can the prediction accuracy of risk prediction models achieve 50% (measured by R^2)?

Now we consider the following model:

$$\mathbf{y} = \mathbf{X}\beta + \mathbf{W}\mathbf{u} + \mathbf{e}, \tag{39}$$

where $\mathbf{X}$ is an $n \times 1$ vector of ones; $\beta = 1$ is the intercept; $\mathbf{W}$ is the standardized version of $\mathbf{G} = [g_{im}] \in \mathbb{R}^{n \times M}$, where each column of $\mathbf{G}$ is generated independently, $g_{im} \in \{0, 1, 2\}$ is sampled according to its allele frequency p_m with probability p_m^2, $2p_m(1 - p_m)$ and $(1 - p_m)^2$; $\mathbf{u} = [\mathbf{u}_{(1)}, \mathbf{0}^T]^T$ are the random effects, where $\mathbf{u}_{(1)}$ includes the first nz components of $\mathbf{u}$ and $\mathbf{0}$ is the $(M - nz) \times 1$ vector of zeros, $\mathbf{u}_{(1)} \sim N(0, \sigma_{u_{(1)}}^2 \mathbf{I})$; $\mathbf{e} \sim N(0, \sigma_e^2 \mathbf{I})$. In our simulation study, we set $n = 200$, $M = 500$, $\sigma_e^2 = 1$, then varied nz and $\sigma_{u_{(1)}}^2$ such that heritability h_g^2 is 50%, that is, the signal-noise-ratio was 1:1.

First, LMM was applied to simulated data sets for heritability estimation. The results are shown in Figure 3.3, which suggests that LMM works well for heritability estimation no matter whether the underlying model is sparse or dense. In fact, our recent theoretical analysis has shown that the REML estimator of LMM is still consistent for heritability estimation even though LMM is mis-specified (Jiang *et al.*, 2013).

Second, we consider the power of association mapping. As shown in Figure 3.4, when there are only $nz = 10$ SNPs jointly contributing all heritability, it is easy to identify these markers because their effect sizes are large (each one can account for 5% phenotypic variance on average). When $nz = 20$, fewer

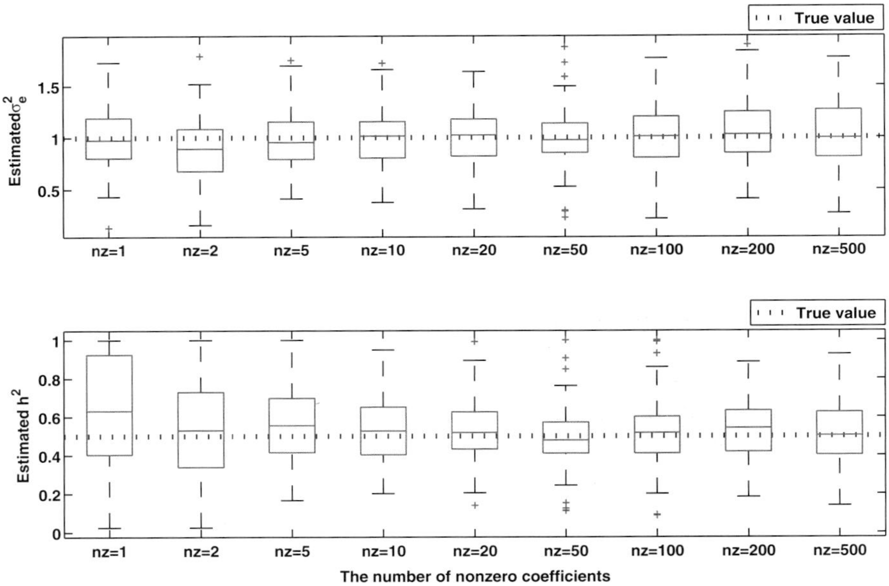

Figure 3.3. Performance of LMM for variance component and heritability estimation. Upper panel: the estimation of σ_e^2 under different *nz*. Lower panel: heritability estimation (h_g^2) under different *nz*. The results were summarized on 100 replications. We can see that LMM works well for heritability estimation under different *nz*. Although bias estimation of h_g^2 has been observed when $nz = 1$, it will disappear as the sample size tends to infinity. Here the sample size is $n = 200$. A black and white version of this figure will appear in some formats. For the color version, please refer to the plate section.

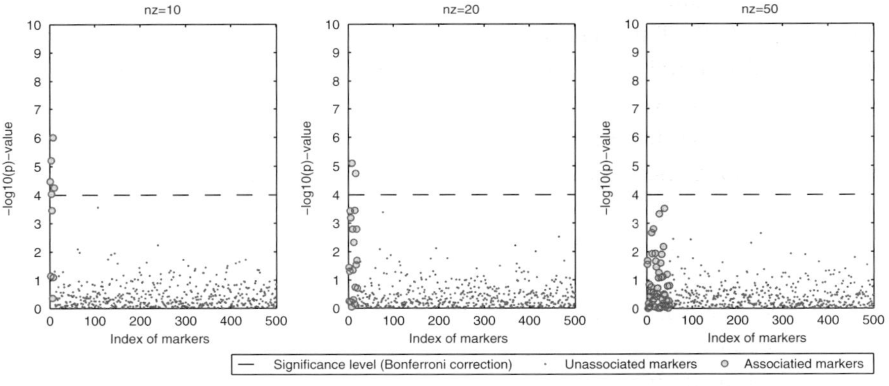

Figure 3.4. Illustration of association mapping ($n = 200$, $M = 500$) for different *nz*. The results from left to right are corresponding to $nz = 10$, 20, and 50, respectively. A black and white version of this figure will appear in some formats. For the color version, please refer to the plate section.

markers can be identified at the significance level after Bonferroni correction.[7] When $nz = 50$, none of the truly associated markers can be identified, because each marker can only explain 1% phenotypic variance on average,

[7] Because all markers are generated independently, Bonferroni correction is a good strategy to control false positives.

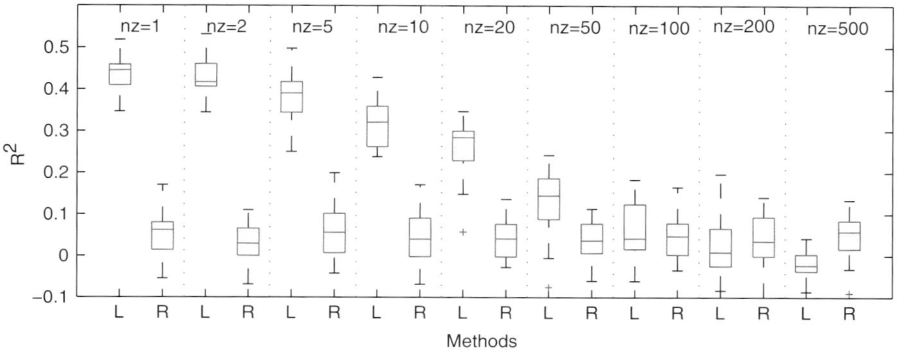

Figure 3.5. Comparison of lasso (L) and ridge (R) regression for different nonzero coefficients ($nz = 1, 2, 5, 10, 20, 50, 100, 200, 500$). The results were summarized based on 20 replications. A black and white version of this figure will appear in some formats. For the color version, please refer to the plate section.

and the limited sample size $n = 200$ does not allow them to be distinguishable from unassociated SNPs. Combining the results shown in Figures 3.3 and 3.4, we can see that although the overall effect (heritability) can be estimated ($n = 200$ and $M = 500$), the power of association mapping largely depends on the number of markers associated with the phenotype. In a real GWAS, typically, the sample size n can be a few thousands, and the total number of SNPs M can be more than one million. For some complex traits, the associated number of SNPs nz has been estimated to be around 10^3–10^4 (Stahl *et al.*, 2012). This makes identification of risk variants very challenging, although chip-heritability estimation can be relatively accurate. Clearly, this leads to the gap between h_g^2 and h_{gwas}^2.

Third, we applied both lasso and ridge regression to the simulated data. The predicted performance of these methods is shown in Figure 3.5. As expected, lasso worked better than ridge regression when nz was small. In particular, lasso nearly achieved $R^2 = 0.5$, which was the limit of prediction accuracy due to $h_g^2 = 50\%$. For large nz, none of these methods worked well, although ridge regression was slightly better for $nz \geq 100$. The reason for better performance of ridge regression was that the underlying model became denser, which matched the assumption of ridge regression. A natural question arises: why can ridge regression (equivalent to LMM) do a very good job of heritability estimation, when its prediction is far from the limit, i.e., $R^2 = 0.5$? The answer to this question is that they are essentially different problems: heritability estimation only requires the estimate of two variance components, which can be considered as "overall" effects of genetic markers, while risk prediction requires the estimate of posterior mean (Equation (36)) of random effects $\mathbf{u} = [u_1, \ldots, u_M]$, where u_m ($m = 1, \ldots, M$) corresponds to the "individual" effect of the m-th marker. A few thousands of samples are enough for estimating the overall effect (i.e., heritability), while a much larger sample size is needed to estimate those individual effects

(Chatterjee *et al.*, 2013; Wray *et al.*, 2013). In a word, unsatisfying prediction accuracy of ridge regression (i.e., the big gap between h_g^2 and R^2) is due to the limited sample size and the mismatched model structure ($nz \neq M$). Given the matched model, statistical theory implies that the prediction accuracy of ridge regression can approach its limit ($R^2 = 0.5$) as $n \to \infty$.

3.6. Future directions

The major challenge of GWAS data analysis is the polygenic genetic architecture of complex diseases: in the presence of numerous variants with small or modest effects, a large sample size is needed for both association mapping and risk prediction. However, sample recruitment may be expensive and time-consuming. In this section, we consider some alternative ways to boost statistical power of GWAS data analysis.

3.6.1. Exploration of common genetic risk factors shared in multiple GWAS

In fact, accumulating evidence suggests that different complex human traits are genetically correlated, that is, multiple traits share common genetic bases, which is formally known as "pleiotropy." In a systematic analysis of the National Human Genome Research Institute (NHGRI) catalog (http://www.genome.gov/gwastudies/) of published GWAS, 16.9% genes and 4.6% SNPs have been reported to show pleiotropic effects (Sivakumaran *et al.*, 2011). The power to detect schizophrenia-associated genetic variants could be greatly improved by exploiting the pleiotropy between schizophrenia and cardiovascular disease (Andreassen *et al.*, 2013). More recently, a study on genome-wide SNP data for five psychiatric disorders in 33,332 cases and 27,888 controls identified four significant loci (p-value $< 5 \times 10^{-8}$) affecting multiple psychiatric disorders (Cross-Disorder Group of the Psychiatric Genomics Consortium, 2013b). Further analysis showed that the genetic correlation between schizophrenia and bipolar disorder was high (0.68 ± 0.04 s.e.; Cross-Disorder Group of the Psychiatric Genomics Consortium, 2013a). Significant genetic correlations between several metabolic syndrome traits were also reported (Vattikuti *et al.*, 2012). A large-scale collaborative oncological gene–environment study also highlighted the existence of "carcinogenic pleiotropy," that is, the overlap between loci that confer genetic susceptibility to multiple types of tumor (Sakoda *et al.*, 2013).

All these findings are very exciting because they suggest that exploring the genetic correlations between phenotypes could be a promising strategy to improve the accuracy of risk prediction and the power of association mapping (Hartley *et al.*, 2012; Hartley and Sebastiani, 2013). Suppose we have two GWAS data sets, and their phenotypes and genotypes are given by $\mathbf{y} \in \mathbb{R}^{n_k \times 1}$ and $\mathbf{G} \in \mathbb{R}^{n_k \times M}$, where n_k is the sample size of the k-th GWAS, and $k = 1, 2$. Here we consider bivariate LMM for risk prediction by integrative analysis of these two data sets:

$$\begin{aligned}\mathbf{y}^{(1)} &= \mathbf{G}^{(1)}\mathbf{u}^{(1)} + \mathbf{e}^{(1)},\\ \mathbf{y}^{(2)} &= \mathbf{G}^{(2)}\mathbf{u}^{(2)} + \mathbf{e}^{(2)}.\end{aligned} \tag{40}$$

Let $u_m^{(k)}$ be the m-th element of $\mathbf{u}^{(k)}$. To model the genetic correlation, we assume that

$$\begin{bmatrix}\mathbf{u}_m^{(1)}\\ \mathbf{u}_m^{(2)}\end{bmatrix} \sim N(\mathbf{0}, \mathbf{\Sigma}_u), \quad \begin{bmatrix}\mathbf{e}^{(1)}\\ \mathbf{e}^{(2)}\end{bmatrix} \sim N(\mathbf{0}, \mathbf{\Sigma}_e) \tag{41}$$

where

$$\mathbf{\Sigma}_u = \begin{bmatrix}\sigma_{u_1}^2 & \rho_u\sigma_{u_1}\sigma_{u_2}\\ \rho_g\sigma_{u_1}\sigma_{u_2} & \sigma_{u_2}^2\end{bmatrix}, \quad \mathbf{\Sigma}_e = \begin{bmatrix}\sigma_{e_1}^2\mathbf{I} & 0\\ 0 & \sigma_{e_2}^2\mathbf{I}\end{bmatrix}. \tag{42}$$

Accordingly, the bivariate BLUP is given as follows:

$$\begin{bmatrix}\hat{\mathbf{u}}^{(1)}\\ \hat{\mathbf{u}}^{(2)}\end{bmatrix} = ((\mathbf{\Sigma}_{\mathbf{u}} \otimes \mathbf{I})^{-1} + \mathbf{W}^T\mathbf{\Sigma}_e^{-1}\mathbf{W})^{-1}\mathbf{W}^T\mathbf{\Sigma}_e^{-1}\begin{bmatrix}\mathbf{y}^{(1)}\\ \mathbf{y}^{(2)}\end{bmatrix}, \tag{43}$$

where $\mathbf{W} = \begin{bmatrix}\mathbf{G}^{(1)} & \mathbf{0}\\ \mathbf{0} & \mathbf{G}^{(2)}\end{bmatrix}$ and $\otimes$ stands for the Kronecker product.

In our simulation, we considered two quantitative traits, with each trait measured on 2000 individuals. Genotypes for these individuals on 20,000 SNPs were first simulated, with 2000 SNPs causal for each trait and heritability $h_g^2 = 60\%$. We varied the proportion of shared causal SNPs between the two traits, γ, to model different levels of genetic factor sharing. We simulated the effect sizes of the

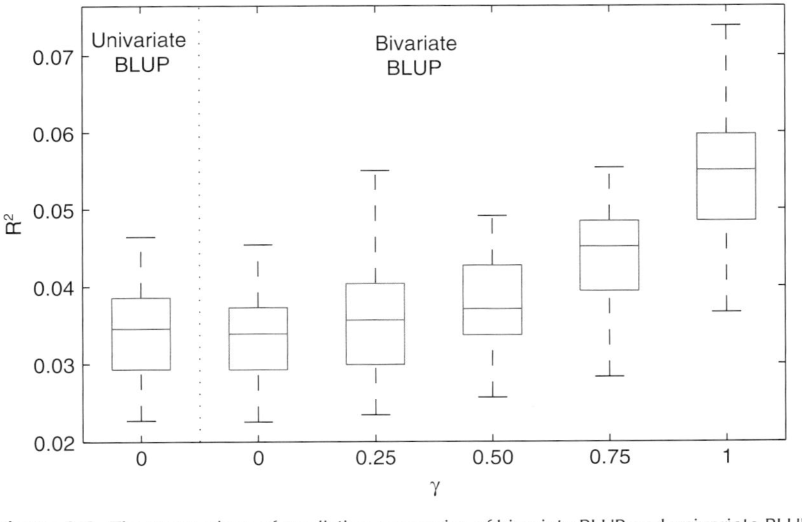

Figure 3.6. The comparison of prediction accuracies of bivariate BLUP and univariate BLUP. Sample size for each phenotype is 2000. Total number of SNPs is 20,000. Number of causal SNPs is 2000 for each phenotype. The parameter γ represents the proportion of shared causal SNPs. As γ increases, bivariate BLUP achieves better prediction accuracy than univariate BLUP. A black and white version of this figure will appear in some formats. For the color version, please refer to the plate section.

shared causal SNPs from a bivariate normal distribution with a correlation coefficient of 0.8. Then we used Equation (43) to predict the phenotypes based on the genotypes on the 20,000 SNPs. Prediction accuracy (measured by R^2 based on cross-validation) is shown in Figure 3.6. The results suggest that the prediction accuracy can indeed be improved by borrowing information between two genetically correlated traits. A comprehensive study based on this approach can be found in C. Li *et al.* (2013).

3.6.2. Integrative analysis with other sources of information

Integration of GWAS with related sources of information has proven to be a powerful and effective approach to boost the statistical power of GWAS data analysis. By integrating additional sources of information related to the phenotype of interest, potentially biologically important SNPs can be prioritized and the search space can be effectively reduced for association mapping and risk prediction. Furthermore, such integration also allows us to understand a biological link relating genetic variants to changes in the phenotype, which could be translated into clinical benefits.

One popular strategy for SNP prioritization is to map genetic variants to its target gene and prioritize genes using related information (Cantor *et al.*, 2010). Such mapping is relatively straightforward for the SNPs located in exonic, intronic, and promoter regions. The SNPs in intergenic regions are more challenging to map and they are often assigned to the closest genes, using a predefined window. There are three key challenges in such mapping. First, among the identified risk-associated SNPs, a large proportion of them are located in non-coding regions. For example, Cowper-Sal-lari *et al.* (2012) analyzed the SNPs associated with breast cancer risk from the NHGRI GWAS catalog (http://www.genome.gov/gwastudies/) and found that among the 44 reported risk-associated SNPs, only one of them maps to coding exons, while 25 map to introns. Second, the target gene of a risk-associated SNP in a non-coding region is often not the closest gene. For example Cowper-Sal-lari *et al.* (2012) showed that the rs4784227 SNP associated with breast cancer risk is located 18 kb away from its target gene, *TOX3*, instead located in the intronic region of LOC643714. Third, the criterion to assign SNPs to genes is often arbitrary, and a SNP can be mapped to multiple genes, which in turn affects prioritization results (Cantor *et al.*, 2010).

When SNPs are located in protein-coding regions, they could be interpretable as they change amino acids in proteins. Similarly, genetic variants in promoter regions could affect transcription of protein-coding sequences by changing the affinity of transcription factors (TF) and RNA polymerase to the promoter regions. After these SNPs are mapped to their target genes, they could be analyzed at either the gene or the pathway level (see Cantor *et al.* (2010) for a more detailed review). Related sources of information include annotation for gene structure (Chen and Witte, 2007; Veyrieras *et al.*, 2008; S.-I. Lee *et al.*, 2009), expression quantitative trait loci (eQTL) (Veyrieras *et al.*, 2008; S.-I. Lee *et al.*, 2009), evolutionary conservation (Chen and Witte, 2007; Veyrieras *et al.*, 2008; S.-I. Lee *et al.*, 2009), and pathway information (S.-I. Lee et al., 2009; Cantor *et al.*, 2010).

Contrary to the genetic variants in protein-coding regions, SNPs in non-coding regions do not directly modify amino acids in protein. As a result, it is more challenging to understand SNPs in non-coding regions (see Ward and Kellis (2012b) for more extensive reviews on related topics and issues). Regulatory annotations potentially provide valuable information to explain SNPs in non-coding regions from functional perspectives. For example, Cowper-Sal-lari *et al.* (2012) showed that allele T at the rs4784227 SNP, previously identified as associated with breast cancer risk, increases the affinity of *FOXA1* transcription factor to the site containing this SNP. Increased affinity of *FOXA1* factor in turn represses *TOX3* expression and upon repression of *TOX3*, the proliferation of breast cancer cells increases. Such examples imply potentials of regulatory annotations for prioritization and interpretation of SNPs in non-coding regions.

Recently, the Encyclopedia of DNA Elements (ENCODE) Consortium has generated vast amounts of data annotating gene expression and regulation from a large number of cell lines. Specifically, the ENCODE database provides experimental data on gene expression (RNA-seq), DNA methylation status (RRBS-seq), chromatin modifications (ChIP-seq), chromatin accessibility (DNase-seq and FAIRE-seq), TF binding sites (ChIP-seq), and long-range chromatin interactions (ChIA-PET, Hi-C, and 5 C). As of September 2012, more than 1600 data from 147 cell lines have been produced, including 2.89 million unique, non-overlapping DNase I hypersensitivity sites (DHSs) in 125 cell lines using DNase-seq and 630,000 binding regions of 119 DNA-binding proteins in 72 cell lines using ChIP-seq, among many (The ENCODE Project Consortium, 2012).

The vast amount of ENCODE data provides previously unattainable opportunities to investigate SNPs from functional or regulatory perspectives and accumulating evidence suggests that GWAS variants are enriched in ENCODE regulatory annotations. For example, The ENCODE Project Consortium (2012) examined 4492 risk-associated SNPs from the NHGRI GWAS catalog and found that 12% overlap TF binding regions and 34% overlap DHSs. Furthermore, integration with ENCODE data also has the potential to improve the mapping of risk-associated SNPs to their target genes. Specifically, ChIA-PET data measure long-range chromatin interaction and allow the prediction of various enhancer–promoter and promoter–promoter interactions, which can be used to link risk-associated SNPs to their target genes.

Integration of GWAS and ENCODE data is currently being investigated actively from various perspectives. RegulomeDB (Boyle *et al.*, 2012) and HaploReg (Ward and Kellis, 2012a) are valuable resources to annotate SNPs using regulatory annotations including ENCODE data. Specifically, RegulomeDB utilizes a large collation of eQTL, DNase-seq predictions, predicted TF binding sites, and positional weight matrices. Research on the prioritization of SNPs in non-coding regions using ENCODE and related data is still in its early stage, with few available approaches, including M. Li *et al.* (2013).

There are several issues to resolve for successful integration of GWAS and related information, especially for the SNPs in non-coding regions. First, genes are regulated and expressed in cell-type specific manner and genetic variants affect a phenotype through various genetic pathways. Hence, diverse data for different types of regulatory mechanisms from a large number of cell lines are required to be considered for such integration. Second, computationally, such integration with large-scale genomic data requires consideration of combinatorial patterns of high-dimensional data and identification of cell lines and data types that are more relevant to the phenotype of interest (Schaub *et al.*, 2012). When considering the large scale of the ENCODE database, it is almost impossible to assign weight to each cell line and data type manually. Hence, it is desirable to automatically determine weight for each ENCODE data in a data-driven way, based on their relevance to the phenotype of interest. Third, the SNPs identified in the GWAS might not be functional, but instead are in strong LD with functional SNPs. Hence, in such integration, larger sets of genetic variants need to be considered by taking the LD structure into account (Schaub *et al.*, 2012). Finally, although the ENCODE database provides large-scale, high-quality data for gene expression and regulation, a catalog for regulatory elements on the genome is still far from completion. As of September 2012, the ChIP-seq data generated by the ENCODE Consortium includes only 119 TFs, which is only a small fraction among 1800 known ones (The ENCODE Project Consortium, 2012). As the ENCODE and related projects are enlarged and integrated, available information could be richer and the accuracy and efficiency of SNP prioritization and interpretation could also be significantly improved.

3.7. Conclusion

In this chapter, we have provided a brief introduction of recent progress in GWAS, including heritability estimation, association mapping, and risk prediction. Heritability is a global measure which quantifies the overall contribution of genetic factors to a phenotype. There are several related concepts of heritability: broad-sense heritability (H^2), narrow-sense heritability (h^2), chip-heritability (h_g^2), and heritability explained by SNPs at the genome-wide significance level (h_{gwas}^2). Their relationship could be summarized as $H^2 \geq h^2 \geq h_g^2 \geq h_{\text{gwas}}^2$. The missing heritability problem is referred to as the gap between h^2 and h_{gwas}^2. As we discussed in this chapter, this gap can be further decomposed into two parts: $h^2 - h_g^2$, and $h_g^2 - h_{\text{gwas}}^2$. The former is due to genetic information lost in the genotyping platform, that is, the imperfect LD between genotyped variants and ungenotyped causal variants, and the latter could be attributed to the small-effect variants remaining undiscovered. Identification of those risk genetic variants is the central task of association mapping, in which confounding effects need to be adjusted to avoid spurious association findings. Risk prediction aims at translating these findings into clinical usage for identifying individuals at higher risk, which may greatly facilitate personalized disease prevention.

From the statistical point of view, we presented a unified framework for heritability estimation, association mapping, and risk prediction. Based on the connection between LMM and some other related approaches (e.g., ridge regression and PCA-based approaches), these three problems can be viewed as different applications of LMM: for heritability estimation, interpretation and estimation of variance components are the main focus, while inference on the fixed effects and estimation of the posterior mean of the random effects are of main interest for association mapping and risk prediction, respectively. In the simulation study, we have used examples to illustrate how these concepts are connected. As discussed before, accurate heritability estimation is possible even when only a few thousands of samples are available, but this sample size may not be enough for identifying those weak-effect variants and accurate risk prediction due to the polygenic genetic architecture of complex traits/diseases.

Facing the challenge presented by the polygenic genetic architecture and limited sample size, we have discussed several possible directions for further exploration of the GWAS data, including combining multiple genetically related GWAS data and incorporating other sources of information (e.g., eQTL and ENCODE data). Accumulating evidence suggests that integrative analysis of genomic data may be a promising direction in the post-GWAS era.

REFERENCES

Allen, H.L., Estrada, K., Lettre, G., *et al.* (2010). Hundreds of variants clustered in genomic loci and biological pathways affect human height. *Nature*, **467**(7317), 832–838.

Andreassen, O.A., Djurovic, S., Thompson, W.K., *et al.* (2013). Improved detection of common variants associated with schizophrenia by leveraging pleiotropy with cardiovascular-disease risk factors. *Am. J. Hum. Genet.*, **92**(2), 97–109.

Asimit, J. and Zeggini, E. (2010). Rare variant association analysis methods for complex traits. *Annu. Rev. Genet.*, **44**, 293–308.

Balding, D. (2006). A tutorial on statistical methods for population association studies. *Nature Rev. Genet.*, **7**(10), 781–791.

Bansal, V., Libiger, O., Torkamani, A. and Schork, N. J. (2010). Statistical analysis strategies for association studies involving rare variants. *Nature Rev. Genet.*, **11**(11), 773–785.

Bishop, C.M. and Nasrabadi, N.M. (2006). *Pattern Recognition and Machine Learning* (Vol. 1). Springer, New York.

Boyle, A., Hong, E., Hariharan, M., *et al.* (2012). Annotation of functional variation in personal genomes using RegulomeDB. *Genome Res.*, **22**(9), 1790–1797.

Browning, S.R. and Browning, B.L. (2013). Identity-by-descent-based heritability analysis in the northern Finland birth cohort. *Hum. Genet.*, **132**(2), 129–138.

Cantor, R., Lange, K. and Sinsheimer, J. (2010). Prioritizing GWAS results: a review of statistical methods and recommendations for their application. *Am. J. Hum. Genet.*, **86**(1), 6–22.

Chatterjee, N., Wheeler, B., Sampson, J., *et al.* (2013). Projecting the performance of risk prediction based on polygenic analyses of genome-wide association studies. *Nature Genet.*, **45**(4), 400–405.

Chen, G. and Witte, J. (2007). Enriching the analysis of genomewide association studies with hierarchical modeling. *Am. J. Hum. Genet.*, **81**(2), 397–404.

Cordell, H.J. (2009). Detecting gene–gene interactions that underlie human diseases. *Nature Rev. Genet.*, 10, 392–404.

Cowper-Sal-lari, R., Zhang, X., Wright, J., *et al.* (2012). Breast cancer risk-associated SNPs modulate the affinity of chromatin for *FOXA1* and alter gene expression. *Nature Genet.*, **44**(11), 1191–1200.

Cross-Disorder Group of the Psychiatric Genomics Consortium. (2013a). Genetic relationship between five psychiatric disorders estimated from genome-wide SNPs. *Nature Genet.*, **45**(9), 984–994.

Cross-Disorder Group of the Psychiatric Genomics Consortium. (2013b). Identification of risk loci with shared effects on five major psychiatric disorders: a genome-wide analysis. *Lancet*, **381**(9875), 1371–1379.

de los Campos, G., Gianola, D. and Allison, D.B. (2010). Predicting genetic predisposition in humans: the promise of whole-genome markers. *Nature Rev. Genet.*, **11** (12), 880–886.

de los Campos, G., Vazquez, A. I., Fernando, R., Klimentidis, Y.C. and Sorensen, D. (2013). Prediction of complex human traits using the genomic best linear unbiased predictor. *PLoS Genet.*, **9**(7), e1003608.

Devlin, B. and Roeder, K. (1999). Genomic control for association studies. *Biometrics*, **55**(4), 997–1004.

Donoso, L.A., Vrabec, T. and Kuivaniemi, H. (2010). The role of complement Factor H in age-related macular degeneration: a review. *Surv. Ophthalmol.*, **55**(3), 227–246.

Falconer, D.S. (1996). *Introduction to Quantitative Genetics* (2nd ed.). Longman, London.

Fisher, R.A. (1918). The correlation between relatives on the supposition of Mendelian inheritance. *Trans. R. Soc. Edinb.*, **52**(2), 399–433.

Goddard, M.E., Wray, N.R., Verbyla, K. and Visscher, P.M. (2009). Estimating effects and making predictions from genome-wide marker data. *Statist. Sci.*, **24**(4), 517–529.

Golan, D. and Rosset, S. (2013). Narrowing the gap on heritability of common disease by direct estimation in case-control GWAS. *arXiv preprint arXiv:1305.5363.*

Hartley, S.W. and Sebastiani, P. (2013). PleioGRiP: genetic risk prediction with pleiotropy. *Bioinformatics*, **29**(8), 1086–1088.

Hartley, S.W., Monti, S., Liu, C.-T., Steinberg, M.H. and Sebastiani, P. (2012). Bayesian methods for multivariate modeling of pleiotropic SNP associations and genetic risk prediction. *Front. Genet.*, **3**, 176.

Hastie, T. and Tibshirani, R. (2004). Efficient quadratic regularization for expression arrays. *Biostatistics*, **5**(3), 329–340.

Hastie, T., Tibshirani, R. and Friedman, J. (2009). *The Elements of Statistical Learning: Data Mining, Inference, and Prediction* (2nd ed.). Springer, New York.

Hindorff, L., Sethupathy, P., Junkins, H., *et al.* (2009). Potential etiologic and functional implications of genome-wide association loci for human diseases and traits. *Proc. Natl Acad. Sci. USA*, **106**(23), 9362–9367.

Hopper, J. (1993). Variance components for statistical genetics: applications in medical research to characteristics related to human diseases and health. *Statist. Meth. Med. Res.*, **2**(3), 199–223.

Hopper, J. and Mathews, J.D. (1982). Extensions to multivariate normal models for pedigree analysis. *Ann. Hum. Genet.*, **46**(4), 373–383.

Hopper, J. and Mathews, J.D. (1983). Extensions to multivariate normal models for pedigree analysis: II. Modeling the effect of shared environment in the analysis of variation in blood lead levels. *Am. J. Epidemiol.*, **117**(3), 344–355.

Jiang, J., Li, C., Debashis, P., Yang, C. and Zhao, H. (2013). High dimensional genome-wide association study and mis-specified mixed model analysis. *arXiv preprint: arXiv.1404.2355 [math.ST].*

Kang, H.M., Sul, J.H., Zaitlen, N.A., *et al.* (2010). Variance component model to account for sample structure in genome-wide association studies. *Nature Genet.*, **42**(4), 348–354.

Klein, R., Zeiss, C., Chew, E., *et al.* (2005). Complement factor H polymorphism in age-related macular degeneration. *Science*, **308**(5720), 385–389.

Korte, A., Vilhjálmsson, B.J., Segura, V., *et al.* (2012). A mixed-model approach for genome-wide association studies of correlated traits in structured populations. *Nature Genet.*, **44**(9), 1066–1071.

Lander, E., Linton, L., Birren, B., *et al.* (2001). Initial sequencing and analysis of the human genome. *Nature*, **409**(6822), 860–921.

Lange, K., WestLake, J. and Spence, M. (1976). Extensions to pedigree analysis III. Variance components by the scoring method. *Ann. Hum. Genet.*, **39**(4), 485–491.

Lee, S.H., Wray, N.R., Goddard, M.E. and Visscher, P.M. (2011). Estimating missing heritability for disease from genome-wide association studies. *Am. J. Hum. Genet.*, **88**(3), 294–305.

Lee, S.H., DeCandia, T.R., Ripke, S., *et al.* (2012). Estimating the proportion of variation in susceptibility to schizophrenia captured by common SNPs. *Nature Genet.*, **44**(3), 247–250.

Lee, S.-I., Dudley, A., Drubin, D., *et al.* (2009). Learning a prior on regulatory potential from eQTL data. *PLoS Genet.*, **5**(1), e1000358.

Li, C., Yang, C., Gelernter, J. and Zhao, H. (2013). Improving genetic risk prediction by leveraging pleiotropy. *arXiv preprint arXiv:1304.7417*.

Li, M., Wang, L., Xia, Z., Sham, P. and Wang, J. (2013). GWAS3D: detecting human regulatory variants by integrative analysis of genome-wide associations, chromosome interactions and histone modifications. *Nucl. Acids Res.*, **41**, W150–W158.

Lippert, C., Listgarten, J., Liu, Y., *et al.* (2011). Fast linear mixed models for genome-wide association studies. *Nature Meth.*, **8**(10), 833–835.

Lippert, C., Quon, G., Kang, E.Y., *et al.* (2013). The benefits of selecting phenotype-specific variants for applications of mixed models in genomics. *Sci. Rep.*, 3.

Lynch, M. and Walsh, B. (1998). *Genetics and Analysis of Quantitative Traits*. Sinauer Associates, Sunderland, MA.

Maher, B. (2008). Personal genomes: the case of the missing heritability. *Nature*, **456**(7218), 18–21.

Manolio, T. (2010). Genomewide association studies and assessment of the risk of disease. *New Engl. J. Med.*, **363**(2), 166–176.

Manolio, T.A., Collins, F.S., Cox, N.J., *et al.* (2009). Finding the missing heritability of complex diseases. *Nature*, **461**(7265), 747–753.

Marchini, J., Cardon, L.R., Phillips, M.S. and Donnelly, P. (2004). The effects of human population structure on large genetic association studies. *Nature Genet.*, **36**(5), 512–517.

Morris, A.P., Voight, B.F., Teslovich, T.M., *et al.* (2012). Large-scale association analysis provides insights into the genetic architecture and pathophysiology of type 2 diabetes. *Nature Genet.*, **44**(9), 981–990.

Price, A.L., Patterson, N.J., Plenge, R.M., *et al.* (2006). Principal components analysis corrects for stratification in genome-wide association studies. *Nature Genet.*, **38**(8), 904–909.

Price, A.L., Zaitlen, N.A., Reich, D. and Patterson, N. (2010). New approaches to population stratification in genome-wide association studies. *Nature Rev. Genet.*, **11**(7), 459–463.

Purcell, S., Neale, B., Todd-Brown, K., *et al.* (2007). PLINK: a tool set for whole-genome association and population-based linkage analyses. *Am. J. Hum. Genet.*, **81**(3), 559–575.

Risch, N. and Merikangas, K. (1996). The future of genetic studies of complex human diseases. *Science*, **273**(5281), 1516–1517.

Sachidanandam, R., Weissman, D., Schmidt, S., *et al.* (2001). A map of human genome sequence variation containing 1.42 million single nucleotide polymorphisms. *Nature*, **409**(6822), 928–933.

Sakoda, L.C., Jorgenson, E. and Witte, J.S. (2013). Turning of COGS moves forward findings for hormonally mediated cancers. *Nature Genet.*, **45**(4), 345–348.

Schaub, M., Boyle, A., Kundaje, A., Batzoglou, S. and Snyder, M. (2012). Linking disease associations with regulatory information in the human genome. *Genome Res.*, **22**, 1748–1759.

Searle, S.R., Casella, G. and McCulloch, C.E. (2006). *Variance Components*. Wiley-Interscience, New York, NY.

Sivakumaran, S., Agakov, F., Theodoratou, E., *et al.* (2011). Abundant pleiotropy in human complex diseases and traits. *Am. J. Hum. Genet.*, **89**(5), 607–618.

Speed, D., Hemani, G., Johnson, M.R. and Balding, D.J. (2012). Improved heritability estimation from genome-wide SNPs. *Am. J. Hum. Genet.*, **91**(6), 1011–1021.

Stahl, E.A., Wegmann, D., Trynka, G., *et al.* (2012). Bayesian inference analyses of the polygenic architecture of rheumatoid arthritis. *Nature Genet.*, **44**(5), 483–489.

Sul, J.H. and Eskin, E. (2013). Mixed models can correct for population structure for genomic regions under selection. *Nature Rev. Genet.*, **14**(4), 300.

Svishcheva, G.R., Axenovich, T.I., Belonogova, N.M., Duijn, C.M. van and Aulchenko, Y.S. (2012). Rapid variance components-based method for whole-genome association analysis. *Nature Genet.*, **44**(10), 1166–1170.

Tabor, H., Risch, N. and Myers, R. (2002). Candidate-gene approaches for studying complex genetic traits: practical considerations. *Nature Rev. Genet.*, **3**(5), 391–397.

The ENCODE Project Consortium. (2012). An integrated encyclopedia of DNA elements in the human genome. *Nature*, **489**, 57–74.

The Wellcome Trust Case Control Consortium. (2007). Genome-wide association study of 14,000 cases of seven common diseases and 3,000 shared controls. *Nature*, **447**(7145), 661–678.

Thompson, R. (1977a). The estimation of heritability with unbalanced data: II. Data available on more than two generations. *Biometrics*, **33**(3), 497–504.

Thompson, R. (1977b). The estimation of heritability with unbalanced data: I. observations available on parents and offspring. *Biometrics*, **33**(3), 485–495.

Tipping, M.E. (2001). Sparse Bayesian learning and the relevance vector machine. *J. Mach. Learn. Res.*, **1**, 211–244.

Tipping, M.E. and Faul, A.C. (2003). Fast marginal likelihood maximisation for sparse Bayesian models. In C.M. Bishop and M. Frey (Eds), *Proceedings of the Ninth International Workshop on Artificial Intelligence and Statistics* (Vol. **1**), Jan 3–6, Key West, FL.

Vattikuti, S., Guo, J., and Chow, C.C. (2012). Heritability and genetic correlations explained by common SNPs for metabolic syndrome traits. *PLoS Genet.*, **8**(3), e1002637.

Venter, J., Adams, M., Myers, E., *et al.* (2001). The sequence of the human genome. *Science*, **291**(5507), 1304–1351.

Veyrieras, J.-B., Kudaravalli, S., Kim, S., *et al.* (2008). High-resolution mapping of expression-QTLs yields insight into human gene regulation. *PLoS Genet.*, **4**(10), e1000214.

Visscher, P.M. (2008). Sizing up human height variation. *Nature Genet.*, **40**(5), 489–490.

Visscher, P.M., Hill, W.G. and Wray, N.R. (2008). Heritability in the genomics era 0150 concepts and misconceptions. *Nature Rev. Genet.*, **9**(4), 255–266.

Visscher, P.M., Brown, M.A., McCarthy, M.I. and Yang, J. (2012). Five years of GWAS discovery. *Am. J. Hum. Genet.*, **90**(1), 7–24.

Wan, X., Yang, C., Yang, Q., *et al.* (2010). Predictive rule inference for epistatic interaction detection in genome-wide association studies. *Bioinformatics*, 26, 30–37.

Wang, W., Barratt, B., Clayton, D. and Todd, J. (2005). Genome-wide association studies: theoretical and practical concerns. *Nature Rev. Genet.*, **6**(2), 109–118.

Ward, L. and Kellis, M. (2012a). HaploReg: a resource for exploring chromatin states, conservation, and regulatory motif alterations within sets of genetically linked variants. *Nucl. Acids Res.*, **40**(D1), D930–D934.

Ward, L. and Kellis, M. (2012b). Interpreting noncoding genetic variation in complex traits and human disease. *Nature Biotechnol.*, **30**, 1095–1106.

Wray, N.R., Yang, J., Hayes, B.J., *et al.* (2013). Pitfalls of predicting complex traits from SNPs. *Nature Rev. Genet.*, **14**(7), 507–515.

Yang, J., Benyamin, B., McEvoy, B.P., *et al.* (2010). Common SNPs explain a large proportion of the heritability for human height. *Nature Genet.*, **42**(7), 565–569.

Yang, J., Manolio, T.A., Pasquale, L.R., *et al.* (2011a). Genome partitioning of genetic variation for complex traits using common SNPs. *Nature Genet.*, **43**(6), 519–525.

Yang, J., Weedon, M.N., Purcell, S., *et al.* (2011b). Genomic inflation factors under polygenic inheritance. *Eur. J. Hum. Genet.*, **19**(7), 807–812.

Zaitlen, N., Kraft, P., Patterson, N., *et al.* (2013). Using extended genealogy to estimate components of heritability for 23 quantitative and dichotomous traits. *PLoS Genet.*, **9**(5), e1003520.

Zhou, X. and Stephens, M. (2012). Genome-wide efficient mixed-model analysis for association studies. *Nature Genet.*, **44**(7), 821–824.

Zhu, X., Zhang, S., Zhao, H. and Cooper, R. S. (2002). Association mapping, using a mixture model for complex traits. *Genet. Epidemiol.*, **23**(2), 181–196.

4 GWAS replicability across time and space

Urko M. Marigorta, Juan Antonio Rodriguez and Arcadi Navarro

4.1. Introduction

The key step to validating associations between genetic variants and complex human diseases is the replication of findings in independent samples. This was, perhaps, the main lesson learned by the community from the candidate–gene association studies that were performed prior to the era dominated by genome-wide association studies (GWAS). Since the mid-1990s, thousands of papers had been published describing new associations between candidate variants and complex diseases (Ioannidis *et al.*, 2001). However, the actual worth of many of these publications was inherently constrained by small sample sizes, among many other factors, which imposed hard limits to statistical power; by a poor characterization of the structure of genomic variability in human populations, which generated many false positives; and by a focus on common alleles discovered in peoples of European ancestry, with frequencies usually above 5%, which resulted in a strong ascertainment bias. Due to these powerful reasons, and despite their enormous popularity, associations reported during the pre-GWAS era frequently failed to replicate in independent studies (Ioannidis *et al.*, 2001). For instance, out of the 166 most widely studied associations by 2002, only six had been positively replicated three or more times (Lohmueller *et al.*, 2003). This plethora of promising but eventually failed associations seriously undermined the credibility of the whole association-mapping approach, but, on the bright side, made researchers aware that they needed to do better.

Many of the problems were indeed addressed by the design of GWAS. In sharp contrast with previous association studies, the GWAS era has been characterized by much larger sample sizes, an extensive coverage of human genomic diversity, careful control of the effects of population stratification, more stringent significance thresholds to avoid false positives due to multiple testing, and, in many publications, built-in replication samples (McCarthy *et al.*, 2008). What has been

Genome-Wide Association Studies: From Polymorphism to Personalized Medicine, ed. K. Appasani. Published by Cambridge University Press.

the impact of these improvements? Do associations discovered by GWAS replicate, and, whatever the answer to these questions, can we learn anything from replication attempts? In what follows, we analyze the degree and patterns of replicability of disease-associated variants discovered by GWAS during the last 10 years. We first summarize the main patterns of GWAS replicability considering the time at which discoveries were made. We study these patterns paying special attention to differences observed according to disease classes, the strength of the reported association, as well as the statistical significance in the discovery GWAS. Finally, we compare the degree of replicability of GWAS results across human populations and discuss its implications in terms of the relative influence of common versus rare variants in explaining the genetic associations reported so far. All these analyses illustrate the kind of inferences about the genetic architecture of complex traits that can be gained from assessing GWAS replicability.

Replication of newly reported associations is usually considered the most reliable validation of GWAS discoveries. A successful replication strongly vouches in favor of the idea that the newly observed association is real and does not constitute a false-positive finding. In principle, several clear and stringent criteria are used to validate the status of an association after a replication attempt (NCI-NHGRI Working Group on Replication in Association Studies, 2007); among others, it is necessary to test the newly discovered risk variant in independent cohorts of patients and controls, and if possible, using alternative genotyping platforms (McCarthy *et al.*, 2008). However, it is difficult to assess whether researchers really adhere to these criteria, because GWAS publications tend to focus on the description of new variants. Thus, patterns of replicability remain an understudied issue.

To obtain an overview of replication patterns, we conducted an analysis of the NHGRI GWAS Catalog (Hindorff *et al.*, 2009; Welter *et al.*, 2014), and studied the main patterns of discovery and replication for GWAS risk variants. Specifically, we focus on 46 disease traits presenting the highest number of reported associations in the *Catalog*. In total, we considered a total of 2950 associations, with an average of 64 associations per disease (ranging from 16 associations for migraine and polycystic ovary syndrome to 210 associations for type 2 diabetes and Crohn's disease, respectively).

Because different standards are used in different contexts, we first need to define the exact criteria we use to determine positive replications. When GWAS present a discovery and a replication phase within the same work, replication is always studied on a SNP-by-SNP basis. In general terms, after an initial GWAS of hundreds of thousands or even millions of SNPs, only a few of them are taken to the validation phase in an independent, in many cases smaller, set of individuals. Given the smaller number of SNPs to be tested, the significance thresholds required for a replication to be considered positive are not as stringent as the genome-wide consensus of $p < 5 \times 10^{-8}$ of the discovery phase (which also explains why smaller sample sizes are required). Usually, a Bonferroni correction based on the number of SNPs taken forward to the GWAS replication stage is considered, but often a simple replication p-value threshold of 0.05 suffices. In other cases, GWAS try to ascertain whether they can replicate discoveries reported in previous, independent publications. Because in these cases the relevant GWAS might have used different arrays and because imputation techniques are not always applied, replication is referred to genes or loci, rather than to individual SNPs.

To use either of these criteria, or combinations thereof, the *p*-values of every SNP in every study need to be known. Unfortunately, GWAS publication policies are far from standardized, and painstaking manual efforts are required to carefully examine the original publication or the limited released data available (Marigorta and Navarro, 2013). Considering that genome-wide *p*-values often remain unavailable, we based our analyses on the GWAS Catalog, a weekly updated central repository of published GWAS that gathers associations that present *p*-values $\leq 1 \times 10^{-5}$ (Hindorff *et al.*, 2009). In the present analysis, we used a loci-based replication criterion that considers an association discovered by a GWAS as positively replicated if a GWAS published in a subsequent date reports the same SNP or another SNP less than 200 kb away from the discovery SNP as associated with the same trait. We are aware that our criteria are far from ideal, because, on one hand, we underestimate the number of replications (because we cannot take SNPs with *p*-values $> 1 \times 10^{-5}$ into account; and, on the other hand, we add false replications whenever we accept SNPs that are not in high LD with the discovery SNP). Still, our results show that we obtain a good approximation to replicability patterns.

4.2. Replicability across time

The first GWAS were performed using relatively low numbers of cases and controls, resulting in smaller total sample sizes when compared to the average of the GWAS period. For instance, the median size of GWAS published before mid-2007 was of 3900 individuals, while the correspondent figure for the whole GWAS period scales up to 6400 individuals. Studies previous to 2007 were characterized by a low rate of success in describing new associations. For instance, the first successful genome-wide scan studied age-related macular degeneration (AMD), one of the leading causes of blindness. Despite being performed only upon 96 AMD cases and 50 controls, a large-effect variant (odds ratio (OR): 4.6) that would account for ~70% of the population-attributable risk for this condition was discovered (Klein *et al.*, 2005). However, the large effect size observed for AMD is quite specific of that condition, and for other diseases much larger sample sizes have been necessary to increase the total number of discovered associations. This aspect was first exemplified by the landmark study by the Wellcome Trust Case Control Consortium (WTCCC) in summer 2007. After studying ~2000 cases for seven different diseases and a set of ~3000 shared controls, totaling an impressive set of ~17,000 individuals, the WTCCC yielded a total of 23 new associations at the genome-wide significance level (at the time $p < 5 \times 10^{-7}$) and a total of 50 SNPs with a *p*-value $< 10^{-5}$ (Wellcome Trust Case Control Consortium, 2007). Interestingly, and despite testing the same sample size for each disease, 80% of the new associations corresponded to autoimmune disorders, while a single locus was described for bipolar disorder at genome-wide significance levels. Enormous heterogeneity across different types of traits in terms of discovery and replicability has been a constant ever since.

One of the most important issues with the replicability of GWAS results is that for a large proportion of variants a simple scheme of discovery/replication analysis is not easy to apply. Because huge sample sizes are required for the discovery of many low-effect risk variants, a considerable proportion of associations have been discovered through meta-analysis that pool results from different GWAS. In these cases a replication sample is either not necessary or, if deemed so, would need a larger sample size than those available in current GWAS.

Interestingly, the first large-scale meta-analysis publications were performed during a very skeptical period, leading many researchers to judge GWAS findings unsatisfactory. For almost every trait, the number of discovered loci and, most importantly, the total heritability explained, had been below the expectations garnered during the first years of GWAS. In that sense, large meta-analyses have turned out to be reassuring due to the sharp increase in both newly associated and replicated risk variants. For instance, regarding discoveries, a recent meta-analysis for inflammatory bowel disease raised known loci from 71 to 163. As to replications, a good example comes from the original WTCCC discoveries. The 100% of the genome-wide significant associations reported by the WTCCC have been replicated, in many cases by meta-analyses. This trend can be observed in Figure 4.1, as an ever-increasing proportion of GWAS results are not novel discoveries, but positive replications of previous findings. Indeed, a large proportion of the replication evidence available in the Catalog accumulates for loci that have been replicated by more than one GWAS (Figure 4.2, the "*repeated replica*" category). Overall, the rising numbers of loci abundantly replicated by prospective studies constitute strong evidence for the true nature of GWAS loci.

4.2.1. New discoveries across time

Interesting patterns emerge when associations are clustered according to the effect size of SNPs in the discovery studies. In Figure 4.3, we classify associated variants into three categories of risk alleles; namely, large-effect size (1.5 > OR > 2.0), medium-effect size (1.2 > OR > 1.5), and low-effect size (OR < 1.2). If we focus on the earliest GWAS, most discovered SNPs belong to the group with the largest OR. This is expected, because these have been the least-powered studies. For instance, the average OR of pre-2008 GWAS discoveries equals 1.52, while the average effect has since dropped to OR = 1.33. Interestingly, the variants of largest effect are the less numerous when we consider the whole range of published GWAS (Figure 4.3a), which confirms that the bulk of associated variants present OR lower than 1.5. An additional interesting observation is that these large-effect variants rarely replicate after their discovery, which would be unexpected if the described effect were true (because larger GWAS should have been empowered to replicate them). Instead, the observed low replicability could indicate that these large effects are inflated due to a winner's curse phenomenon, where low sample size GWAS overestimate the effect size of associated variants.

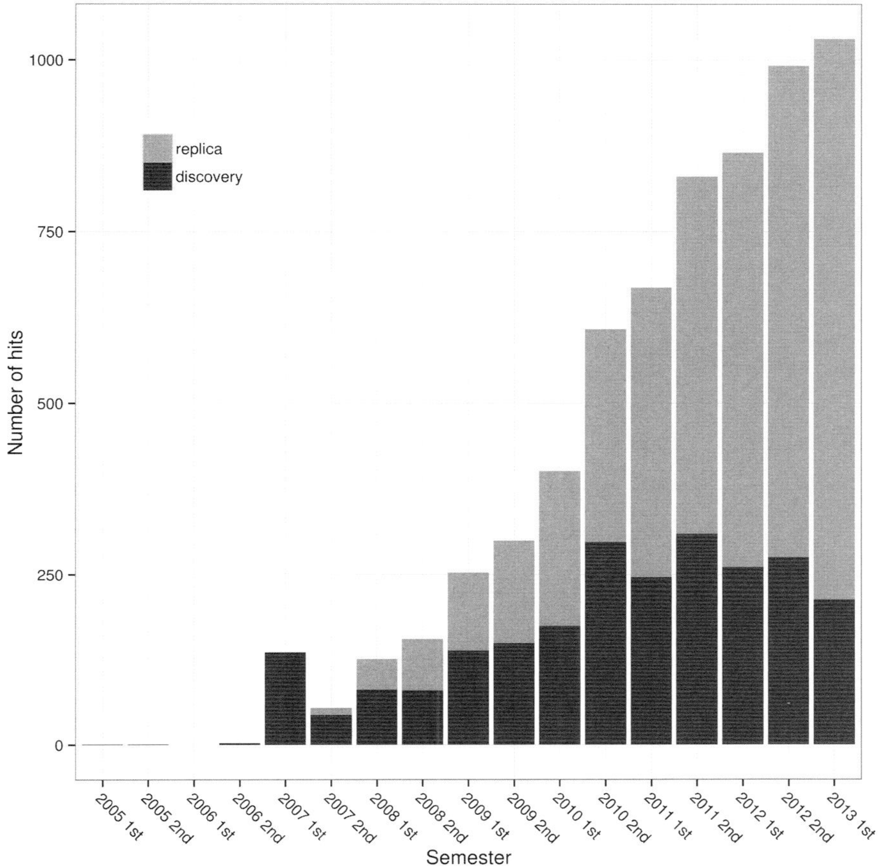

Figure 4.1. Number of discoveries of new loci and replication events of previously discovered loci according to the semester of GWAS publication (from 2005 to 2013) for the 46 diseases under study. The replication figure for a given semester corresponds to the total cumulative number of replication evidence since 2005.

Associations with intermediate and lowest effect sizes show a different picture in terms of replicability (Figure 4.3b and Figure 4.3c). A noticeable proportion of these associations have been replicated, which vouches in favor of them being real. However, the bulk of these variants have been discovered recently. This pattern is most obvious for variants with the lowest effects (OR < 1.2), discovered mostly by large meta-analyses published after the second semester of 2010. Although it is too early to inquire about the true nature of these low-effects variants because not many replication attempts have been published, plenty of replication evidence for these infinitesimal effects is starting to accumulate.

As mentioned above, the GWAS era has been characterized by an extreme caution to avoid the false-positive findings that were so widespread in candidate–gene studies. In this regard, the WTCCC study was very influential and established a standard criterion for significance cutoff for discovery at p-value 5×10^{-7}. In recent years, GWAS have scaled up to millions of variants tested after imputation and the most commonly used p-value cutoff has become 5×10^{-8}. The

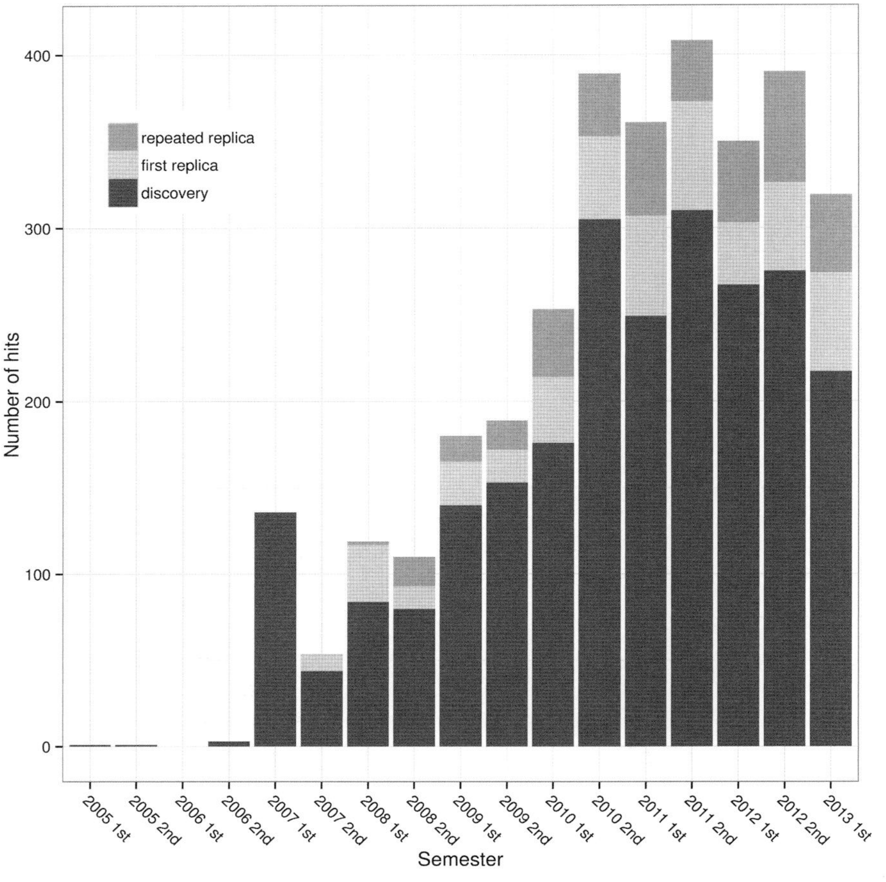

Figure 4.2. Number of discoveries of new loci and replication events of previously discovered loci according to the semester of GWAS publication (from 2005 to 2013) for the 46 diseases under study. Replication events are separated into "first replica" and "repeated replica" depending on whether the replication event constitutes the first evidence for replication of a given locus or if the locus had been previously replicated.

availability of associations at p-value 10^{-5} in the GWAS Catalog permits checking of the replicability of associations discovered under different significance thresholds. Specifically, we distinguish between variants discovered under a liberal significance threshold ($10^{-5} \geq p$-value $\geq 10^{-7}$, Figure 4.4a), variants at strict genome-wide significance level ($10^{-7} > p$-value $\geq 10^{-10}$, Figure 4.4b), and variants at a very stringent significance level ($10^{-10} > p$-value $\geq 10^{-15}$, Figure 4.4c). Overall, Figure 4.4 shows that an overwhelming majority of variants recorded by the GWAS Catalog have not achieved an appropriate genome-wide significance threshold. Indeed, these variants present the poorest rates of replicability, even if many were described a relatively long time ago, during the first GWAS years (e.g., 2007–2008). This result indicates that many records available at the Catalog are of dubious nature, and should be considered with caution when using the GWAS Catalog as a repository of disease-associated variants. On the contrary, risk variants associated at genome-wide significance levels in the discovery GWAS

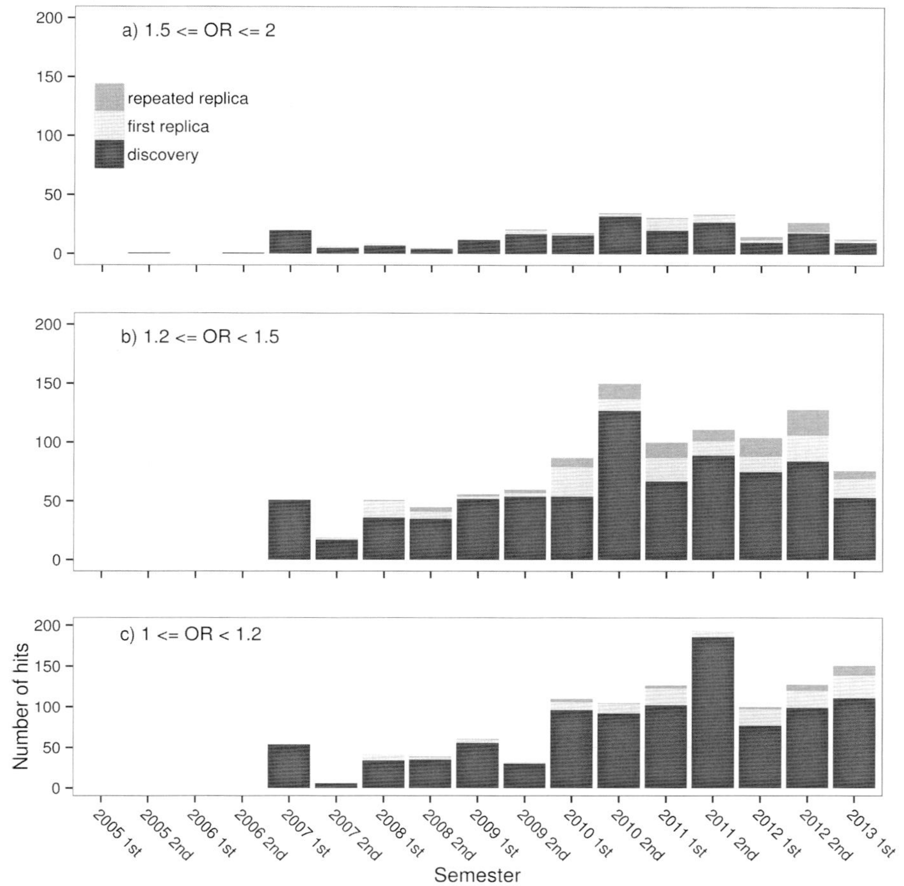

Figure 4.3. Numbers of discovered loci, first replicas, and repeated replicas for the 46 diseases according to the effect size of the original association in the discovery GWAS. (a) OR between 1.5 and 2.0 ($n = 231$); (b) OR between 1.2 and 1.5 ($n = 950$); (c) Hits with OR below 1.2 ($n = 1170$).

replicate more often (Figure 4.4b,c), ensuring the true nature of these disease-associated loci.

4.2.2. Disease replicability across time

One of the most interesting observations on GWAS discovery and replication patterns is that they highlight differences in the genetic architectures of different diseases. The risk for some conditions, such as AMD, is strongly influenced by a few loci of relatively large effect, which allowed early discoveries, whereas other conditions seem to be far more polygenic and thus discoveries have had to wait until sample sizes have been large enough.

In Figure 4.5, we consider three different groups of diseases. The first group is composed of central nervous systems (CNS) disorders, specifically schizophrenia and bipolar disorder. In the second group, we pool together several autoimmune diseases such as asthma, rheumatoid arthritis, or Crohn's disease. Finally, we

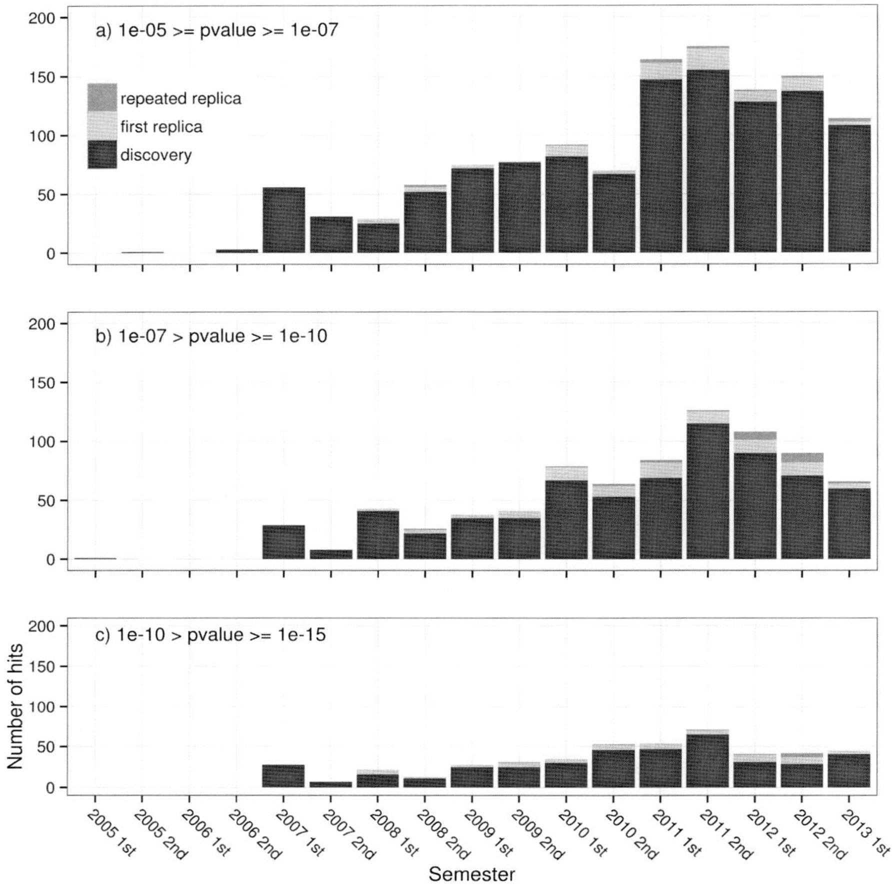

Figure 4.4. Numbers of discovered loci, first replicas, and repeated replicas for the 46 diseases according to the *p*-value in the discovery GWAS. (a) Hits discovered at *p*-value between 10^{-5} and 10^{-7} ($n = 1237$); (b) at *p*-value between 10^{-7} and 10^{-10} ($n = 805$); (c) at *p*-value between 10^{-10} and 10^{-15} ($n = 470$).

consider the joined patterns of replicability for 11 different malignancies such as prostate and breast cancer. Overall, the three groups show very divergent patterns of discovery and replicability. For instance, diseases related to the CNS show the lowest number of discovered associations, and replication evidence for these variants was not achieved until recently, from 2010 onwards (Figure 4.5a). These patterns confirm the infinitesimal character of the genetic architecture of psychiatric conditions, influenced by a large number of variants with low effect (International Schizophrenia Consortium *et al.*, 2009). On the contrary, immune-mediated conditions form the most successful groups of diseases in terms of discovered and replicated variants (Figure 4.5b). For instance, 32 variants explaining ~20% of the genetic susceptibility to Crohn's disease had been discovered as early as 2008 (Barrett *et al.*, 2008), and several examples of translation from GWAS to clinical practice are available (Visscher *et al.*, 2012). Finally, the figure for cancer diseases exemplifies the large payoff of meta-analysis efforts as several peaks of

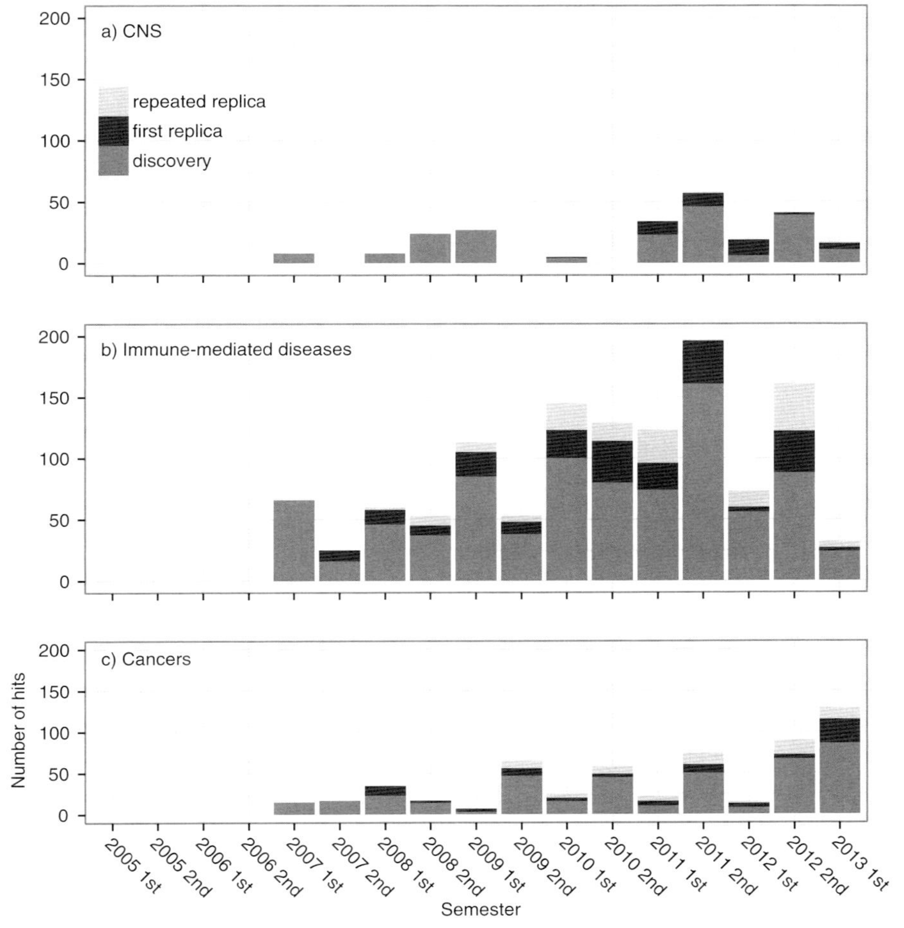

Figure 4.5. Numbers of discovered loci, first replicas, and repeated replicas for three different categories of disease. (a) Two disorders of the CNS, namely schizophrenia and bipolar disorder; (b) immune-mediated disorders ($n = 17$); (c) cancer ($n = 11$).

discovered and replicated variants coincide with the publication of high-power GWAS published by several large international consortia (Figure 4.5c).

4.3. Replicability across space

The demographic history of our species has led to a non-homogeneous apportionment of genetic diversity in human populations (International HapMap Consortium, 2003). Hence, the ease with which risk variants are discovered or replicated can vary across populations due to differences in allele frequencies as well as to divergent linkage disequilibrium (LD) patterns between causal variants and genotyped tag SNPs (Rosenberg *et al.*, 2010). Therefore, the characterization of the genetic architecture of complex disease in humans can benefit from a cross-population angle, because risk variants may be missed if not all populations are studied. A population perspective is also necessary for the long-term goal to achieve personalized medicine, as genetic ancestry encloses a fraction of the genetic risk present in any individual (Rosenberg *et al.*, 2010). In this section, we summarize the replicability of GWAS across geographical space.

4.3.1. Challenges to the replication of discoveries across populations

Although it is important to address the genetic risk of complex disease in all human groups, the GWAS period has been characterized by an extraordinary bias in terms of the genetic ancestry under study. The first major GWAS, a seminal paper on the genetic risk of myocardial infarction, was performed upon individuals of Japanese origin (Ozaki *et al.*, 2002), but GWAS using European individuals have been preponderant ever since. This disproportion was more evident during the initial years of the genome-wide era, as fewer than 10% of papers published until 2009 made exclusive use of individuals of non-European descent (Rosenberg *et al.*, 2010). Only from 2010 onwards did this tendency start changing, with the publication of GWAS performed mostly upon individuals of Asian, and to a lesser extent African and Hispanic, ancestries. As of December 2013, European ancestry still accounted for 69% of the published GWAS, and only 7% and 3% of GWAS included samples of African and Hispanic genetic ancestry, respectively (Adeyemo and Rotimi, 2014). The case of type 2 diabetes (T2D) exemplifies the extent of this phenomenon: up to 27 (87%) of the 31 T2D GWAS studying >1000 subjects have been performed on individuals of either European or East Asian ancestry (Figure 4.6).

A further aspect related to sample size makes it difficult to ascertain the sharing and replicability of variants across populations. As shown in Figure 4.6, the sample size of most T2D GWAS has not gone beyond the few thousands of individuals, and only a minority of studies has achieved the statistical power necessary to discover common variants of the smallest effect (OR<1.2). For instance, we can easily calculate the sample size required to detect, at genome-wide significance levels of p-value $< 5 \times 10^{-8}$, a hypothetical risk allele with an allele frequency of 0.2 and that increases the risk of developing T2D by 20% (OR = 1.2, multiplicative effect). Only the five largest T2D GWAS fulfill the minimum sample size (~7000 individuals) required to achieve >80% power to declare such a variant as genome-wide significant. It is obvious that even for a widely "GWASed" disease such as T2D, the different pictures regarding the susceptibility of the disease across populations remain incomplete.

Interestingly, a few diseases at low prevalence in Europeans have been studied mostly on peoples of non-European descent. For instance, nasopharyngeal and hepatocellular cancers, which present higher incidences in East Asians (Tse *et al.*, 2009), have been the subject of genome-wide studies only in that ancestry (four and five GWAS, respectively). This has also been the case for conditions related to the burden of infectious disease present in Africa, such as malaria (Jallow *et al.*, 2009).

Besides the deficient representation of human population diversity, the portability at the SNP level of GWAS in space is inherently challenged by

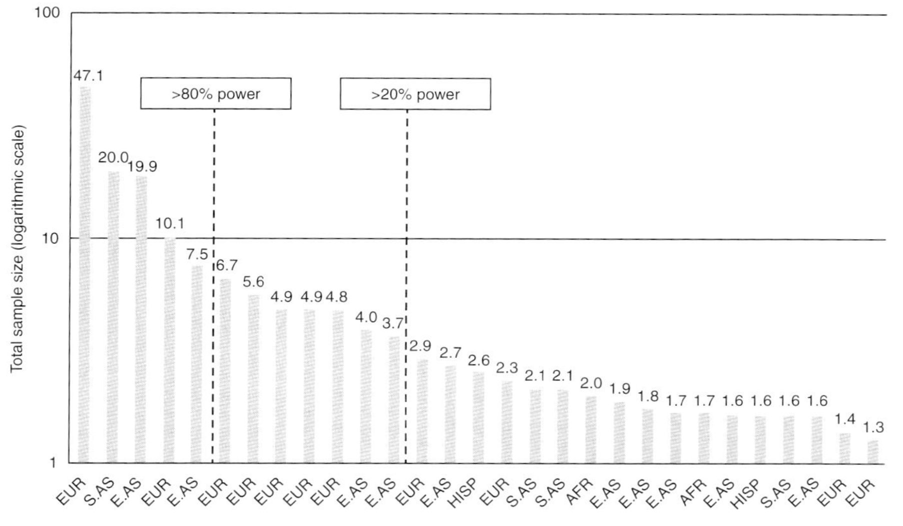

Figure 4.6. Total sample size (in thousands) of the 31 GWAS for type 2 diabetes studying more than 1000 individuals, ranked according to the number of individuals studied. The genetic ancestry of each GWAS is shown on the *x*-axis; EUR (European), S.AS (South Asian), E.AS (East Asian), HISP (Hispanic), and AFR (African). The two vertical bars indicate the sample size threshold needed to achieve 80% and 20% power to detect, at the genome-wide threshold of *p*-value 5×10^{-8}, a risk allele with OR 1.2 and risk allele frequency of 0.20.

ascertainment bias and the considerable variation in LD patterns found across human populations. The focus on Europeans in both SNP discovery and array design has typically resulted in sets of markers (i.e., in commercial arrays) that provide a less-extensive genomic coverage in non-European populations, particularly for Africans (Bhangale *et al.*, 2008). Recent developments in the accuracy and portability of imputation methods have allowed the scale up of the number of interrogated variants, thus mitigating the poorer coverage of non-European GWAS (Rosenberg *et al.*, 2010). Interestingly, differences in LD patterns across populations turn out to be advantageous for the fine mapping of causal variants underlying GWAS associations (Teo *et al.*, 2010).

4.3.2. Replicating discoveries across populations

When replication is attempted with populations from the same continental origin, GWAS discoveries are almost invariably reproduced, even at the SNP level. This high within-continent replicability is observed across ancestries and diseases, as shown in an exhaustive study of GWAS replicability at the SNP level for 28 different diseases (Marigorta and Navarro, 2013). For instance, and after accounting for statistical power, European GWAS achieve a ~98% rate of replication at p-value<0.05 when examining SNPs that had been previously discovered by previous GWAS in Europeans. A similar figure of 84% is observed within East Asian replication attempts, and it strengthens to ~100% when considering only SNPs associated at $p < 10^{-10}$ in the East Asian discovery GWAS. This pattern indicates that most results reported by GWAS are true positives, and confirm that an internal replication phase coupled with strict significance thresholds constitute an excellent barrier to avoid false-positive reports.

Although more challenging, replication attempts across ancestries tend to be successful too. In 2009, two initial studies focused on single traits by the PAGE Consortium (Population Architecture using Genomics and Epidemiology) showed encouraging replicability rates. Specifically, they examined the portability of T2D and prostate cancer variants that had been discovered in European populations (Waters *et al.*, 2009, 2010), and found widespread concordances in the direction of effects across diverse ancestries were reported, including African Americans, Latinos, and Native Hawaiians.

More recently, several studies have considered a wider range of traits to investigate the portability of discoveries made with European samples (Ntzani *et al.*, 2012; Marigorta and Navarro, 2013). Because of the more extensive availability of GWAS, replicability between Europeans and East Asians has been studied in more depth. Marigorta and Navarro (2013) observed that 78% of the associations discovered in Europeans were positively replicated at the same SNP in East Asians, with associations of similar strength. Moreover, differences in LD patterns account for the heterogeneity in those associations that present differences in effect size between both ancestries. A similar tendency was described after inspecting the replicability in Europeans of 43 associations initially discovered in East Asians (Fu *et al.*, 2011). Interestingly, widespread transportability has been reported for several risk variants that present large differences in allele frequencies. This was the case of the association of a SNP within *KCNQ1* (rs2237892) and T2D, which had been missed by the first European GWAS probably due to its low allele frequency of ~5%. However, this susceptibility locus arose as the strongest association by the first GWAS performed upon East Asians and it was shown to replicate in Swedish cohorts (Yasuda *et al.*, 2008). This example highlights that a fraction of risk variants are shared, but may remain unnoticed if GWAS are not attempted across populations.

Even if the bulk of disease susceptibility appears to be shared across populations (de Candia *et al.*, 2013), the generalization beyond Eurasians is more challenging. The PAGE Consortium inspected 68 variants associated with five conditions (including lipid traits, body mass index, and T2D) in ~73,000 individuals from six different ancestries (Carlson *et al.*, 2013). An overwhelming consistency across all ancestries was observed. This result confirms that most causal variants underlying GWAS associations are shared across all human populations. Even so, caution is warranted when using the effect sizes reported in Europeans for risk prediction in individuals of other ancestries. For instance, 25% of the associations scrutinized by the latter study present significant differences in the magnitude of the effects in at least one ancestral group, although the dilution in effect size can be attributed to differences in LD patterns across populations (Carlson *et al.*, 2013; Marigorta and Navarro, 2013). Importantly, differences in LD bring an opportunity for fine mapping of GWAS loci because the comparison of local *p*-value patterns across populations may allow successful narrowing down of the causal variant present in the susceptibility locus (Wu *et al.*, 2013).

In summary, the low availability of GWAS performed upon populations of non-European ancestry has not precluded noticing a widespread replication of risk variants across populations. Indeed, large inter-population replicability rates allow researchers to reject a scenario of susceptibility variants being continental-specific, and rather favor a scenario of causal variants being common and shared across human populations. These results are shadowed by a poor and unbalanced availability of across population data, but the extension of GWAS across ancestries remains as one of the best tactics to finally unveil the location of causal variants in susceptibility loci.

REFERENCES

Adeyemo, A. and Rotimi, C. (2014). What does genomic medicine mean for diverse populations? *Mol. Genet. Genom. Med.*, **2**, 3–6.

Barrett, J.C., Hansoul, S., Nicolae, D.L., *et al.* (2008). Genome-wide association defines more than 30 distinct susceptibility loci for Crohn's disease. *Nature Genet.*, **40**, 955–962.

Bhangale, T.R., Rieder, M.J. and Nickerson, D.A. (2008). Estimating coverage and power for genetic association studies using near-complete variation data. *Nature Genet.*, **40**, 841–843.

Carlson, C.S., Matise, T.C., North, K.E., *et al.* (2013). Generalization and dilution of association results from European GWAS in populations of non-European ancestry: the PAGE study. *PLoS Biol.*, **11**, e1001661.

de Candia, T.R., Lee, S.H., Yang, J., *et al.* (2013). Additive genetic variation in schizophrenia risk is shared by populations of African and European descent. *Am. J. Hum. Genet.*, **93**, 463–470.

Fu, J., Festen, E.A. and Wijmenga, C. (2011). Multi-ethnic studies in complex traits. *Hum. Mol. Genet.*, **20**, R206–213.

Hindorff, L.A., Macarthur, K., Morales, J., *et al.* (2009). *A Catalog of Published Genome-Wide Association Studies*. Available at: www.genomes.gov/gwastudies. Accessed September 2013.

International HapMap Consortium. (2003). The International HapMap Project. *Nature*, **426**, 789–796.

International Schizophrenia Consortium, Purcell, S.M., Wray, N.R., *et al.* (2009). Common polygenic variation contributes to risk of schizophrenia and bipolar disorder. *Nature*, **460**, 748–752.

Ioannidis, J.P., Ntzani, E.E., Trikalinos, T.A. and Contopoulos-Ioannidis, D.G. (2001). Replication validity of genetic association studies. *Nature Genet.*, **29**, 306–309.

Jallow, M., Teo, Y.Y., Small, K.S., *et al.* (2009). Genome-wide and fine-resolution association analysis of malaria in West Africa. *Nature Genet.*, **41**, 657–665.

Klein, R.J., Zeiss, C., Chew, E.Y., *et al.* (2005). Complement factor H polymorphism in age-related macular degeneration. *Science*, **308**, 385–389.

Lohmueller, K.E., Pearce, C.L., Pike, M., Lander, E.S. and Hirschhorn, J.N. (2003). Meta-analysis of genetic association studies supports a contribution of common variants to susceptibility to common disease. *Nature Genet.*, **33**, 177–182.

Marigorta, U.M. and Navarro, A. (2013). High trans-ethnic replicability of GWAS results implies common causal variants. *PLoS Genet.*, **9**, e1003566.

Mccarthy, M.I., Abecasis, G.R., Cardon, L.R., *et al.* (2008). Genome-wide association studies for complex traits: consensus, uncertainty and challenges. *Nature Rev. Genet.*, **9**, 356–369.

NCI-NHGRI Working Group On Replication In Association Studies. (2007). Replicating genotype-phenotype associations. *Nature*, **447**, 655–660.

Ntzani, E.E., Liberopoulos, G., Manolio, T.A. and Ioannidis, J.P. (2012). Consistency of genome-wide associations across major ancestral groups. *Hum. Genet.*, **131**, 1057–1071.

Ozaki, K., Ohnishi, Y., Iida, A., *et al.* (2002). Functional SNPs in the lymphotoxin-alpha gene that are associated with susceptibility to myocardial infarction. *Nature Genet.*, **32**, 650–654.

Rosenberg, N.A., Huang, L., Jewett, E.M., *et al.* (2010). Genome-wide association studies in diverse populations. *Nature Rev. Genet.*, **11**, 356–366.

Teo, Y.Y., Small, K.S. and Kwiatkowski, D.P. (2010). Methodological challenges of genome-wide association analysis in Africa. *Nature Rev. Genet.*, **11**, 149–160.

Tse, K.P., Su, W.H., Chang, K.P., *et al.* (2009). Genome-wide association study reveals multiple nasopharyngeal carcinoma-associated loci within the HLA region at chromosome 6p21.3. *Am. J. Hum. Genet.*, **85**, 194–203.

Visscher, P.M., Brown, M.A., McCarthy, M.I. and Yang, J. (2012). Five years of GWAS discovery. *Am. J. Hum. Genet.*, **90**, 7–24.

Waters, K.M., Le Marchand, L., Kolonel, L.N., *et al.* (2009). Generalizability of associations from prostate cancer genome-wide association studies in multiple populations. *Cancer Epidemiol. Biomark. Prev.*, **18**, 1285–1289.

Waters, K.M., Stram, D.O., Hassanein, M.T., *et al.* (2010). Consistent association of type 2 diabetes risk variants found in Europeans in diverse racial and ethnic groups. *PLoS Genet.*, **6**, e1001078.

Wellcome Trust Case Control Consortium. (2007). Genome-wide association study of 14,000 cases of seven common diseases and 3,000 shared controls. *Nature*, **447**, 661–678.

Welter, D., MacArthur, J., Morales, J., *et al.* (2014). The NHGRI GWAS Catalog, a curated resource of SNP-trait associations. *Nucl. Acids Res.*, Jan; **42** (Database issue); D1001–D1006.

Wu, Y., Waite, L.L., Jackson, A.U., *et al.* (2013). Trans-ethnic fine-mapping of lipid loci identifies population-specific signals and allelic heterogeneity that increases the trait variance explained. *PLoS Genet.*, **9**, e1003379.

Yasuda, K., Miyake, K., Horikawa, Y., *et al.* (2008). Variants in *KCNQ1* are associated with susceptibility to type 2 diabetes mellitus. *Nature Genet.*, **40**, 1092–1097.

Part II

Genome-wide studies in disease biology

5 Genome-wide association studies of body mass index

Tuomas O. Kilpeläinen

5.1. Introduction

Body mass index (BMI), the ratio of body weight to height squared, is the most commonly used index of adiposity and the diagnostic criterion for obesity, a leading risk factor for type 2 diabetes, cardiovascular disease, cancer, and premature death (WHO 2000; Zimmet *et al.*, 2001). Approximately half of BMI is heritable (Hemani *et al.*, 2013), and unraveling the specific genetic variation that contributes to this heritability is vital for understanding the biological mechanisms that regulate adiposity. Despite the evident heritability of BMI, identification of BMI-increasing genetic loci has long been unsuccessful. This has been largely due to our limited knowledge of the underlying molecular pathways and of the architecture of the human genome, as well as small, insufficiently powered studies (Li and Loos, 2008). The implementation of the genome-wide association study (GWAS) approach has dramatically increased the speed of gene discovery for BMI. Since 2006, more than 100 novel BMI loci have been identified, providing valuable novel insights into the biological pathways and molecular mechanisms that lead to the development of obesity. In this chapter, I will review these recent discoveries and discuss their biological and public health implications.

5.2. Discovery of more than 100 novel loci in GWAS of BMI

5.2.1. The first discovery – *FTO*

The first GWAS of BMI, published in 2006, included 694 participants of the Framingham Heart Study and identified a common variant in the *INSIG2* gene to be associated with BMI (Herbert *et al.*, 2006). The *INSIG2* variant was not, however, confirmed at the genome-wide significant statistical thresholds that are applied today, and subsequent large-scale studies have not been able to replicate the findings (Dina *et al.*, 2007; Loos *et al.*, 2007).

Genome-Wide Association Studies: From Polymorphism to Personalized Medicine, ed. K. Appasani. Published by Cambridge University Press.

The first robustly confirmed BMI locus was identified a year later when two independent GWAS discovered variants in the first intron of the *FTO* gene as being unequivocally associated with BMI (Frayling *et al.*, 2007; Scuteri *et al.*, 2007). The initial discovery of *FTO* did not, however, occur in a GWAS of BMI, but in a GWAS of type 2 diabetes. This study identified a common variant in the first intron of *FTO* to be strongly associated with type 2 diabetes by comparing 1924 diabetic patients and 2938 healthy individuals. Adjustment for BMI, however, abolished the association, suggesting that the association was mediated through increased adiposity (Frayling *et al.*, 2007). Indeed, the association of *FTO* with BMI was subsequently confirmed in 38,759 adults and children, where each risk allele of the *FTO* variant increased adult BMI by 0.34–0.46 kg/m^2 (~1 kg of body weight). Soon after, a GWAS of BMI among 4741 Sardinians and a replication in 2355 Europeans and Hispanic Americans further confirmed *FTO* as the first locus robustly associated with common obesity (Scuteri *et al.*, 2007).

At the time of discovery, the function of *FTO* was largely unknown, and it was not clear whether the identified obesity variant affected BMI by modifying the function of *FTO* or another gene nearby. *FTO* had been originally identified in mice as one of six genes included in a 1.6 Mbp deletion on chromosome 8, which was responsible for the fused toes (Ft) phenotype (van der Hoeven *et al.*, 1994). Mice heterozygous for the deletion displayed fused toes and enlargement of the thymus, whereas homozygous mice died during mid-gestation. No changes in adiposity or leanness of the mice had been reported.

The confirmation of *FTO* as a BMI locus inspired an extensive follow up to gain insight into the potential mechanisms linking *FTO* to adiposity. Sequence analysis and *in vitro* studies revealed that the protein product of *FTO* is a 2-oxoglutarate-dependent nucleic acid demethylase (Gerken *et al.*, 2007), suggesting a potential role for *FTO* in nucleic acid repair or modification. Gene expression studies demonstrated that even if *FTO* is expressed ubiquitously, the levels of expression are particularly high in the hypothalamic nuclei regulating energy balance and are regulated by feeding and fasting (Frayling *et al.*, 2007; Gerken *et al.*, 2007). A mouse knockout model for *FTO* was developed, showing postnatal growth retardation and a lean phenotype due to increased energy expenditure despite decreased levels of spontaneous activity (Fischer *et al.*, 2009). Mice overexpressing *FTO*, on the other hand, showed an increased fat mass as a result of increased energy intake, independent of whether the mice were fed a standard or a high-fat diet (Church *et al.*, 2010). *FTO* was also shown to have a role in cellular sensing of amino acid levels (Gulati *et al.*, 2013), a critical regulator of appetite-regulating pathways in the brain (Cota *et al.*, 2006), and most human epidemiological studies suggest a subtle effect of the *FTO* locus on food intake and preference (Cecil *et al.*, 2008; Speakman *et al.*, 2008; Wardle *et al.*, 2008, 2009; Haupt *et al.*, 2009). However, evidence of a direct link between single nucleotide changes in *FTO* and alteration in the gene function or expression levels has been lacking, and most recently, findings based on chromatin conformation capture suggest that rather than affecting *FTO* itself, variations in *FTO* may

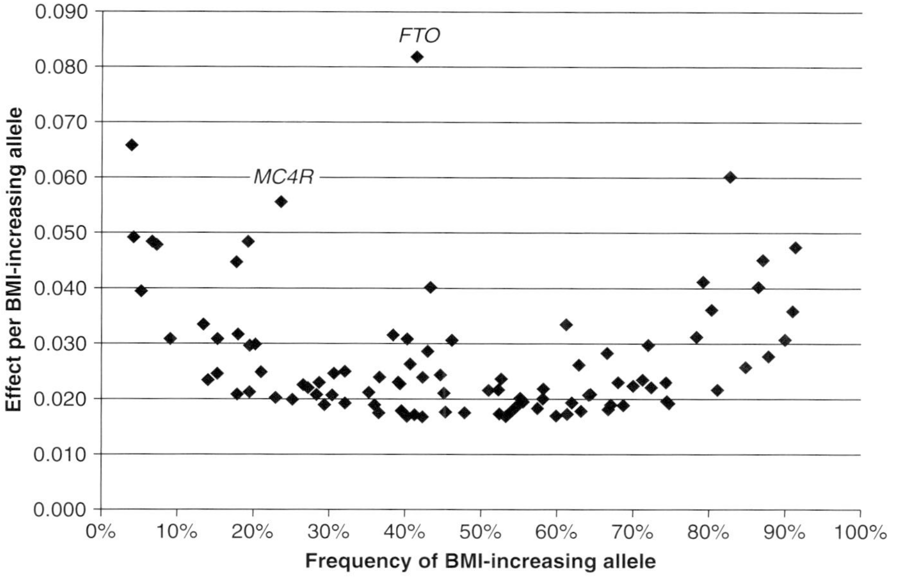

Figure 5.1. Relationship between the effect size (in units of standard deviation of BMI) and allele frequency of the BMI-increasing alleles of the 97 loci identified in the GIANT consortium meta-analysis in 2015 (Locke *et al.*, 2015). The two first discovered loci, *FTO* (Frayling *et al.*, 2007) and *MC4R* (Loos *et al.*, 2008), are indicated.

influence body weight by regulating the expression of a nearby transcription factor-coding gene *IRX3* (Smemo *et al.*, 2014). Thus, it still remains to be determined whether the genetic mechanism linking *FTO* to human obesity involves disrupted regulation of *FTO, IRX3*, or perhaps both.

5.2.2. The second BMI locus – *MC4R*

In the initial GWAS of adiposity traits, several other variants were taken forward, but only *FTO* was replicated in subsequent analyses. This suggested that *FTO* was the "lowest hanging fruit," and larger sample sizes would be needed to discover additional BMI loci of smaller effect (Figure 5.1). In 2008, an international collaboration named the GIANT (Genomic Investigation of ANthropometric Traits) consortium was launched to facilitate larger meta-analyses of BMI and other anthropometric traits. By pooling together GWAS data from 16,876 adults of European descent and a replication in 60,352 additional adults, the consortium identified a novel BMI signal near the melanocortin-4 receptor (*MC4R*) gene that explained 0.14% of the variance in adult BMI.

Unlike for *FTO*, the role of the *MC4R* gene in obesity development had already been well established at the time of discovery. *MC4R* is expressed in the arcuate nucleus of the hypothalamus where the protein product is activated by the alpha-melanocyte stimulating hormone, a cleavage product of pro-opiomelanocortin (POMC), leading to a decrease in appetite (De Jonghe *et al.*, 2011). In mice, *MC4R* deficiency leads to hyperphagia, hyperinsulinemia, hyperglycemia, and mature-onset obesity (Huszar *et al.*, 1997). In humans, loss-of-function variants in *MC4R* are the most common cause of severe monogenic obesity (Farooqi *et al.*, 2003).

5.2.3. New loci emphasize the role of the central nervous system in regulating BMI

In 2009, two large-scale GWAS meta-analyses of BMI were carried out by the GIANT consortium and deCODE Genetics. The GIANT consortium combined data from 32,387 adults of European descent and identified six novel BMI loci in or near the *GNPDA2, KCTD15, MTCH2, NEGR1, SH2B1*, and *TMEM18* genes (Willer *et al.*, 2009). At the same time, deCODE Genetics carried out a GWAS meta-analysis of 31,392 adults of mainly Icelandic origin, identifying four of the loci discovered by the GIANT consortium, but also four new loci in or near the *BCDIN3D, BDNF, ETV5*, and *SEC16B* genes (Thorleifsson *et al.*, 2009).

The new discoveries underscored the role of the central nervous system (CNS) in the predisposition to obesity. Most of the likely causal genes were highly expressed in the brain and hypothalamus and suggested to act through the CNS (Thorleifsson *et al.*, 2009; Willer *et al.*, 2009). These findings were consistent with the results from animal models where neuronal genes, particularly those expressed in the hypothalamus, had been shown to be involved in the regulation of appetite and energy balance and to have a major role in determining obesity (Huszar *et al.*, 1997; Ren *et al.*, 2007; O'Rahilly and Farooqi, 2008).

5.2.4. Extended meta-analyses reveal 90 more BMI loci

The third meta-analysis of the GIANT consortium, published in 2010, involved 123,865 individuals of European descent with genome-wide variant data and a follow-up of the strongest loci in 125,931 additional individuals (Speliotes *et al.*, 2010). This meta-analysis confirmed all 12 loci identified in the previous GWASs and enabled the identification of 20 novel BMI-associated loci. In total, the 32 loci explained 1.45% of the variance in BMI (Speliotes *et al.*, 2010).

Success in GWAS of populations of European ancestry was followed by large-scale meta-analyses in populations of other ancestries, in whom differences in patterns of linkage disequilibrium (LD), effect sizes, and allele frequencies may facilitate the discovery of novel loci (Rosenberg *et al.*, 2010). In 2012, two large GWAS meta-analyses of BMI in 26,620 and 27,715 individuals of South-Asian ancestry were carried out, identifying four novel BMI loci in or near the *CDKAL1, KLF9, PCSK1*, and *GP2* genes (Okada *et al.*, 2012; Wen *et al.*, 2012). A year later, a meta-analysis of 39,144 individuals of African ancestry and a replication in 32,268 Africans identified a new locus in *GALNT10*. Another locus was identified near the *NFE2L3* gene when the African-ancestry data were meta-analyzed with the data of 123,865 European-ancestry individuals from the meta-analysis of the GIANT consortium (Speliotes *et al.*, 2010; Monda *et al.*, 2013). In 2013, a trans-ethnic meta-analysis in 108,912 adults of European, African-American, Hispanic, or East

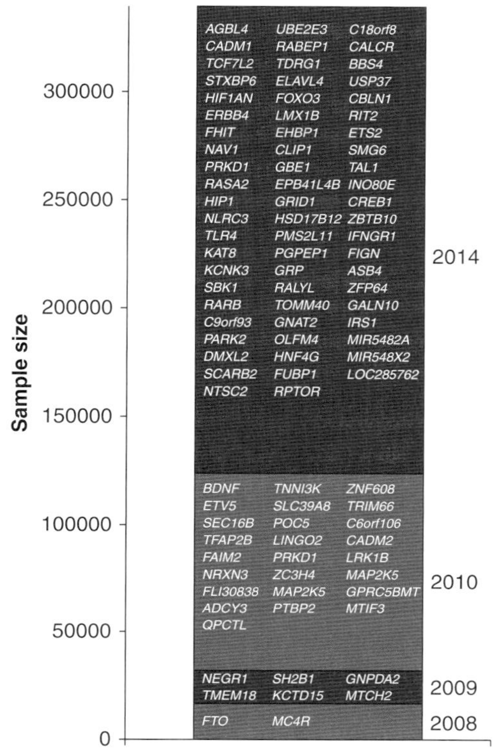

Figure 5.2. Sample size of the discovery stage (number of individuals with genome-wide data) and the new loci identified in the GIANT consortium meta-analyses of BMI in 2008 (Loos *et al.*, 2008), 2009 (Willer *et al.*, 2009), 2010 (Speliotes *et al.*, 2010), and 2014 (Locke *et al.*, 2015).

Asian ancestry, utilizing a gene-centric chip across ~2100 metabolic and cardiovascular-related loci, identified yet another novel adiposity locus, *TOMM40* (Guo *et al.*, 2013).

The GIANT consortium published its most recent meta-analysis in 2015. The meta-analysis included 236,231 individuals with genome-wide data and 103,047 individuals genotyped with the Metabochip – a custom-selected genotyping array designed to cover the loci that fell just below genome-wide significant thresholds in previous GWAS of anthropometric and metabolic traits (Voight *et al.*, 2012; Locke *et al.*, 2015). Of the 339,226 individuals that were included in this meta-analysis, 322,154 were of European descent. A meta-analysis of the European-ancestry individuals identified 77 loci reaching genome-wide significance. Ten additional loci were identified when 17,072 non-European-descent individuals were included, and 10 further loci were unraveled in analyses of men only, women only, or population-based studies only. The total number of identified loci thus reached 97 (Figure 5.2), of which 56 were novel and 41 had been identified in previous meta-analyses of adiposity traits (Speliotes *et al.*, 2010; Kilpelainen *et al.*, 2011; Bradfield *et al.*, 2012; Berndt *et al.*, 2013; Guo *et al.*, 2013; Monda et al., 2013).

To summarize, GWAS meta-analyses have discovered more than 100 loci robustly associated with BMI. The effect sizes of all loci are, however, small, and in total the 97 loci identified in the latest GIANT consortium meta-analysis (Locke *et al.*, 2015) explain only 2.7% of variance in BMI (Figure 5.1). The more recently identified loci generally have smaller effect sizes and/or minor allele frequencies than the loci identified in the early meta-analyses, and explain a smaller proportion of variance in BMI. Many more loci are expected to be identified in extended meta-analyses in the years to come.

5.3. Follow-up of the GWAS-identified BMI loci

5.3.1. Biological pathways implicated by the BMI loci

Discoveries in GWAS of BMI have underscored the role of central nervous system in the development of obesity. The lead variants in many of the strongest BMI-increasing loci map near genes that are highly expressed in the brain, and several of these genes are known to be key hypothalamic regulators of energy balance (Locke *et al.*, 2015). Several examples also exist where rare loss-of-function mutations in the key hypothalamic genes, such as *MC4R* (Farooqi *et al.*, 2003), *POMC* (Krude *et al.*, 1998), *BDNF* (Gray *et al.*, 2006) and *PCSK1* (Jackson *et al.*, 1997), lead to severe early-onset obesity, and common variants in the same genes show an association with BMI in the general population (Speliotes *et al.*, 2010). Pathway analyses of GWAS meta-analysis results have highlighted CNS-associated gene sets involved in, for example, synaptic function and neurotransmitter signaling (Locke *et al.*, 2015). The conclusion from the current evidence is therefore that the genetic basis of obesity is primarily derived from neuronal mechanisms. Nevertheless, the most recent GWAS meta-analyses of BMI suggest that other mechanisms, such as energy metabolism, polyphagia, and insulin secretion and action, are also involved (Locke *et al.*, 2015).

5.3.2. Associations of BMI loci with cardiovascular and metabolic traits

BMI correlates with a number of cardio-metabolic traits, such as high-density lipoprotein cholesterol, triglycerides and insulin, and increased risk of type 2 diabetes. Consistently, GWAS-identified BMI loci also show associations with cardio-metabolic traits more often than expected by chance (Locke *et al.*, 2015). These findings corroborate the common pathophysiology between BMI and cardio-metabolic traits. Intriguingly, however, for some BMI-increasing loci, the associations with cardio-metabolic traits follow a pattern that is directionally discordant with the correlation between BMI and other traits. For example, the BMI-increasing allele of the locus at *HHIP* is associated with decreased risk of type 2 diabetes and increased levels of high-density lipoprotein cholesterol (Locke *et al.*, 2015). Another locus showing directionally unexpected pattern of associations is a locus near the insulin substrate-1 (*IRS1*) gene. The BMI-increasing allele of the *IRS1* locus is associated with an improved cardiovascular risk profile, including lower fasting insulin and triglycerides, increased high-density lipoprotein cholesterol and adiponectin, and lower risk of coronary heart disease, type 2 diabetes, and diabetic nephropathy (Kilpelainen *et al.*, 2011; Locke *et al.*, 2015). The unexpected pattern of associations may, however, be explained by the fact that the BMI-increasing allele of *IRS1* is also associated with a higher ratio of subcutaneous adipose tissue to visceral adipose tissue, suggesting that this allele may convey beneficial cardio-metabolic effects by increasing subcutaneous deposition of fat and decreasing fat deposition viscerally and ectopically (Kilpelainen *et al.*, 2011).

Several other BMI-increasing loci show directionally discordant effects as compared to the epidemiological correlations between traits (Locke *et al.*, 2015). Many of these unexpected findings may be due to genetic pleiotropy, where a gene simultaneously influences multiple traits and disorders (Solovieff *et al.*, 2013).

5.3.3. Further characterization of the established BMI loci

Further characterization of the identified BMI loci is necessary for understanding the underlying biological mechanisms. Follow-up of the BMI loci in epidemiological studies for associations with more refined body composition traits, such as body fat mass, lean mass, or specific fat tissue compartments, and with intermediary traits known to induce weight gain, such as physical activity behavior and dietary intake, may give more refined phenotypic understanding of the observed associations. Follow-up in prospective study samples with repeated measurement of BMI is important to elucidate the effects of the identified loci during the lifecourse. Analyses of gene–gene interactions, gene–environment interactions, gene- or pathway-based analyses, and analyses of genetic pleiotropy may also provide important mechanistic insights.

The discovery of a novel BMI locus indicates that one or more genetic variants within the locus have biological functions that drive the observed association. A crucial aim after the initial observation is to move from the associated variants to identify the causal variants and their target genes. Few of the identified BMI variants are predicted to disrupt protein-coding regions with the vast majority being located in non-coding, intergenic, or intronic regions, likely to influence the regulation of nearby genes. Most GWAS-identified BMI loci contain several genes that could explain the observed associations and the causal genes have not been unequivocally established. Some loci span very large regions in the genome and sometimes with a complex genetic architecture, making the identification of the causal genes and variants challenging.

Analyzing the LD between the identified lead variants and potential functional variants may, in some cases, give clues of the causal genes and variants. For example, the *NEGR1* and *GEPRC5B* loci harbor nearby deletions that may lead to the observed association with BMI (Willer *et al.*, 2009; Speliotes *et al.*, 2010). Differences in LD patterns between ethnicities also make it possible to perform fine-mapping studies to narrow down the BMI-associated genomic regions likely to contain the causal variant. In the follow-up of the most recent GWAS results of the GIANT consortium, fine-mapping across ancestries decreased the number of possible causal SNPs for 10 of the 27 loci that were examined (Voight *et al.*, 2012; Locke *et al.*, 2015). The fine-mapping was particularly successful for the *SEC16B* and *FTO* loci that were narrowed down to a single variant.

After the number of likely causal variants has been reduced by epidemiological methods, the next step is to elucidate the molecular mechanisms by which the candidate variants could influence the function of nearby genes. For non-coding SNPs, regulatory function can be predicted computationally using publicly available genome-wide data sets such as ENCODE, Nuclear Receptor Cistrome, and the NIH Roadmap Epigenomics project, which can be used to annotate variants based on histone modifications, chromatin interactions, expression quantitative trait loci, transcription factor binding motifs, or other regulatory mechanisms. These predictions help in designing experimental analyses, which are ultimately necessary to validate each gene and to identify the fundamental mechanisms that lead to the association with increased BMI.

Either *in vitro* or *in vivo* models can be used to study the effects of overexpression, knock-down, or knockout of a certain gene. Recently, promising new tools have also become available to study the effect of targeted DNA variation in human cell cultures or animal models by using programmable DNA nucleases that induce site-specific double-strand breaks (Urnov *et al.*, 2010; Cong *et al.*, 2013). Studies in animal models have often been performed in rodents, but given the large number of new loci and the advent of high-throughput technologies, simpler model systems such as zebrafish and *Drosophila* have gained popularity in recent years to prioritize genes or variants for testing in more complex model organisms.

5.4. Concluding remarks

Genome-wide association studies of BMI have been tremendously successful in identifying novel susceptibility loci and have greatly increased our understanding of the genetic basis of common obesity. Since 2006, GWAS of BMI have discovered more than 100 independent susceptibility loci, and many more loci will be identified in larger meta-analyses in the future. The effect sizes of the identified loci are small and they explain a small proportion of the phenotypic variability of BMI. Nevertheless, these new loci have given invaluable biological and mechanistic insights that may open up new avenues for the prevention and treatment of obesity. The translation of the numerous novel discoveries to benefit public health is now an area of major focus.

REFERENCES

Berndt, S.I., Gustafsson, S., Magi, R., *et al.* (2013). Genome-wide meta-analysis identifies 11 new loci for anthropometric traits and provides insights into genetic architecture. *Nature Genet.*, **45**, 501–512.

Bradfield, J.P., Taal, H.R., Timpson, N.J., *et al.* (2012). A genome-wide association meta-analysis identifies new childhood obesity loci. *Nature Genet.*, **44**, 526–531.

Cecil, J.E., Tavendale, R., Watt, P., Hetherington, M.M. and Palmer, C.N. (2008). An obesity-associated *FTO* gene variant and increased energy intake in children. *New Engl. J. Med.*, **359**, 2558–2566.

Church, C., Moir, L., McMurray, F., *et al.* (2010). Overexpression of *FTO* leads to increased food intake and results in obesity. *Nature Genet.*, **42**, 1086–1092.

Cong, L., Ran, F.A., Cox, D., *et al.* (2013). Multiplex genome engineering using CRISPR/Cas systems. *Science*, **339**, 819–823.

Cota, D., Proulx, K., Smith, K.A., *et al.* (2006). Hypothalamic mTOR signaling regulates food intake. *Science*, **312**, 927–930.

De Jonghe, B.C., Hayes, M.R. and Bence, K.K. (2011). Melanocortin control of energy balance: evidence from rodent models. *Cell. Molec. Life Sci.*, **68**, 2569–2588.

Dina, C., Meyre, D., Samson, C., *et al.* (2007). Comment on "A common genetic variant is associated with adult and childhood obesity". *Science*, **315**, 187; author reply 187.

Farooqi, I.S., Keogh, J.M., Yeo, G.S., *et al.* (2003). Clinical spectrum of obesity and mutations in the melanocortin 4 receptor gene. *New Engl. J. Med.*, **348**, 1085–1095.

Fischer, J., Koch, L., Emmerling, C., *et al.* (2009). Inactivation of the *FTO* gene protects from obesity. *Nature*, **458**, 894–898.

Frayling, T.M., Timpson, N.J., Weedon, M.N., *et al.* (2007). A common variant in the *FO* gene is associated with body mass index and predisposes to childhood and adult obesity. *Science*, **316**, 889–894.

Gerken, T., Girard, C.A., Tung, Y.C., *et al.* (2007). The obesity-associated *FTO* gene encodes a 2-oxoglutarate-dependent nucleic acid demethylase. *Science*, **318**, 1469–1472.

Gray, J., Yeo, G.S., Cox, J.J., *et al.* (2006). Hyperphagia, severe obesity, impaired cognitive function, and hyperactivity associated with functional loss of one copy of the brain-derived neurotrophic factor (BDNF) gene. *Diabetes*, **55**, 3366–3371.

Gulati, P., Cheung, M.K., Antrobus, R., *et al.* (2013). Role for the obesity-related *FTO* gene in the cellular sensing of amino acids. *Proc. Natl Acad. Sci.*, **110**, 2557–2562.

Guo, Y., Lanktree, M.B., Taylor, K.C., *et al.* (2013). Gene-centric meta-analyses of 108 912 individuals confirm known body mass index loci and reveal three novel signals. *Hum. Molec. Genet.*, **22**, 184–201.

Haupt, A., Thamer, C., Staiger, H., *et al.* (2009). Variation in the *FTO* gene influences food intake but not energy expenditure. *Exp. Clin. Endocrinol. Diab.*, **117**, 194–197.

Hemani, G., Yang, J., Vinkhuyzen, A., *et al.* (2013). Inference of the genetic architecture underlying BMI and height with the use of 20,240 sibling pairs. *Am. J. Hum. Genet.*, **93**, 865–875.

Herbert, A., Gerry, N.P., McQueen, M.B., *et al.* (2006). A common genetic variant is associated with adult and childhood obesity. *Science*, **312**, 279–283.

Huszar, D., Lynch, C.A., Fairchild-Huntress, V., *et al.* (1997). Targeted disruption of the melanocortin-4 receptor results in obesity in mice. *Cell*, **88**, 131–141.

Jackson, R.S., Creemers, J.W., Ohagi, S., *et al.* (1997). Obesity and impaired prohormone processing associated with mutations in the human prohormone convertase 1 gene. *Nature Genet.*, **16**, 303–306.

Kilpelainen, T.O., Zillikens, M.C., Stancakova, A., *et al.* (2011). Genetic variation near *IRS1* associates with reduced adiposity and an impaired metabolic profile. *Nature Genet.*, **43**, 753–760.

Krude, H., Biebermann, H., Luck, W., *et al.* (1998). Severe early-onset obesity, adrenal insufficiency and red hair pigmentation caused by *POMC* mutations in humans. *Nature Genet.*, **19**, 155–157.

Li, S. and Loos, R.J. (2008). Progress in the genetics of common obesity: size matters. *Curr. Opin. Lipidol.*, **19**, 113–121.

Locke, A.E., Kahali, B., Berndt, S.I. and Justice, A.E. (2015). Large-scale genetic studies of body mass index provide insight into the biological basis of obesity. *Nature*, 518, 197–206.

Loos, R.J., Barroso, I., O'Rahilly, S. and Wareham, N.J. (2007). Comment on "A common genetic variant is associated with adult and childhood obesity". *Science*, **315**, 187; author reply 187.

Loos, R.J., Lindgren, C.M., Li, S., *et al.* (2008). Common variants near *MC4R* are associated with fat mass, weight and risk of obesity. *Nature Genet.*, **40**, 768–775.

Monda, K.L., Chen, G.K., Taylor, K.C., *et al.* (2013). A meta-analysis identifies new loci associated with body mass index in individuals of African ancestry. *Nature Genet.*, **45**, 690–696.

O'Rahilly, S. and Farooqi, I. S. (2008). Human obesity as a heritable disorder of the central control of energy balance. *Int. J. Obes.*, **32** (Suppl 7), S55–S61.

Okada, Y., Kubo, M., Ohmiya, H., *et al.* (2012). Common variants at *CDKAL1* and *KLF9* are associated with body mass index in East Asian populations. *Nature Genet.*, **44**, 302–306.

Ren, D., Zhou, Y., Morris, D., *et al.* (2007). Neuronal *SH2B1* is essential for controlling energy and glucose homeostasis. *J. Clin. Investig.*, **117**, 397–406.

Rosenberg, N.A., Huang, L., Jewett, E.M., *et al.* (2010). Genome-wide association studies in diverse populations. *Nature Rev. Genet.*, **11**, 356–366.

Scuteri, A., Sanna, S., Chen, W.M., *et al.* (2007). Genome-wide association scan shows genetic variants in the *FTO* gene are associated with obesity-related traits. *PLoS Genet.*, **3**, e115.

Smemo, S., Tena, J.J., Kim, K.H., *et al.* (2014). Obesity-associated variants within *FTO* form long-range functional connections with *IRX3*. *Nature*, **507**, 371–375.

Solovieff, N., Cotsapas, C., Lee, P.H., Purcell, S.M. and Smoller, J.W. (2013). Pleiotropy in complex traits: challenges and strategies. *Nature Rev. Genet.*, **14**, 483–495.

Speakman, J.R., Rance, K.A. and Johnstone, A.M. (2008). Polymorphisms of the *FTO* gene are associated with variation in energy intake, but not energy expenditure. *Obesity*, **16**, 1961–1965.

Speliotes, E.K., Willer, C.J., Berndt, S.I., *et al.* (2010). Association analyses of 249,796 individuals reveal 18 new loci associated with body mass index. *Nature Genet.*, **42**, 937–948.

Thorleifsson, G., Walters, G.B., Gudbjartsson, D.F., *et al.* (2009). Genome-wide association yields new sequence variants at seven loci that associate with measures of obesity. *Nature Genet.*, **41**, 18–24.

Urnov, F.D., Rebar, E.J., Holmes, M.C., Zhang, H.S. and Gregory, P.D. (2010). Genome editing with engineered zinc finger nucleases. *Nature Rev. Genet.*, **11**, 636–646.

Van Der Hoeven, F., Schimmang, T., Volkmann, A., *et al.* (1994). Programmed cell death is affected in the novel mouse mutant Fused toes (Ft). *Development*, **120**, 2601–2607.

Voight, B.F., Kang, H.M., Ding, J., *et al.* (2012). The Metabochip, a custom genotyping array for genetic studies of metabolic, cardiovascular, and anthropometric traits. *PLoS Genet.*, **8**, e1002793.

Wardle, J., Carnell, S., Haworth, C.M., *et al.* (2008). Obesity associated genetic variation in *FTO* is associated with diminished satiety. *J. Clin. Endocrinol. Metab.*, **93**, 3640–3643.

Wardle, J., Llewellyn, C., Sanderson, S. and Plomin, R. (2009). The *FTO* gene and measured food intake in children. *Int. J. Obes.*, **33**, 42–45.

Wen, W., Cho, Y.S., Zheng, W., *et al.* (2012). Meta-analysis identifies common variants associated with body mass index in east Asians. *Nature Genet.*, **44**, 307–311.

WHO. (2000). Obesity: preventing and managing the global epidemic. Report of a WHO consultation. *World Health Organization Technical Report Series*, 894, i–xii, 1–253.

Willer, C.J., Speliotes, E.K., Loos, R.J., *et al.* (2009). Six new loci associated with body mass index highlight a neuronal influence on body weight regulation. *Nature Genet.*, **41**, 25–34.

Zimmet, P., Alberti, K.G. and Shaw, J. (2001). Global and societal implications of the diabetes epidemic. *Nature*, **414**, 782–787.

6 Identification of myocardial infarction-susceptible genes and their functional analyses

Kouichi Ozaki and Toshihiro Tanaka

6.1. Introduction

Coronary artery disease (CADs), including myocardial infarction (MI), have been the major cause of mortality and morbidity among late-onset diseases in many industrialized countries with a Western lifestyle (Braunwald, 1997; Breslow, 1997). MI often occurs without any preceding clinical signs and is followed by severe complications, especially ventricular fibrillation and cardiac rupture, which might result in sudden death. Although recent advances in treatment and diagnosis have greatly improved the quality of life for patients after MI, its morbidity is still high. MI is a disease of the vessel that feeds the cardiac muscle, called the coronary artery. Abrupt occlusion of the coronary artery results in irreversible damage to cardiac muscle. Plaque rupture with thrombosis is a well-established critical factor in the pathogenesis of MI (Falk *et al.*, 1995; Libby, 1995). Although the detailed mechanisms of plaque rupture are unknown, inflammation is thought to play an important role in its pathogenesis (Ross, 1999). Inflammatory mediators like cytokines are involved in atheroma formation; rapid evolution of the atheromatous injury, leading to rupture of the plaque; and intraluminal thrombosis (Ross, 1999). Epidemiologic studies reveal that coronary risk factors include type 2 diabetes mellitus, hypercholesterolemia, hypertension, and obesity. Some studies report a genetic factor; one reported that first-degree relatives of patients who have had an acute MI before age 55 have a 2–7-times higher risk of MI (Lusis *et al.*, 2004). A twin study indicated an eightfold increase in risk of death from MI when a first twin dies of MI before age 55 (Marenberg *et al.*, 1994). Common genetic variants are believed to contribute to genetic risk of disease (Lander, 1996; Risch and Merikangas, 1996; Collins *et al.*, 1997). In this context, we started genome-wide association studies (GWAS) for MI using nearly 90,000 gene-based SNPs (http://snp.ims.u-tokyo.ac.jp/; Haga *et al.*, 2002) by high-throughput multiplex polymerase chain reaction (PCR) invader

Genome-Wide Association Studies: From Polymorphism to Personalized Medicine, ed. K. Appasani. Published by Cambridge University Press.

assay system (Ohnishi *et al.*, 2001), and identified several genes associated with the risk of MI including *LTA* (Ozaki *et al.*, 2002; Ishii *et al.*, 2006; Ebana *et al.*, 2007; Aoki *et al.*, 2011). Although the detailed roles of these susceptible genes in MI pathogenesis are under investigation, these findings showed the potent power of GWAS, which is hypothesis-free, to identify unexpected anchors to further understand the disease. Through examining the LTA cascade by a combination of molecular biological and genetic analyses, we have identified additional MI-susceptible genes (Ozaki *et al.*, 2004, 2006, 2009), so far. In this chapter, we focus on our genetic association studies and show that our initial hypothesis-free strategy unexpectedly revealed the importance of vascular inflammation in the pathogenesis of MI.

6.2. The world's first genome-wide association study

In 2000, we started genotyping with 94 MI patients with our high-throughput multiplex PCR-invader assay method (Ohnishi *et al.*, 2001) using 92,788 gene-based SNPs (Haga *et al.*, 2002) and compared the results with the allelic frequencies found in 658 members of the general Japanese population as a first step in a comprehensive association study. The method was based on a combination of multiplex PCR and invader assay, and allows high-throughput, low-cost screening that requires as little as 0.1 ng of genomic DNA per single SNP (Ohnishi *et al.*, 2001). The success rate for this genotyping was 70.8% (65,671 SNPs), and distribution of the allelic frequency was largely even (Ozaki *et al.*, 2002). The cutoff P value of 0.01 for association in either recessive or dominant models was used for this screening; approximately 99% of the SNP loci that were successfully genotyped were excluded from the second screening. We further genotyped SNPs that showed P values less than 0.01 in a larger replication panel of MI patient and control subjects and found that most of the loci eventually showed the lack of association at this screening phase of the project. From these observations, we think one should carefully handle the association results using fewer (~100) samples for common diseases.

However, through this GWAS, one SNP in the *LTA*, encoding an inflammatory cytokine lymphotoxin-α, (6p21.3) was identified as a candidate susceptibility locus for MI (Ozaki *et al.*, 2002). Following linkage disequilibrium (LD), haplotype mapping and further functional analyses revealed that two functional SNPs (*LTA* intron 1 252A>G and exon 3 804 C>A) were in complete LD in this locus and conferred the risk of MI. Recently, we have further confirmed an association between MI and *LTA* exon 3 804 C>A SNP (rs1041981) using larger sample sizes (approximately 3400 case and 3500 control subjects), and we obtained a similar association result ($P < 0.001$; recessive association model; Table 6.1). Furthermore, among white Europeans (in the Precocious Coronary Artery Disease [PROCARDIS] study), a transmission disequilibrium test analysis of 447 trio families with CAD demonstrated that the *LTA* 804 C allele (26 N-LTA) was excessively transmitted to affected offspring ($\chi^2 = 8.44$, $P = 0.002$, recessive association model) (PROCARDIS Consortium, 2004).

Table 6.1. Association study of *LTA* exon3 SNP (rs1041981) and MI.

	MI		CO	
Genotype	Number	%	Number	%
Exon 3 804 C/A, T26N				
CC	1028	36.3	1333	39.2
CA	1318	46.5	1630	48
AA	487	17.2	436	12
Total	2833	100	3399	100
Statistics	χ^2	*P*	Odds ratio	95% CI
(Recessive model)	23.3	0.0000014	1.41	1.23–1.62

6.3. *LGALS2*, encoding galectin-2 that interacts with LTA, associated with MI

After identifying *LTA* as a novel genetic risk factor for MI, we searched for proteins that interact with LTA to better understand its role in the pathogenesis of this disease. Using both the *Escherichia coli* two-hybrid system and a phage display method, we identified a protein, galectin-2, as a binding partner of LTA (Ozaki *et al.*, 2004). Because galectin-2 was shown to bind to LTA, we examined whether variations on *LGALS2* (encoding galectin-2) were also associated with susceptibility to MI. We found one SNP (3279C>T) in intron 1 of *LGALS2*; this substitution represses the level of galectin-2 expression and shows a significant association with MI (Ozaki *et al.*, 2004). This genetic substitution seemed to affect the transcriptional level of galectin-2, which led to altered secretion of LTA, thereby affecting the degree of inflammation. We also found that galectin-2 binds to tubulins, which are important components of microtubules, suggesting a role in intracellular trafficking. It is likely that LTA is another molecule that uses the microtubule cytoskeleton network for translocation, and galectin-2 mediates LTA trafficking through binding to microtubules (Ozaki and Tanaka, 2005; Tanaka and Ozaki, 2006), although the precise role of galectin-2 in this trafficking machinery complex has yet to be elucidated (Figure 6.1).

6.4. *PSMA6*, encoding the intercellular LTA signaling molecule, is also associated with MI

Because LTA binding to its receptor strongly activates nuclear factor κB (NFκB) by proteasomal degradation of its inhibitory partner (IκB) protein (Figure 6.2; Beinke and Ley, 2004), we hypothesized that the variation(s) in the genes encoding proteasomal proteins could confer risk of MI. The 20S proteasome, which is composed of seven α- and 10 β-subunits, is the core particle for 26S proteasome system (Coux *et al.*, 1996). We selected tagSNPs with minor allele frequencies of more than 10% that covered most of the haplotypes in the genomic region of genes encoding these subunits and found that one SNP (rs1048990) in the

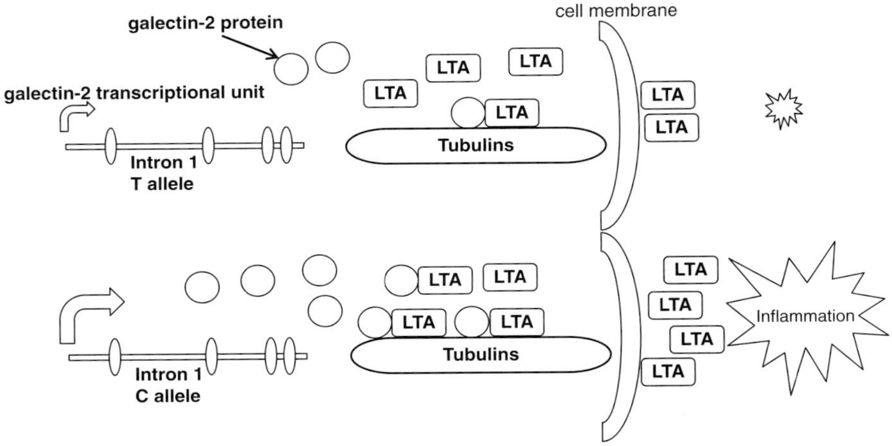

Figure 6.1. Hypothetical roles of the SNP in *LGALS2* in inflammatory process in the pathogenesis of MI.

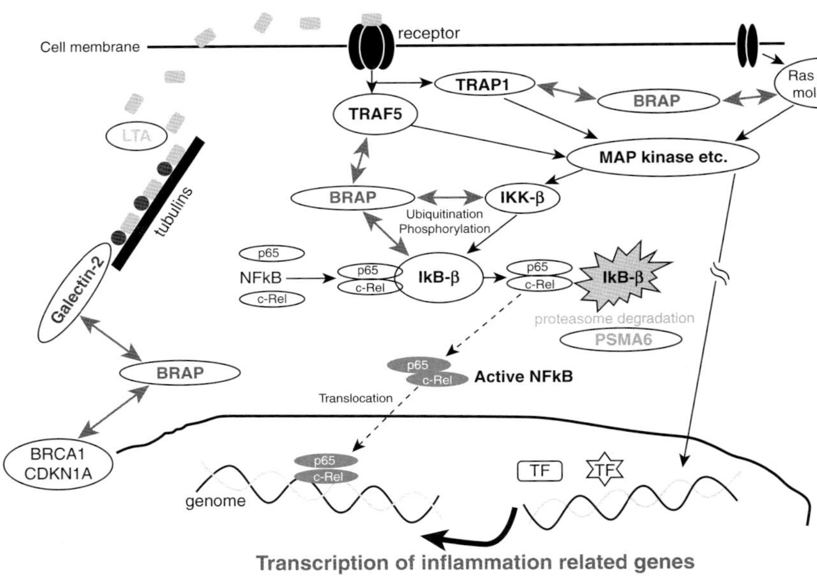

Figure 6.2. MI-related inflammatory cascade. Red arrows indicate direct interaction of BRAP. TRAF5, TNF receptor associated factor 5; TRAP1, TNF receptor-associated protein 1; IκK-β, I kappa kinase β; IκB-β; I kappa B-β; BRCA1, breast cancer 1; CDKN1A, cyclin-dependent kinase inhibitor 1A; TF, transcription factor. A black and white version of this figure will appear in some formats. For the color version, please refer to the plate section.

5′-untranslated region of exon 1 (5′UTR –8 C>G) of *PSMA6*, encoding proteasome subunit, alpha type, 6 was significantly associated with MI (Table 6.2; Ozaki *et al.*, 2006). This association was replicated by nearly the same effect with Japanese in a large Chinese cohort (Liu *et al.*, 2009). The SNP, located within 5′UTR of exon 1 in this gene, enhanced the transcriptional level of *PSMA6*. Moreover, suppression of *PSMA6* expression level using siRNA in cultured coronary vascular endothelial cells as well as a T-lymphocyte cell line reduced activation of NFκB, a central

Table 6.2. Association study of *PSMA6* exon1 SNP (rs1048990) and MI.

	MI		CO	
Genotype	Number	%	Number	%
Exon 1 –8C/G				
CC	1134	43.8	1382	48.5
CG	1137	43.9	1216	42.7
GG	321	12.4	253	8.9
Total	2592	100	2851	100
Statistics	χ^2	*P*	Odds ratio	95% CI
(Allele frequency)	21.1	0.0000044	1.21	1.11–1.31

mediator of inflammation, by stabilizing phosphorylated IκB. Thus, the levels of PSMA6 protein influence the degree of inflammation (Figure 6.2), indicating that *PSMA6* SNP is a novel genetic risk factor for MI (Ozaki *et al.*, 2006).

6.5. *BRAP*, encoding galectin-2 binding protein, associated with the risk of MI

To further understand the molecular mechanism that confers risk of MI, we searched proteins that interact with galectin-2. By means of tandem affinity purification, followed by matrix-assisted laser desorption/ionization–time of flight (MALDI/TOF) mass spectrometry analyses, we identified BRAP, *BRCA1*-associated protein, as a possible binding partner of galectin-2 (Ozaki *et al.*, 2009) We confirmed their interaction by co-immunoprecipitation experiments, and examined whether the genetic variation in BRAP was associated with susceptibility to MI by resequencing genomic DNA, and then an association study of selected tag SNPs. As an initial association study, we compared genotype frequencies of these tag SNPs in about 450 individuals with MI and 450 controls and found that one SNP, rs3782886, in exon 5 (90A>G, R241 R) of *BRAP* was significantly associated with MI (P = 0.0014). Haplotypes based on these tag SNPs showed a less significant association with MI. rs11066001 was in very strong LD (r^2 = 0.96) with rs3782886, so we examined these two SNPs in 2475 cases and 2778 controls, and found strong associations with MI (χ^2 = 83.6, $P = 3.0 \times 10^{-18}$, OR = 1.48, by comparison of allele frequency of rs11066001, Table 6.3). To further confirm the association, we examined two additional panels, 862 cases and 1113 controls from the Japanese population and 349 cases and 994 controls from the Taiwanese population, and confirmed the associations of rs11066001 and rs3782886 with MI in these two sets (Table 6.3).

According to HapMap data (http://www.hapmap.org; International HapMap Consortium, 2005), the minor allele frequencies of rs3782886 were 0.239 in Japanese in Tokyo and 0.148 in Han Chinese in Beijing, but this allele was observed in neither CEPH individuals (Utah residents with ancestry from

Table 6.3. Association results of the two *BRAP* SNPs (rs11066001 and rs3782886) with MI in East Asian population.

	rs11066001				rs3782886			
	MAF*		Allele frequency		MAF*		Allele frequency	
Study population	Cases	Controls	Odds ratio (95% CI)	*P* value**	Cases	Controls	Odds ratio (95% CI)	*P* value**
Japanese								
First panel	0.34	0.26	1.48 (1.36–1.61)	3.0×10^{-18}	0.35	0.28	1.42 (1.31–1.54)	2.8×10^{-15}
Replication panel	0.34	0.26	1.46 (1.27–1.67)	4.4×10^{-6}	0.36	0.27	1.50 (1.31–1.71)	1.8×10^{-7}
Combined	0.34	0.26	1.47 (1.37–1.56)	1.3×10^{-24}	0.35	0.27	1.44 (1.34–1.55)	7.0×10^{-23}
Taiwanese	0.33	0.27	1.31 (1.09–1.58)	4.7×10^{-3}	0.33	0.28	1.26 (1.05–1.52)	1.5×10^{-2}

* Minor allele frequency; ** adjusted for Bonferroni's correction in Japanese cohorts.

northern and western Europe) nor Yoruba individuals from Ibadan, Nigeria. No information was available for rs11066001. We additionally examined a panel of 50 CEPH individuals and found there was no variation at these two SNP loci. These results indicate that these SNPs are likely to be present only in Asian populations. However, the possibility cannot be excluded that other variations in this gene confer risk of MI in other populations. Because rs1041981 in *LTA* and rs7291467 in *LGALS2* were associated with MI as described above, we also performed logistic regression analysis for the combinatorial effect of rs11066001 (*BRAP*), rs1041981 (*LTA* exon 3 804 C>A SNP) and rs7291467 (*LGALS2*) on MI susceptibility. We did not find any evidence of gene–gene interactions, as addition of a statistical interaction term showed no significance. The combinatorial effect was consistent with a multiplicative odds ratio model (Ozaki *et al.*, 2009). We also examined the possibility of confounding effects by age, gender, and classical risk factors within cases, and found no obvious relationships between genotype and these factors. These results indicated that the significant SNP in *BRAP* is an independent risk factor of MI (Ozaki *et al.*, 2009). Using antibodies directly labeled with fluorescein (α-galectin-2) or rhodamine (α-BRAP), we examined subcellular localization of native galectin-2 and BRAP in human coronary artery smooth muscle cells (HCASMC) and found that these proteins co-localized in the cytoplasm and nucleus. We also investigated whether the BRAP protein is in fact expressed in the myocardial infarction lesion, that is, the atherosclerotic lesion of the coronary artery. Immunoreactivities for BRAP were detected in the smooth muscle cells (SMCs) and macrophages in atherosclerotic plaques. Co-expression of BRAP and galectin-2 was also observed in the majority of polymorphic SMCs and activated macrophages by double-labeled immunohistochemistry (Ozaki *et al.*, 2009).

The two SNPs in *BRAP* showing very strong associations did not cause amino acid substitutions. Therefore, we investigated whether these SNPs, rs11066001 (intron3 270A>G) and rs3782886 (exon5 90A>G; R241 R), would affect *BRAP* expression by reporter gene analysis. A clone containing the intron3 270A allele showed approximately half of the transcriptional activity of the 270 G allele or that of the *BRAP* promoter only. No allelic difference was observed in constructs containing the exon5 SNP (Ozaki *et al.*, 2009). To confirm these results, we cloned three tandem copies of the genomic segment including these SNPs, and obtained similar results. These results indicate that the substitution in intron3, but not the one in exon5, affected the transcription level of *BRAP*. We subsequently looked for nuclear factor(s) that might bind to oligonucleotides corresponding to genomic sequences of the 270A allele. No known protein was predicted to bind to this DNA segment by TFSEARCH program (http://www.cbrc.jp/research/db/TFSEARCHJ.html) based on the TRANSFAC database. Using nuclear extracts from HCASMC, we observed one band in the lane corresponding to the A allele, indicating binding of a nuclear protein(s) to the A allele (Ozaki *et al.*, 2009). This result suggested that an unidentified nuclear factor(s) interacting with this genomic region might suppress the transcription of *BRAP* and thereby play a role in the pathogenesis of MI. BRAP was originally identified as a protein that binds to the signal peptide of breast cancer suppressor protein BRCA1 (Li *et al.*, 1998), and is known to be an E3 ubiquitin ligase that associates with Ras and modulates mitogen-activated protein (MAP) kinase signaling through regulation of the scaffolding activity of KSR (kinase suppressor of ras; Figure 6.2; Matheny *et al.*, 2004). The MAP kinase signaling pathway is well known as a regulator of cell survival, growth, differentiation, transformation, and production of proinflammatory factors (Ory and Morrison, 2004; O'Neill, 2006). Galectin-2 was also implicated in the inflammatory pathway through interactions with lymphotoxin-α and tubulin proteins (Figures 6.1 and 6.2; Ozaki *et al.*, 2004). Thus, we examined whether the cellular level of BRAP protein could influence NFkB, a central transcription factor of inflammation (Karin and Delhase, 2000). An siRNA against *BRAP* significantly suppressed *BRAP* mRNA, and resulted in inhibition of NFκB activation in *human coronary artery endothelial cells* (HCAEC; Ozaki *et al.*, 2009). This result implied that altered expression of BRAP may influence the expression levels of genes involved in the NFκB-dependent inflammatory pathway (Figure 6.2). Furthermore, we found that the BRAP protein binds several molecules related with inflammation and cell proliferation (Figure 6.2), such as major components of the IκK-signalosome (Liao *et al.*, 2011; and unpublished data). Thus, it is conceivable that a higher expression level of *BRAP* with an minor allele of intron3 (G allele) might enhance the degree of inflammation through activation of the NFκB–IκK signalosome protein, thereby playing an important role in the pathogenesis of MI, although the relationship of BRAP protein with activation of HCAECs involved in plaque instability and/or rupture remains to be tested.

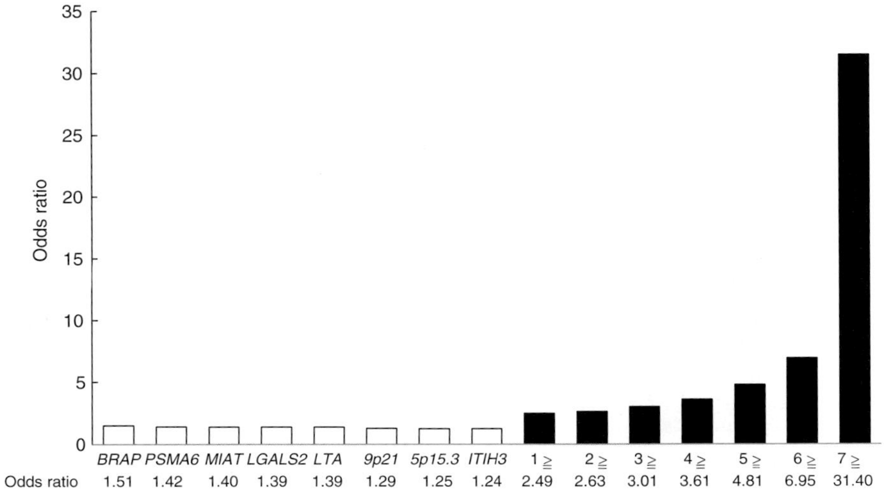

Figure 6.3. Odds ratios according to the number of genetic risk factors.

6.6. Conclusion

To date, we have identified another three loci (SNPs in *MIAT, ITIH3*, chromosome 5p15.3; Ebana *et al.*, 2007; Ishii *et al.*, 2006; Aoki *et al.*, 2011) by GWAS that confer increased risk of MI, and also have replicated the association for SNP locus on 9p21 previously identified by European GWAS (Helgadottir *et al.*, 2007; McPherson *et al.*, 2007). Each odds ratio of these genetic risk factors described here is relatively small; however, by combination of these genetic risk factors, the odds ratio has risen by about 30 (Figure 6.3). This result suggests that the combination of genetic risk factors contributes to the pathogenesis of MI. Coronary artery disease attributable to atherosclerosis is a leading cause of death in many countries. We believe that knowledge of genetic factors contributing to its pathogenesis provides a useful clue for the development of diagnostic methods, treatments, and preventive measures through combinations of risk variants (for diagnostic methods) and clarification of the molecular mechanism in the pathogenesis of causative genes (for therapeutics) for this common but serious disorder.

REFERENCES

Aoki, A., Ozaki, K., Sato, H., *et al.* (2011). SNPs on 5p15.3 associated with myocardial infarction in Japanese population. *J. Hum. Genet.*, **56**, 47–51.

Beinke, S. and Ley, S.C. (2004). Functions of NF-kappaB1 and NF-kappaB2 in immune cell biology. *Biochem. J.*, **382**, 393–409.

Braunwald, E. (1997). Shattuck lecture – cardiovascular medicine at the turn of the millennium: triumphs, concerns and opportunities. *New Engl. J. Med.*, **337**, 1360–1369.

Breslow, JW. (1997). Cardiovascular disease burden increases, NIH funding decreases. *Nature Med.*, **3**, 600–601.

Collins, FS., Guyer, MS. and Charkravarti, A. (1997) Variations on a theme: cataloging human DNA sequence variation. *Science*, **278**, 1580–1581.

Coux, O., Tanaka, K. and Goldberg, AL. (1996). Structure and functions of the 20S and 26S proteasomes. *Annu. Rev. Biochem.*, **65**, 801–847.

Ebana, Y., Ozaki, K., Sato, H., *et al.* (2007) A functional SNP in *ITIH3* is associated with susceptibility to myocardial infarction. *J. Hum. Genet.*, **52**, 220–229.

Falk, E., Shah, P.K. and Fuster, V. (1995) Coronary plaque disruption. *Circulation*, **92**, 657–671.

Haga, H., Yamada, R., Ohnishi, Y., Nakamura, Y. Tanaka, T. (2002). Gene-based SNP discovey as part of the Japanese Millennium Genome project: identification of 190,562 genetic variations in the human genome. *J. Hum. Genet.*, **47**, 605–610.

Helgadottir, A., Thorleifsson, G., Manolescu, A., *et al.* (2007). A common variant on chromosome 9p21 affects the risk of myocardial infarction. *Science*, **316**, 1491–1493.

International HapMap Consortium. (2005) A haplotype map of the human genome. *Nature*, **437**, 1299–1320.

Ishii, N., Ozaki, K., Sato, H., *et al.* (2006). Identification of a novel non-coding RNA, *MIAT*, that confers risk of myocardial infarction. *J. Hum. Genet.*, **51**, 1087–1099.

Karin, M. and Delhase, M. (2000). The I kappa B kinase (IKK) and NF-kappa B: key elements of proinflammatory signalling. *Seminars in Immunology*, **12**, 85–98.

Lander, ES. (1996). The new genomics: global views of biology. *Science* **274**, 536–539.

Li, S., Ku, C.Y., Farmer, A.A., *et al.* (1998). Identification of a novel cytoplasmic protein that specifically binds to nuclear localization signal motifs. *J. Biol. Chem.*, **273**, 6183–6189.

Liao, Y.C., Wang, Y.S., Guo, Y.C., *et al.* (2011). BRAP activates the inflammatory cascades and increases the risk for carotid atherosclerosis. *Molec. Med.*, **17**, 1065–1074.

Libby, P. (1995). Molecular bases of the acute coronary syndromes. *Circulation*, **91**, 2844–2850.

Liu, X., Wang, X., Shen, Y., *et al.* (2009). The functional variant rs1048990 in *PSMA6* is associated with susceptibility to myocardial infarction in a Chinese population. *Atherosclerosis*, **206**(1), 199–203.

Lusis, A.J., Mar, R. and Pajukanta, P. (2004). Genetics of atherosclerosis. *Annu. Rev. Genom. Hum. Genet.*, **5**, 189–218.

Marenberg, M.E., Risch, N., Berkman, L.F., Floderus, B. and de Faire, U. (1994). Genetic susceptibility to death from coronary heart disease in a study of twins. *New Engl. J. Med.*, **330**, 1041–1046.

Matheny, S.A., Chen, C., Kortum, R.L., *et al.* (2004). Ras regulates assembly of mitogenic signalling complexes through the effector protein IMP. *Nature*, **427**, 256–260.

McPherson, R., Pertsemlidis, A., Kavaslar, N., *et al.* (2007). A common allele on chromosome 9 associated with coronary artery disease. *Science*, **316**, 1488–1491.

Ohnishi, Y., Tanaka, T., Ozaki, K., *et al.* (2001). A high-throughput SNP typing system for genomewide association studies. *J. Hum. Genet.*, **46**, 471–477.

O'Neill, L.A. (2006). Targeting signal transduction as a strategy to treat inflammatory diseases. *Nature Rev. Drug Discov.*, **5**, 549–563.

Ory, S. and Morrison, DK. (2004). Signal transduction: implications for Ras-dependent ERK signaling. *Curr. Biol.*, **14**, R277–R278.

Ozaki, K. and Tanaka, T. (2005). Genome-wide association study to identify SNPs conferring risk of myocardial infarction and their functional analyses. *Cell. Molec. Life Sci.*, **62**, 1804–1813.

Ozaki, K., Ohnishi, Y., Iida, A., *et al.* (2002). Functional SNPs in the lymphotoxin-alpha gene that are associated with susceptibility to myocardial infarction. *Nature Genet.*, **32**, 650–654.

Ozaki, K., Inoue, K., Sato, H., *et al.* (2004). Functional variation in *LGALS2* confers risk of myocardial infarction and regulates lymphotoxin-alpha secretion in vitro. *Nature*, **429**, 72–75.

Ozaki, K., Sato, H., Iida, A., *et al.* (2006). A functional SNP in *PSMA6* confers risk of myocardial infarction in the Japanese population. *Nature Genet.*, **38**, 921–925.

Ozaki, K., Sato, H., Inoue, K., *et al.* (2009). SNPs in *BRAP* associated with risk of myocardial infarction in Asian populations. *Nature Genet.*, **41**, 329–333.

PROCARDIS Consortium. (2004). A trio family study showing association of the lymphotoxin-alpha N26 (804A) allele with coronary artery disease. *Eur. J. Hum. Genet.*, **12**, 770–774.

Risch, N. and Merikangas, K. (1996). The future of genetic studies of complex human diseases. *Science*, **273**, 1516–1517.

Ross, R. (1999). Atherosclerosis – an inflammatory disease. *New Engl. J. Med.*, **340**, 115–126.

Tanaka, T. and Ozaki, K. (2006). Inflammation as a risk factor for myocardial infarction. *J. Hum. Genet.*, **51**, 595–604.

7 Admixture mapping for disease gene discovery

Randall C. Johnson, Cheryl A. Winkler and Meredith Yeager

7.1. Introduction

Mapping disease genes by admixture linkage disequilibrium (MALD) is a method that exploits observed associations between disease and ancestry. The sources of these observed associations generally fall into two categories: (1) environmental exposures, including factors such as diet, cultural practices and pathogens, and (2) heritable genetic risk modifiers. In a genome-wide association study (GWAS), association between a genetic marker and the disease are sought for, and any statistically significantly associated marker is inferred to be near a risk-modifying genetic variant. By contrast, a MALD study maps disease genes by identifying associations between ancestry and disease. Thus, loci with a statistically significantly different ancestral origin, when compared to the rest of the genome or to a control group, will be inferred to harbor a risk-modifying genetic variant. Additional research can then be carried out to identify the causal variant responsible for the observed association. The size of the associated region in a GWAS or MALD study is dependent upon the extent of genetic linkage at the locus in question.

Genetic linkage is fundamental to genetic association studies as a means to narrowing the search for causal variants, by identifying a chromosomal region associated with disease. When two markers are physically near each other on a chromosome, they are more likely to be inherited together, because there is a smaller chance of a crossover between the two during meiosis. The initial source of LD is mutation, in that a new allele arising at a locus by mutation necessarily occurs on a single chromosome, and is thus associated with all alleles carried on that specific chromosome (Bateson and Kilby, 1905; Morgan, 1910, 1911). In successive generations recombination breaks up this original chromosome, but even after 5000 generations – roughly the age of fully modern humans – chromosome segments of an average length of 20 kb will be inherited unbroken with probability less than 0.0001 (Matise *et al.*, 2007). Random drift of allele

Genome-Wide Association Studies: From Polymorphism to Personalized Medicine, ed. K. Appasani. Published by Cambridge University Press.

frequencies, and selection for advantageous alleles against deleterious alleles, contribute to this process, in a complex and extensively studied pattern (Ohta, 1982; Sober, 1993; Keightley and Otto, 2006; Palaisa *et al.*, 2004). As populations age, they acquire more variants and have more opportunity for recombination, which results in shorter blocks (or haplotypes) of LD, while younger populations tend to have longer LD blocks and fewer common variants (International HapMap Consortium *et al.*, 2007). Populations that are geographically separated, or reproductively isolated for other reasons, will acquire different LD patterns over time, forming genetically distinct subpopulations with distinctive phenotypic characteristics. This population substructure is evident in the genetics of Europeans, for example, with individuals from different geographic regions exhibiting slightly different allele frequencies and patterns of LD (Price *et al.*, 2004).

LD structure between disease-causing variants and nearby loci allows the researcher to infer the presence of the causal variant by virtue of its linkage to a nearby marker. This key postulate of genetics is at the core of candidate gene and GWAS studies, which are only possible because of the extensive LD blocks that are shared among individuals within a population (Altshuler *et al.*, 2008). When a significant association is found between a genetic marker and the phenotype under study, an inference can be made that one or more causal variants in LD with the genetic marker have a functional effect on phenotype.

7.1.1. Case/control studies

In Mendelian diseases, where the penetrance is nearly complete, that is, inheritance of a mutation in the appropriate genetic mode (e.g., autosomal dominant, recessive or X-linked) absolutely causes disease, family studies provide a powerful tool for disease gene discovery. Due to the relatively small number of recombination events that occur over a few generations, extended linkage blocks in these families allow simple, yet statistically powerful disease gene-mapping methods to be used. These methods, however, are most powerful in diseases with high penetrance, that is, when inheritance of a mutation from a parent nearly always causes disease; however, Mendelian diseases tend to be rare and account for only a small fraction of morbidity and mortality. Common diseases such as infectious and non-infectious diseases (e.g., HIV, cardiovascular disease and cancers) account for most of the burden to human disease. Case/control studies, on the other hand, can be used to search for genetic variants involved in complex traits, but rely only on local LD for linkage. This reliance on local LD comes with the requirement that many more markers must be tested in order to fully interrogate the entire human genome. Because of this and the relative high cost of genotyping, GWAS were not practical until SNP genotyping chips with non-redundant, haplotype-tagging SNPs became available and of sufficient density (0.6–2.5 million markers). Prior to this, candidate genes were identified and a minimal set of SNPs in and around the gene of interest were carefully chosen. Post GWAS analysis has moved from testing candidate gene hypotheses to hypothesis generation, with subsequent independent replication for validation of SNP association followed by fine-mapping and functional studies to identify causal variants responsible for the association.

Identification of linkage between genes and disease is much less straightforward in diseases with lower penetrance, usually because of the multifactorial (e.g., gene–environment and gene–gene interactions) nature of these diseases. Most have environmental components that affect disease outcome and are impacted by multiple biological systems and pathways. When a candidate gene is identified, a careful analysis of the genetic variants in the region is undertaken in cases and controls to identify possible modifiers of disease outcome.

7.1.2. Candidate gene studies

Prior to the current genotyping and sequencing technologies, great care was taken to minimize the number of markers to be genotyped. Haplotypes, defined as a haploid, multi-locus genotype, are often used to help optimize this minimal marker set. Haplotype blocks can be defined in a number of ways, but the unifying principle is that they have a low inter-block recombination rate. As such, they have strong LD across the block and can be represented by very few genetic markers, referred to as haplotype tagging SNPs. Haplotype tagging is used both to increase power and decrease genotyping costs in candidate gene studies (Zhao *et al.*, 2003). These haplotype blocks can also be very useful in inference of population history (Tishkoff *et al.*, 1996).

Data from the International HapMap Project is critical when identifying haplotype-tagging SNPs (International HapMap Consortium, 2005), but may not be sufficient in populations where good population data are lacking. Many populations in Africa, for example, retain a good deal of variability, which has been lost by populations migrating out to Europe, requiring information not found in HapMap. Other endeavors are being pursued to capture additional variability in less well-studied populations, including the 1000 Genomes project and the African Diaspora SNP Chip (1000 Genomes Project Consortium *et al.*, 2010; Herrera-Paz, 2014). In these populations, investigators can infer haplotype blocks from sequence of a subset of the sample. Once haplotype blocks are identified, rather than genotyping the entire cohort using the full set of all identified SNPs, the smallest subset of SNPs uniquely identifying all haplotypes is chosen to represent the region (Martin *et al.*, 2000). This allows the researcher to sample a larger group of individuals, reduce the number of multiple comparisons, and save time in narrowing down the search for disease-causing variants. The HapMap project increased the efficiency of this method significantly by providing the LD profiles of various populations across the genome, and current bioinformatics tools exploiting these data can deliver good haplotype-tagging SNPs for most regions with relatively minimal effort.

Candidate gene studies have fallen out of favor because they are limited by prior knowledge and they are subject to an unusually high type 1 error rate (Hirschhorn *et al.*, 2002). These studies are still performed, particularly when trying to identify the disease-causing variant associated with a GWAS finding, but haplotype tagging as a means of minimizing the number of markers genotyped is becoming less common with the increased economy of targeted sequencing and increased availability of whole-genome sequence.

7.1.3. Genome-wide studies

Genome-wide studies have been carried out for decades in family studies of Mendelian diseases with high penetrance. As genotyping technology continued to advance, genotyping chips became available with increasing numbers of SNPs. The first SNP chip with genome-wide coverage was made available by the mid-2000s, with the first studies to use this technology published in 2005 (Klein *et al.*, 2005). As of July, 2014 there had been 1927 peer-reviewed GWAS studies, with 17,080 findings reported (Hindorf *et al.*, 2014).

Whole exome and genome sequencing data are particularly well suited for the study of consanguineous families and rare Mendelian diseases, and these data are increasingly making personalized medicine a reality (Rabbani *et al.*, 2014). Additionally, many have proposed that rare variants sampled in exome and whole-genome sequence may bridge the missing heritability gap observed in GWASs (Manolio *et al.*, 2009; Schork *et al.*, 2009; Eichler *et al.*, 2010). A recent study of type 2 diabetes (T2D), for example, identified 63 SNPs associated with T2D which only accounted for 5.7% of the variance in disease susceptibility. Even when modeling the effect of all common variants in the genome, the variance explained by this full set of GWAS SNPs was only 50–60% (Morris *et al.*, 2012). Common variants increasing disease risk generally have low penetrance and fall into one of three categories: (1) they affect late onset of disease, making them evolutionarily neutral; (2) they undergo balancing selection due to gene pleiotropy; or (3) the direction of selection has changed due to recent environmental changes (e.g., a high-fat diet coupled with a sedentary lifestyle). Rare variants generally have a neutral effect and increase or decrease in frequency by genetic drift. Those affecting disease risk, however, can have moderate to high penetrance, but need to have a mild enough phenotype so as to ensure passage to the next generation, because those with more deleterious phenotypes will be strongly selected against and will be less likely to be observed in more than a few cases (Saint Pierre and Génin, 2014). Sequencing studies are able to sample these rare variants, but suffer considerably from a lack of power when analyzed with traditional GWAS methods. This lack of power stems primarily from two sources: (1) because of their rarity, the number of individuals with a given variant will be small, even with relatively large sample sizes; and (2) the number of rare variants sampled using next-generation sequencing techniques increases roughly linearly with the number of individuals sampled, resulting in many more statistical comparisons being performed (Ionita-Laza *et al.*, 2009; Sham and Purcell, 2014). Other methods (e.g., gene burden testing, the sequence kernel association test, and the C-alpha test) analyze rare variant patterns on a gene- or region-wide basis (Moutsianas and Morris, 2014), and have been used to identify a number of genes affecting disease (see Conley and Casanova, 2014, and Ning *et al.*, 2014, for reviews of some of these findings).

MALD is an alternative GWAS method, which uses long-range linkage extant in recently admixed populations (e.g., African Americans). The linkage blocks are smaller than in family studies, but significantly larger than the local LD blocks found in non-admixed populations typically studied in case-control studies, resulting in a significant power advantage (Charkraborty and Weiss, 1988). MALD studies also retain the advantage over family studies in their ability to study complex disease associations.

7.2. Admixture linkage

Recently admixed populations have much longer, extended LD blocks that form a mosaic of the ancestral populations (e.g., African Americans are primarily descended from West Africans with approximately 15–20% European admixture; Parra *et al.*, 1998). Extended haplotypes descend from each ancestral population and are slowly broken up through recombination during meiosis in each subsequent generation (see Figure 7.1). Haplotype blocks consist of an entire chromosome inherited from each ancestral parental chromosome in the initial admixed generation. The length of these blocks as they decay with each generation is a function of genetic distance, measured in centimorgans (cM). After one generation, for example, each extended haplotype block will vary, but the average block size will be approximately 100 cM, or roughly one hundred million bases. The correlation of markers within these extended haplotype blocks is referred to as admixture linkage disequilibrium (ALD). Within these extended haplotype blocks, however, ALD is confounded by extant local LD retained from ancestral populations, because each extended haplotype is descended from one ancestral population (Tang *et al.*, 2006; Sundquist *et al.*, 2008). A complex correlation structure arises from ALD and local LD, in which knowing the allele at one locus can give significant information about alleles of neighboring loci, and knowing the alleles in one haplotype block can give significant information about the ancestry of that locus.

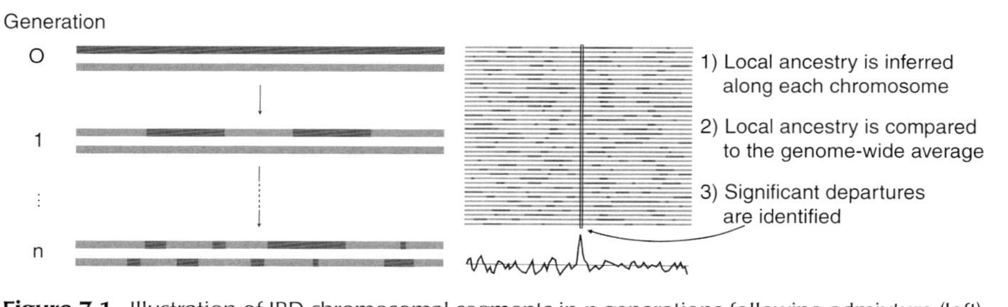

Figure 7.1. Illustration of IBD chromosomal segments in *n* generations following admixture (left) and strategy for mapping of a disease locus by admixture linkage disequilibrium (right). A black and white version of this figure will appear in some formats. For the color version, please refer to the plate section.

Table 7.1. Resolution and cost, in numbers of markers needed and multiple comparisons burden, of various genetic association study approaches (see International HapMap Consortium *et al.*, 2007; Winkler *et al.*, 2010).

	Family linkage	Candidate gene study	GWAS	MALD
# Markers	< 1000	10–500	1–2 million	1500–3000
Bonferroni threshold (Study-wide a = 0.05)	5 × 10	5 × 10	1 × 10	3 × 10
Resolution	20–30 cM	0.1–5 cM	0.1–5 cM	1–10 cM

ALD can also add meaningful information to the analysis of association studies using admixed populations. In addition to performing a genetic study for association between markers and a phenotype, one can test for linkage between an extended haplotype block derived from one or the other ancestral populations with a trait or phenotype that differs in frequency between the two ancestral parental populations. In other words, we anticipate that the causal variant is more likely to be harbored on the same ancestral chromosome as the ancestral population showing the higher disease prevalence (Stephens *et al.*, 1994). This is analogous to family studies that use long linkage blocks to identify disease mutations in affected families segregating a disease gene (McKeigue, 1997). These extended ancestral haplotype blocks can also be inferred with many fewer markers, resulting in fewer statistical comparisons than a traditional GWAS. A GWAS, however, has a higher resolution, resulting in a smaller genomic region associated with the trait (see Table 7.1).

7.2.1. Admixture linkage disequilibrium

Linkage disequilibrium (LD) is the non-random association of the alleles from a group of loci, due to shared inheritance of the alleles over many generations, while admixture linkage disequilibrium (ALD) discussed above is the non-random association of alleles from each ancestral population over a few generations following admixture. Regions of high LD, often referred to as LD blocks, are not generally recombined during meiosis and are inherited together. This is most often due to the loci in the LD block being in close physical proximity. These regions vary somewhat in size between populations, mainly due to the size of the founding population and the number of generations since the population's founding. Africans, for example, have an overall lower level of LD than Europeans.

When there is admixture between two populations, individuals of the initially admixed generation (Generation 0) inherit one entire chromosome from each parent and thus inherit one entire chromosome from each ancestral population. The LD structure across each of these chromosomes can be quite different, making them easily distinguishable from each other. In the first generation after admixture (Generation 1) the contiguous chromosomal segments from each ancestral population are 100 cM long on average, which is quite large in comparison to the background identical-by-descent (IBD) segments originating from each of the ancestral populations. These chromosomal segments form extended haplotypes that are unique to each ancestral population, which can be described by local LD patterns specific to each ancestral population, resulting in considerable ALD in addition to the local LD present in the ancestral populations. Through succeeding generations, the IBD chromosomal segments resulting from admixture will become smaller by the process of genetic recombination (see Figure 7.1), but ALD continues to be significant and measurable for many generations (Lautenberger *et al.*, 2000; Patterson *et al.*, 2012).

7.2.2. Inference of local ancestry

Earlier software was limited to an analysis of sparse marker panels, with 1000–3000 markers, because the algorithms did not account for local LD extant from the ancestral populations (McKeigue *et al.*, 2000a; Pritchard *et al.*, 2000; Patterson *et al.*, 2004). With the increasing availability of GWAS data, other algorithms supporting dense marker data have been developed (see Table 7.2). The LAMP (Sankararaman *et al.*, 2008b) and WINPOP (Pasaniuc *et al.*, 2009) packages do not model local LD, but rely instead on clustered windows of markers to avoid bias. The SABER (Tang *et al.*, 2006) and SWITCH-MHMM (Sankararaman *et al.*, 2008a) packages model first-order local LD (i.e., one-way LD between pairs of markers), but this has been shown to allow some bias from higher-order local LD into the model (Price *et al.*, 2008). The most accurate modeling of local LD is done by HAPMIX (Price *et al.*, 2009), but the increased computational requirements associated with its complex modeling and the fact that it will only support two-way admixed populations limit its usability (Seldin *et al.*, 2011). Other software packages either indirectly model local LD or model it with approximate distributions.

Another distinguishing feature provided by few admixture software packages is the support of statistical tests to map disease genes by admixture linkage disequilibrium (MALD). Of the seven MALD studies published in 2013–2014, four used older sparse marker software (Divers *et al.*, 2013; Molineros *et al.*, 2013; Bensen *et al.*, 2014; Kim-Howard *et al.*, 2014), and at least one of these actually thinned GWAS data to accommodate the software used (Kim-Howard *et al.*, 2014). While MALD can be performed using local ancestry estimates from any appropriately used admixture software package, ALDsuite is the only package that offers this statistical analysis along with local ancestry estimates (Johnson *et al.*, 2015).

7.2.3. Mapping by admixture linkage disequilibrium

Ancestry plays a critical role in our genetic makeup and affects distinct phenotypic traits in various racial groups. This can negatively impact genetic studies through unrecognized substructure, but it can also be exploited to map genes

Table 7.2. Admixture inference softare.

Software	Dense markers	Local ancestry	Local LD modeled	MALD	>2 populations	References
STRUCTURE		✓			✓	Pritchard *et al.*, 2000; Falush *et al.*, 2003
ADMIXMAP		✓		✓		McKeigue *et al.*, 2000a, 2013
ANCESTRYMAP		✓		✓		Patterson *et al.*, 2004
FRAPPE	✓	Global only			✓	Tang *et al.*, 2005
SABER+	✓	✓	1st order		✓	Tang *et al.*, 2010
LAMP-LD	✓	✓			✓	Baran *et al.*, 2012
HAPAA	✓	✓	✓		✓	Sundquist *et al.*, 2008
SWITCH-MHMM	✓	✓	1st order		✓	Sankararaman *et al.*, 2008a
WINPOP	✓	✓			✓	Pasaniuc *et al.*, 2009
HAPMIX	✓	✓	✓			Price *et al.*, 2009
ADMIXTURE	✓	Global only			✓	Alexander *et al.*, 2009
PCAdmix	✓	✓			✓	Brisbin *et al.*, 2012
MULTIMIX	✓	✓	✓		✓	Churchhouse and Marchini, 2013
SEQMIX	✓	✓				Hu *et al.*, 2013
ALDER	✓	✓			✓	Loh *et al.*, 2013
RFMix	✓	✓			✓	Maples *et al.*, 2013
ALLOY	✓	✓	✓		✓	Rodriguez *et al.*, 2013
EILA	✓	✓			✓	Yang *et al.*, 2013
DBM-Admix	✓	✓	✓		✓	Zhang, 2013
MaCH-Admix	✓	✓	✓		✓	Liu *et al.*, 2013
ELAI	✓	✓	✓		✓	Guan, 2014
ALDsuite	✓	✓	✓	✓	✓	Johnson *et al.*, 2015

under the right circumstances. For several decades admixed populations have been recognized for their potential to elucidate genetic causes of observed phenotypic differences or disease disparities between racial groups (Thoday, 1969; MacLean and Workman, 1973), but it was not until relatively recently that genotyping and computing technologies have evolved sufficiently to truly harness the capabilities of this method.

The basic premise of MALD is to infer the location of disease genes by way of association between local ancestry and phenotype (McKeigue, 1998). Power to observe statistically significant associations between ancestry and phenotype increases as a function of the magnitude of the incidence rate ratio of disease between the ancestral populations, but even in populations with a large difference in incidence rates a genetic cause is not guaranteed. In many instances, an environmental factor can be responsible for some or all of the observed difference in incidence rates (Risch *et al.*, 2002).

Ancestry at each sampled locus in the genome is typically inferred using one of the methods discussed above. When the local ancestry at a locus under study is significantly different than in controls, or when it is significantly different from global ancestry in a case only study design, it can be inferred that a variant specific to that ancestral population is nearby, and that this variant is responsible for a change in prevalence or character of the phenotype of interest (see Figure 7.1). Statistical power is greatly increased when dense marker sets are used to infer local ancestry, primarily due to the increased accuracy of inferred local ancestry, but care must be taken to correctly account for local LD. The two most common consequences of failing to appropriately handle local LD are that the model will tend to overestimate the divergence of the admixing populations and it will tend to infer significant admixture in unadmixed individuals (Falush *et al.*, 2003). An ongoing area of research is how best to efficiently and accurately model local LD to make use of this increase in power (Seldin *et al.*, 2011).

7.2.4. Disease association statistics

The first test proposed for MALD was a transmission disequilibrium test (TDT) utilizing parent–child trios (McKeigue, 1997). The TDT is a powerful method, but requires additional resources to sample both parents of each case. Testing unrelated individuals avoids this problem, and is also a valid test for linkage when done properly (McKeigue, 1998).

One major advantage in admixture mapping is that a case-only study design may be employed. This is not normally an option in traditional genetic association studies, but in MALD studies the genome of each case can be effectively used as a substitute for a more traditional control population. In these case-only tests, local ancestry is compared to genome-wide ancestry, with large deviations between the two resulting in statistically significant test results. Two examples of this class of test statistics are a Bayes factor proposed by Patterson *et al.* (2004) and a non-parametric statistic proposed by Montana and Pritchard (2004).

Given the i-th individual's global ancestry proportion of the risk population: A_i, the estimated probability of the number of chromosomes from the risk population at the locus being tested, $\gamma_{i0} = P(0 \text{ chromosomes})$, $\gamma_{i1} = P(1 \text{ chromosome})$ and $\gamma_{i2} = P(2 \text{ chromosomes})$; and the prior risk increase associated with one or two risk chromosomes, ψ_1 and ψ_2, the Bayes factor testing association of the locus and disease is defined in Equation (7.1) (Patterson *et al.*, 2004).

$$Bf = \Pi_1 \frac{\gamma_{i0} + \psi_1\gamma_{i1} + \psi_2\gamma_{i2}}{(1 - A_i)^2 + \psi_1 2A_i(1 - A_i) + \psi_2 2A_i} \qquad (7.1)$$

Alternately, the non-parametric test statistic defined in Equation (7.2) is defined in terms of the sample averages for local ancestry at the locus of interest, $\bar{\gamma}$, and global ancestry, $\bar{A}$ (Montana and Pritchard, 2004).

$$T = \frac{\bar{\gamma} - \bar{A}}{\mathrm{SD}(\bar{\gamma})} \qquad (2)$$

Case/control statistics are typically explored independently of case-only statistics. Pasaniuc *et al.* (2011), however, have proposed a χ^2 test, merging both case/control GWAS results and case-only admixture mapping results.

Accounting for confounding variables and testing for associations with continuous traits has also been modeled using generalized linear regression techniques (Hoggart *et al.*, 2003). Linear models typically take the form of

$$f(Y_i) = \beta_0 + \overline{A}_i\beta_i + g(\gamma_i) + X_1\beta_4 + \ldots \varepsilon_i, \quad (3)$$

where $f()$ is the link function, $\overline{A}_i$ is the global ancestry of the risk population, $g()$ is a function of the local ancestry probabilities (e.g., under an additive model $g(\gamma) = \gamma_1 + 2\gamma_2$), and X is a vector of environmental and/or confounding variables. There have also been concerns that differences in admixture between chromosomes inherited from each parent can also introduce bias in disease association. While this bias is likely to be very small, it has been shown that the product of each parent's global ancestry proportion, $\overline{A}_{1i}\overline{A}_{2i}$, is sufficient to account for this potential bias (Redden *et al.*, 2006):

$$f(Y_i) = \beta_0 + \overline{A}_i\beta_1 + \overline{A}_{1i}\overline{A}_{2i}\beta_3 + g(\gamma_i)\beta_4 + X_1\beta_5 + \ldots + \varepsilon_i. \quad (4)$$

7.3. MALD applications

An underused option for increasing power in GWAS is MALD, which combines the power gained from long range admixture linkage disequilibrium (ALD) similar to that found in family studies with the ability of case-control studies to model complex disease (McKeigue, 1997). It is important to note that MALD assumes the existence of a causal genetic factor, more frequent in one ancestral population compared to the other ancestral population(s) contributing to the admixture. Importantly, computer simulations exploring the effect of the difference in disease allele frequency between ancestral populations indicates that statistical power remains stable with as little as a 20% difference in disease prevalence between populations (Stephens *et al.*, 1994).

A further benefit of MALD is that the association sought after is not between phenotype and a specific marker in linkage disequilibrium (LD) with the functional genetic variant, but rather linkage between phenotype and local ancestry at the locus harboring the functional genetic variant (Chakraborty and Weiss, 1986). Notably, the *APOL1* variants profoundly affecting kidney disease were discovered only after an admixture linkage study identifying the C22 q13.1 region with glomerular disease and non-diabetic end-stage renal disease. The causal alleles are only found in Africans, where they are under selection by trypanosomes, and in admixed populations with recent African ancestry (Freedman *et al.*, 2010, 2011; Genovese *et al.*, 2010; Kopp *et al.*, 2011). This gene is strongly associated with end-stage renal disease (OR = 7), focal segmental glomerulosclerosis (OR = 17), and HIV-associated nephropathy (OR = 29), as well as increased rate of progression to end-stage renal disease

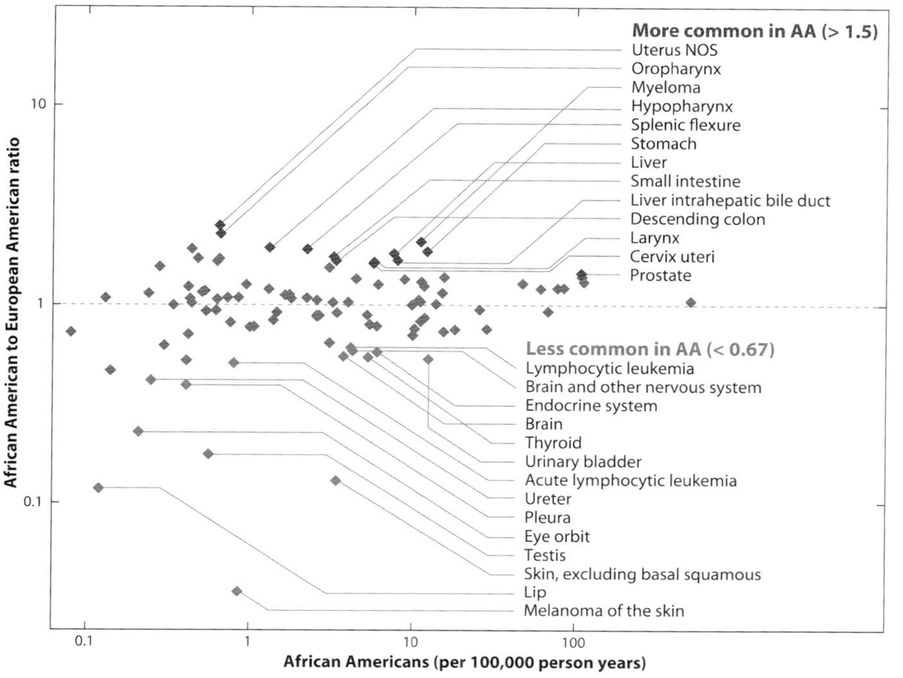

Figure 7.2. Cancer incidence rate ratios between African Americans and European Americans for various cancers (Winkler *et al.*, 2010).

in persons with chronic kidney disease (Kopp *et al.*, 2011). This locus, however, was not found by prior GWAS studies for chronic and end-stage kidney disease enrolling many more people because the causal variants were not tagged by markers included in the earlier GWAS (Bostrom *et al.*, 2010; Genovese *et al.*, 2010; McDonough *et al.*, 2011). The locus was easily identified using MALD because it showed an excess of African ancestry compared to the rest of the genome in cases and when compared to the control group at that locus using under 200 cases (Kopp *et al.*, 2008).

Phenotypes with large differences in incidence between ancestral populations that make up an admixed population provide the greatest power in MALD studies. Given the genetic diversity that exists between continental populations, it is not surprising that a number of diseases are more common in one or another population. Figure 7.2 illustrates an analysis of the SEER database tracking cancer in the United States over a five-year period, for example, which identified 13 cancers with higher incidence in those with African ancestry and 14 cancers with higher incidence in those with European ancestry (Winkler *et al.*, 2010).

Prostate cancer, the most prevalent of the cancers identified in Figure 7.2, was the first MALD study published (M.L. Freedman *et al.*, 2006). Many reports of disease-causing variants have been made over the ensuing seven years in diseases as diverse as renal disease, asthma, and neutrophil count (see Table 7.3 for a more complete list). MALD has not only been applied to the study of human

Table 7.3. Sample of genetic associations identified using MALD.

Disease	Locus	References
Multiple sclerosis	1	Reich *et al.*, 2005
Prostate cancer	8q24.21	Freedman *et al.*, 2006; Bensen *et al.*, 2014
Kidney disease	*APOL1*	Kopp *et al.*, 2008; Kao *et al.*, 2008; Freedman *et al.*, 2010
White cell count	1q	Nalls *et al.*, 2008
Obesity	2p23.3	Cheng *et al.*, 2010
Lung cancer	1 and 3q	Schwartz *et al.*, 2011
Asthma	6q15 and 8q12	Torgerson *et al.*, 2012
Bone mineral density and fat mass	19q13 and Xp22	Ochs-Balcom *et al.*, 2013
Blood lipid levels	Various	Coram *et al.*, 2013
Lupus	NCF2	Molineros *et al.*, 2013; Kim-Howard *et al.*, 2014
Forced expiratory volume	12q14.1	Parker *et al.*, 2014
Type II diabetes	Various	Jeff *et al.*, 2014
Hepatitis C	15q24 and 20p12	Wojcik *et al.*, 2014

populations, but also to organisms of agricultural importance and in ecological modeling (Mezaka *et al.*, 2012; Rogers, 2012).

Only recently admixed populations have reliably distinguishable contiguous chromosome segments from identifiable ancestral populations (Hoggart *et al.*, 2004). With newer methods, however, this condition is being relaxed and characterizations of anciently admixed populations are also shedding light on the history of human migration (A.L. Price *et al.*, 2009; Patterson *et al.*, 2012). Admixture mapping is also being used to refine the human reference sequence. Using admixture mapping techniques, Genovese *et al.* (2013) were able to map the locations of 70 previously unmapped scaffolds, including a number of protein-coding genes and eight inter-chromosomal segmental duplications, which span four million base pairs. The examples discussed in this section highlight the importance and utility of including admixture linkage disequilibrium in genetic studies, and they provide clear evidence of the scientific value of admixed populations.

REFERENCES

1000 Genomes Project Consortium, Abecasuis, G.R., Altshuler, D., *et al.* (2010). A map of human genome variation from population-scale sequencing. *Nature*, **467**(7319), 1061–1073.

Alexander, D.H., Novembre, J. and Lange, K. (2009). Fast model-based estimation of ancestry in unrelated individuals. *Genome Res.*, **19**(9), 1655–1664.

Altshuler, D., Daly, M.J. and Lander, E.S. (2008). Genetic mapping in human disease. *Science*, **322**(5903), 881–888.

Baran, Y., Pasaniuc, B., Sankararaman, S., *et al.* (2012). Fast and accurate inference of local ancestry in Latino populations. *Bioinformatics*, **28**(10), 1359–1367.

Bateson, W. and Kilby, H. (1905). Experimental studies in the physiology of heredity. *R. Soc. Rep. Evol. Comm.*, **2**, 1–55.

Bensen, J.T., Xu, Z., McKeigue, P.M., *et al.* (2014). Admixture mapping of prostate cancer in African Americans participating in the North Carolina–Louisiana Prostate Cancer Project (PCaP). *The Prostate*, **74**(1), 1–9.

Bostrom, M.A., Lu, L., Chou, J., *et al.* (2010). Candidate genes for non-diabetic ESRD in African Americans: a genome-wide association study using pooled DNA. *Hum. Genet.*, **128**(2), 195–204.

Brisbin, A., Bryc, K., Zakharia, F., *et al.* (2012). PCAdmix: principal components-based assignment of ancestry along each chromosome in individuals with admixed ancestry from two or more populations. *Hum. Biol.*, **84**(4), 343–364.

Chakraborty, R. and Weiss, K.M. (1986). Frequencies of complex diseases in hybrid populations. *Am. J. Phys. Anthropol.*, **70**(4), 489–503.

Chakraborty, R. and Weiss, K.M. (1988). Admixture as a tool for finding linked genes and detecting that difference from allelic association between loci. *Proc. Natl Acad. Sci. USA*, **85**(23), 9119–9123.

Cheng, C.-Y., Reich, D., Coresh, J., *et al.* (2010). Admixture mapping of obesity-related traits in African Americans: the Atherosclerosis Risk in Communities (ARIC) Study. *Obesity (Silver Spring, Md.)*, **18**(3), 563–572.

Churchhouse, C. and Marchini, J. (2013). Multiway admixture deconvolution using phased or unphased ancestral panels. *Genet. Epidemiol.*, **37**(1), 1–12.

Conley, M.E. and Casanova, J.-L. (2014). Discovery of single-gene inborn errors of immunity by next generation sequencing. *Curr. Opin. Immunol.*, **30**C, 17–23.

Coram, M.A., Duan, Q., Hoffmann, T.J., *et al.* (2013). Genome-wide characterization of shared and distinct genetic components that influence blood lipid levels in ethnically diverse human populations. *Am. J. Hum. Genet.*, **92**(6), 904–916.

Divers, J., Palmer, N.D., Lu, L., *et al.* (2013). Admixture mapping of coronary artery calcified plaque in African Americans with type 2 diabetes mellitus. *Cardiovasc. Genet.*, **6**(1), 97–105.

Eichler, E.E., Flint, J., Gibson, G., *et al.* (2010). Missing heritability and strategies for finding the underlying causes of complex disease. *Nature Rev. Genet.*, **11**(6), 446–450.

Falush, D., Stephens, M. and Pritchard, J.K. (2003). Inference of population structure using multilocus genotype data: linked loci and correlated allele frequencies. *Genetics*, **164**(4), 1567–1587.

Freedman, B.I., Kopp, J.B., Langefeld, C.D., *et al.* (2010). The apolipoprotein L1 (*APOL1*) gene and nondiabetic nephropathy in African Americans. *J. Am. Soc. Nephrol.*, **21**(9), 1422–1426.

Freedman, B.I., Langefeld, C.D., Lu, L., *et al.* (2011). Differential effects of *MYH9* and *APOL1* risk variants on *FRMD3* Association with diabetic ESRD in African Americans. *PLoS Genet.*, **7**(6), e1002150.

Freedman, M.L., Haiman, C.A., Patterson, N., *et al.* (2006). Admixture mapping identifies 8q24 as a prostate cancer risk locus in African-American men. *Proc. Natl Acad. Sci. USA*, **103**(38), 14068–14073.

Genovese, G., Friedman, D.J., Ross, M.D., *et al.* (2010). Association of trypanolytic *APOL1* variants with kidney disease in African Americans. *Science*, **329**(5993), 841–845.

Genovese, G., Handsaker, R.E., Li, H., Kenny, E.E. and McCarroll, S.A. (2013). Mapping the human reference genome's missing sequence by three-way admixture in Latino genomes. *Am. J. Hum. Genet.*, **93**(3), 411–421.

Guan, Y. (2014). Detecting structure of haplotypes and local ancestry. *Genetics*, **196**(3), 625–642.

Herrera-Paz, E.-F. (2014). The African Diaspora Power SNP Chip Developed via the CAAPA Consortium. Available at: http://www.academia.edu/7233528/The_African_Diaspora_Power_SNP_Chip_Developed_via_the_CAAPA_Consortium.

Hindorff, L.A., *et al.*, A Catalog of Published Genome-Wide Association Studies. Available at: http://www.genome.gov/gwastudies/. Accessed July 16, 2014.

Hirschhorn, J.N., Lohmueller, K., Byrne, E. and Hirschhorn, K. (2002). A comprehensive review of genetic association studies. *Genet. Med.*, **4**(2), 45–61.

Hoggart, C.J., Parra, E.J., Shriver, M.D., *et al.* (2003). Control of confounding of genetic associations in stratified populations. *Am. J. Hum. Genet.*, **72**(6), 1492–1504.

Hoggart, C.J., Shriver, M.D., Kittles, R.A., *et al.* (2004). Design and analysis of admixture mapping studies. *Am. J. Hum. Genet.*, **74**(5), 965–978.

Hu, Y., Willer, C., Zhan, X., Kang, H.M. and Abecasis, G.R. (2013). Accurate local-ancestry inference in exome-sequenced admixed individuals via off-target sequence reads. *Am. J. Hum. Genet.*, **93**(5), 891–899.

International HapMap Consortium. (2005). A haplotype map of the human genome. *Nature*, **437**(7063), 1299–1320.

International HapMap Consortium, Frazer, K.A., Ballinger, D.G., *et al.* (2007). A second generation human haplotype map of over 3.1 million SNPs. *Nature*, **449**(7164), 851–861.

Ionita-Laza, I., Lange, C. and M Laird, N. (2009). Estimating the number of unseen variants in the human genome. *Proc. Natl Acad. Sci. USA*, **106**(13), 5008–5013.

Jeff, J.M., Armstrong, L.L., Ritchie, M.D., *et al.* (2014). Admixture mapping and subsequent fine-mapping suggests a biologically relevant and novel association on chromosome 11 for type 2 diabetes in African Americans. *PloS ONE*, **9**(3), e86931.

Johnson, R.C., Nelson, G.W., Zagury, J.F. and Winkler, C.A. (2015). ALDsuite: dense marker MALD using principal components of ancestral linkage disequilibrium. *Bioinformatics*, **16**(1), 23.

Kao, W.H.L., Klag, M.J., Meoni, L.A., *et al.* (2008). *MYH9* is associated with nondiabetic end-stage renal disease in African Americans. *Nature Genet.*, **40**(10), 1185–1192.

Keightley, P.D. and Otto, S.P. (2006). Interference among deleterious mutations favours sex and recombination in finite populations. *Nature*, **443**(7107), 89–92.

Kim-Howard, X., Sun, C., Molineros, J.E., *et al.* (2014). Allelic heterogeneity in *NCF2* associated with systemic lupus erythematosus (SLE) susceptibility across four ethnic populations. *Hum. Molec. Genet.*, **23**(16), 1656–1668.

Klein, R.J., Zeiss, C., Chow, E.J., *et al.* (2005). Complement factor H polymorphism in age-related macular degeneration. *Science*, **308**(5720), 385–389.

Kopp, J.B., Smith, M.W., Nelson, G.W., *et al.* (2008). *MYH9* is a major-effect risk gene for focal segmental glomerulosclerosis. *Nature Genet.*, **40**(10), 1175–1184.

Kopp, J.B., Nelson, G.W., Sampath, K., *et al.* (2011). *APOL1* genetic variants in focal segmental glomerulosclerosis and HIV-associated nephropathy. *J. Am. Soc. Nephrol.*, **22** (11), 2129–2137.

Lautenberger, J.A., Stephens, J.C., O'Brien, S.J. and Smith, M.W. (2000). Significant admixture linkage disequilibrium across 30 cM around the FY locus in African Americans. *Am. J. Hum. Genet.*, **66**(3), 969–978.

Liu, E.Y., Li, M., Wang, W. and Li, Y. (2013). MaCH-admix: genotype imputation for admixed populations. *Genet. Epidemiol.*, **37**(1), 25–37.

Loh, P.-R., Lipson, M., Patterson, N., *et al.* (2013). Inferring admixture histories of human populations using linkage disequilibrium. *Genetics*, **193**(4), 1233–1254.

MacLean, C.J. and Workman, P.L. (1973). Genetic studies on hybrid populations. I. Individual estimates of ancestry and their relation to quantitative traits. *Ann. Hum. Genet.*, **36**(3), 341–351.

Manolio, T.A., Collins, F.S., Cox, N.J., *et al.* (2009). Finding the missing heritability of complex diseases. *Nature*, **461**(7265), 747–753.

Maples, B.K., Gravel, S., Kenny, E.E. and Bustamante, C.D. (2013). RFMix: A discriminative modeling approach for rapid and robust local-ancestry inference. *Am. J. Hum. Genet.*, **93**(2), 278–288.

Martin, E.R., Lai, E.H., Gilbert, J.R., *et al.* (2000). SNPing away at complex diseases: analysis of single-nucleotide polymorphisms around APOE in Alzheimer disease. *Am. J. Hum. Genet.*, 67(2), 383–394.

Matise, T.C., Chen, F., Chen, W., *et al.* (2007). A second-generation combined linkage physical map of the human genome. *Genome Res.*, **17**(12), 1783–1786.

McDonough, C.W., Palmer, N.D., Hicks, P.J., *et al.* (2011). A genome-wide association study for diabetic nephropathy genes in African Americans. *Kidney Int.*, **79**(5), 563–572.

McKeigue, P.M. (1997). Mapping genes underlying ethnic differences in disease risk by linkage disequilibrium in recently admixed populations. *Am. J. Hum. Genet.*, **60**(1), 188.

McKeigue, P.M. (1998). Mapping genes that underlie ethnic differences in disease risk: methods for detecting linkage in admixed populations, by conditioning on parental admixture. *Am. J. Hum. Genet.*, **63**, 241–251.

McKeigue, P.M., Carpenter, J.R., Parra, E.J. and Shriver, M.D. (2000a). Estimation of admixture and detection of linkage in admixed populations by a Bayesian approach: application to African-American populations. *Ann. Hum. Genet.*, **64**(Pt 2), 171–186.

McKeigue, P.M., Colombo, M., Agakov, F., *et al.* (2013). Extending admixture mapping to nuclear pedigrees: application to sarcoidosis. *Genet. Epidemiol.*, **37**(3), 256–266.

Mezaka, I., Legzdina, L., Waugh, R., Close, T.J. and Rostoks, N. (2012). Genetic diversity in Latvian spring barley association mapping population. In G. Zhang, C. Li and X. Liu (Eds), *Advances in Barley Science: Proceedings of the 11th International Barley Genetics Symposium*. Dordrecht: Springer Netherlands, pp. 25–35.

Molineros, J.E., Maiti, A.K., Sun, C., *et al.* (2013). Admixture mapping in lupus identifies multiple functional variants within *IFIH1* associated with apoptosis, inflammation, and autoantibody production. *PLoS Genet.*, **9**(2), e1003222.

Montana, G. and Pritchard, J.K. (2004). Statistical tests for admixture mapping with case-control and cases-only data. *Am. J. Hum. Genet.*, **75**(5), 771–789.

Morgan, T.H. (1910). Sex limited inheritance in drosophila. *Science*, **32**(812), 120–123.

Morgan, T.H. (1911). Random segregation versus coupling in medellian inheritance. *Science*, **34**(873), 384.

Morris, A.P., Voight, B.F., Teslovich, T.M., *et al.* (2012). Large-scale association analysis provides insights into the genetic architecture and pathophysiology of type 2 diabetes. *Nature Genet.*, **44**(9), 981–990.

Moutsianas, L. and Morris, A.P. (2014). Methodology for the analysis of rare genetic variation in genome-wide association and re-sequencing studies of complex human traits. *Brief. Funct. Genom.*, **13**(5), 362–370.

Nalls, M.A., Wilson, J.G., Patterson, N.J., *et al.* (2008). Admixture mapping of white cell count: genetic locus responsible for lower white blood cell count in the Health ABC and Jackson Heart studies. *Am. J. Hum. Genet.*, **82**(1), 81–87.

Ning, B., Su, Z., Mei, N., *et al.* (2014). Toxicogenomics and cancer susceptibility: advances with next-generation sequencing. *J. Environ. Sci. Health. Part C, Environ. Carcin. Ecotoxicol. Rev.*, **32**(2), 121–158.

Ochs-Balcom, H.M., Preus, L., Wactawski-Wende, J., *et al.* (2013). Association of DXA-derived bone mineral density and fat mass with African ancestry. *J. Clin. Endocrinol. Metab.*, **98**(4), E713–717.

Ohta, T. (1982). Linkage disequilibrium due to random genetic drift in finite subdivided populations. *Proc. Natl Acad. Sci. USA*, **79**(6), 1940–1944.

Palaisa, K., Morgante, M., Tingey, S. and Rafalski, A. (2004). Long-range patterns of diversity and linkage disequilibrium surrounding the maize *Y1* gene are indicative of an asymmetric selective sweep. *Proc. Natl Acad. Sci. USA*, **101**(26), 9885–9890.

Parker, M.M., Foreman, M.G., Abel, H.J., *et al.* (2014). Admixture mapping identifies a quantitative trait locus associated with *FEV1/FVC* in the COPD Gene Study. *Genet. Epidemiol.*, **38**(7), 652–659.

Parra, E.J., Marcini, A., Akey, J., *et al.* (1998). Estimating African American admixture proportions by use of population-specific alleles. *Am. J. Hum. Genet.*, **63**(6), 1839–1851.

Pasaniuc, B., Sankararaman, S., Kimmel, G. and Halperin, E. (2009). Inference of locus-specific ancestry in closely related populations. *Bioinformatics*, **25**(12), i213–221.

Pasaniuc, B., Zaitlen, N., Lettre, G., *et al.* (2011). Enhanced statistical tests for GWAS in admixed populations: assessment using African Americans from CARe and a Breast Cancer Consortium. *PLoS Genet.*, **7**(4), e1001371.

Patterson, N., Hattangadi, N., Lane, B., *et al.* (2004). Methods for high-density admixture mapping of disease genes. *Am. J. Hum. Genet.*, **74**(5), 979–1000.

Patterson, N., Moorjani, P., Luo, Y., *et al.* (2012). Ancient admixture in human history. *Genetics*, **192**(3), 1065–1093.

Price, A.L., Weale, M.E., Patterson, N., *et al.* (2008). Long-range LD can confound genome scans in admixed populations. *Am. J. Hum. Genet.*, **83**(1), 132–135.

Price, A.L., Tandon, A., Patterson, N., *et al.* (2009). Sensitive detection of chromosomal segments of distinct ancestry in admixed populations. *PLoS Genet.*, **5**(6), e1000519.

Price, P., James, I., Fernandez, S. and French, M.A. (2004). Alleles of the gene encoding IL-1α may predict control of plasma viraemia in HIV-1 patients on highly active antiretroviral therapy. *AIDS*, **18**(11), 1495–1501.

Pritchard, J.K., Stephens, M. and Donnelly, P. (2000). Inference of population structure using multilocus genotype data. *Genetics*, **155**(2), 945–959.

Rabbani, B., Tekin, M. and Mahdieh, N. (2014). The promise of whole-exome sequencing in medical genetics. *J. Hum. Genet.*, **59**(1), 5–15.

Redden, D.T., Divers, J., Vaughan, L.K., *et al.* (2006). Regional admixture mapping and structured association testing: conceptual unification and an extensible general linear model. *PLoS Genet.*, **2**(8), e137.

Reich, D., Patterson, N., De Jager, P.L., *et al.* (2005). A whole-genome admixture scan finds a candidate locus for multiple sclerosis susceptibility. *Nature Genet.*, **37**(10), 1113–1118.

Risch, N., Burchard, E., Ziv, E. and Tang, H. (2002). Categorization of humans in biomedical research: genes, race and disease. *Genome Biol.*, **3**(7), comment2007.

Rodriguez, J.M., Bercovici, S., Elmore, M. and Batzoglou, S. (2013). Ancestry inference in complex admixtures via variable-length Markov chain linkage models. *J. Comput. Biol.*, **20**(3), 199–211.

Rogers, S.M. (2012). Mapping the genomic architecture of ecological speciation in the wild: does linkage disequilibrium hold the key? *Molec. Ecol.*, **21**(21), 5155–5158.

Saint Pierre, A. and Génin, E. (2014). How important are rare variants in common disease? *Brief. Funct. Genom.*, **13**(5), 353–361.

Sankararaman, S., Kimmel, G., Halperin, E. and Jordan, M.I. (2008a). On the inference of ancestries in admixed populations. *Genome Res.*, 18(4), 668–675.

Sankararaman, S., Sridhar, S. and Kimmel, G. (2008b). Estimating local ancestry in admixed populations. *Am. J. Hum. Genet.*, **82**(2), 290–303.

Schork, N.J., Murray, S.S., Frazer, K.A. and Topol, E.J. (2009). Common vs. rare allele hypotheses for complex diseases. *Curr. Opin. Genet. Develop.*, **19**(3), 212–219.

Schwartz, A.G., Wenzlaff, A.S., Bock, C.H., *et al.* (2011). Admixture mapping of lung cancer in 1812 African-Americans. *Carcinogenesis*, **32**(3), 312–317.

Seldin, M.F., Pasaniuc, B. and Price, A.L. (2011). New approaches to disease mapping in admixed populations. *Nature Rev. Genet.*, **12**(8), 523–528.

Sham, P.C. and Purcell, S.M. (2014). Statistical power and significance testing in large-scale genetic studies. *Nature Rev. Genet.*, **15**(5), 335–346.

Sober, E. (1993). *The Nature of Selection: Evolutionary Theory in Philosophical Focus*. University of Chicago Press, Chicago, IL.

Stephens, J.C., Briscoe, D. and O'Brien, S.J. (1994). Mapping by admixture linkage disequilibrium in human populations: limits and guidelines. *Am. J. Hum. Genet.*, **55**, 809–824.

Sundquist, A., Fratkin, E., Do, C.B. and Batzoglou, S. (2008). Effect of genetic divergence in identifying ancestral origin using HAPAA. *Genome Res.*, **18**(4), 676–682.

Tang, H., Peng, J., Wang, P. and Risch, N.J. (2005). Estimation of individual admixture: analytical and study design considerations. *Genet. Epidemiol.*, **28**(4), 289–301.

Tang, H., Coram, M., Wang, P., Zhu, X. and Risch, N. (2006). Reconstructing genetic ancestry blocks in admixed individuals. *Am. J. Hum. Genet.*, **79**(1), 1–12.

Tang, H., Siegmund, D.O., Johnson, N.A., Romieu, I. and London, S.J. (2010). Joint testing of genotype and ancestry association in admixed families. *Genet. Epidemiol.*, **34**(8), 783–791.

Thoday, J.M. (1969). Limitations to genetic comparison of populations. *J. Biosoc. Sci.*, Suppl. 1, 3–14.

Tishkoff, S.A., Dietzch, E., Speed, W., *et al.* (1996). Global patterns of linkage disequilibrium at the CD4 locus and modern human origins. *Science*, **271**(5254), 1380–1387.

Torgerson, D.G., Gignoux, C.R., Galanter, J.M., *et al.* (2012). Case-control admixture mapping in Latino populations enriches for known asthma-associated genes. *J. Allergy Clin. Immunol.*, **130**(1), 76–82.e12.

Winkler, C.A., Nelson, G.W. and Smith, M.W. (2010). Admixture mapping comes of age. *Annu. Rev. Genom. Hum. Genet.*, 11, 65–89.

Wojcik, G.L., Thio, C.L., Kao, W.H., *et al.* (2014). Admixture analysis of spontaneous hepatitis C virus clearance in individuals of African descent. *Genes Immun.*, **15**(4), 241–246.

Yang, J.J., Li, J., Buu, A. and Williams, L.K. (2013). Efficient inference of local ancestry. *Bioinformatics*, **29**(21), 2750–2756.

Zhang, Y. (2013). *De novo* inference of stratification and local admixture in sequencing studies. *BMC Bioinformatics*, **14**(Suppl. 5), S17.

Zhao, H., Pfeiffer, R. and Gail, M.H. (2003). Haplotype analysis in population genetics and association studies. *Pharmacogenomics*, **4**(2), 171–178.

8 Genome-wide association analysis in schizophrenia

Sven Stringer, Dorien H. Nieman, René S. Kahn, and Eske M. Derks

8.1. Introduction

Schizophrenia is a severe mental disorder with a typical onset in adolescence or young adulthood. Global lifetime prevalence is about 0.3–0.7% (van Os and Kapur, 2009). Symptoms can be divided into positive symptoms (e.g., delusions and hallucinations), negative (deficit) symptoms (e.g., anhedonia, blunted affect, and avolition), and disorganization symptoms (e.g. disorganized speech). In addition, the majority of schizophrenia patients show cognitive dysfunctioning. In general, schizophrenia patients have deficits in most cognitive domains (e.g., attention, memory, and executive functioning) approximately one standard deviation below the normative mean (Mesholam-Gately *et al.*, 2009). However, there is no specific cognitive profile that distinguishes schizophrenia patients from patients with other Diagnostic and Statistical Manual of Mental Disorders (DSM) (American Psychiatric Association, 2000) diagnoses. Schizophrenia is often preceded by a prodromal period of months to years in which mild psychotic and other symptoms can occur and psycho-social functioning deteriorates.

A short case example:

Michael is a 20-year-old philosophy student who has been skipping a lot of classes lately. In the class room he hears his name being whispered by fellow students in the front row, although the distance is too far to be able to hear them. During the breaks, he hears other students talk and laugh about him. Sometimes he thinks they are conspiring to kill him, especially because he also hears them talking about how they are going to get him when he is alone in his room. He is unable to concentrate on what the professor says in the classroom. It is as if he cannot extract the meaning of what is being said. He has suffered from that problem for several years. His grades have decreased during this period and he will probably drop out from university. He also experiences a feeling of emptiness, which started years ago. Nothing seems to get through to him. Even the birth of

Genome-Wide Association Studies: From Polymorphism to Personalized Medicine, ed. K. Appasani. Published by Cambridge University Press.

his niece left him cold. He experiences a loss of identity. Lately, he sometimes has the feeling that someone else is putting thoughts in his head or moves his limbs outside his own will.

Schizophrenia has remained a mental disorder with an unknown etiology, unchanged prevalence, and disabling outcomes for the vast majority of the patients. Sustained recovery occurs in less than 14% within the first five years following a psychotic episode and in an additional 16% in a later phase (Harrison *et al.*, 2001; Robinson *et al.*, 2004). Throughout Europe, less than 20% of people with schizophrenia are employed (Marwaha *et al.*, 2007). Various treatments, especially pharmacological, have been tried to improve its disabling lifetime course (Lieberman *et al.*, 2005). Antipsychotic medication reduces the positive symptoms, but the negative symptoms often remain, sometimes even worsened by antipsychotic medication. One of the causes of lack of progress in scientific understanding and treatment may be that the DSM-IV category of schizophrenia is very broad and therefore psychopathology is too heterogeneous to find a biological substrate (Miller, 2010). Wessman et al. (2009) performed unsupervised clustering of individuals from Finnish schizophrenia families, based on extensive clinical and neuropsychological data, including Structured Clinical Interview for DSM-IV information (SCID; First *et al.*, 1997). The sample consisted of 904 individuals from 288 families with at least one member with schizophrenia. Wessman *et al.* (2009) found several subgroups. One group was characterized by psychotic and mood symptoms and an association with allelic variants of the *DISC1* gene. The other group showed mainly negative and cognitive symptoms and a strong association to several allelic variants in the *DTNBP1* gene. Other examples include the studies of Derks and colleagues (2010, 2012, 2013) and Fanous *et al.* (2012), which suggest that genetic associations may be particularly strong for negative and disorganization symptoms. Summarizing, modeling phenotypic heterogeneity within schizophrenia may provide increased insight into the biological substrate of the diverse symptoms of schizophrenia.

8.2. Twin and family studies

The risk of developing schizophrenia is increased in relatives of schizophrenia patients, indicating that familial or genetic factors influence disease risk. In the early 1990s, Gottesman (1991) showed that disease probability is more strongly increased in first-degree relatives of schizophrenia patients (6–17%) compared to second-degree relatives (2–6%). In agreement with a large genetic component for schizophrenia, the probability of developing schizophrenia is 48% in the monozygotic twin of a proband with schizophrenia. The fact that disease risk is more strongly increased in relatives, who are genetically more alike, indicates that genetic factors are important in causing schizophrenia. Indeed, a meta-analysis of twin studies indicates that the heritability of liability to schizophrenia is 81% (Sullivan *et al.*, 2003). Since the publication of this meta-analysis, Lichtenstein and colleagues investigated the heritability of schizophrenia based on multigeneration register data including information on psychiatric inpatient admissions in Sweden (Lichtenstein *et al.*, 2009). This sample comprised over nine million unique individuals, including 35,985 probands with schizophrenia. The heritability of schizophrenia was estimated at 64%, which is lower compared to previous estimates. However, the latter estimate is consistent with the study of Wray and Gottesman (2012), who reported a heritability estimate of 67% based on Danish population-based cohort data.

Lichtenstein and colleagues did not limit the Swedish registry study to schizophrenia only, but also aimed to investigate whether schizophrenia and bipolar disorder share a common genetic etiology. In relatives of schizophrenia probands, the risk to develop bipolar disorder was higher than the risk in relatives of healthy probands. Furthermore, in relatives of probands with bipolar disorder, an increased risk of schizophrenia was reported. Based on the relative risks, Lichtenstein and colleagues concluded that the comorbidity between schizophrenia and bipolar disorder was for a large percentage (63%) due to shared genetic effects.

The genetic architecture of schizophrenia is complex. Early linkage and association studies were challenged by the fact that twin and family data are not in agreement with a single major gene effect (McGue and Gottesman, 1989). Indeed, the identification of causal genetic variants has been less successful than expected. However, genome-wide association (GWA) studies in large samples of patients with schizophrenia and healthy controls have recently resulted in conclusive evidence for the role of several genetic loci (section 8.4). These studies have also increased our knowledge about the genetic architecture of this severe disorder (section 8.5).

8.3. Genome-wide association studies

In 2009, the first GWA study on schizophrenia was published (Kirov *et al.*, 2009). In this study, DNA pooling was used to estimate allele frequencies in 574 schizophrenia patients, their parents, and 605 unaffected controls. Although genetic variants in the genes *CCDC60* and *RBP1* were highlighted, no genome-wide significant findings were reported. Kirov *et al.* predicted that thousands of cases and controls would have to be included to reach statistical significance. Indeed, since then, large consortia have been formed to increase sample size. In 2009, three consortia presented their data in *Nature*. Stefansson and colleagues, representing the SGENE-plus consortium, included data of 2663 cases and 13,498 controls (Stefansson *et al.*, 2009). The International Schizophrenia Consortium (ISC) collected data in 3322 schizophrenia cases and 3587 controls from European ancestry (Purcell *et al.*, 2009). Finally, the Molecular Genetics of Schizophrenia (MGS) sample included a European-ancestry sample (2681 cases and 2652 controls) and an African-American sample (1286 cases and 973 controls) (Shi *et al.*, 2009).

Even though these studies included thousands of cases and controls, very few genome-wide significant findings were reported for each of the individual samples. To further increase statistical power, the three consortia exchanged GWA study summary results. Meta-analyses provided strong support for the involvement of the major histocompatibility complex (MHC) (Purcell *et al.*, 2009; Shi *et al.*, 2009; Stefansson *et al.*, 2009), and for involvement of genetic variants in *NRGN* and *TCF4* (Stefansson *et al.*, 2009). The associations with *NRGN* and *TCF4* point to a causal role of pathways involved in brain development, memory and cognition. The MHC region (6p21.32–p22.1) is a gene-dense region, characterized by high levels of linkage disequilibrium. The MHC region comprises 0.3% of the genome and contains hundreds of genes (i.e., 1.5% of the genes in Online Mendelian Inheritance in Man (OMIM)) (McKusick, 2007). It is enriched for genome-wide significant SNP associations, while 6.4% of the associations reported in the NHGRI GWAS catalog are located in this region (Hindorff *et al.*, 2009; Ripke *et al.*, 2013).

Due to the high level of linkage equilibrium in the MHC region, multiple SNPs show significant associations with schizophrenia and any of the genes in this region could be functionally related to schizophrenia. Walters *et al.* studied the association between four SNPs in the MHC region and cognitive deficits, a core component of schizophrenia (Kahn and Keefe, 2013; Walters *et al.*, 2013). The G allele of SNP rs6904071 was found to be associated with delayed episodic memory and decreased hippocampal volume in schizophrenia patients and controls (Walters *et al.*, 2013), which supports a causal relation between genetic variants in the MHC region and schizophrenia. However, until the location of the causal genetic variants is more precisely identified, it is impossible to distinguish between several theoretically plausible mechanisms.

Because of the large samples required to detect genetic variants with small effects, in 2011 the samples of the three consortia were combined into a single mega-analysis by the Psychiatric Genomics Consortium (PGC) (Ripke *et al.*, 2011). This analysis included a stage I discovery sample of 9394 cases and 12,462 controls and a stage II replication sample of 8442 cases and 21,397 controls. Ten independent SNPs located in seven different loci were significantly associated with case-control status. Two loci were previously implicated in schizophrenia: 6p21.32–p22.1 (MHC) and 18q21.1 (*CCDC68* and *TCF4*). One of the novel loci is located within intron three of *AK094607*, which contains the primary transcript for microRNA-137 (*MIR137*; Bemis *et al.*, 2008). *MIR137* has been implicated in regulating adult neurogenesis and neuronal maturation (see Silber *et al.*, 2008; Smrt *et al.*, 2010; Szulwach *et al.*, 2010; Ripke *et al.*, 2011), and could therefore contribute to brain development abnormalities. This study was followed up in 2013, by including GWAs data from an additional 6454 subjects from Sweden (Ripke *et al.*, 2013). Further evidence was provided for eight loci previously implicated in schizophrenia, one locus previously implicated in bipolar disorder, and 13 new risk loci (see Table 8.1). Two loci (*CACNA1C* and *CACNB2*) implicate a role for calcium signaling in the etiology of schizophrenia. To further test the role of calcium channels, which are involved in learning, memory and synaptic plasticity, Ripke and colleagues performed a gene-set test which indeed showed enrichment of smaller *p*-values in genes encoding calcium channel subunits (Ripke *et al.*, 2013).

Table 8.1. Description of the 22 genome-wide significant loci in the combined analysis (from Ripke *et al.*, 2013).

Chromosomal region	*P* value	Previous association[a]	Candidate gene in relation to index SNP[b]	Other genes in genomic region defined by LD[c]	eQTL[d]	Disease associations[e]
Chr. 6: 31,596,138–32,813,768	9.14×10^{-14}	SCZ	*HLA-DRB9*	MHC class II, many other genes, lincRNA	Many	Many
Chr. 10: 104,487,871–105,245,420	3.68×10^{-13}	SCZ	*C10orf32-AS3MT*	*CALHM1, CALHM2, CALHM3, CNNM2, CYP17A1, INA, MIR1307, NT5C2, PCGF6, PDCD11, SFXN2, ST13P13, TAF5, USMG5, WBP1L*	*ACTR1A, ARL3, AS3MT, C10orf32, C10orf78, NT5C2, TMEM180, TRIM8, WBP1L*	GWAS: blood pressure, CAD, aneurysm
Chr. 7: 1,827,717–2,346,115	5.93×10^{-13}	No	*MAD1L1*	*FTSJ2, NUDT1, SNX8*	*C7orf27, FTSJ2, MAD1L1, NUDT1*	
Chr. 1: 98,141,112–98,664,991	1.72×10^{-12}	SCZ	(*MIR137*, 37 kb)	*DPYD*, lincRNA	*DPYD*	*DPYD*: mental retardation
Chr. 12: 2,285,731–2,440,464	5.22×10^{-12}	SCZ, BPD	*CACNA1C*	–	No data	*CACNA1 C*: autism, Timothy syndrome, Brugada syndrome 3
Chr. 10: 18,601,928–18,934,390	1.27×10^{-10}	5 disorders	*CACNB2*	*NSUN6*	No data	*CACNB2*: Brugada syndrome 4; GWAS: blood pressure

Chr. 8: 143,297,312–143,410,423	2.19×10^{-10}	No	*TSNARE1*	–	No data	
Chr. 1: 73,275,828–74,099,273	3.64×10^{-10}	No	(x10NST00000415686.1, 4 kb)	lincRNA	No data	
Chr. 11: 130,706,918–130,894,976	1.83×10^{-9}	No	(*SNX19*, 31 kb)	lincRNA	SNX19	
Chr. 5: 151,888,959–152,835,304	2.65×10^{-9}	No	ENST00000503048.1	lincRNA (*GRIA1*)	No data	
Chr. 5: 152,505,453–152,707,306	4.12×10^{-8}	No				
Chr. 19: 19,354,937–19,744,079	3.44×10^{-9}	BPD	(*MAU2*, 4 kb)	*CILP2, GATAD2A, GMIP, HAPLN4, LPAR2, MIR640, NCAN, NDUFA13, PBX4, SUGP1, TM6SF2, TSSK6, YJEFN3*	No data	GWAS: lipid levels
Chr. 2: 37,422,072–37,592,628	6.78×10^{-9}	No	*QPCT*	C2orf56, *CEBPZ, PRKD3, SULT6B1 lincRNA*	No eQTL	
Chr. 5: 101,581,848–101,870,822	9.03×10^{-9}	No	*SLCO6A1*	lincRNA	No data	
Chr. 3: 52,215,002–53,175,017	1.16×10^{-8}	SCZ, BPD	*ITIH3*	*ALAS1, ALDOAP1, BAP1, C3orf78, DNAH1, GLT8D1, GLYCTK, GNL3, ITIH1, ITIH4, MIR135A1, MIRLET7G, MUSTN1, NEK4, NISCH, NT5DC2,*	No data (ITIH1-ITIH3-ITIH4)	*GLYCTK*: D-glyceric aciduria, mental retardation; *RTF1*: mental retardation; GWAS: adiponectin, height, waist–hip ratio

Table 8.1. (cont.)

Chromosomal region	*P* value	Previous association[a]	Candidate gene in relation to index SNP[b]	Other genes in genomic region defined by LD[c]	eQTL[d]	Disease associations[e]
				PBRM1, PHF7, PPM1M, RFT1, SEMA3G, SFMBT1, SPCS1, STAB1, TLR9, TMEM110, TNNC1, TWF2, WDR82, lincRNA		
Chr. 2: 145,139,727–145,214,607	1.19×10^{-8}	No	*ZEB2*	–	No eQTL	*ZEB2*: Mowat–Wilson syndrome, mental retardation
Chr. 2: 200,628,118–201,293,421	1.21×10^{-8}	No	*FONG*	*C2orf47, C2orf69, SPATS2L, TYW5,* lincRNA	No data	GWAS: osteoporosis
Chr. 18: 52,722,378–52,827,668	1.22×10^{-8}	No	(ENST00000565991.1, 21 kb)	lincRNA (*TCF4*)	No data	
Chr. 2: 233,550,961–233,808,241	1.51×10^{-8}	No	*C2orf82*	*GIGYF2, KCNJ13, NGEF*	No data	
Chr. 1: 243,593,066–244,025,999	1.80×10^{-8}	No	*AKT3*	CEP170	AKT3	
Chr. 1: 243,418,063–243,627,135	2.53×10^{-8}	Yes	*SDCCAG8*		SDCCAG8	

Chr. 12: 123,447,928–123,913,433	2.28×10^{-8}	No	*C12orf65*	*ABCB9, ARL6IP4, CDK2AP1, MIR4304, MPHOSPH9, OGFOD2, PITPNM2, RILPL2, SBNO1, SETD8, lincRNA*	ARL6IP4, CDK2AP1, OGFOD2, SBNO1	*C12orf65*: mental retardation; GWAS: HDL, height, head size
Chr. 8: 89,188,454–89,761,163	3.33×10^{-8}	SCZ	Intergenic	*MMP16*, lincRNA	MMP16	
Chr. 5: 60,484,179–60,843,706	3.78×10^{-8}	No	ENST00000506902.1	*ZSWIM6*, C5orf43, lincRNA	C5orf43, ZSWIM6	

[a] Regions reported to meet genome-wide significance thresholds of association for schizophrenia (SCZ) or bipolar disorder (BPD).
[b] The gene within which an index SNP is located is given. For intergenic index SNPs, the nearest gene is given in parentheses.
[c] Other named genes in the genomic interval.
[d] SNP-transcript associations with $q < 0.05$ in peripheral blood. eQTLs with the SNP with the strongest association are shown in bold.
[e] Data from the NHGRI GWAS catalog (Hindorff *et al.*, 2009), OMIM (McKusick, 2007) and a compilation of genes related to autism (Betancur, 2011) and mental retardation (McKusick, 2007; Chiurazzi *et al.*, 2008; Inlow and Restifo, 2004). No data means no Affymetrix U219 probe sets or low expression in peripheral blood. The *CACNB2* association emerged when considering attention deficit/hyperactivity disorder (ADHD), autism, bipolar disorder, major depressive disorder and schizophrenia as affected (Smoller *et al.*, 2013). CAD, coronary artery disease; HDL, high-density lipoprotein.

Summarizing, current GWAs data provide convincing evidence for several loci. Recent, yet unpublished, analyses of even larger samples have resulted in an increase in the number of significant hits and we can safely conclude that the GWAs approach has been successful in identification of risk loci that can be followed up in functional studies. At present, GWAs data support the role of *MIR137* and calcium signaling in the etiology of schizophrenia. Genetic variants in the MHC region are strongly associated with schizophrenia, but the biological mechanisms are as yet unknown. Furthermore, large sample sizes were required to detect a relatively small number of common genetic polymorphisms affecting schizophrenia, as effect sizes were small (OR~1.1). This raises the question how many other SNPs affect schizophrenia and with what effect size. In the next section, we therefore discuss what GWA studies have revealed so far about the genetic architecture of schizophrenia.

8.4. Genetic architecture

Although subsequent PGC schizophrenia studies have identified an increasingly large number of genome-wide significant SNPs due to increased sample sizes, Visscher and colleagues estimated the contribution of SNPs robustly associated with schizophrenia to be less than 1% (Visscher *et al.*, 2012). Despite the fact that additional loci have been found since then, this small percentage is in stark contrast with the 64–81% heritability range estimated by twin studies (Sullivan *et al.*, 2003; Lichtenstein *et al.*, 2009; Wray and Gottesman, 2012); a discrepancy which has been coined the missing heritability (Maher, 2008). This section examines several explanations for the missing heritability and discusses their implications for the genetic architecture of schizophrenia.

One explanation for the missing heritability is model misspecification (Gibson, 2012; Zuk *et al.*, 2012). According to this view, the additive models often used to analyze genetic data are oversimplifications of the complex genetic architecture of schizophrenia. Additive models assume that the cumulative genetic effect of SNPs is the sum of the individual genetic effects. This assumption is convenient as it implies that heritability is the sum of genetic effects of individual SNPs. It also implies that no epistasis is present; that is, no (statistical) interactions between genetic effects are assumed (Frazer *et al.*, 2009; Manolio *et al.*, 2009; Eichler *et al.*, 2010; Zuk *et al.*, 2012; Hemani *et al.*, 2013). However, if epistasis is present, heritability estimates based on family and twin studies could be overestimated instead of missing (Zuk *et al.*, 2012). In this view, missing heritability is a problem of overestimated heritability in family and twin studies.

The role of epistasis in missing heritability has been much debated (Hill *et al.*, 2008; Frazer *et al.*, 2009; Manolio *et al.*, 2009; Eichler *et al.*, 2010; Zuk *et al.*, 2012; Hemani *et al.*, 2013; Stringer *et al.*, 2013). The problem with this debate is that it is currently impossible to empirically quantify the amount of epistasis because most epistatic models are too complex to estimate. The epistasis debate is therefore foremost a philosophical debate. However, even if we do not believe the additive model to be strictly true, there are pragmatic reasons for using it. The additive model is currently the most parsimonious model to investigate genetic effects and has taught important lessons about the genetic architecture of complex diseases, as we will discuss next.

Most other explanations for missing heritability assume the additive model to be valid and attribute the discrepancy in heritability either to unobserved variants or a lack of statistical power. First, a proportion of the causal variants might not be properly tagged by SNPs on current genotype platforms, which makes it difficult to detect their effects (Manolio et al., 2009; Gibson, 2012). This is especially true for SNPs with (very) rare variants (minor allele frequency < 1%). Extremely large sample sizes are required to reliably detect effects of rare variants even for relatively large effect sizes. Second, hundreds or thousands of common SNPs could have effect sizes too small to identify individually, but which cumulatively could explain a sizeable portion of the heritability (Manolio et al., 2009; Gibson, 2012). Although the common and rare variant hypotheses are often contrasted, both might be better viewed as two extremes of a continuum (Visscher *et al.*, 2012). Ultimately it is the combination of sample size, allele frequency, effect size, and linkage equilibrium structure which determines the power to detect an individual causal variant.

So far, only few rare variants affecting schizophrenia have been identified and none of those were disease-specific (Sullivan *et al.*, 2012). In contrast, several approaches have shown that many common variants with small effect sizes indeed comprise a sizeable proportion of the heritability. For example, risk score analysis, has shown the importance of (currently) non-significant SNPs. In polygenic risk score analysis (Purcell *et al.*, 2009), the SNP effect sizes (i.e., logistic regression weights) from a discovery sample are used to compute additive genetic risk scores and to predict disease status in an independent target sample. The effect sizes from a GWA analysis of 8831 schizophrenia cases and 12,067 controls (the PGC-1 schizophrenia sample) showed that a risk score based on all SNPs with a p-value below 0.1 predicts 6% of the variance in a Swedish case-control target sample (based on Nagelkerke pseudo-R^2), whereas a score based on SNPs with a p-value below 0.001 only resulted in a pseudo-R^2 of 3% (Ripke *et al.*, 2013). Given that the p-value threshold for genome-wide significance is 5×10^{-8}, these results imply that many non-significant SNPs contribute to schizophrenia. In other words the missing heritability is, at least partly, hidden in thousands of non-significant SNPs with small effects.

Although polygenic risk score analysis provides evidence for the importance of non-significant common SNPs, the pseudo-R^2 does not provide an estimate of the total heritability contributed by these common SNPs. Genome-wide complex trait analysis (GCTA), a second approach, was explicitly designed to estimate the total heritability (on the liability scale) of complex traits based on common SNPs (Yang *et al.*, 2010, 2011; Lee *et al.*, 2012). GCTA does so by implementing a linear mixed model (LMM). The LMM is an additive model with the additional assumption that the effect sizes on the liability scale of SNPs come from a normal distribution with mean zero. In other words, most SNPs are assumed to have a near-zero effect size. Under these assumptions, GCTA infers the total heritability contributed by all observed common SNPs, both significant and non-significant. For the PGC-1 schizophrenia sample, GCTA estimated the heritability at 27% (s.e. of 2%) assuming a population risk of 0.4% (Ripke *et al.*, 2013). In the Swedish target sample, the heritability estimate was even 32% (s.e. 3%), possibly due to higher genetic homogeneity in this sample (Ripke *et al.*, 2013).

A third approach which illustrated the importance of common variants is approximate Bayesian polygenic analysis (ABPA) (Wegmann *et al.*, 2010; Stahl *et al.*, 2012; Ripke *et al.*, 2013). ABPA is an approximate Bayesian method which, like GCTA, estimates the heritability, but uses polygenic risk score analysis as a model. Note that in Bayesian statistics the credible interval is analogous but not equal to the well-known confidence interval in frequentist statistics. ABPA yielded a heritability estimate of 34% (95% credible interval 31–37%) assuming a population prevalence of 0.4% for the PGC and Swedish samples combined (Ripke *et al.*, 2013). In GWA studies it is customary to impute unobserved SNPs using a reference panel in which these SNPs are observed. In this case the Utah residents with ancestry from northern and Western Europe (CEU) reference panel from HapMap 3 (Altshuler *et al.*, 2010) was used for imputation. However, when the much denser CEU 1000 Genomes reference panel (Abecasis *et al.*, 2010) was used and a 1% population prevalence of schizophrenia was assumed, the estimated heritability was even 50% (95% credible interval 45–54%) (Ripke *et al.*, 2013). Furthermore, unlike GCTA, ABPA can also estimate other parameters of interest. For example, it was estimated that 8300 (95% credible interval 6300–10,200) common independent SNPs contributed to the 50% heritability estimate (Ripke *et al.*, 2013). Moreover, compared to other diseases such as rheumatic arthritis and celiac disease, the estimated effect sizes per SNP are smaller in schizophrenia (Ripke *et al.*, 2013).

In conclusion, all three approaches demonstrate that under the additive model a large part of the missing heritability can be explained. Depending on method, population prevalence, sample (size), and reference panel used for imputation, 52–78% of the missing heritability can be explained (Ripke *et al.*, 2013). Hence, GWA studies have provided clear evidence that a sizeable proportion of the heritability in schizophrenia can be attributed to common SNPs, while most individual effect sizes are too small to be detected with current sample sizes. It is therefore expected that increasing sample sizes even further will continue to produce additional significant hits. Identifying the many genetic loci that contribute to schizophrenia is only a first step, however. The next challenge will be to investigate to what extent these hundreds or thousands of schizophrenia SNPs point towards common pathways (Sullivan *et al.*, 2012).

8.5. Genetic overlap with other disorders

GWA studies not only support the contribution of genetic factors to the risk of developing schizophrenia, but also point towards pleiotropy, a shared genetic basis with other diseases. For example, a cross-disorder study by PGC identified four different genome-wide significant loci (see Figure 8.1) associated with disease

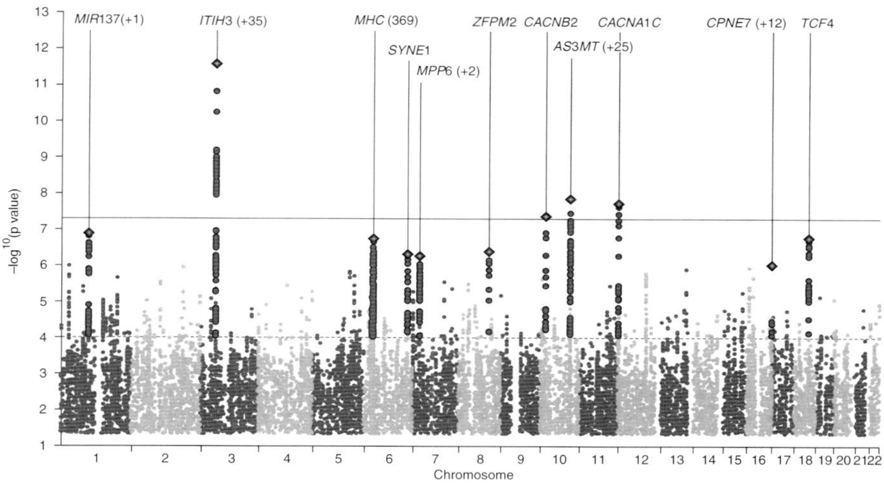

Figure 8.1. Manhattan plot of cross-disorder meta-analysis. Horizontal line represents genome-wide significance threshold ($p < 5 \times 10^{-8}$) (from Smoller *et al.*, 2013). A black and white version of this figure will appear in some formats. For the color version, please refer to the plate section.

in a large cross-disorder meta-analysis in which schizophrenia, bipolar disorder, major depressive disorder, autism spectrum disorder, and attention deficit–hyperactivity disorder PGC samples were combined (Smoller *et al.*, 2013). The associated SNPs were located on chromosome 3p21 and 10q24, and on two L-type voltage gated calcium-channel subunits (*CACNA1C* and *CACNB2*). Earlier, Huang *et al.* (2010) also performed a cross-disorder GWA analysis of schizophrenia, bipolar disorder, and depression and identified a genome-wide significant SNP (rs6484218) near the adrenomedullin (*ADM*) gene which was most strongly associated with bipolar disorder II. These cross-disorder findings are in line with the evidence from family studies that, for example, schizophrenia and bipolar disorder have a common genetic basis (Lichtenstein *et al.*, 2009).

However, due to the relatively small number of significant hits in schizophrenia GWA studies, it is no surprise that a cross-disorder meta-analysis only results in a few genome-wide significant associations. Polygenic risk score analysis can also be applied to estimate the extent of genetic overlap between disorders. For example, the cross-disorder group of the psychiatric genetics consortium performed a risk score analysis using the PGC-I schizophrenia sample as a discovery sample and the PGC samples for bipolar disorder (BPD), major depression disorder (MDD), autism spectrum disorder (ASD), and attention deficit–hyperactivity disorder (ADHD) as target samples (cross-disorder PGC, 2013). If there is no genetic overlap between schizophrenia and bipolar disorder, polygenic risk scores based on a schizophrenia sample should not predict disease status in BPD. However, if many SNPs affect the liability for both schizophrenia and BPD, a polygenic risk score based on schizophrenia liability will predict BPD.

This is indeed what has been found (Smoller *et al.*, 2013). Schizophrenia risk scores based on SNPs with *p*-values < 0.3 explained more than 2% of the variance in BPD status (Nagelkerke R^2), which was highly significant ($p < 10^{-16}$). The same analysis for MDD resulted in a predictive value of 0.8%, which was also highly significant ($p < 10^{-16}$). Early-onset psychiatric diseases such as ASD and ADHD showed much less genetic overlap with schizophrenia: 0.1% for ASD ($p < 0.05$), and 0.0% for ADHD ($p > 0.05$). Note that these results refer to overall genome-wide genetic overlap. For example, the low genetic overlap between schizophrenia and ASD does not preclude genetic overlap of small effect size at specific sites such as reported by Voineskos and colleagues (2011). Although a genome-wide analysis of genetic overlap does not specify the location of contributing SNPs, a pathway analysis for all five diseases combined revealed significant enrichment for a set of calcium channel activity genes. More precise mapping of gene pathways affected in schizophrenia is expected as increasingly large sample sizes allow better distinction between relevant and non-relevant SNPs.

Similar to polygenic risk score analysis, GCTA can also be used for cross-disorder genetic studies (Lee *et al.*, 2013). Generalizing the LMM, not only the heritability of a single disease can be estimated, but also the coheritability of two diseases. Note that coheritability is a measure of genetic overlap which depends not only on the correlation between genetic effects, but also on the individual heritabilities. In other words, a low coheritability can be due to low correlation in genetic effects as well as low heritabilities in either disease. The coheritabilities of schizophrenia and BPD (68%, s.e. 4%), MDD (43%, s.e. 6%), ASD (16%, s.e. 6%), and ADHD (non-significant) estimated by Lee *et al.* (2013) are shown in Figure 8.2, and followed the same pattern as the results from polygenic risk score analysis.

The evidence of a genetic overlap between schizophrenia and BPD that has been provided by family studies (Lichtenstein *et al.*, 2009) has been confirmed in cross-disorder GWA studies. However, analogous to identifying the loci contributing to the heritability of schizophrenia, the evidence for pleiotropy in psychiatric diseases provides a challenge to identify the loci involved and to establish which common biological pathways these diseases have in common (Solovieff *et al.*, 2013).

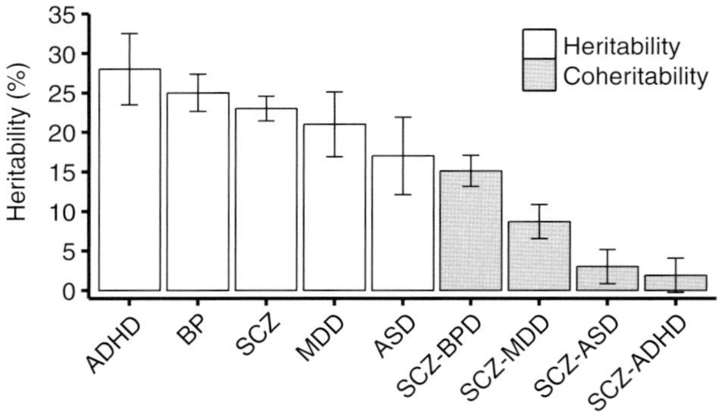

Figure 8.2. SNP-based heritabilities and coheritabilities for five psychiatric disorders: attention deficit hyperactivity disorder (ADHD), bipolar disorder (BPD), schizophrenia (SCZ), major depression disorder (MDD), and autism spectrum disorder (ASD) (adapted from Lee *et al.*, 2013).

8.6. Conclusions

Family, twin, and GWA studies have provided clear evidence that genetic factors play an important role in the onset of schizophrenia. Moreover, genetic studies have contributed hugely to a better appreciation of the complexity of the genetic architecture of schizophrenia. Not only is it now evident that many different loci affect schizophrenia; many of these loci are expected to contribute to the risk of other psychiatric diseases as well. However, many challenges need to be overcome for a thorough understanding of the etiology of schizophrenia. First, many more loci affecting schizophrenia need to be identified. Second, to understand the biological processes behind schizophrenia it is important to establish the common pathways that contribute to schizophrenia. Third, to biologically differentiate schizophrenia from, for example, bipolar disorder it is necessary to further investigate the similarities and dissimilarities in biological pathways between both diseases.

Nonetheless, the coordinated effort of PGC to continuously increase sample sizes promises to mitigate some of the statistical challenges in genetic studies. Although it is not feasible to identify all loci, Ripke *et al.* (2013) propose a goal for the field to identify the top 2000 loci for schizophrenia. Furthermore, especially for follow-up studies, next-generation sequencing allows more precise mapping of the location of causal common SNPs (Koboldt *et al.*, 2013). Similarly, family studies based on next-generation sequencing might reveal large rare effect loci which may not contribute much to the heritability, but could provide important clues about the biological processes behind schizophrenia. Finally, other types of genomic information, such as copy number variations (Malhotra and Sebat, 2012), expression data (Kunii *et al.*, 2014), and epigenetic factors (Nair and Howard, 2013), might provide additional biological insight into the etiology of schizophrenia.

Due to the unknown etiology of psychiatric diseases, classification of psychiatric disorders has traditionally been based on clinical observation. By solving the genetic puzzle behind schizophrenia and other psychiatric diseases, genetic studies would complement the traditional classification with an alternative classification based on etiological coherence instead of phenotypic coherence. Such a breakthrough would allow the development of drugs that target specific pathways identified for schizophrenia. Therefore, genetic studies will continue to play an important and exciting role in schizophrenia research.

REFERENCES

Abecasis, G.R., Altshuler, D., Auton, A., *et al.* (2010). A map of human genome variation from population-scale sequencing. *Nature*, **467**, 1061–1073.

Altshuler, D.M., Gibbs, R.A., Peltonen, L., *et al.* (2010). Integrating common and rare genetic variation in diverse human populations. *Nature*, **467**, 52–58.

American Psychiatric Association (2000). *Diagnostic and Statistical Manual of Mental Disorders: DSM-IV-TR*. American Psychiatric Association, Arlington, VA.

Bemis, L.T., Chen, R., Amato, C.M., *et al.* (2008). MicroRNA-137 targets microphthalmia-associated transcription factor in melanoma cell lines. *Cancer Res.*, **68**, 1362–1368.

Betancur, C. (2011). Etiological heterogeneity in autism spectrum disorders: more than 100 genetic and genomic disorders and still counting. *Brain Res.*, **1380**, 42–77.

Chiurazzi, P., Schwartz, C.E., Gecz, J. and Neri, G. (2008). XLMR genes: update 2007. *Eur. J. Hum. Genet.*, **16**(4), 422–434.

Derks, E.M. and Ophoff, R.A., Genetic Risk and Outcome of Psychosis (GROUP) (2013). Replication and refinement of the role of rs548181 in schizophrenia: results from a family based study. *Am. J. Med. Genet. B Neuropsych. Genet.*, **162B**, 75–77.

Derks, E.M., Allardyce, J., Boks, M.P., *et al.* (2010). Kraepelin was right: a latent class analysis of symptom dimensions in patients and controls. *Schizophr. Bull.*, **38**, 495–505.

Derks, E.M., Vorstman, J.A., Ripke, S., Kahn, R.S., Schizophrenia Psychiatric Genomic Consortium and Ophoff, R.A. (2012). Investigation of the genetic association between quantitative measures of psychosis and schizophrenia: a polygenic risk score analysis. *PLoS ONE*, **7**, e37852.

Eichler, E.E., Flint, J., Gibson, G., *et al.* (2010). Missing heritability and strategies for finding the underlying causes of complex disease. *Nature Rev. Genet.*, **11**, 446–450.

Fanous, A.H., Zhou, B., Aggen, S.H., *et al.* (2012). Genome-wide association study of clinical dimensions of schizophrenia: polygenic effect on disorganized symptoms. *Am. J. Psych.*, **169**, 1309–1317.

First, M.B., Spitzer, R.L., Gibbon, M. and Williams, J.B. (1997). *Structured Clinical Interview for DSM-IV Axis I Disorders (SCID-I), Clinician Version, Administration Booklet.* American Psychiatric Association, Arlington, VA.

Frazer, K.A., Murray, S.S., Schork, N.J. and Topol, E.J. (2009). Human genetic variation and its contribution to complex traits. *Nature Rev. Genet.*, **10**, 241–251.

Gibson, G. (2012). Rare and common variants: twenty arguments. *Nature Rev. Genet.*, **13**, 135–145.

Gottesman, I. (1991). *Schizophrenia Genesis: The Origin of Madness.* Freeman, New York, NY.

Harrison, G., Hopper, K., Craig, T., *et al.* (2001). Recovery from psychotic illness: a 15- and 25-year international follow-up study. *Br. J. Psych.*, **178**, 506–517.

Hemani, G., Knott, S. and Haley, C. (2013). An evolutionary perspective on epistasis and the missing heritability. *PLoS Genet.*, **9**, e1003295.

Hill, W.G., Goddard, M.E.and Visscher, P.M. (2008). Data and theory point to mainly additive genetic variance for complex traits. *PLoS Genet.*, **4**, e1000008.

Hindorff, L.A., Sethupathy, P., Junkins, H.A., *et al.* (2009). Potential etiologic and functional implications of genome-wide association loci for human diseases and traits. *Proc. Natl Acad. Sci. USA*, **106**, 9362–9367.

Huang, J., Perlis, R.H., Lee, P.H., *et al.* (2010). Cross-disorder genomewide analysis of schizophrenia, bipolar disorder, and depression. *Am. J. Psych.*, **167**, 1254–1263.

Inlow, J.K. and Restifo, L.L. (2004). Molecular and comparative genetics of mental retardation. *Genetics*, **166**(2), 835–881.

Kahn, R.S. and Keefe, R. S. (2013). Schizophrenia is a cognitive illness: time for a change in focus. *JAMA Psych.*, **70**, 1107–1112.

Kirov, G., Zaharieva, I., Georgieva, L., *et al.* (2009). A genome-wide association study in 574 schizophrenia trios using DNA pooling. *Molecul. Psych.*, **14**, 796–803.

Koboldt, D.C., Steinberg, K.M., Larson, D.E., Wilson, R.K. and Mardis, E.R. (2013). The Next-Generation Sequencing Revolution and Its Impact on Genomics. *Cell*, **155**, 27–38.

Kunii, Y., Hyde, T.M., Ye, T., *et al.* (2014). Revisiting DARPP-32 in postmortem human brain: changes in schizophrenia and bipolar disorder and genetic associations with t-DARPP-32 expression. *Molec. Psych.*, **19**, 192–199.

Lee, S.H., Decandia, T.R., Ripke, S., *et al.* (2012). Estimating the proportion of variation in susceptibility to schizophrenia captured by common SNPs. *Nature Genet.*, **44**, 831.

Lee, S.H., Ripke, S., Neale, B.M., *et al.* (2013). Genetic relationship between five psychiatric disorders estimated from genome-wide SNPs. *Nature Genet.*, **45**, 984–994.

Lichtenstein, P., Yip, B.H., Bjork, C., *et al.* (2009). Common genetic determinants of schizophrenia and bipolar disorder in Swedish families: a population-based study. *Lancet*, **373**, 234–239.

Lieberman, J.A., Stroup, T.S., McEvoy, J.P., *et al.* (2005). Effectiveness of antipsychotic drugs in patients with chronic schizophrenia. *New Engl. J. Med.*, **353**, 1209–1223.

Maher, B. (2008). Personal genomes: the case of the missing heritability. *Nature*, **456**, 18–21.

Malhotra, D. and Sebat, J. (2012). CNVs: harbingers of a rare variant revolution in psychiatric genetics. *Cell*, **148**, 1223–1241.

Manolio, T.A., Collins, F.S., Cox, N.J., *et al.* (2009). Finding the missing heritability of complex diseases. *Nature*, **461**, 747–753.

Marwaha, S., Johnson, S., Bebbington, P., *et al.* (2007). Rates and correlates of employment in people with schizophrenia in the UK, France and Germany. *Br. J. Psych.*, **191**, 30–37.

McGue, M. and Gottesman, I.I. (1989). Genetic linkage in schizophrenia: perspectives from genetic epidemiology. *Schizophr. Bull.*, **15**, 453–464.

McKusick, V.A. (2007). Mendelian Inheritance in Man and its online version, OMIM. *Am. J. Hum. Genet.*, **80**, 588–604.

Mesholam-Gately, R.I., Giuliano, A.J., Goff, K.P., Faraone, S.V. and Seidman, L.J. (2009). Neurocognition in first-episode schizophrenia: a meta-analytic review. *Neuropsychology*, **23**, 315.

Miller, G. (2010). Beyond DSM: seeking a brain-based classification of mental illness. *Science*, **327**, 1437.

Nair, A. and Howard, R. (2013). ENCODE and a new landscape for psychiatric genetics. *Br. J. Psych.*, **203**, 84–85.

Purcell, S.M., Wray, N.R., Stone, J.L., *et al.* (2009). Common polygenic variation contributes to risk of schizophrenia and bipolar disorder. *Nature*, **460**, 748–752.

Ripke, S., Sanders, A.R., Kendler, K.S., *et al.* (2011). Genome-wide association study identifies five new schizophrenia loci. *Nature Genet.*, **43**, 969–976.

Ripke, S., O'Dushlaine, C., Chambert, K., *et al.* (2013). Genome-wide association analysis identifies 13 new risk loci for schizophrenia. *Nature Genet.*, **45**, 1150–1159.

Robinson, D.G., Woerner, M.G., McMeniman, M., Mendelowitz, A. and Bilder, R.M. (2004). Symptomatic and functional recovery from a first episode of schizophrenia or schizoaffective disorder. *Am. J. Psych.*, **161**, 473–479.

Shi, J., Levinson, D.F., Duan, J., *et al.* (2009). Common variants on chromosome 6p22.1 are associated with schizophrenia. *Nature*, **460**, 753–757.

Silber, J., Lim, D.A., Petritsch, C., *et al.* (2008). miR-124 and miR-137 inhibit proliferation of glioblastoma multiforme cells and induce differentiation of brain tumor stem cells. *BMC Med.*, **6**, 14.

Smoller, J.W., Craddock, N., Kendler, K., *et al.* (2013). Identification of risk loci with shared effects on five major psychiatric disorders: a genome-wide analysis. *Lancet*, **381**, 1371–1379.

Smrt, R.D., Szulwach, K.E., Pfeiffer, R.L., *et al.* (2010). MicroRNA miR-137 regulates neuronal maturation by targeting ubiquitin ligase mind bomb-1. *Stem Cells*, *28*, 1060–1070.

Solovieff, N., Cotsapas, C., Lee, P.H., Purcell, S.M. and Smoller, J.W. (2013). Pleiotropy in complex traits: challenges and strategies. *Nature Rev. Genet.*, **14**, 483–495.

Stahl, E.A., Wegmann, D., Trynka, G., *et al.* (2012). Bayesian inference analyses of the polygenic architecture of rheumatoid arthritis. *Nature Genet.*, **44**, 483–489.

Stefansson, H., Ophoff, R.A., Steinberg, S., *et al.* (2009). Common variants conferring risk of schizophrenia. *Nature*, **460**, 744–747.

Stringer, S., Derks, E.M., Kahn, R.S., Hill, W.G. and Wray, N.R. (2013). Assumptions and properties of limiting pathway models for analysis of epistasis in complex traits. *PLoS ONE*, **8**, e68913.

Sullivan, P.F., Kendler, K.S. and Neale, M.C. (2003). Schizophrenia as a complex trait: evidence from a meta-analysis of twin studies. *Arch. Gen. Psych.*, **60**, 1187–1192.

Sullivan, P.F., Daly, M.J. and O'Donovan, M. (2012). Genetic architectures of psychiatric disorders: the emerging picture and its implications. *Nature Rev. Genet.*, **13**, 537–551.

Szulwach, K.E., Li, X., Smrt, R.D., *et al.* (2010). Cross talk between microRNA and epigenetic regulation in adult neurogenesis. *J. Cell Biol.*, **189**, 127–141.

van Os, J. and Kapur, S. (2009). Schizophrenia. *Lancet*, **374**, 635–645.

Visscher, P.M., Goddard, M.E., Derks, E.M. and Wray, N.R. (2012). Evidence-based psychiatric genetics, AKA the false dichotomy between common and rare variant hypotheses. *Molec. Psych.*, **17**, 474–485.

Voineskos, A.N., Lett, T.A., Lerch, J.P., *et al.* (2011). Neurexin-1 and frontal lobe white matter: an overlapping intermediate phenotype for schizophrenia and autism spectrum disorders. *PloS ONE*, **6**, e20982.

Walters, J.T., Rujescu, D., Franke, B., *et al.* (2013). The role of the major histocompatibility complex region in cognition and brain structure: a schizophrenia GWAS follow-up. *Am. J. Psych.*, **170**, 877–885.

Wegmann, D., Leuenberger, C., Neuenschwander, S. and Excoffier, L. (2010). Abctoolbox: a versatile toolkit for approximate bayesian computations. *BMC Bioinform.*, **11**, 116.

Wessman, J., Paunio, T., Tuulio-Henriksson, A., *et al.* (2009). Mixture model clustering of phenotype features reveals evidence for association of *DTNBP1* to a specific subtype of schizophrenia. *Biol. Psych.*, **66**, 990–996.

Wray, N.R. and Gottesman, I.I. (2012). Using summary data from the Danish national registers to estimate heritabilities for schizophrenia, bipolar disorder, and major depressive disorder. *Front. Genet.*, **3**, 118.

Yang, J., Benyamin, B., McEvoy, B.P., *et al.* (2010). Common SNPs explain a large proportion of the heritability for human height. *Nature Genet.*, **42**, 565–569.

Yang, J., Lee, S.H., Goddard, M.E. and Visscher, P.M. (2011). GCTA: a tool for genome-wide complex trait analysis. *Am. J. Hum. Genet.*, **88**, 76–82.

Zuk, O., Hechter, E., Sunyaev, S.R. and Lander, E.S. (2012). The mystery of missing heritability: genetic interactions create phantom heritability. *Proc. Natl Acad. Sci. USA*, **109**, 1193–1198.

9 Epigenome-wide association studies in neurodevelopmental disorders

Takeo Kubota, Kunio Miyake and Takae Hirasawa

9.1. Introduction

The brain is a gene-dosage sensitive organ in which either under- or overexpression of genes encoding proteins associated with brain function results in a range of congenital neurodevelopmental disorders, suggesting that the brain is extremely sensitive to perturbations in gene regulation, and further indicating the importance of a proper gene expression control in the brain.

Intrinsic epigenetic mechanisms are involved in the controls of gene expression, and are essential for normal development during embryogenesis and for differentiation of neural cells (Takizawa *et al.*, 2001). It has been reported that abnormalities in epigenetic mechanisms can result in a number of congenital neurodevelopmental disorders.

Advances in methodologies for epigenetic analysis, such as Beadchip microarrays and next-generation sequencing, now enable the investigation of the epigenetic status at individual loci, multiple loci, or the whole genome. These new approaches also enable epigenome-wide association studies (EWAS).

Several lines of evidence suggest that epigenetic abnormalities can be induced by environmental factors. Thus, clinical epigenetic research not only needs to target congenital disorders, but must also investigate acquired chronic diseases including common mental and neurodevelopmental disorders, in which epigenomic abnormalities may reside at multiple genomic loci.

We are now in the process of identifying environmentally induced epigenomic changes that can be used as "epigenomic disease signature," that is, predictive markers for chronic diseases. Realization of this goal will ensure the start of "personalized medicine" or "preemptive medicine."

In this chapter, we describe epigenetic and epigenomic (genome-wide epigenetic) abnormalities associated with congenital neurodevelopmental disorders. Additionally, we describe environment-induced epigenetic abnormalities, and

Genome-Wide Association Studies: From Polymorphism to Personalized Medicine, ed. K. Appasani. Published by Cambridge University Press.

discuss EWAS on various diseases including neurodevelopmental disorders. We also discuss personalized medicine as a goal of EWAS.

9.2. Congenital neurodevelopmental disorders with epigenetic abnormalities

9.2.1. Genomic imprinting disorders

Genomic imprinting is an epigenetic phenomenon that was initially discovered in mammals, and results in the monoallelic, parent-of-origin expression of some genes. These inherited maternal and paternal imprints are erased in the germ line and a new imprinting pattern is established according to the sex of the individual. For example, analysis of the mouse insulin-like growth factor 2 (*Igf2*) gene, which is expressed from the maternally inherited allele in the soma, showed that it contains a differentially methylated region 2 (DMR2) that is demethylated in primordial gene cells (PGCs) by day (E) 13.5 of embryonic development days; the maternal or paternal DNA methylation patterns that contribute to the monoallelic expression in the next generation are then established (Sato *et al.*, 2003; Yamaguchi *et al.*, 2013). Therefore, a defect in the expression allele of the imprinted gene results in the loss of expression. This phenomenon underlies a subset of neurodevelopmental disorders such as Prader–Willi syndrome and Angelman syndrome (Kubota *et al.*, 1997; Figure 9.1A).

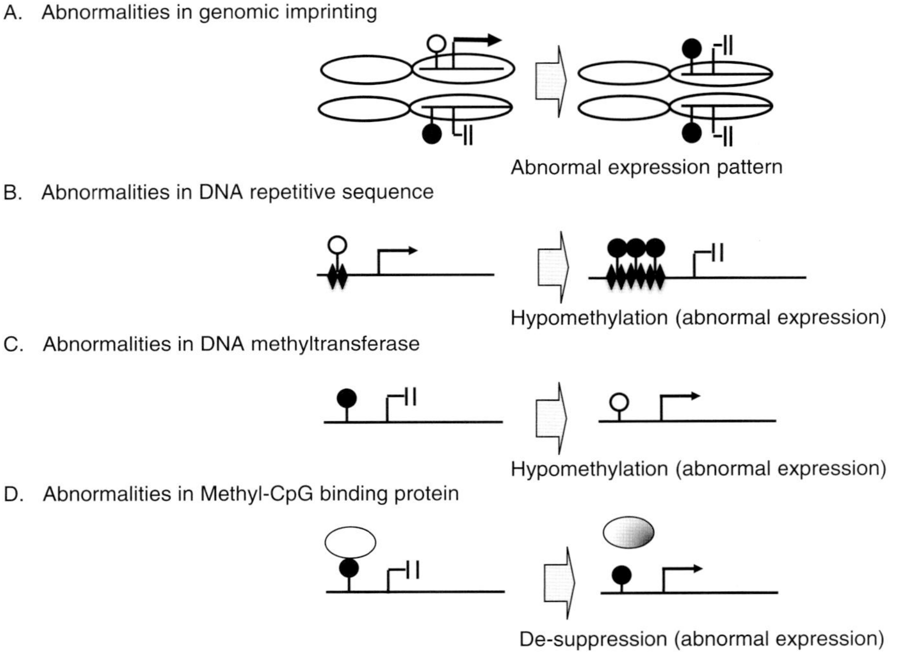

Figure 9.1. Epigenetic abnormalities in congenital neurodevelopmental disorders.

9.2.2. DNA sequence-based epigenetic disorder

The human genome contains a large number of repeat sequences, such as the unstable CGG triplet repeat within exon 1 of the Fragile X Mental Retardation Syndrome 1 gene (*FMR1*) in the X chromosome. When the number of CGG trinucleotides is increased, CpG dinucleotides within CGG triplet repeat sequence become hypermethylated; that suppresses *FMR1* expression, causing the autistic Fragile X syndrome (Figure 9.1B; Verkerk *et al.*, 1991). Fragile X mental retardation protein (FMRP) encoded by *FMR1* regulates neuronal genes, and thus, decreased FMRP expression by hypermethylation of *FMR1* may cause these mental abnormalities including autism and schizophrenia (Fromer *et al.*, 2014). The example of Fragile X syndrome indicates that abnormalities in DNA sequence can lead to aberrant epigenetic status.

9.2.3. Epigenetic molecule disorders

DNA methyltransferases (DNMTs) mediate addition of a methyl group (CH_3) to CpG dinucleotides. Mutations in *DNMT3B*, a *DNMT* gene, causes ICF syndrome, which is characterized by immunodeficiency, centromere instability and facial anomalies, and mild mental retardation (Kubota *et al.*, 2004). Thus, dysregulation of genes expression as a result of a DNMT mutation may cause features in this syndrome (Figure 9.1C). Methyl-CpG-binding domain proteins (MBDs) also have an important role in epigenetic gene regulation. The methyl-CpG-binding protein 2 gene, *MECP2*, encodes an MBD; mutation of *MECP2* causes Rett syndrome, which is characterized by seizures, ataxic gait, language dysfunction, and autistic behavior (Amir *et al.*, 1999). Recent studies have shown that the MeCP2 protein controls expression of neuronal genes (Miyake *et al.*, 2011), suggesting that dysregulation of multiple neuronal genes may cause the neurological features of this syndrome (Figure 9.1D). The examples briefly described above provide a clear indication of how defects in epigenetic molecular mechanisms can induce an aberrant epigenetic status.

9.3. Congenital neurodevelopmental disorders with epigenomic (genome-wide epigenetic) abnormalities

9.3.1. Abnormal X-chromosome inactivation disorder

In mammals, the X chromosome is much larger than the Y chromosome, so the X chromosome carries substantially more active genes. Consequently, females potentially have greater gene expression from their two X chromosomes than do males, who have a single X. This potential imbalance between females and males is prevented by the epigenetic inactivation of one of two X chromosomes in females (Kubota *et al.*, 2002). If X chromosome inactivation (XCI) does not occur properly, it can cause lethality in the affected female embryo; this effect is evident in mouse embryonic clones produced by somatic nuclear transfer in which a majority of clones abort due to failure of XCI (Nolen *et al.*, 2005), although this effect has not been demonstrated yet in humans (Figure 9.2A). When one of the

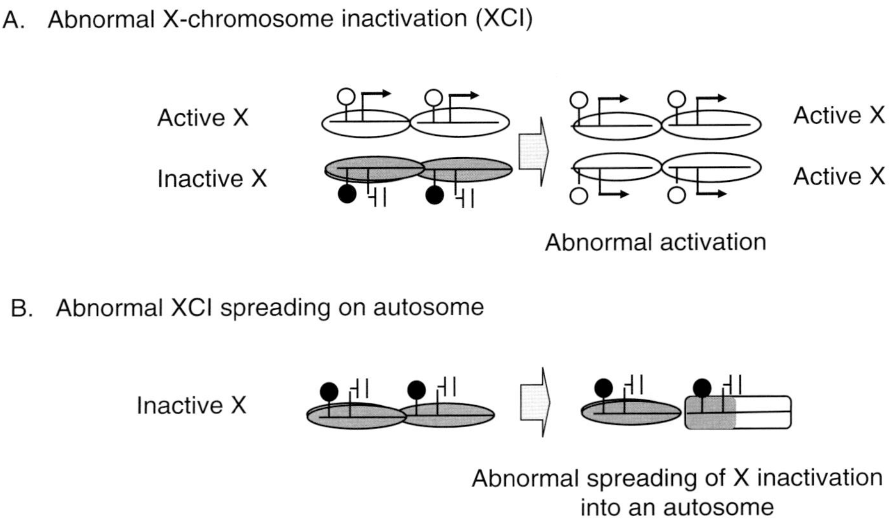

Figure 9.2. Epigenomic (genome-wide epigenetic) abnormalities in congenital neurodevelopmental disorders.

X chromosomes in a female is very tiny, as a result of a chromosome rearrangement that generates a small ring-shape X chromosome with a centromere, it may escape XCI. In this case, the female has a normal and a small ring X chromosome that are active but nevertheless do not cause embryonic lethality. However, such females generally show extremely severe neurodevelopmental delay (Kubota *et al.*, 2002), indicating that proper epigenetic chromosome inactivation is essential for normal development.

9.3.2. X-chromosome inactivation spreading disorder

Various types of chromosome rearrangement can generate an abnormal X chromosome attached to an autosomal chromosome region. In these rearranged chromosomes, XCI can spread from the X chromosome in the autosomal chromosome, and the affected autosomal genes are hypermethylated as a consequence of XCI (Figure 9.2B). The effect of such an inappropriate spread of XCI is illustrated by a patient with a translocation between the X chromosome and chromosome 15, which caused hypermethylation and downregulation of the *OCA2* gene on chromosome 15. The patient showed hypopigmentation (thin skin and hair color) as a result of the XCI-mediated suppression of *OCA2* (Sakazume *et al.*, 2012).

9.4. Environmental stress-induced epigenetic changes and disorders

9.4.1. Short-term mental stress-induced epigenetic changes

Epigenetic mechanisms are involved in the selection of the appropriate genes for expression in a particular cell type. Initially, it was believed that this system is

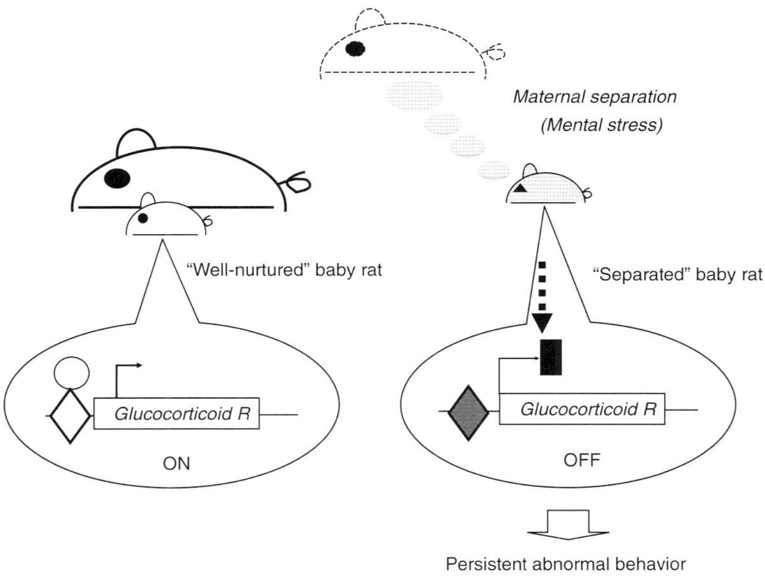

Figure 9.3. Mental stress-induced DNA methylation in the brain.

stable and did not change through the lifetime; thus, cell identity was also assumed to be maintained throughout the lifetime. The only exception to this expected stability was the change associated with carcinogenesis in which disorder of the epigenetic gene selection system is induced by long-term environmental stimuli.

However, it has been demonstrated that epigenetic status in the brains can be altered by short-term mental stress. The level of DNA methylation is normally low in the promoter region of the glucocorticoid receptor (*Gr*) (also known as *NR3C1*) gene in the hippocampus of the brains of the offspring given high maternal care during the first week (Figure 9.3, left). By contrast, rat newborns separated from their mothers during the first week showed aberrant hypermethylation of the promoter and repression of expression in the hippocampal tissue (Figure 9.3, right; Weaver *et al.*, 2004). This finding was suggested to provide a putative animal model for childhood neglect and maltreatment in humans during the early period of life. The ability of stress to induce long-term changes was supported by a postmortem brain analysis of suicide victims with a history of childhood abuse in which hypermethylation of the neuron-specific promoter of *NR3C1* as well as reduced expression was observed in the hippocampus (McGowan *et al.*, 2009). These findings further indicate that mental stress in early life can alter personality and this effect can persist throughout the lifetime as a consequence of long-lasting epigenetic changes (Murgatroyd *et al.*, 2009).

9.4.2. Malnutrition-induced epigenetic changes

Current epidemiological studies of populations affected by famines in the Netherlands and China have shown that the generation of children born to the affected population had lower birth weights and an increased risk of metabolic disorders (e.g., obesity, diabetes mellitus) and mental disorders (St Clair *et al.*, 2005; Painter *et al.*, 2006). A similar phenomenon appears to be occurring in Japan, where birth weights have decreased over the last 30 years due to intentional dieting in young Japanese women, which has resulted in fatal malnutrition (Gluckman *et al.*, 2007).

This association of fatal circumstances and susceptibility of adult diseases is referred to as the Developmental Origin of Health and Diseases (DOHaD) and stress-related epigenetic changes by malnutrition during the fetal period underlie this association. Recent studies in rats demonstrated that malnutrition during the fetal period decreases DNA methylation and increases expression of the peroxisome proliferator-activated receptor alpha (*PPARa*) gene in the liver (Lillycrop *et al.*, 2008). Additionally, an increase in DNA methylation that inhibits the binding of the CTCF protein to an *Igf2* enhancer causes an increase in expression of this gene in the adipose tissue. Similarly, DNA methylation changes have been identified in the peripheral blood tissues of individuals who suffered malnutrition during a period of famine in the Netherlands (Tobi *et al.*, 2009).

It has also been reported that assisted reproductive technologies (ART; e.g., *in vitro* fertilization and intracytoplasmic sperm injection), which are now widely used due to increases in maternal and paternal ages, lead to a decrease in the DNA methylation status at multiple maternally methylated imprinted loci (Lim *et al.*, 2009).

Therefore, it is interesting to speculate that two social factors in modern societies – namely, DOHaD and ART – might contribute to the recent increase in children with mild neurodevelopmental disorders in Japan and other countries due to an altered epigenetic status in the brain and metabolic tissues (Kim *et al.*, 2011).

9.4.3. Evidence for environmental stress-induced epigenetic changes in humans

Although recent animal studies have provided convincing evidence that environmental stresses can induce epigenetic changes, information of this topic from human studies is generally more sparse and indirect. It has been reported that genome-wide epigenetic differences are larger in older monozygotic twins than in younger twins; one possible explanation for this is that the epigenetic status may be altered during the lifetime by environmental stresses (Fraga *et al.*, 2005; Figure 9.4). Likewise, in monozygotic twins with discordant severity of Rett syndrome, the DNA methylation patterns and expression of neuronal genes can differ (Miyake *et al.*, 2013). These reports indicate that environmental factors may affect the human epigenome and that epigenomic differences induced by environmental factors may contribute to mental and neurological status.

9.4.4. Transgenerational inheritance of environmental stress-induced epigenetic changes

In mammals, methylation marks should be removed by demethylating factors such as the cytidine deaminases (e.g., AID, APOBEC1) and re-established in each

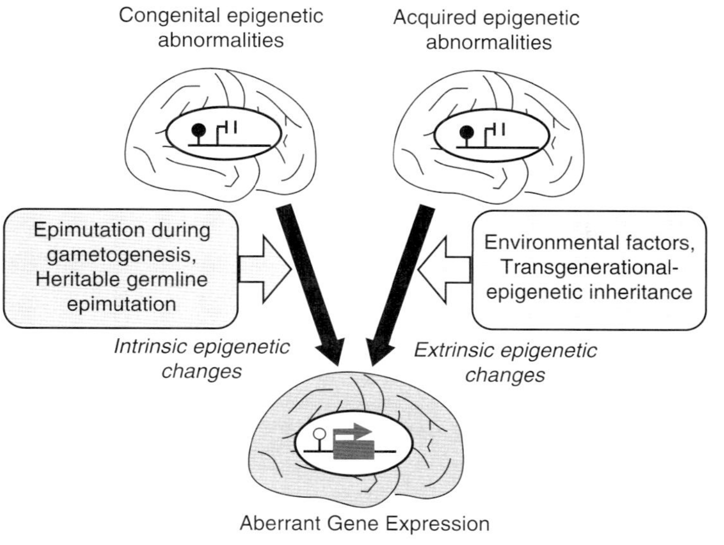

Figure 9.4. Current understanding of intrinsic and extrinsic mechanisms of neurodevelopmental and mental disorders. Congenital epigenetic disorders are caused by intrinsic mechanisms through mutations in proteins associated with epigenetic gene regulation or epimutation (e.g., *de novo* DNA methylation) during spermatogenesis and oogenesis, or by a "heritable germline epimutation" mechanism. Acquired epigenetic disorders are caused either by an extrinsic mechanism via various environmental factors, or by a "transgenerational-epigenetic inheritance" mechanism.

generation. However, this clearance may be incomplete at some loci due to a deficiency of demethylating factors. Recently, it was shown that rare regulatory elements escape systematic DNA demethylation in PGCs, providing a potential mechanistic basis for transgenerational epigenetic inheritance (Hackett *et al.*, 2013). These observations provide a biological basis of the hypothesis "transgenerational epigenetic inheritance," which suggests the germline transmission of epigenetic marks (Horsthemke, 2007; Daxinger and Whitelaw, 2010) and may give support to the concept of "Lamarckism," in which acquired characteristics are postulated to be heritable.

Any discussion of the heritability of epigenetic marks needs to distinguish between "transgenerational epigenetic inheritance" and "heritable germline epimutation." Transgenerational epigenetic inheritance is considered to be independent of the DNA sequence, whereas heritable germline epimutation occurs as a direct consequence of a *cis*-acting epigenetic alteration (Horsthemke, 2007; Daxinger and Whitelaw, 2010). The best example of a heritable germline epimutation in humans is the CGG trinucleotides repeat expansion-driven DNA methylation observed in the mother and son in a family with Fragile X syndrome, as described above (Chong *et al.*, 2007).

As described above, mental stress can induce epigenetic changes in the brain during infancy (Weaver *et al.*, 2004). A recent study in mice showed that such epigenetic marks could be transmitted to the next generation to generate an abnormal phenotype (Franklin *et al.*, 2010). Abnormal behavior was observed in the third generation, and altered DNA methylation in several neuronal genes was observed in the sperm of the F_1 generation males and in the brains of F_2 subjects. It has also been reported that female rat pups suffering from chronic maternal separation show increased levels of cytosine methylation of the estrogen receptor (Er)-alpha1b gene promoter; as variation in ER-alpha1b expression is associated with differences in maternal behavior, then there will be an epigenetically transmitted effect on behavior from the mother to her female offspring (Champagne *et al.*, 2006). These findings provide support for the brief that environmental stress-induced neurodevelopmental disorders can be transmitted to succeeding generations. The possibility of the inheritance of epigenetic marks is also supported by recent studies in which it was shown in mice that a low paternal dietary intake of folate could affect the epigenomic pattern (DNA methylation and histone H3 methylation) in the sperm (Lambrot *et al.*, 2013).

9.5. Epigenome-wide association study on chronic diseases including neurodevelopmental disorders

9.5.1. Genome-wide approaches for chronic diseases

It is known that both congenital genetic predisposition and exposure to environmental factors affect chronic diseases, such as diabetes mellitus, obesity, and mental disorders. In recent years, the numbers of patients suffering from such chronic diseases have been reported to be increasing in Asian countries, including Japan (Nguyen *et al.*, 2013).

In order to identify the genetic predisposition to chronic diseases, many genome-wide association studies (GWAS) have been performed. These studies have identified various primary DNA sequence variations that influence gene expression and that are unique to each chronic disease (Lahiri *et al.*, 2012). Such primary DNA sequence variations are referred to as single-nucleotide polymorphisms (SNPs).

However, SNPs only account for the relative risk for a disease, and individuals who carry these SNPs do not necessarily develop the corresponding disease. SNPs also suffer from the limitation that they are based on a single sampling point from a subject's lifetime and, therefore, do not reflect the lifetime influence of environmental factors and the longitudinal aspect of the disease (Lahiri *et al.*, 2012). Because the association of disease occurrence with SNPs is not very high, SNPs are not generally recognized as a valuable disease-predicting marker. For this reason, other genetic markers with higher prediction accuracies have been sought (Xu *et al.*, 2013).

Due to improvements in methodologies and facilities for epigenetic analysis, information on the epigenetic state can now be obtained from multiple points on the genome (e.g., we can collect DNA methylation levels at 450,000 CpG sites on the human genome using BeadChip microarrays). These new technologies enable EWAS, which identifies disease-risk epigenetic sites and disease-specific epigenetic changes. For example, a recent EWAS study demonstrated that DNA methylation at specific CpG sites was associated with blood metabolic traits (Petersen *et al.*, 2014).

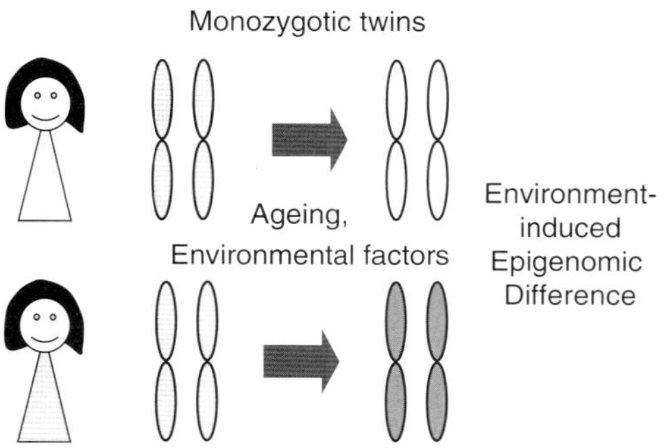

Figure 9.5. Epigenomic differences between monozygotic twins.

9.5.2. Epigenotype influenced by genotype

To date, a subset of EWAS have aimed at identifying disease-specific DNA methylation sites (DNA methylation positions, or DMPs; Kerkel *et al.*, 2008; Xu *et al.*, 2013; Liu *et al.*, 2013; Hidalgo *et al.*, 2014). A comparative GWAS and EWAS study on rheumatoid arthritis (RA) have revealed that most RA-associated DMPs are located in the same genomic region (MHC cluster region), suggesting that epigenotype may be influenced by genotype (Liu *et al.*, 2013). Genotype-dependent DNA methylation has been reported in another study (Kerkel *et al.*, 2008). Furthermore, allele-specific methylation (or heritable germline epimutation) has been demonstrated in a three-generation family and the heritable germline epimutation is found to occur at more loci than genomic imprinted loci (Gertz *et al.*, 2011; Figure 9.5, left).

9.5.3. Epigenotype independent of genotype

External environmental factor can influence the epigenotype, presumably in a non-genotype-dependent fashion (Petersen *et al.*, 2014). Actually, epigenotype including genomic imprinting loci can be altered by various nutritional factors (Choi *et al.*, 2013) and smoking (Breitling *et al.*, 2011; Shenker *et al.*, 2013; Figure 9.5, right).

DNA methylation patterns also change with age in peripheral blood and brain tissues (Florath *et al.*, 2014); DNA methylation levels are higher in newborn children of older parents than those of younger parents (Adkins *et al.*, 2011). As described above, DNA methylation patterns can differ between monozygotic twins (Tsai *et al.*, 2012; Miyake *et al.*, 2013). The alterations in DNA methylation patterns associated with these various factors indirectly suggest that DNA changes are due to external environmental factors.

Based on these findings, one can speculate that environmental factors can induce epigenomic changes and, thereby, increase the risk of diseases. Some support for this proposition comes from the report that DNA methylation changes increased breast cancer risk independent of genetic risk (*BRCA1* and *BRCA2* mutations; Xu *et al.*, 2013), and that changes to the DNA methylation patterns in peripheral blood are associated with an increased risk of head and neck squamous cell carcinoma (Langevin *et al.*, 2012).

9.5.4. Future medicine based on EWAS studies

GWAS studies have identified SNPs associated with disease risks and EWAS has identified DMPs associated with various diseases (Kerkel *et al.*, 2008; Liu *et al.*, 2013; Xu *et al.*, 2013; Hidalgo *et al.*, 2014). SNPs and DMPs differ in that the latter are affected by environmental factors whereas SNPs are independent of environmental effects. SNPs are stable disease markers that reflect differences between individuals, while DMPs are changeable markers that reflect environment, age, disease condition, and effectiveness of therapies. Therefore, SNPs may be suitable for personalized medicine and DMPs may be suitable for preemptive medicine in which disease risk has been identified before onset, because DMPs potentially have higher prediction accuracy than do SNPs (Xu *et al.*, 2013). Ultimately, epigenomic risk detection in preemptive medicine may require epigenomic scanning using the Gurhrie papers (heel-prick blood spots) collected from newborn infants to identify future disease risk (Cruickshank *et al.*, 2012).

Further studies will be necessary to answer the following fundamental questions as follows:

(1) Which environmental factors alter the epigenotype?
(2) Which genomic regions are epigenetically susceptible to the environmental factor that is associated with alteration of the gene expression pattern, resulting in a disease or pre-disease condition (Figure 9.6)?
(3) Which tissues (organs) are vulnerable to environmental factors from an epigenetic viewpoint?
(4) Which ages (periods of life) are vulnerable to environmental factors from an epigenetic viewpoint?

By answering these questions, fundamental knowledge will be obtained for preemptive medicine. Recently, an NIH research project grant was funded for this purpose (http://grants.nih.gov/grants/guide/pa-files/PAR-13–385.html#_Part_1._Overview/; Feb 26, 2014).

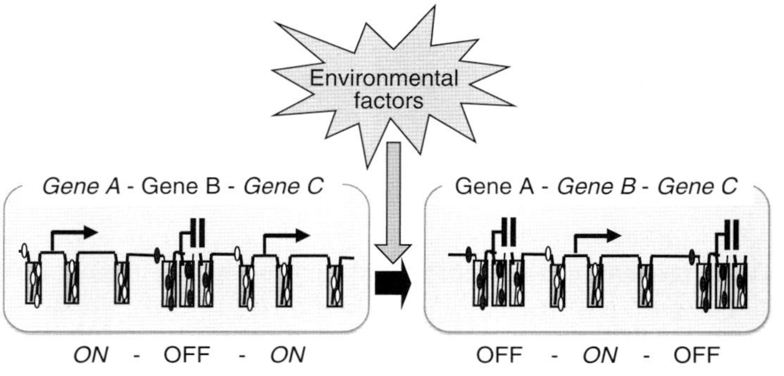

Figure 9.6. Environmental factors change epigenotype, leading to change of gene expression patterns.

EWAS studies have mainly been performed for cancer (Langevin *et al.*, 2012; Xu *et al.*, 2013) and diabetes mellitus (Verma *et al.* 2012; Hidalgo *et al.*, 2014). Because the choice of appropriate tissue is important for an EWAS (Rakyan *et al.*, 2011) and it is important to understand the predictive value of epigenomic changes in peripheral blood tissues for neurodevelopmental disorder, further understanding of the linkage between epigenomic changes in the brain and those in peripheral blood tissues will be essential for EWAS of neurodevelopmental disorders. To achieve this goal, comparative studies between brain and peripheral tissues (Kaminsky *et al.*, 2012), combined with use of real-time imaging systems for visualizing the epigenetic status in the brains of patients with mental disorders (Wang *et al.*, 2013), will be necessary.

REFERENCES

Adkins, R.M., Thomas, F., Tylavsky, F.A. and Krushkal, J. (2011). Parental ages and levels of DNA methylation in the newborn are correlated. *BMC Med. Genet.*, **12**, 47.

Amir, R.E., Van den Veyver, I.B., Wan, M., *et al.* (1999). Rett syndrome is caused by mutations in X-linked MECP2 encoding methyl-CpG-binding protein 2. *Nature Genet.*, **23**, 185–188.

Breitling, L.P., Yang, R., Korn, B., Burwinkel, B. and Brenner, H. (2011). Tobacco-smoking-related differential DNA methylation: 27K discovery and replication. *Am. J. Hum. Genet.*, **88**, 450–457.

Champagne, F.A., Weaver, I.C., Diorio, J., *et al.*(2006). Maternal care associated with methylation of the estrogen receptor-alpha1b promoter and estrogen receptor-alpha expression in the medial preoptic area of female offspring. *Endocrinology*, **147**, 2909–2915.

Choi, S.W., Claycombe, K.J., Martinez, J.A., Friso, S. and Schalinske, K.L. (2013). Nutritional epigenomics: a portal to disease prevention. *Adv. Nutr.*, **4**, 530–532.

Chong, S., Youngson, N.A. and Whitelow, E. (2007). Heritable germline epimutation is not the same as transgenerational epigenetic inheritance. *Nature Genet.*, **39**, 574–575.

Cruickshank, M.N., Pitt, J. and Craig, J.M. (2012). Going back to the future with Guthrie-powered epigenome-wide association studies. *Genome Med.*, **4**, 83.

Daxinger, L. and Whitelaw, E. (2010). Transgenerational epigenetic inheritance: more questions than answers. *Genome Res.*, **20**, 1623–1628.

Florath, I., Butterbach, K., Müller, H., Bewerunge-Hudler, M. and Brenner, H. (2014). Cross-sectional and longitudinal changes in DNA methylation with age: an epigenome-wide analysis revealing over 60 novel age-associated CpG sites. *Hum. Molec. Genet.*, **23**, 1186–1201.

Fraga, M.F., Ballestar, E., Paz, M.F., *et al.* (2005). Epigenetic differences arise during the lifetime of monozygotic twins. *Proc. Natl Acad. Sci. USA*, **102**, 10604–10609.

Franklin, T.B., Russig, H., Weiss, I.C., *et al.* (2010). Epigenetic transmission of the impact of early stress across generations. *Biol. Psych.*, **68**, 408–415.

Fromer, M., Pocklington, A.J., Kavanagh, D.H., *et al.* (2014). *De novo* mutations in schizophrenia implicate synaptic networks. *Nature*, **506**, 179–184.

Gertz, J., Varley, K.E., Reddy, T.E, *et al.* (2011). Analysis of DNA methylation in a three-generation family reveals widespread genetic influence on epigenetic regulation. *PLoS Genet.*, **7**, e1002228.

Gluckman, P.D., Seng, C.Y., Fukuoka, H., Beedle, A.S. and Hanson, M.A. (2007). Low birthweight and subsequent obesity in Japan. *Lancet*, **369**, 1081–1082.

Hackett, J.A., Sengupta, R., Zylicz, J.J., *et al.* (2013). Germline DNA demethylation dynamics and imprint erasure through 5-hydroxymethylcytosine. *Science*, **339**, 448–452.

Hidalgo, B., Irvin, M.R., Sha, J., *et al.* (2014). Epigenome-wide association study of fasting measures of glucose, insulin, and HOMA-IR in the genetics of lipid lowering drugs and diet network study. *Diabetes*, **63**, 801–807.

Horsthemke, B. (2007). Heritable germline epimutations in humans. *Nature Genet.*, **39**, 573–574.

Kaminsky, Z., Tochigi, M., Jia, P., *et al.* (2012). A multi-tissue analysis identifies HLA complex group 9 gene methylation differences in bipolar disorder. *Molec. Psych.*, **17**, 728–740.

Kerkel, K., Spadola, A., Yuan, E., *et al.* (2008). Genomic surveys by methylation-sensitive SNP analysis identify sequence-dependent allele-specific DNA methylation. *Nature Genet.*, **40**, 904–908.

Kim, Y.S., Leventhal, B.L., Koh, Y.J., *et al.* (2011). Prevalence of autism spectrum disorders in a total population sample. *Am. J. Psych.*, **168**, 904–912.

Kubota, T., Das, S., Christian, S.L., *et al.* (1997). Methylation-specific PCR simplifies imprinting analysis. *Nature Genet.*, **16**, 16–17.

Kubota, T., Wakui, K., Nakamura, T., *et al.* (2002). Proportion of the cells with functional X disomy is associated with the severity of mental retardation in mosaic ring X Turner syndrome females. *Cytogenet. Genome Res.*, **99**, 276–284.

Kubota, T., Furuumi, H., Kamoda, T., *et al.* (2004). ICF syndrome in a girl with DNA hypomethylation but without detectable DNMT3B mutation. *Am. J. Med. Genet. A*, **129A**, 290–293.

Lahiri, D.K. and Maloney, B. (2012). Gene × environment interaction by a longitudinal epigenome-wide association study (LEWAS) overcomes limitations of genome-wide association study (GWAS). *Epigenomics*, **4**, 685–699.

Lambrot, R., Xu, C., Saint-Phar, S., *et al.* (2013). Low paternal dietary folate alters the mouse sperm epigenome and is associated with negative pregnancy outcomes. *Nature Commun.*, **4**, 2889.

Langevin, S.M., Koestler, D.C., Christensen, B.C., *et al.* (2012). Peripheral blood DNA methylation profiles are indicative of head and neck squamous cell carcinoma: an epigenome-wide association study. *Epigenetics*, **7**, 291–299.

Lillycrop, K.A., Phillips, E.S., Torrens, C., *et al.* (2008). Feeding pregnant rats a protein-restricted diet persistently alters the methylation of specific cytosines in the hepatic PPAR alpha promoter of the offspring. *Br. J. Nutr.*, **100**, 278–282.

Lim, D., Bowdin, S.C. and Tee, L. (2009). Clinical and molecular genetic features of Beckwith–Wiedemann syndrome associated with assisted reproductive technologies. *Hum. Reprod.*, **24**, 741–747.

Liu, Y., Aryee, M.J., Padyukov, L., *et al.* (2013). Epigenome-wide association data implicate DNA methylation as an intermediary of genetic risk in rheumatoid arthritis. *Nature Biotechnol.*, **31**, 142–147.

McGowan, P.O., Sasaki, A., D'Alessio, A.C., *et al.* (2009). Epigenetic regulation of the glucocorticoid receptor in human brain associates with childhood abuse. *Nature Neurosci.*, **12**, 342–348.

Miyake, K., Hirasawa, T., Soutome, M., *et al.* (2011). The protocadherins, PCDHB1 and PCDH7, are regulated by MeCP2 in neuronal cells and brain tissues: implication for pathogenesis of Rett syndrome. *BMC Neurosci.*, **12**, 81.

Miyake, K., Yang, C., Minakuchi, Y., *et al.* (2013). Comparison of genomic and epigenomic expression in monozygotic twins discordant for Rett syndrome. *PLoS ONE*, **8**, e66729.

Murgatroyd, C., Patchev, A.V., Wu, Y., *et al.* (2009). Dynamic DNA methylation programs persistent adverse effects of early-life stress. *Nature Neurosci.*, **12**, 1559–1566.

Nguyen, H.N., Fujiyoshi, A., Abbott, R.D. and Miura, K. (2013). Epidemiology of cardiovascular risk factors in Asian countries. *Circul. J.*, **77**, 2851–2859.

Nolen, L.D., Gao, S., Han, Z., *et al.* (2005). X chromosome reactivation and regulation in cloned embryos. *Develop. Biol.*, **279**, 525–540.

Painter, R.C., de Rooij, S.R., Bossuyt, P.M., *et al.* (2006). Early onset of coronary artery disease after prenatal exposure to the Dutch famine. *Am. J. Clin. Nutr.*, **84**, 322–327.

Petersen, A.K., Zeilinger, S., Kastenmüller, G., *et al.* (2014). Epigenetics meets metabolomics: an epigenome-wide association study with blood serum metabolic traits. *Hum. Molec. Genet.*, **23**, 534–545.

Rakyan, V.K., Down, T.A., Balding, D.J. and Beck, S. (2011). Epigenome-wide association studies for common human diseases. *Nature Rev. Genet.*, **12**, 529–541.

Sakazume, S., Ohashi, H., Sasaki, Y., *et al.* (2012) Spread of X-chromosome inactivation into chromosome 15 is associated with Prader–Willi syndrome phenotype in a boy with a t(X;15)(p21.1;q11.2) translocation. *Hum. Genet.*, **131**, 121–130.

Sato, S., Yoshimizu, T., Sato, E. and Matsui, Y. (2003). Erasure of methylation imprinting of Igf2r during mouse primordial germ-cell development. *Molec. Reprod. Devel.*, **65**, 41–50.

Shenker, N.S., Polidoro, S., van Veldhoven, K., *et al.* (2013). Epigenome-wide association study in the European Prospective Investigation into Cancer and Nutrition (EPIC-Turin) identifies novel genetic loci associated with smoking. *Hum. Molec. Genet.*, **22**, 843–851.

St Clair, D., Xu, M., Wang, P., *et al.* (2005). Rates of adult schizophrenia following prenatal exposure to the Chinese famine of 1959–1961. *J. Am. Med. Ass.*, **294**, 557–562.

Takizawa, T., Nakashima, K., Namihira, M., *et al.* (2001). DNA methylation is a critical cell-intrinsic determinant of astrocyte differentiation in the fetal brain *Develop. Cell*, **1**, 749–758.

Tobi, E.W., Lumey, L.H., Talens, R.P., *et al.* (2009). DNA methylation differences after exposure to prenatal famine are common and timing- and sex-specific. *Hum. Molec. Genet.*, **18**, 4046–4053.

Tsai, P.C., Spector, T.D. and Bell, J.T. (2012). Using epigenome-wide association scans of DNA methylation in age-related complex human traits. *Epigenomics*, **4**, 511–526.

Verkerk, A.J., Pieretti, M., Sutcliffe, J.S., *et al.* (1991). Identification of a gene (FMR-1) containing a CGG repeat coincident with a breakpoint cluster region exhibiting length variation in fragile X syndrome. *Cell*, **65**, 905–914.

Verma, M. (2012). Epigenome-Wide Association Studies (EWAS) in cancer. *Curr. Genom.*, **13**, 308–313.

Wang, Y., Zhang, Y.L., Hennig, K., *et al.* (2013). Class I HDAC imaging using [(3)H]CI-994 autoradiography. *Epigenetics*, **8**, 756–764.

Weaver, I.C., Cervoni, N., Champagne, F.A., *et al.* (2004). Epigenetic programming by maternal behavior. *Nature Neurosci.*, **7**, 847–854.

Xu, Z., Bolick, S.C., DeRoo, L.A., *et al.* (2013). Epigenome-wide association study of breast cancer using prospectively collected sister study samples. *J. Natl Cancer Inst.*, **105**, 694–700.

Yamaguchi, S., Shen, L., Liu, Y., Sendler, D. and Zhang, Y. (2013). Role of Tet1 in erasure of genomic imprinting. *Nature*, **504**, 460–464.

Part III

Single nucleotide polymorphisms, copy number variants, haplotypes and eQTLs

10 Finding SNPs that affect microRNA regulation in disease-associated genomic regions

Laurent F. Thomas and Pål Sætrom

10.1. Introduction

MicroRNAs (miRNAs) are small, single-stranded RNAs of about 22 nucleotides involved in gene regulation by binding to 3′ untranslated regions (UTRs) of messenger RNAs (mRNAs) (Bartel, 2004). By bringing the RNA-induced silencing complex (RISC) to the mRNAs, they enable gene expression inhibition (gene silencing), either by affecting protein translation or by destabilizing mRNAs through deadenylation or decapping (Fabian *et al.*, 2010). Target mRNAs are recognized by miRNAs through Watson–Crick matching between the nucleotides two to seven of the 5′ end of miRNAs (seed sequences), and complementary sequences called seed sites in the 3′ UTR of mRNAs (Bartel, 2009). Gene silencing by miRNAs is an important mechanism in physiological processes, and its deregulation can lead to complex diseases such as cancer (Garzon *et al.*, 2006).

Complex diseases that are heritable are commonly analysed by studying genomic DNA variants, such as single nucleotide polymorphisms (SNPs), which are a change of one nucleotide that occurs in more than 1% in a population (Frazer *et al.*, 2009). A SNP can take several forms called alleles. Because recombination events between closely located SNPs are less likely than recombination between SNPs that are located far apart on a chromosome, alleles at close SNPs often co-occur, or correlate, in which case the SNPs are said to be in linkage disequilibrium (LD; Reich *et al.*, 2001; Clague *et al.*, 2010).

SNPs in the coding sequence of mRNAs have been well studied for their role in changing the amino-acid chain, as it may result in protein isoforms with affected function, leading to phenotypic differences and also disease. Nevertheless, SNPs can also occur in non-coding regions of the genome such as the 3′ UTR of mRNAs, which harbors many functional sequence elements involved in gene regulation. One type of functional element that can be disrupted by SNPs is the miRNA target site. SNPs in miRNA target sites (miRSNPs) can change the affinity between the

Genome-Wide Association Studies: From Polymorphism to Personalized Medicine, ed. K. Appasani. Published by Cambridge University Press.

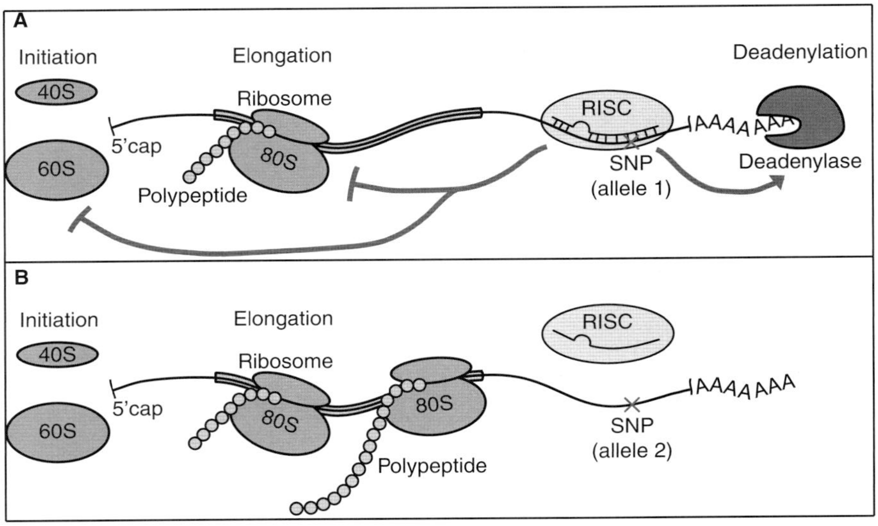

Figure 10.1. SNPs in miRNA target sites. An mRNA depicted by a coding region (in gray), its 5′ and 3′ UTRs, a 5′ cap and a polyA tail, is being translated by the ribosome (depicted by the 40S and 60S subunits and the 80S complex) into a protein (shown by the polypeptide chain in green). Panel (A) shows an mRNA with the allele 1 at a SNP located in a miRNA target site in the 3′ UTR of the mRNA. The miRNA can bind to the target site with this allele, guiding the RISC complex to the target mRNA, which results in gene downregulation by translation inhibition at initiation or elongation of the polypeptide, or by activating deadenylation of the mRNA, through deadenylase, destabilizing the transcript. Panel (B) shows an mRNA with the alternative allele at the SNP, which disrupts the affinity between the miRNA and the mRNA, resulting in the absence of miRNA silencing and a higher gene expression compared to the other allele. A black and white version of this figure will appear in some formats. For the color version, please refer to the plate section.

miRNA seed sequence and its target mRNA, resulting in deregulation of gene expression (Figure 10.1), and possibly in phenotype differences and diseases (Sethupathy and Collins, 2008).

One classic example that a miRSNP determines phenotype is the single nucleotide change in the myostatin gene (*GDF8*) of Texel sheep (Clop *et al.*, 2006). The Texel-specific allele creates a target site for miRNAs *miR-1* and *miR-206* within the myostatin 3′ UTR. As both miRNAs are highly expressed in skeletal muscle, this Texel-specific target site results in myostatin being downregulated and thereby increasing the muscle mass in Texel sheep (Clop *et al.*, 2006). Another example is the SNP rs1434536 located in the 3′ UTR of the gene Bone Morphogenic Receptor 1b (*BMPR1b*). This SNP, which is associated with breast cancer susceptibility, reduces the binding affinity of miRNA *miR-125b*, and this altered binding leads to *BMPR1b* upregulation (Saetrom *et al.*, 2009).

Further evidence for miRSNPs functional importance comes from examining SNP densities at miRNA target sites. Because the SNP-based disruption of the seed sequence of miRNA target sites may have important effects on gene regulation, seed sequences within 3′ UTRs harbor a lower SNP density due to high selection pressure (Chen and Rajewsky, 2006; Saunders *et al.*, 2007; Mu *et al.*, 2011).

Although SNPs can affect miRNA regulation by disrupting or creating miRNA target sites, 3′ UTRs have other functional elements that, if disrupted by SNPs, can also affect miRNA regulation. Specifically, SNPs within the sequence elements involved in determining polyadenylation sites may affect polyadenylation and thereby 3′ UTR length and the presence of miRNA target sites. The mRNA polyadenylation process consists of cleaving the precursor mRNA at its 3′ region and adding a poly(A) tail – a sequence of several consecutive A nucleotides. The poly (A) tail is involved in mRNA export into the cytoplasm, increase of mRNA stability in the cytoplasm, and translation efficiency (Lutz, 2008). Important sequence elements that affect polyadenylation cleavage sites are the polyadenylation signal (PAS) located 30 nucleotides upstream of the cleavage site, and the GU-rich region located 20–40 nucleotides downstream of the cleavage site (Colgan and Manley, 1997).

When a 3′ UTR harbors several PAS, the transcript can have several possible polyadenylation cleavage sites, and therefore longer or shorter 3′ UTRs (Lutz, 2008). This concept is called alternative polyadenylation (APA), and it is estimated that about 54% of human mRNAs have several polyadenylation sites (Tian *et al.*, 2005). Shorter 3′ UTRs may lose functional elements such as miRNA binding sites, which may result in gene upregulation (Di Giammartino *et al.*, 2011).

Similarly to SNPs in miRNA target sites, SNPs in the PAS or in the GU-rich region can affect the polyadenylation process (Figure 10.2). Such SNPs (called APA-SNPs) disrupt or increase affinity at binding sites of the polyadenylation machinery, resulting in different 3′ UTR lengths, with gain or loss of regulatory elements such as miRNA target sites, affected gene expression, and increased disease risk (Uitte de Willige *et al.*, 2007; Danckwardt *et al.*, 2008).

Genome-wide association studies (GWAS) can identify the association between DNA variants and traits/diseases. However, associated variants are not necessarily causative, but may co-occur with the causative variants, because of the absence of recombination events between them (variants in LD). Therefore, potentially functional SNPs such as miRSNPs or APA-SNPs that are in LD with disease-associated SNPs may be the causative variants.

This chapter presents a computational framework for identifying and predicting the regulatory effect of miRSNPs and APA-SNPs, and demonstrates how this framework can be used to identify functional candidates within genomic regions found to be associated with disease. We first explain an LD mapping method that identifies candidate functional DNA variants linked to known disease-associated variants. Second, we define two SNP variables that we use to prioritize the candidate SNPs of interest: (i) the miRSNP score, predicting the SNP effect on miRNA-based regulation, and (ii) the APA score, predicting the effect of APA-SNPs on miRNA regulation through loss of miRNA binding sites in shorter 3′ UTRs. Third, we present different results, including a comparison between our predicted miRSNP effects and measured effects from experiments, the effect of different APA variables on gene expression, and LD mapping results of the GWAS catalog mapped to predicted miRSNPs and APA-SNPs.

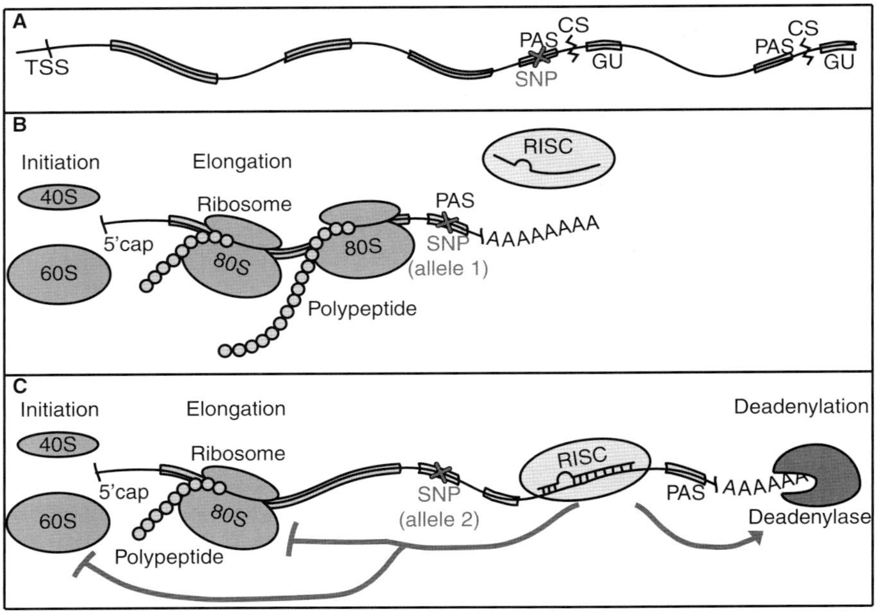

Figure 10.2. SNPs in polyadenylation signals. Panel (A) shows a gene on a DNA strand, including its transcription start site (TSS), its 5′ UTR, its coding region (in gray) divided by two introns, and its 3′ UTR containing two polyadenylation cleavage sites (CS) together with their respective PAS and GU-rich region. A SNP is located in the proximal PAS. Panel (B) shows the transcribed mRNA of the gene in panel A, with allele 1 at the SNP, which increases affinity of the polyadenylation machinery to the proximal PAS, resulting in cleavage at the proximal polyadenylation site, a shorter 3′ UTR, and loss of potential downstream miRNA binding sites. Panel (C) shows the transcript with allele 2 at the SNP, which disrupts the proximal PAS, and results in the usage of the distal one. The longer 3′ UTR contains miRNA target sites that can trigger gene downregulation by miRNAs. A black and white version of this figure will appear in some formats. For the color version, please refer to the plate section.

10.2. Methods

We present here the LD mapping method, and the different variables for prioritizing miRSNPs and APA-SNPs.

10.2.1. Linkage disequilibrium mapping

To identify miRSNPs or APA-SNPs in LD with significant SNPs from GWAS, we took LD data from human haplotype map Hapmap r27 (Frazer *et al.*, 2007), which consists of D' and r^2 values of SNP pairs that are relatively closely collocated on the same chromosomes. For SNP pairs of GWAS SNPs and functional SNPs that have LD data available and are closely collocated, we quantified the LD level between the two SNPs by computing a block score, which is based on D' values of the block between the two SNPs (Figure 10.3; Thomas *et al.*, 2011). That block score belongs to the interval [0; 2], and we filtered SNPs that have a block score greater than 1.988 (top 10%).

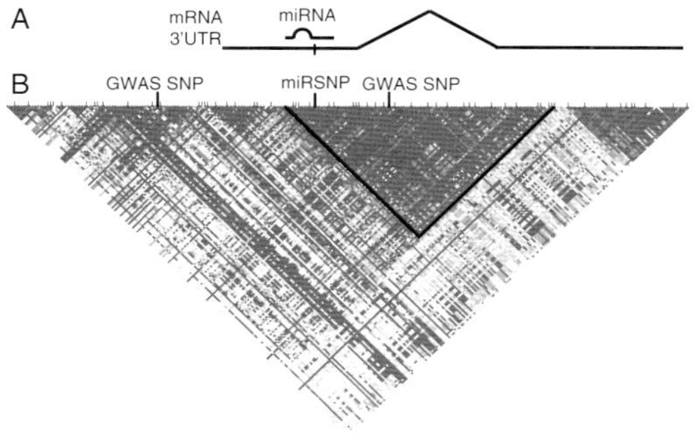

Figure 10.3. LD mapping of potentially functional SNPs to associated SNPs from GWAS. Panel (A) shows the 3′ UTR of an mRNA that contains a SNP in a miRNA target site. Panel (B) shows the LD between the different SNPs in the region. Dark means high LD while light means low LD. Among all the SNPs in the region, two GWAS SNPs and the miRSNP are represented. One GWAS SNP and the miRSNP are located in a SNP block with strong LD, shown by the dark triangle. Figure modified from Thomas *et al.*, 2011.

10.2.2. Regulatory scores for SNPs in miRNA target sites

To score the regulatory effect of miRSNPs on mRNA expression, we used our miRNA target site prediction tool (Saito and Saetrom, 2010), which is based on a machine learning technique involving a two-step classification by support vector machines (SVM). In this tool, the first step predicts whether a candidate miRNA seed site is a likely miRNA target site, whereas the second step predicts whether an mRNA that contains predicted miRNA target sites is likely to be downregulated by a given miRNA. Specifically, the first SVM step classifies each target site based on the local features of the site, such as pairing at the seed site and at the 3′ end of the miRNA, the AU context of the target site, its position in the 3′ UTR, and its distance to other neighboring target sites. For a given miRNA/mRNA pair, the second SVM step then considers the different target sites that were predicted at the first step and scores how much the given miRNA may regulate the given target mRNA. Features used in this scoring include the number and predicted strength of the target sites, the 3′ UTR length, and the number of optimally spaced target sites (Saetrom *et al.*, 2007). The global score quantifies how a miRNA regulates an mRNA, and can be used to score 3′ UTR haplotypes for miRNA regulation, by setting the different alleles of the haplotype in the 3′ UTR sequence (Thomas *et al.*, 2011).

If a 3′ UTR sequence has no predicted target sites for a given miRNA, the second SVM step provides no score. For standard miRNA target predictions, this is fine, as no score means that the mRNA is not an miRNA target. For predicting the effect of a miRSNP, however, we have to consider two scenarios: (i) either one of the SNP alleles destroys the miRNA target site, or (ii) none of the SNP alleles destroys the miRNA target site, but one of the alleles may affect the miRNA's binding affinity for the site. In both scenarios the SNP's effect can vary; for example, one SNP can destroy a weak site, thereby having little effect, whereas another SNP can alter a strong site into a weak site, thereby having a large effect. To handle both scenarios consistently, we therefore developed an approach that normalizes the score distribution for 3′ UTRs that contain target sites such that the resulting distribution also can be used to provide consistent scores for 3′ UTRs that have no predicted target sites (Thomas *et al.*, 2011). Using this approach, we compute a normalized score for each 3′ UTR haplotype sequence and compare haplotype differences and normalized score differences to identify which SNPs are responsible for the score differences (Thomas *et al.*, 2011). Finally, because an miRSNP will have a different score for different miRNAs and gene isoforms, we assigned to the SNP its maximum score difference.

10.2.3. SNPs in alternative polyadenylation

Several variables are important for evaluating the effect of APA-SNPs: the PAS motif, the GU content downstream, and the scoring of miRNA target sites lost in APA. There are 13 known polyadenylation signals (AAUAAA, AUUAAA, UAUAAA, AGUAAA, AAGAAA, AAUAUA, AAUACA, CAUAAA, GAUAAA, AAUGAA, UUUAAA, ACUAAA, and AAUAGA), listed here from most to least frequent (Tian *et al.*, 2005). Assuming that the most frequent signals are the strongest in terms of polyadenylation efficiency, the motifs' frequency ranks can be used to evaluate PAS strength. Here, we defined the six most frequent PAS motifs as strong and the seven less frequent as weak.

For scoring GU content, we compute the proportion of G and U in five nucleotide windows downstream of the PAS. The GU score is then computed as the mean GU proportion for windows sliding from the 25th to the 45th window downstream of the PAS.

Analogous to scoring of miRSNP haplotypes, we compute an miRNA score that predicts the effect of APA-SNPs on miRNA regulation by using the two-step SVM tool for miRNA target prediction to score the short and the long 3′ UTR. Similarly to miRSNP scores, miRNA scores for APA-SNPs are normalized to handle cases where the miRNA target site is lost in the short 3′ UTR. Scores for all miRNAs can be combined by weighting them by the proportion of miRNA expression, if available from analyses in particular tissues, or by taking the average score between all miRNAs. The latter approach is relevant for general analyses and for cases where data on tissue-specific miRNA expression are unavailable. After combining miRNA scores for the short and the long 3′ UTRs separately, the score difference between the short and long UTRs gives the APA-SNP's effect on miRNA-based regulation. Specifically, the score difference predicts the effect on miRNA regulation from the loss of miRNA target sites in the shorter UTR version.

10.3. Results

We tested our different miRSNP and APA-SNP variables. Specifically, we found that the predicted miRSNP effects fit measured allelic imbalance, and that APA-SNPs with strong signal, high GU-level downstream, and high miRNA score (loss of miRNA sites in the shorter UTR) gave higher gene expression. Finally, we mapped our miRSNPs and APA-SNPs to trait-/disease-associated SNPs from published GWAS to identify potentially causative SNPs.

10.3.1. Predicted effects of heterozygous miRSNPs fit their measured allelic imbalance

To assess the validity of our scores for predicted miRSNP effects, we compared them with experimental data based on allelic imbalance (AI), which is the ratio of allelic expression at heterozygous SNPs and gives an estimate of the difference of regulation between alleles. Specifically, we used mice allelic imbalance sequencing data of 47 miRSNPs for three miRNAs (Kim and Bartel, 2009). Importantly, the allelic expression levels were measured both in tissues that do and do not express the specific miRNA. Allelic ratios (AR) were defined as the expression of transcripts with the target alleles over those with the non-target alleles. The difference between the log AR in tissues expressing the targeting miRNA (AR_E) and the log AR in tissues that do not express the miRNA (AR_{NE}), $\Delta AR = \log_2(AR_{NE}) - \log_2(AR_E)$, then gave the measured SNP effect on transcript expression. Similarly, the predicted SNP effect was given by the score difference $\Delta S = S_T - S_{NT}$, where S_T and S_{NT} are the scores for the target allele (S_T) and for the non-target allele (S_{NT}). Comparing our predicted SNP effect with the experimental measurements gave a significant correlation (Pearson's correlation: $r = 0.431$, $p = 0.0025$; Spearman's rank correlation: $\rho = 0.524$, $p = 0.00019$; Figure 10.4A).

Alternative scoring strategies, such as changes in minimum free energy or TargetScan scores, did not fit the allelic imbalance data as well as our scores (Figure 10.4B,C). Minimum free energy (MFE) is often used to assess RNA binding and can be computed by RNAHybrid 2.1 (Rehmsmeier *et al.*, 2004). However, comparing the AI measurements with differences of MFE scores gave an insignificant correlation. Comparing the allelic imbalance measurements with context scores from the target site prediction tool TargetScan 5.0 (Lewis *et al.*, 2005) gave a significant but lower correlation than for our scoring strategy. Consequently, our miRSNPs scores fit AI data well and can therefore be used to assess miRSNP effects.

10.3.2. Effects of APA-SNPs in RNA-seq data

Based on data from the HapMap project (Frazer *et al.*, 2007), we identified 412 SNPs possibly creating polyA signals (APA-SNPs) in the 3′ UTRs of human coding genes (Thomas and Saetrom, 2012). We then mapped RNA-seq data from 19 samples of diverse tissue types from the Burge Lab (Wang *et al.*, 2008) to these 412 loci, and computed the AI at the heterozygous APA-SNPs, defined as the log ratio of the APA allele read count over read for the other allele (log AR). We found that the PAS rank (strength) negatively correlated with log AR, so that top signal ranks (strong PAS) had overexpressed APA alleles compared to the other allele ($r = -0.144$, $p = 0.013$, $n = 300$). Furthermore, considering the PAS that have a GU-rich region downstream also gave a significant correlation ($r = -0.195$, $p = 0.032$, $n = 122$), while those without GU-rich regions were not significant ($p = 0.17$).

A linear regression of our three-variable based model showed that the signal rank, the GU-level, and the miRNA score could explain 12.39% of the log AR

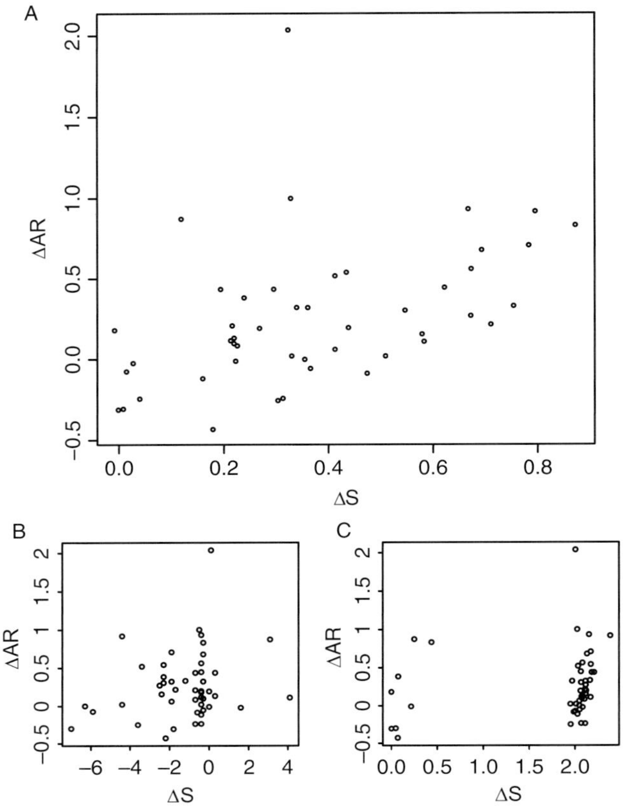

Figure 10.4. Correlations between measured and predicted miRSNP effects. The measured allelic ratio difference (ΔAR) and the predicted score difference (ΔS) for (A) our SVM-based miRSNP scores, (B) MFE differences, and (C) TargetScan context score differences. Figure from Thomas *et al.*, 2011.

variance ($p = 0.04$). These results indicate that APA-SNPs creating strong signals with high GU level and loss of miRNA target sites increase the expression of the APA allele compared to the other allele.

We further grouped SNPs by signal strength (strong/weak), GU levels (high/low), miRNA scores (high/low), and by log *AR* sign (positive/negative), and computed the median SNP expression (read count at the SNP position) in each group. Median SNP expression differences between positive and negative log *AR* were computed for each subgroup, together with 95% confidence intervals (CI) of the median differences, based on bootstrapping (Figure 10.5). Only one group had a median difference significantly different from 0: the SNP group with strong signals, high GU level, and high miRNA scores (loss of miRNA target sites). In other words, the group of SNPs that likely resulted in an APA site that had a strong effect on miRNA regulation had a significantly higher expression of the APA allele, suggesting that increased expression is due to the APA allele and the APA site use.

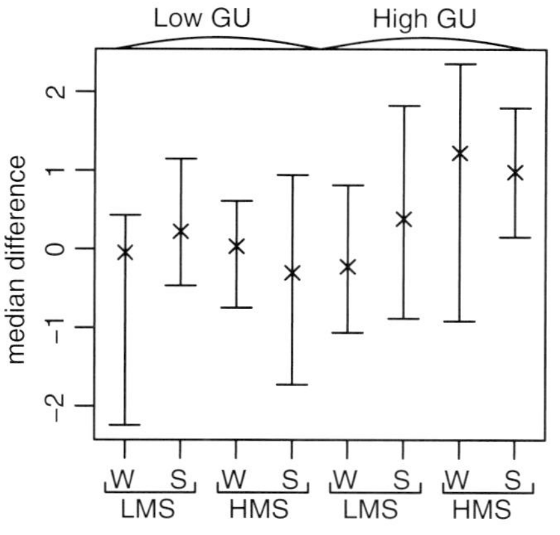

Figure 10.5. Median SNP expression differences between positive and negative log *AR*. SNP expressions are grouped by strong (S) and weak (W) signals, high and low GU level, high and low miRNA score, and the log *AR* sign. Median differences between positive and negative log *AR* for each subgroup together with 95% CI of median differences, shows that only SNPs with strong signal, high GU level, and high miRNA scores (loss of miRNA target sites), have a significantly higher SNP expression for SNPs with a higher APA allele proportion compared to those with higher proportion of the other allele. This suggests that the increased expression comes from the APA allele, when there are strong sites and loss of miRNA target sites. Figure from Thomas *et al.*, 2012.

10.3.3. Analysis of GWAS data

To identify associations between functional miRSNPs or APA-SNPs and disease, we used European ancestry LD data from Hapmap (CEU) (Frazer *et al.*, 2007) to map miRSNPs and APA-SNPs to GWAS SNPs from the NHGRI GWAS catalog (Hindorff *et al.*, 2009)(www.genome.gov/gwastudies; accessed Sept. 19, 2013) – particularly to the 7810 associated SNPs from GWAS based on European ancestry-related samples. For this analysis, we used the 24,620 miRSNPs with maximum score difference greater than 0.15, the 22,606 miRSNPs of conserved miRNAs, and the 412 APA-SNPs. We looked at functional-SNP/GWAS-SNP pairs that have a non-null minimum allele frequency (MAF) and filtered the results through different variables, such as the miRSNP score (greater than 0.45), and APA variables (strong signal rank (top six ranks), GU level greater than 0.55, and miRNA score greater than the median 0.0725). The miRSNP scores and the miRNA scores for APA-SNPs were computed based on miRNAs from miRBase v20 (Griffiths-Jones *et al.*, 2006). We found that there are about 102 miRSNPs with a score greater than 0.45 that overlap with trait- or disease-associated SNPs from CEU-based GWA studies and that about 1088 miRSNPs are in strong LD blocks with 682 GWAS SNPs (Table 10.1, Figure 10.6).

From the 79 miRSNPs from conserved miRNAs that are also found in the GWAS catalog of CEU-related studies, seven miRSNPs were disrupting an 8mer site and had a miRSNP score greater than 0.9. Two SNPs were associated with metabolic traits: the SNP rs15285 associated with triglycerides–blood pressure (Kraja *et al.*, 2011) disrupted the target site of miR-571 on the *LPL* gene and the SNP rs8396 associated with metabolite level (Illig *et al.*, 2010; Suhre *et al.*, 2011) disrupted an miR-376a-3p site on gene *PPID*. Two SNPs were associated

Table 10.1. Counts of miRSNPs and their corresponding GWAS SNPs.

		SNP count				
SNPs	miRNAs	All *BS*	*BS* ≥ 1.9	*BS* ≥ 1.988	*BS* = 2	Same SNP
miRSNPs	All	7754	2498	1088	833	102
	miR Fam	5912	1951	861	665	79
GWAS SNPs	All	3788	1555	682	522	102
	miR Fam	3593	1408	584	448	79

The columns show the SNP type (miRSNP or GWAS SNP), the set of miRNAs (all miRNAs or conserved miRNAs from miRNA families), and SNP counts with different block score (*BS*) filtering. All*BS*shows the SNP count without block score filtering. The Same SNP column shows the number of SNP that are both miRSNPs and GWAS SNPs.

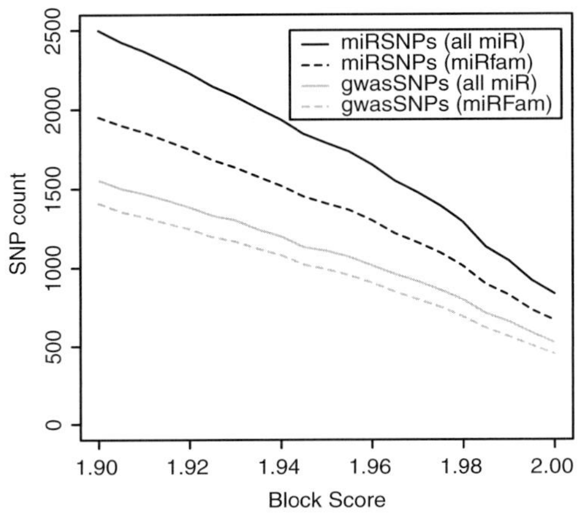

Figure 10.6. SNP count of miRSNPs and GWAS SNPs with different LD block score filtering. The *x*-axis shows the LD block score thresholds (greater than) used to count SNPs (inverse cumulative distribution) for strong LD block scores (from block scores 1.9 to 2.0, where 2.0 is the maximum). Black and gray lines show, respectively, the miRSNP and GWAS SNP counts that are in LD (block score greater than the threshold). Full and dashed lines show, respectively, the SNP counts computed from all miRNAs and from conserved miRNAs (miRNA families).

with hematological phenotypes: the SNP rs9253 associated with red blood cell count (Yang *et al.*, 2007) disrupted an miR-634 site on gene *MEAF6*, and the SNP rs4132601 associated with lymphoblastic leukemia (Papaemmanuil *et al.*, 2009) disrupted an miR-525-3p site on gene *IKZF1*. The SNP rs8040009 associated with alcoholism (Heath *et al.*, 2011) disrupted an miR-586 site on gene *C15orf32*. Two SNPs were associated with neurodegenerative diseases: the SNP rs2275336 associated with Parkinson's disease (Pankratz *et al.*, 2012) created an miR-524-5p site on gene *CNKSR3* and the SNP rs2275336 associated with Alzheimer's disease (Hollingworth *et al.*, 2011) affected an miR-6876-3p site on gene *MS4A6A*.

Five APA-SNPs had strong signals and miRNA score, and mapped to GWAS SNPs with high block score. Two of those APA-SNPs, rs3737977 and rs3731389 on genes *TGFB2* and *CDK6*, were in LD with height-associated SNPs. Two other APA-SNPs rs15563 and rs9261533 on genes *UBE2Z* and *TRIM10*, were in LD with cardiac-related GWAS SNPs, and the last APA-SNP rs3776448 on gene *NSUN2* was in LD with a visceral fat-associated SNP.

10.4. Conclusion

Genome-wide association studies can identify trait- or disease-associated SNPs, but the SNPs identified by such studies are usually not the causative ones. Whereas it is relatively easy to identify SNPs affecting the coding sequence and the final protein product, it is more challenging to identify SNPs affecting gene regulation, such as miRSNPs and APA-SNPs. We have presented a methodology for identifying such regulatory SNPs and mapped them to published trait- and disease-associated SNPs. Our methodology can be used to prioritize regulatory SNPs for subsequent functional validation.

REFERENCES

Bartel, D.P. (2004). MicroRNAs: genomics, biogenesis, mechanism, and function. *Cell*, **116**, 281–297.

Bartel, D.P. (2009). MicroRNAs: target recognition and regulatory functions. *Cell*, **136**, 215–233.

Chen, K. and Rajewsky, N. (2006). Natural selection on human microRNA binding sites inferred from SNP data. *Nature Genet.*, **38**, 1452–1456.

Clague, J., Lippman, S.M., Yang, H., *et al.* (2010). Genetic variation in microRNA genes and risk of oral premalignant lesions. *Molec. Carcinog.*, **49**, 183–189.

Clop, A., Marcq, F., Takeda, H., *et al.* (2006). A mutation creating a potential illegitimate microRNA target site in the myostatin gene affects muscularity in sheep. *Nature Genet.*, **38**, 813–818.

Colgan, D.F. and Manley, J.L. (1997). Mechanism and regulation of mRNA polyadenylation. *Genes Develop.*, **11**, 2755–2766.

Danckwardt, S., Hentze, M.W. and Kulozik, A.E. (2008). 3′ end mRNA processing: molecular mechanisms and implications for health and disease. *EMBO J.*, **27**, 482–498.

Di Giammartino, D.C., Nishida, K. and Manley, J.L. (2011). Mechanisms and consequences of alternative polyadenylation. *Molec. Cell.*, **43**, 853–866.

Fabian, M.R., Sonenberg, N. and Filipowicz, W. (2010). Regulation of mRNA translation and stability by microRNAs. *Annu. Rev. Biochem.*, **79**, 351–379.

Frazer, K.A., Ballinger, D.G., Cox, D.R., *et al.* (2007). A second generation human haplotype map of over 3.1 million SNPs. *Nature*, **449**, 851–861.

Frazer, K.A., Murray, S.S., Schork, N.J. and Topol, E.J. (2009). Human genetic variation and its contribution to complex traits. *Nature Rev. Genet.*, **10**, 241–251.

Garzon, R., Fabbri, M., Cimmino, A., Calin, G.A. and Croce, C.M. (2006). MicroRNA expression and function in cancer. *Trends Molec. Med.*, **12**, 580–587.

Griffiths-Jones, S., Grocock, R.J., van Dongen, S., Bateman, A. and Enright, A.J. (2006). miRBase: microRNA sequences, targets and gene nomenclature. *Nucleic Acids Res.*, **34**, D140–144.

Heath, A.C., Whitfield, J.B., Martin, N.G., *et al.* (2011). A quantitative-trait genome-wide association study of alcoholism risk in the community: findings and implications. *Biol. Psych.*, **70**, 513–518.

Hindorff, L.A., Sethupathy, P., Junkins, H.A., *et al.* (2009). Potential etiologic and functional implications of genome-wide association loci for human diseases and traits. *Proc. Natl Acad. Sci. USA*, **106**, 9362–9367.

Hollingworth, P., Harold, D., Sims, R., *et al.* (2011). Common variants at *ABCA7, MS4A6A/MS4A4E, EPHA1, CD33* and *CD2AP* are associated with Alzheimer's disease. *Nature Genet.*, **43**, 429–435.

Illig, T., Gieger, C., Zhai, G., *et al.* (2010). A genome-wide perspective of genetic variation in human metabolism. *Nature Genet.*, **42**, 137–141.

Kim, J. and Bartel, D.P. (2009). Allelic imbalance sequencing reveals that single-nucleotide polymorphisms frequently alter microRNA-directed repression. *Nature Biotechnol.*, **27**, 472–477.

Kraja, A.T., Vaidya, D., Pankow, J.S., *et al.* (2011). A bivariate genome-wide approach to metabolic syndrome: STAMPEED consortium. *Diabetes*, **60**, 1329–1339.

Lewis, B.P., Burge, C.B. and Bartel, D.P. (2005). Conserved seed pairing, often flanked by adenosines, indicates that thousands of human genes are microRNA targets. *Cell*, **120**, 15–20.

Lutz, C.S. (2008). Alternative polyadenylation: a twist on mRNA 3′ end formation. *ACS Chem. Biol.*, **3**, 609–617.

Mu, X.J., Lu, Z.J., Kong, Y., Lam, H.Y. and Gerstein, M.B. (2011). Analysis of genomic variation in non-coding elements using population-scale sequencing data from the 1000 Genomes Project. *Nucleic Acids Res.*, **39**, 7058–7076.

Pankratz, N., Beecham, G.W., DeStefano, A.L., *et al.* (2012). Meta-analysis of Parkinson's disease: identification of a novel locus, RIT2. *Ann. Neurol.*, **71**, 370–384.

Papaemmanuil, E., Hosking, F.J., Vijayakrishnan, J., *et al.* (2009). Loci on 7p12.2, 10q21.2 and 14q11.2 are associated with risk of childhood acute lymphoblastic leukemia. *Nature Genet.*, **41**, 1006–1010.

Rehmsmeier, M., Steffen, P., Hochsmann, M. and Giegerich, R. (2004). Fast and effective prediction of microRNA/target duplexes. *RNA*, **10**, 1507–1517.

Reich, D.E., Cargill, M., Bolk, S., *et al.* (2001). Linkage disequilibrium in the human genome. *Nature*, **411**, 199–204.

Saetrom, P., Heale, B.S., Snove, O., Jr., *et al.* (2007). Distance constraints between microRNA target sites dictate efficacy and cooperativity. *Nucleic Acids Res.*, **35**, 2333–2342.

Saetrom, P., Biesinger, J., Li, S.M., *et al.* (2009). A risk variant in an miR-125b binding site in BMPR1B is associated with breast cancer pathogenesis. *Cancer Res.*, **69**, 7459–7465.

Saito, T. and Saetrom, P. (2010). A two-step site and mRNA-level model for predicting microRNA targets. *BMC Bioinform.*, **11**, 612.

Saunders, M.A., Liang, H. and Li, W.H. (2007). Human polymorphism at microRNAs and microRNA target sites. *Proc. Natl Acad. Sci. USA*, **104**, 3300–3305.

Sethupathy, P. and Collins, F.S. (2008). MicroRNA target site polymorphisms and human disease. *Trends Genet*, **24**, 489–497.

Suhre, K., Shin, S.Y., Petersen, A.K., *et al.* (2011). Human metabolic individuality in biomedical and pharmaceutical research. *Nature*, **477**, 54–60.

Thomas, L.F. and Saetrom, P. (2012). Single nucleotide polymorphisms can create alternative polyadenylation signals and affect gene expression through loss of microRNA-regulation. *PLoS Comput. Biol.*, **8**, e1002621.

Thomas, L.F., Saito, T. and Saetrom, P. (2011). Inferring causative variants in microRNA target sites. *Nucleic Acids Res.*, **39**, e109.

Tian, B., Hu, J., Zhang, H. and Lutz, C.S. (2005). A large-scale analysis of mRNA polyadenylation of human and mouse genes. *Nucleic Acids Res.*, **33**, 201–212.

Uitte de Willige, S., Rietveld, I.M., De Visser, M.C., Vos, H.L. and Bertina, R.M. (2007). Polymorphism 10034C>T is located in a region regulating polyadenylation of FGG transcripts and influences the fibrinogen gamma′/gammaA mRNA ratio. *J. Thromb. Haemost.*, **5**, 1243–1249.

Wang, E.T., Sandberg, R., Luo, S., *et al.* (2008). Alternative isoform regulation in human tissue transcriptomes. *Nature*, **456**, 470–476.

Yang, Q., Kathiresan, S., Lin, J.P., Tofler, G.H. and O'Donnell, C.J. (2007). Genome-wide association and linkage analyses of hemostatic factors and hematological phenotypes in the Framingham Heart Study. *BMC Med. Genet.*, **8**(Suppl 1), S12

11 From linkage to complex associations: the role of *GABRA2* as a risk factor for alcohol use

Sandra M. Villafuerte, Elisa M. Trucco and Margit Burmeister

11.1. Introduction

Alcohol use disorder (AUD) is partially under genetic control due to common variants in several genes, each of small effect. Most relevant to AUD, in addition to genetic risk, environmental exposures play important roles in the development of AUD. It is increasingly recognized that genetic and environmental factors depend on each other, suggesting that the expression of genetic liability depends on environmental factors. Furthermore, risk factors known as intermediate phenotypes are under the same gene–environment influence and may provide important clues to further understand this heterogeneous disorder. With this in mind, we review more than a decade of interdisciplinary research on the role of the gamma-aminobutyric acid-Aα2 (*GABRA2*) gene on the risk of developing AUD. The initial evidence came from wide genome studies in alcoholic families, followed by electrophysiological measures and replication studies in case-control designs. Going further, the study of *GABRA2* extended to risk factors (i.e., impulse control) imaging and environment during development. This plethora of evidence and interdisciplinary research emphasizes the importance of converging different lines of investigation to understand the interplay between genetic risk (or resilience) and environmental factors in the development of AUD.

11.2. *GABRA2* structure and function

The gamma-aminobutyric acid (GABA) system is the major inhibitory system in the central nervous system (CNS) with a major role in homeostatic neuronal plasticity processes. Voltage-dependent conductance and fine-tuning of synaptic strength are two mechanisms by which plasticity operates in neurons (Onori *et al.*, 2010).

Genome-Wide Association Studies: From Polymorphism to Personalized Medicine, ed. K. Appasani. Published by Cambridge University Press.

GABA type A (GABA-A) receptors can be found in the GABA synapse or periphery (mediate tonic inhibition). Synaptic receptors mediate a phasic, high-chloride conductance GABA response and respond to benzodiazepine agonist stimulation. GABA-A receptors are pentamers with combinations of seven different subunit classes. Typically, *GABRA2* is found in the synaptic receptors. Sensitivity to the ligand GABA depends on the α subunit present in receptors that contain α, β, and γ subunits. The order of sensitivity is $\alpha6<\alpha1<\alpha2<\alpha4<\alpha5<<\alpha3$ (Reich *et al.*, 1998).

GABRA2 mRNA shows different splicing forms and alternative promoter use resulting in many isoforms present in human brain (Tian *et al.*, 2005). Genetic differences may account for this complexity with different complex patterns of expression in brain regions.

The GABA receptor encoded by the genes on chromosome 4p13-q11 (α_2, α_4, β_1, γ_1) is mainly expressed in the reward pathway. In the adult rat, the mRNA from these genes are highly expressed in the hippocampus, dopamine neurons in the substantia nigra, and the ventral tegmental area (VTA), while in the rat embryo these genes were ubiquitously expressed (Steiger and Russek, 2004), indicating they undergo changes in expression. Interestingly, environmental manipulations such as maternal separation have lasting effects on stress-related behavior on GABA receptors in adult animals (Hsu *et al.*, 2003). The fast synaptic inhibition role of GABA receptors may mediate this sensitivity to the environment. These data suggest that these genes may be involved in addiction influenced by environmental exposure and by drugs with different effects. *GABRA2* in particular has binding sites for alcohol and benzodiazepines (anxiolytics).

11.3. Linkage and genome-wide association studies

Initial genetic efforts to identify susceptibility genes for AUD included linkage studies in families with several affected members with the intent to increase the power to detect chromosomal regions with genes involved in AUD. Among several susceptibility chromosomal regions, a locus on chromosome 4 containing a cluster of GABA-related genes was associated with both protection as well as risk for alcohol dependence (AD) in a large sample of families primarily of European-American descent (Reich *et al.*, 1998). In an independent sample of sib-pairs of Southwestern American Indian tribe, suggestive linkage was observed near one of the GABA genes in the same cluster on chromosome 4 (Long *et al.*, 1998). These studies helped narrow down regions of interest.

AUD is a clinical outcome representing a range of symptoms and level of severity with different combinations of risk factors, including genetic and environmental factors. Moreover, individuals with AUD may differ on their risk and level of severity, making this a heterogeneous condition. One strategy is to focus on the genetics of risk factors or intermediate phenotypes, also known as endophenotypes. Endophenotypes are biologically more stable manifestations of risk factors with a clear genetic connection compared to the actual diagnostic category. Diagnostic categories may have weaker direct genetic effects compared to endophenotypes because they are impacted by other factors, including environmental exposures. One such endophenotype is brain oscillations measured by electroencephalography (EEG) that measure changes in the membrane potential of neurons. It has been suggested that neural disinhibition is a feature of the manifestation of AUD (Begleiter and Porjesz, 1999). Specifically, beta oscillations (B) involve GABA-A receptor action (Gaetz *et al.*, 2011), making them good indicators of synaptic inhibition function mediated by GABA-A receptors. Increased power in the beta (B) frequency band (13–28 Hz) of resting EEG has been observed in the cortex in AUD subjects (Rangaswamy *et al.*, 2002) and their offspring (Rangaswamy *et al.*, 2004) compared to controls. In cortical networks, GABAergic transmission is critical for the maintenance of the excitation–inhibition homeostatic balance (Le Roux *et al.*, 2008). Linkage analysis of EEG-B showed linkage in a region of chromosome 4p spanning a cluster of GABA-related genes (Porjesz *et al.*, 2002). A closer examination narrowed down the region associated with both beta oscillations and alcohol dependence to a region of strong linkage disequilibrium (LD) across the *GABRA2* gene (Edenberg *et al.*, 2004).

Additional support for *GABRA2* as a susceptibility gene involved in AUD comes from a study that converged genome-wide association studies (GWAS) data and hypothesis-driven candidate genes (Olfson and Bierut, 2012). Only one variant from a candidate gene, *GABRA2* SNP rs279858, showed modest association in a large sample from the Study of Addiction: Genetics and Environment (SAGE), a GWAS comparing alcohol-dependent and non-dependent subjects. Validation from different independent family-based genome-wide studies was further confirmed by several association studies as discussed below.

11.4. *GABRA2* and alcohol use-related studies

The initial findings from linkage studies led to several association analyses of the *GABRA2* gene and several different alcohol use outcomes (Table 11.1). *GABRA2* SNPs within the region downstream from intron 3 are in high linkage disequilibrium (LD), resulting in two major yin–yang haplotypes. Depending on study design (e.g., different outcomes, ethnicity, gender) results have implicated either haplotype. Using the same sample as the Edenberg *et al.* (2004) study, Agrawal and colleagues reported an association between both haplotypes and alcoholics with comorbid illicit drug dependence with the major haplotype showing stronger association (Agrawal *et al.*, 2006). The major haplotype was also associated with alcohol dependence and withdrawal symptoms in a German sample of treatment-seeking alcohol-dependent individuals (79% male; Soyka *et al.*, 2008). Family-based association studies in a large Australian sample also demonstrated an association between the major haplotype and quantitative proxy measures of AUD (e.g., neuroticism), but no association was confirmed with dichotomous measures of AUD (Lind *et al.*, 2008).

By contrast, several case-control studies in European-American adults reported associations with alleles from the less-frequent of the two major haplotypes as being more abundant in individuals with AUD. In an effort to narrow down the heterogeneity of the phenotype, Covault and colleagues (2004) found a strong

Table 11.1. *GABRA2* common haplotypes and alleles from SNPS in linkage disequilibrium associated with AUD.

Study/SNPs	rs534459	rs279871	rs279869	rs279858	rs279844	rs279845	rs279826	rs279836	rs279837	rs9291283
Haplotype 1	C	A	C	A	A	T	A	T	A	G
Haplotype 2	T	G	A	G	T	A	G	A	G	A
Covault	T		A	G					G	
Lappalainen			A	G						
Fehr		G		G		T		T		
Enoch				G						
Pierucci-Lagha				G						
Bauer				G						
Li-meta analysis				G						
Edenberg		A				T		A		
Soyka	C		C	A	T				T	G
Agrawal		A		A			A			
Cells in gray depict the studies showing association of haplotype 2 with AUD.										
Tecumseh	A	(Impulsiveness)		A	(Impulsiveness)					
	G	(Angry host., Extra.)		G	(Extraversion)					
MLS	G	(Neurot., Impuls.)		G	(Impulsiveness)					

association between the less-frequent *GABRA2* haplotype among adults with AUD but no comorbid drug dependence or depression. These findings are consistent with another study demonstrating that, in a small sample of Russians, those with the minor haplotype had higher rates of alcohol dependence and admittance for detoxification compared to a control sample (Lappalainen *et al.*, 2005). In Germans, the minor haplotype was associated with alcohol dependence and was strongest among those with a family history of AUD, withdrawal seizures, and early onset of alcohol use (Fehr *et al.*, 2006). In a small sample of social drinkers, those with the minor allele of a tag SNP for this haplotype, rs279858, need to drink more alcohol than those with the major allele A to achieve similar stimulation, and were more likely to drink if alcohol was available (Pierucci-Lagha *et al.*, 2005). Along these lines, subjects with AUD seeking psychosocial treatment revealed that those with the minor G allele of rs279858 had a higher tendency for heavy drinking during the treatment period. Those with the G allele show no differences in response to the different treatments (cognitive behavioral therapy (CBT), motivational enhancement therapy (MET), or twelve-step facilitation (TSF)) compared to those with the A allele who drink less and show differences in outcomes to three different psychosocial treatments (Bauer *et al.*, 2007).

The apparent contradictory results regarding which *GABRA2* haplotype is most strongly related to AUD reveals the complexity of these associations. Additional evidence for this dual finding comes from a study in two different populations of subjects with AUD, Finnish Caucasians and Plains American Indians. Interestingly, those with AUD and high trait anxiety had the highest frequency of the major haplotype, while those with low anxiety had the highest frequency of the minor haplotype; controls had intermediate frequencies (Enoch *et al.*, 2006). It may be that different pathways involving anxiety may be underlying these differences, although race and gender are not completely ruled out as alternative explanations. For example, sex differences may be explained by the effect of neurosteroids (e.g., progesterone) on the GABA system modulating an array of behaviors including anxiety (Engel and Grant, 2001) and therefore the susceptibility to develop AUD.

The *GABRA2* haplotype block extends into the adjacent gene, GABA-A receptor gamma 1 (*GABRG1*). In an attempt to understand the contribution of *GABRG1* gene SNPs to alcohol dependence, Covault *et al.* (2008) found a stronger association in *GABRG1* SNPs in a haplotype that is in moderate LD with the *GABRA2* haplotype implicated in all the studies described above. Drgon *et al.* (2006) reported moderate associations with *GABRA2* SNPs in three small groups of poly-substance abusers and control samples, two from European-American and one of African-American descent. Linkage and family-based association analyses in a small collection of families with multiple members diagnosed with AD and some with other drug use found no association with *GABRA2* (Matthews *et al.*, 2007). No evidence for an association was reported in an Italian case-control study where cases were both alcohol-dependents and -abusers (Onori *et al.*, 2010). The first *GABRA2* meta-analysis study provided evidence of an association with AUD for rs567926 and marginal significance for the G allele of rs279858, a SNP in exon

5, although with a high level of heterogeneity for the latter (Zintzaras, 2012). A recent meta-analysis confirmed the association of these two *GABRA2* SNPs, rs567926 and rs279858, with AUD with samples from all ethnicities and from European-Americans, respectively (Li *et al.*, 2014). Rather than being discouraged by the divergent results from these studies, this is an indication of the complexity and heterogeneity of the disorder with different gene pathways operating under different conditions (e.g., environmental influences, gender, race, severity of the disorder).

Risk for AUD starts early in adolescence and investigating the genetic susceptibility of risk factors including behavior, drinking patterns, and age of onset will provide clues as to how genes shape this risk. A systematic investigation of all genes in the GABA system in a large population-based sample of young individuals (adolescents and young adulthood) did not reveal any association with any of these genes (Irons *et al.*, 2014). Previous efforts have not revealed associations with AUD in younger populations (Dick *et al.*, 2006; Sakai *et al.*, 2010). Together, these studies suggest that heterogeneity from gender, age, population stratification, phenotypes, power to detect effects, and method of analysis are important factors to narrow down the contribution of *GABRA2* to the development of AUD. Furthermore, these divergent results might indicate that *GABRA2* is associated with more severe aspects of alcohol use, such as heavy drinking evident in older and more genetically determined subjects (i.e., family history). Nevertheless, it is clear that more refined studies of *GABRA2*, which focus mainly on factors such as intermediate phenotypes (e.g., behavior, personality) and risk behavior during adolescence that underlie the risk to develop AUD are warranted.

11.5. *GABRA2* and dimensions of impulsivity as risk factor associated with AUD

Problems associated with alcohol use are related to undercontrol-disinhibition behavior (Zucker *et al.*, 2011) largely shaped during development and influenced by social environmental factors (e.g., stress, family cohesiveness, peer influences) and genetic liability. Alcohol dependence overlaps with other drug dependence, antisocial behavior, and childhood conduct disorder due in part to shared genetic factors (Young *et al.*, 2000; Krueger *et al.*, 2002; Kendler *et al.*, 2003). Impaired impulse control, a feature of the undercontrol-disinhibition behavior, is shared among these traits. Impulsivity is a multidimensional behavioral construct with at least four different constructs, each with its own mechanism. The tendency to act rashly under extremely positive or negative mood is based on emotions. Lack of planning and lack of perseverance are deficits of conscientiousness, and the tendency to seek out thrilling stimulation is known as sensation-seeking or extraversion. Various dimensions of impulsivity are related with different stages of AUD risk, including initiation, dependence, and relapse as evidenced by studies in animals, behavioral and self-report measures, and longitudinal studies in humans at different developmental stages (Lejuez *et al.*, 2010). For example, young adult (18-year-old) males with high scores of impulsivity were more likely to increase alcohol consumption and be involved in drunk driving compared to those who drink but have low scores of impulsivity. In this study, impulsivity corresponds to behavior of acting on the spur of the moment with no regard of the consequences (Moan *et al.*, 2013). Genetic variation in *GABRA2* has been associated with a myriad of traits related to impulse control at different stages of development and adulthood. In a sample of children and adolescents (7–17 years old), *GABRA2* was associated with conduct disorder symptoms but not with childhood alcohol-dependence symptoms (Dick *et al.*, 2006). It appears that in adolescence *GABRA2* is associated with subclinical self-reports of externalizing behavior (Zuckerman's Sensation Seeking and NEO Extraversion), but no association was observed with these measures in young adults (Dick *et al.*, 2013).

In a high-risk sample of young adults and adults with AUD, subjects with the high-risk haplotype had a higher probability of endorsing alcohol abuse symptoms, higher scores of the NEO-PI-R Impulsiveness facet, and greater insula activation during anticipation of reward and loss (Villafuerte *et al.*, 2012). Insula activation was also correlated with NEO-PI-R Impulsiveness scores. Furthermore, these effects were stronger in females than in males which were not due to lack of power, as more male participants than females were analyzed (Villafuerte *et al.*, 2012). The insula activation finding provides insights into the mechanism by which these processes are facilitated, integrating emotional and homeostatic information to and from limbic and cortical areas, where the insula has been implicated in negative emotional states such as craving and anxiety. The association of *GABRA2* in this sample with alcohol abuse symptoms was weak compared to another measure that emphasizes problems or difficulties related to alcohol, the Lifetime Alcohol Problems Score (LAPS; Zucker, 1991; Zucker *et al.*, 1997). *GABRA2* SNPs were associated with LAPS and this association was mediated by impulsiveness. On average, 22.6% of the effect of *GABRA2* on LAPS is accounted for by impulsivity (Villafuerte *et al.*, 2013). Interestingly, a plausible explanation for the association being stronger with alcohol-related problems was supported by a study in college students that impulsiveness predicts alcohol problems but not alcohol use or symptoms (Magid and Colder, 2007). The convergence of these findings suggests that variation in *GABRA2* contributes to the risk of AUD (mainly alcohol-related problems) through the influence on impulsive behaviors specific to behavior involving lack of control under distress.

11.6. Modulating effects of *GABRA2* SNPs on the response to alcohol

Sensitivity to alcohol (e.g., stimulation, vigor, mood, sedation) is an endophenotype for alcohol use and represents a series of quantitative traits capturing individual differences on the response to alcohol. Few studies have attempted to study the effect of *GABRA2* genetic variation on feelings of alcohol intoxication tests, namely physiologic, subjective, and pharmacologic responses to alcohol and placebo challenges. Due to the nature of these tests, sample sizes were small and thus findings are not consistent regarding the direction of these associations. Nevertheless, they provide some insights towards a better understanding regarding which aspects of subjective response are operating under *GABRA2* influence. Using exploratory factor analysis Ray *et al.* (2009) proposed three factors to the subjective response: positive reinforcement (e.g., stimulation, vigor, and happiness), sedating and unpleasant effects, and negative reinforcement (i.e., tension and low mood alleviated by alcohol).

GABRA2 SNP rs279858, a synonymous coding SNP in exon 5, has been examined in several studies. In a small sample of 27 healthy social drinkers, those with the major genotype, AA, experienced more subjective effects of alcohol and reduced subjective effects under Finasteride (which enhances the action of GABA at GABA-A receptors) administration than individuals carrying the G allele (Pierucci-Lagha *et al.*, 2005). In addition to the small sample size of the AA group ($N = 7$), these results should be interpreted with caution. It is possible that those carrying the G allele experienced decreased subjective response to alcohol (i.e., need a higher dose of alcohol) and are therefore at risk for alcohol abuse. Consistent with this result, in a group of Japanese social drinkers, those homozygotes for the major allele, including rs279858, showed greater subjective response to alcohol compared to those carrying the minor allele (Roh *et al.*, 2011). In contrast, by using an oral alcohol challenge (acute alcohol administration), subjective responses were evaluated in a group of European-American social drinkers. More recently, two studies evaluated sedation and stimulation scores measured to reflect the extent of stimulating and reward effects from alcohol. Arias and colleagues (2014) reported that those with the rs279858 G allele showed "alcohol-like" feelings before and after drinking alcohol. Specifically, they reported feeling more alcohol effects, liking alcohol more, and feeling "higher." Similarly, decreased subjective responses to alcohol were observed in those with the rs279858 G allele and T allele of SNP rs279844 (Uhart *et al.*, 2013). Based on these studies, it has been suggested that the G (or C, if the complementary strand is considered) allele of rs279858 is associated with measures of positive reinforcement (i.e., increased stimulation, vigor, and positive mood) in response to alcohol challenge (i.e., acute administration of alcohol), but not with the sedative or unpleasant effects of alcohol.

Subjects under alcohol intoxication and placebo conditions were exposed to alcohol drink odors and appetitive control odors during an fMRI (Steiger and Russek, 2004). Subjects with rs279871 AA genotype had a larger response to alcohol than to control odors in the medial frontal cortical areas associated with reward responses, while heterozygote subjects had a larger alcohol drinking effect in the VTA. Although these results need replication, they suggest that genotypic effects may have unique effects in different brain regions, as recently reported in brain-derived neurotrophic factor (BDNF) and depression (Covault *et al.*, 2008).

11.7. *GABRA2* and environment

Although research has consistently demonstrated direct effects of *GABRA2* on alcohol, nicotine, and drug dependence in adulthood (Edenberg *et al.*, 2004; Lappalainen *et al.*, 2005; Fehr *et al.*, 2006; Philibert *et al.*, 2009), findings have largely found null effects among younger samples (Dick *et al.*, 2006; Sakai *et al.*, 2010). For example, both Dick and colleagues (2006) as well as Sakai and colleagues (2010) found weak support for a direct effect of *GABRA2* on adolescent-onset alcohol dependence. One potential reason for these seemingly inconsistent findings is that genetic effects may have stronger associations depending on the developmental period of interest. In fact, a meta-analysis demonstrated that age moderates genetic effects on subsequent antisocial behavior, including alcohol use (Rhee and Waldman, 2002). More specifically, one study demonstrated that at age 14, 18% of the variance in drinking was accounted for by genes, compared to nearly 50% of the variance being accounted for by genetic influences at age 18 (Rose *et al.*, 2001). Yet, the lack of direct effects of *GABRA2* on substance abuse and dependence in adolescence does not preclude the potential for indirect effects.

It is possible that during adolescence, genetic effects on later substance dependence may be non-specific, reflecting broader behavior problems compared to adulthood, where specific genetic risk factors relating to the pharmacologic impact of substance use increases in importance (Kendler *et al.*, 2011). One study examining age-specific effects of *GABRA2* on rule-breaking (i.e., normative transgressions to authority figures such as lying, swearing) and substance use outcomes (i.e., problematic alcohol use and substance abuse symptomatology) across adolescence demonstrated that although the direct effects of *GABRA2* on substance use outcomes were weak, *GABRA2* was significantly associated with rule-breaking behaviors in mid to late adolescence (Trucco *et al.*, 2014c). This is consistent with prior work demonstrating main effects of *GABRA2* on conduct disorder, but weak effects on alcohol dependence (Dick *et al.*, 2006). Moreover, findings demonstrate that *GABRA2* effects on problematic alcohol use and substance abuse symptomatology in late adolescence operate largely (45.3% and 71.1%, respectively, $p < .05$) via rule-breaking in mid-adolescence (Trucco *et al.*, 2014c). This offers support for the mechanistic role of *GABRA2* as an early risk factor for an externalizing pathway to the development of substance use disorders.

Other research has widely supported the indirect effects of genes on outcomes via synergistic effects with environmental contexts, or gene × environment (G×E) interactions. G×E effects are especially important to consider during younger developmental periods given research demonstrating that environmental factors, compared to genetic effects, are especially salient during adolescence. Namely, research indicates that environmental factors tend to have a greater impact on the initiation of alcohol and drug use, while escalation to heavy use and to the onset of abuse/dependence is largely driven by genetic factors (Kendler *et al.*, 2008). It is important to note that G×E interactions are often characterized differently depending on the discipline of the researcher. That is, genotype may be viewed as the main causal factor (i.e., independent variable) of an outcome while the environment is seen as the moderator of this effect, or the environment can be viewed as the main causal factor and the genotype as the moderator of this effect (Moffitt *et al.*, 2005). This is more of a theoretical question, because from a statistical perspective these nuances are undifferentiated. Developmental psychologists tend to adopt the view that a youth's experiences and their social environments are the main causal factors influencing outcomes, with the youth's genotype acting as a moderator of these effects (Kochanska *et al.*, 2011); this is the view that we have adopted.

Another important aspect of G×E interactions concerns differences in the form of interactions. Traditionally, G×E interactions have been conceptualized within a *diathesis–stress* framework, whereby some individuals are believed to be vulnerable to adverse effects of environmental stressors due to genetic predispositions (Zuckermann, 1999). That is, adolescents carrying a "risk genotype" (i.e., diathesis) are viewed as being at greatest risk for developing psychopathology when exposed to environmental adversity (i.e., stress). Yet, more recently, two other models have been hypothesized as characterizing G×E interactions. The *differential susceptibility* hypothesis posits that genotypes traditionally conferring risk may actually be more affected by adverse as well as advantageous environments in a "for-better-and-for-worse" manner (Belsky *et al.*, 2007). More recently, a second hypothesis, *vantage sensitivity*, suggests that some G×E interactions reflect increased susceptibility to exclusively positive or adaptive social contexts (Pluess and Belsky, 2013). Given that these different forms of G×E interactions connote different meanings regarding the potential effect of genes (risk, protective, increased susceptibility), research has advocated for the use of empirical tests in order to accurately characterize the form of G×E interactions. Namely, new methodologies have been constructed specifically to distinguish diathesis–stress, differential susceptibility, and vantage sensitivity interactions, such as regions of significance, the proportion of interaction, and the proportion affected index (see Roisman *et al.*, 2012).

From a developmental perspective, parents have a significant impact on adolescent development. Although adolescent problem behavior has been associated with a number of parenting behaviors, most research has focused on two general parenting dimensions associated with adolescent delinquent outcomes, including substance use: parental warmth/responsiveness, and parental control/demandingness. Of the two, parental control/demandingness has demonstrated more robust effects on subsequent problem behaviors in adolescence (Barnes *et al.*, 2006). More specifically, parental monitoring, which reflects parent's knowledge of the associations, whereabouts, and activities of their child, has been demonstrated to be one of the strongest predictors of adolescent delinquent trajectories (Laird *et al.*, 2003; Shaw *et al.*, 2012). Dick *et al.* (2009) examined the effect of *GABRA2*, parental monitoring, and the interaction between the two on externalizing (i.e., an aggregate of rule-breaking and aggressive behavior) trajectories of individuals between the ages of 12 and 22. These researchers identified two separate linear trajectories: a developmentally limited (low-risk) and a

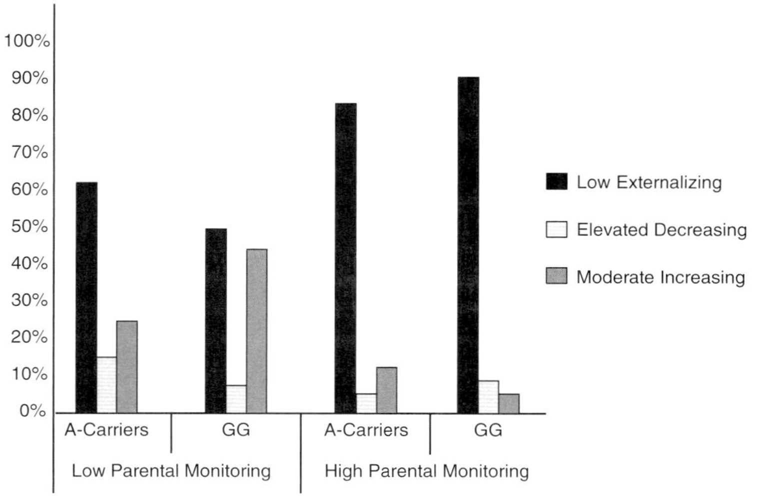

Figure 11.1. *GABRA2* rs279858 and parental monitoring interaction on trajectories of externalizing behavior. (Trucco *et al.*, in press)

persistently elevated (high-risk) externalizing behavior group. The authors frame their findings within a traditional diathesis–stress framework. That is, those individuals carrying the AA genotype of *GABRA2* were more likely to demonstrate a persistently elevated pattern of externalizing behavior, especially in the context of low parental monitoring (Dick *et al.*, 2009). Yet, a more recent replication study found evidence for the differential susceptibility hypothesis (Trucco *et al.*, 2013b). In this study, three unique non-linear externalizing trajectories were estimated (low, decreasing, and high) and *GABRA2* moderated the effects of parental monitoring on externalizing trajectories, uniquely differentiating the lower-risk classes (low and decreasing) and the high-risk class in a cross-over interaction pattern. Namely, A-carriers were largely unaffected by parental monitoring, while at high parental monitoring those with the GG genotype were more likely to be in the low-risk trajectory classes, but less likely to be in the low-risk trajectory classes at low levels of parental monitoring (Figure 11.1).

Another social context that has a significant impact on adolescent development is the peer group. In particular, early adolescence is marked by a change in the quality and amount of time spent with peers, making it a period when adolescents are more susceptible to peer influence (Brown *et al.*, 1997). Although it is assumed that peers have positive and negative influences on adolescents, research has tended to focus on negative peer influence as it relates to problem behaviors (e.g., Jaccard *et al.*, 2005; Trucco *et al.*, 2011, 2014a). Because both peer influence and externalizing behavior are strong predictors of future AUD, we examined the prospective role of *GABRA2* genetic variation on this developmental relationship in the Michigan Longitudinal Study (MLS) adolescent sample. Perceived peer delinquency at age 12–14 predicts higher scores of externalizing behavior at ages 15–17, and this relation was moderated by *GABRA2* variants. Namely, affiliating with peers engaging in delinquent behaviors predicted increased rates of externalizing and rule-breaking behavior among adolescents, and this association was significantly stronger for those with the rs279826 GG genotype compared to A-carriers (Villafuerte *et al.*, 2014).

Few studies have examined the role of positive peer contexts on development, especially as it relates to adaptive outcomes, as opposed to the absence of problem behaviors (e.g., lower rates of aggression; Brown and Klute, 2006; Padilla-Walker and Bean, 2009). Still, there is evidence that peers can have a positive influence on adolescents by promoting proactive adolescent development. For example, positive indirect peer association (i.e., affiliating with close friends who value studying, engage in religious activities, and participate in school organizations) predicted both reductions in maladaptive behavior (delinquency, aggression) as well as increased adaptive outcomes (social initiative, self-esteem) among adolescents (Padilla-Walker and Bean, 2009). Even fewer studies have examined the role of genetic variants as moderating the association between positive peer contexts on adolescent behavior, consistent with the vantage sensitivity hypothesis. A recent study sought to examine whether *GABRA2* (rs279858) moderated the effect of involvement with positive peers on increased adaptive functioning (Trucco *et al.*, 2014b). In preliminary work, findings demonstrate that affiliating with peers that are perceived to do well in academics, to be involved in extracurricular activities, and engage in religious activities, predicted elevations in self-reported behavioral competence, but only among those with the GG genotype and not AA carriers (see Figure 11.2; Trucco *et al.*, 2014b). Taken together, these findings indicate that *GABRA2* variants increase susceptibility to positive social contexts consistent with vantage sensitivity. Thus, *GABRA2* effects on adolescent behavior are likely to be indirect. In addition, conceptualizing *GABRA2* variants as purely risk factors is likely to be inaccurate and mischaracterizes genetic effects on adolescent development and behavior.

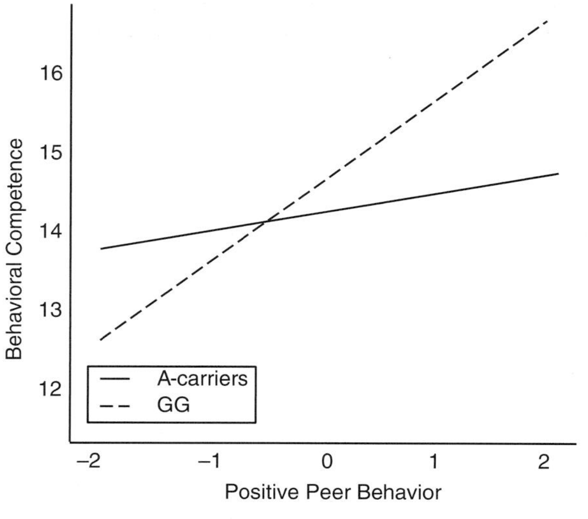

Figure 11.2. *GABRA2* rs279858 moderates the effect of positive peer influence on behavioral competence.

It will be important for future work to consider factors that may be related to mechanisms underlying the increased susceptibility of *GABRA2* variants. For example, preliminary work within the serotonergic and dopaminergic systems have demonstrated that genetic variants associated with increased sensitivity to social contexts are also associated with individual differences to sensitivity to social experiences (Way and Taylor, 2010), reward sensitivity (Roiser *et al.*, 2006), and enhanced attention to emotionally relevant stimuli (Beevers *et al.*, 2009). Given that GABA-A is mainly expressed in brain reward circuitry such as the amygdala, the VTA, and the nucleus acumbens (Pirker *et al.*, 2000), it follows that *GABRA2* variants may reflect predispositions to reward sensitivity. This may be particularly relevant in the context of peers, because adolescents are especially attuned to positive regard and belongingness (Brechwald and Prinstein, 2011).

11.8. Conclusion

The role of the *GABRA2* gene in the development of AUD is supported by several lines of research, replication in independent studies in a myriad of settings from adolescence to adulthood, gender, ethnicities, and endophenotypes. Despite the fact that effect sizes are small, it was identified in large genome studies from linkage to association not only with alcohol-related measures, but also with endophenotypes. Researchers in the field are encouraged to study the effects of this gene on both behavior and environment in early adolescence and further understand their genetic effect during the transition into adulthood in longitudinal studies. The work on the subjective response to alcohol needs to be integrated with personality and behaviors associated with *GABRA2* to further understand their contributions to AUD risk. Despite the advances, no functional polymorphisms have yet been identified. The SNP rs279858 is a silent variation in exon 5 and rs279827 is located next to a splice site. Several *GABRA2* isoforms have been described. Finally, studies on *GABRA2* shed light on the challenges and contributions of genetic variation of small size effects on complex disorders extending as a guide to the scientific community.

REFERENCES

Agrawal, A., Edenberg, H.J., Foroud, T., *et al.* (2006). Association of *GABRA2* with drug dependence in the collaborative study of the genetics of alcoholism sample. *Behav. Genet.*, **36**, 640–650.

Arias, A.J., Covault, J., Feinn, R., *et al.* (2014). A *GABRA2* variant is associated with increased stimulation and "high" following alcohol administration. *Alcohol Alcohol.*, **49**, 1–9.

Barnes, G.M., Hoffman, J.H., Welte, J.W., Farrell, M.P. and Dintcheff, B.A. (2006). Effects of parental monitoring and peer deviance on substance use and delinquency. *J. Marriage Fam.*, **68**, 1084–1104.

Bauer, L.O., Covault, J., Harel, O., *et al.* (2007). Variation in *GABRA2* predicts drinking behavior in project MATCH subjects. *Alcohol. Clin. Exp. Res.*, **31**, 1780–1787.

Beevers, C.G., Wells, T.T., Ellis, A.J. and McGeary, J.E. (2009). Association of the serotonin transporter gene promoter region (5-HTTLPR) polymorphism with biased attention for emotional stimuli. *J. Abnorm. Psychol.*, **118**, 670–681.

Begleiter, H. and Porjesz, B. (1999). What is inherited in the predisposition toward alcoholism? A proposed model. *Alcohol. Clin. Exp. Res.*, **23**, 1125–1135.

Belsky, J., Bakermans-Kranenburg, M.J. and Van Ijzendoorn, M.H. (2007). For better and for worse: differential susceptibility to environmental influences. *Curr. Dir. Psychol. Sci.*, **16**, 300–304.

Brechwald, W.A. and Prinstein, M.J. (2011). Beyond homophily: a decade of advances in understanding peer influence processes. *J. Res. Adolesc.*, **21**, 166–179.

Brown, B.B. and Klute, C. (2006). Friendships, cliques, and crowds. In G.R. Adams and M.D. Berzonsky (Eds.) *Blackwell Handbook of Adolescence*. Blackwell Publishing, Oxford.

Brown, B.B., Dolcini, M.M. and Leventhal, A. (1997). Transformations in peer relationships at adolescence: Implications for health-related behavior. In J. Schulenberg, J.L. Maggs and K. Hurrelman (Eds.) *Health Risks and Developmental Transitions During Adolescence*. Cambridge University Press, New York, NY

Covault, J., Gelernter, J., Hesselbrock, V., Nellissery, M. and Kranzler, H.R. (2004). Allelic and haplotypic association of *GABRA2* with alcohol dependence. *Am. J. Med. Genet. B*, **129B**, 104–109.

Covault, J., Gelernter, J., Jensen, K., Anton, R. and Kranzler, H.R. (2008). Markers in the 5′-region of *GABRG1* associate to alcohol dependence and are in linkage disequilibrium with markers in the adjacent *GABRA2* gene. *Neuropsychopharmacology*, **33**, 837–848.

Dick, D.M., Bierut, L., Hinrichs, A., *et al.* (2006). The role of *GABRA2* in risk for conduct disorder and alcohol and drug dependence across developmental stages. *Behav. Genet.*, **36**, 577–590.

Dick, D.M., Latendresse, S.J., Lansford, J.E., *et al.* (2009). Role of *GABRA2* in trajectories of externalizing behavior across development and evidence of moderation by parental monitoring. *Arch. Gen. Psych.*, **66**, 649–657.

Dick, D.M., Aliev, F., Latendresse, S., *et al.* (2013). How phenotype and developmental stage affect the genes we find: *GABRA2* and impulsivity. *Twin Res. Hum. Genet.*, **16**, 661–669.

Drgon, T., D'Addario, C. and Uhl, G.R. (2006). Linkage disequilibrium, haplotype and association studies of a chromosome 4 GABA receptor gene cluster: candidate gene variants for addictions. *Am. J. Med. Genet. B Neuropsychiatr. Genet.*, **141B**, 854–860.

Edenberg, H.J., Dick, D.M., Xuei, X.L., *et al.* (2004). Variations in *GABRA2*, encoding the alpha 2 subunit of the GABA(A) receptor, are associated with alcohol dependence and with brain oscillations. *Am. J. Hum. Genet.*, **74**, 705–714.

Engel, S.R. and Grant, K.A. (2001). Neurosteroids and behavior. *Int. Rev. Neurobiol.*, **46**, 321–348.

Enoch, M.A., Schwartz, L., Albaugh, B., Virkkunen, M. and Goldman, D. (2006). Dimensional anxiety mediates linkage of gabra2 haplotypes with alcoholism. *Am. J. Med. Genet. B*, **141B**, 599–607.

Fehr, C., Sander, T., Tadic, A., *et al.* (2006). Confirmation of association of the *GABRA2* gene with alcohol dependence by subtype-specific analysis. *Psych. Genet.*, **16**, 9–17.

Gaetz, W., Edgar, J.C., Wang, D.J. and Roberts, T.P. (2011). Relating MEG measured motor cortical oscillations to resting gamma-aminobutyric acid (GABA) concentration. *NeuroImage*, **55**, 616–621.

Hsu, F.C., Zhang, G.J., Raol, Y.S., *et al.* (2003). Repeated neonatal handling with maternal separation permanently alters hippocampal GABAA receptors and behavioral stress responses. *Proc. Natl Acad. Sci. USA*, **100**, 12213–12218.

Irons, D.E., Iacono, W.G., Oetting, W.S., *et al.* (2014). Gamma-aminobutyric acid system genes – no evidence for a role in alcohol use and abuse in a community-based sample. *Alcohol. Clin. Exp. Res.*, 38, 938–947.

Jaccard, J., Blanton, H. and Dodge, T. (2005). Peer influences on risk behavior: an analysis of the effects of a close friend. *Develop. Psychol.*, **41**, 135–147.

Kendler, K.S., Gardner, C. and Dick, D.M. (2011). Predicting alcohol consumption in adolescence from alcohol-specific and general externalizing genetic risk factors, key environmental exposures and their interaction. *Psychol. Med.*, **41**, 1507–1516.

Kendler, K.S., Prescott, C.A., Myers, J. and Neale, M.C. (2003). The structure of genetic and environmental risk factors for common psychiatric and substance use disorders in men and women. *Arch. Gen. Psych.*, **60**, 929–937.

Kendler, K.S., Schmitt, E., Aggen, S.H. and Prescott, C.A. (2008). Genetic and environmental influences on alcohol, caffeine, cannabis, and nicotine use from early adolescence to middle adulthood. *Arch. Gen. Psych.*, **65**, 674–682.

Kochanska, G., Kim, S., Barry, R.A. and Philibert, R.A. (2011). Children's genotypes interact with maternal responsive care in predicting children's competence: diathesis-stress or differential susceptibility? *Develop. Psychopathol.*, **23**, 605–616.

Krueger, R.F., Hicks, B.M., Patrick, C.J., *et al.* (2002). Etiologic connections among substance dependence, antisocial behavior, and personality: modeling the externalizing spectrum. *J. Abnorm. Psychol.*, **111**, 411–424.

Laird, R.D., Pettit, G.S., Bates, J.E. and Dodge, K.A. (2003). Parents' monitoring-relevant knowledge and adolescents' delinquent behavior: evidence of correlated developmental changes and reciprocal influences. *Child Develop.*, **74**, 752–768.

Lappalainen, J., Krupitsky, E., Remizov, M., *et al.* (2005). Association between alcoholism and gamma-amino butyric acid alpha 2 receptor subtype in a Russian population. *Alcohol. Clin. Exp. Res.*, **29**, 493–498.

Le Roux, N., Amar, M., Moreau, A., Baux, G. and Fossier, P. (2008). Impaired GABAergic transmission disrupts normal homeostatic plasticity in rat cortical networks. *Eur. J. Neurosci.*, **27**, 3244–3256.

Lejuez, C.W., Magidson, J.F., Mitchell, S.H., *et al.* (2010). Behavioral and biological indicators of impulsivity in the development of alcohol use, problems, and disorders. *Alcohol. Clin. Exp. Res.*, **34**, 1334–1345.

Li, D., Sulovari, A., Cheng, C., *et al.* (2014). Association of gamma-aminobutyric acid A receptor alpha2 gene (*GABRA2*) with alcohol use disorder. *Neuropsychopharmacology*, **39**, 907–918.

Lind, P.A., MacGregor, S., Agrawal, A., *et al.* (2008). The role of *GABRA2* in alcohol dependence, smoking, and illicit drug use in an Australian population sample. *Alcohol. Clin. Exp. Res.*, **32**, 1721–1731.

Long, J.C., Knowler, W.C., Hanson, R.L., *et al.* (1998). Evidence for genetic linkage to alcohol dependence on chromosomes 4 and 11 from an autosome-wide scan in an American Indian population. *Am. J. Med. Genet.*, **81**, 216–221.

Magid, V. and Colder, C.R. (2007). The UPPS impulsive behavior scale: factor structure and associations with college drinking. *Pers. Indiv. Differ.*, **43**, 1927–1937.

Matthews, A.G., Hoffman, E.K., Zezza, N., Stiffler, S. and Hill, S.Y. (2007). The role of the *GABRA2* polymorphism in multiplex alcohol dependence families with minimal comorbidity: within-family association and linkage analyses. *J. Stud. Alcohol Drugs*, **68**, 625–633.

Moan, I.S., Norstrom, T. and Storvoll, E.E. (2013). Alcohol use and drunk driving: the modifying effect of impulsivity. *J. Stud. Alcohol Drugs*, **74**, 114–119.

Moffitt, T.E., Caspi, A. and Rutter, M. (2005). Strategy for investigating interactions between measured genes and measured environments. *Arch. Gen. Psych.*, **62**, 473–481.

Olfson, E. and Bierut, L.J. (2012). Convergence of genome-wide association and candidate gene studies for alcoholism. *Alcohol. Clin. Exp. Res.*, **36**, 2086–2094.

Onori, N., Turchi, C., Solito, G., *et al.* (2010). *GABRA2* and alcohol use disorders: no evidence of an association in an Italian case-control study. *Alcohol. Clin. Exp. Res.*, **34**, 659–668.

Padilla-Walker, L.M. and Bean, R.A. (2009). Negative and positive peer influence: relations to positive and negative behaviors for African American, European American, and Hispanic adolescents. *J. Adolesc.*, **32**, 323–337.

Philibert, R.A., Gunter, T.D., Beach, S.R., *et al.* (2009). Role of *GABRA2* on risk for alcohol, nicotine, and cannabis dependence in the Iowa Adoption Studies. *Psychiatr. Genet.*, **19**, 91–98.

Pierucci-Lagha, A., Covault, J., Feinn, R., *et al.* (2005). *GABRA2* alleles moderate the subjective effects of alcohol, which are attenuated by finasteride. *Neuropsychopharmacology*, **30**, 1193–1203.

Pirker, S., Schwarzer, C., Wieselthaler, A., Sieghart, W. and Sperk, G. (2000). GABA(A) receptors: immunocytochemical distribution of 13 subunits in the adult rat brain. *Neuroscience*, **101**, 815–850.

Pluess, M. and Belsky, J. (2013). Vantage sensitivity: individual differences in response to positive experiences. *Psychol. Bull.*, **139**, 901–916.

Porjesz, B., Almasy, L., Edenberg, H.J., *et al.* (2002). Linkage disequilibrium between the beta frequency of the human EEG and a GABA(A) receptor gene locus. *Proc. Natl Acad. Sci. USA*, **99**, 3729–3733.

Rangaswamy, M., Porjesz, B., Chorlian, D.B., *et al.* (2002). Beta power in the EEG of alcoholics. *Biol. Psych.*, **52**, 831–842.

Rangaswamy, M., Porjesz, B., Chorlian, D.B., *et al.* (2004). Resting EEG in offspring of male alcoholics: beta frequencies. *Int. J. Psychophysiol.*, **51**, 239–251.

Ray, L.A., Mackillop, J., Leventhal, A. and Hutchison, K.E. (2009). Catching the alcohol buzz: an examination of the latent factor structure of subjective intoxication. *Alcohol. Clin. Exp. Res.*, **33**, 2154–2161.

Reich, T., Edenberg, H.J., Goate, A., *et al.* (1998). Genome-wide search for genes affecting the risk for alcohol dependence. *Am. J. Med. Genet.*, **81**, 207–215.

Rhee, S.H. and Waldman, I.D. (2002). Genetic and environmental influences on antisocial behavior: a meta-analysis of twin and adoption studies. *Psychol. Bull.*, **128**, 490–529.

Roh, S., Matsushita, S., Hara, S., *et al.* (2011). Role of *GABRA2* in moderating subjective responses to alcohol. *Alcohol. Clin. Exp. Res.*, **35**, 400–407.

Roiser, J.P., Rogers, R.D., Cook, L.J. and Sahakian, B.J. (2006). The effect of polymorphism at the serotonin transporter gene on decision-making, memory and executive function in ecstasy users and controls. *Psychopharmacology (Berl.)*, **188**, 213–227.

Roisman, G.I., Newman, D.A., Fraley, R.C., *et al.* (2012). Distinguishing differential susceptibility from diathesis–stress: recommendations for evaluating interaction effects. *Develop. Psychopathol.*, **24**, 389–409.

Rose, R.J., Dick, D.M., Viken, R.J., Pulkkinen, L. and Kaprio, J. (2001). Drinking or abstaining at age 14? A genetic epidemiological study. *Alcohol. Clin. Exp. Res.*, **25**, 1594–1604.

Sakai, J.T., Stallings, M.C., Crowley, T.J., *et al.* (2010). Test of association between *GABRA2* (SNP rs279871) and adolescent conduct/alcohol use disorders utilizing a sample of clinic referred youth with serious substance and conduct problems, controls and available first degree relatives. *Drug Alcohol Depend.*, **106**, 199–203.

Shaw, D.S., Hyde, L.W. and Brennan, L.M. (2012). Early predictors of boys' antisocial trajectories. *Develop. Psychopathol.*, **24**, 871–888.

Soyka, M., Preuss, U.W., Hesselbrock, V., *et al.* (2008). GABA-A2 receptor subunit gene (*GABRA2*) polymorphisms and risk for alcohol dependence. *J. Psychiatr. Res.*, **42**, 184–191.

Steiger, J.L. and Russek, S.J. (2004). GABAA receptors: building the bridge between subunit mRNAs, their promoters, and cognate transcription factors. *Pharmacol. Therapeut.*, **101**, 259–281.

Tian, H.J., Chen, H.J., Cross, T.H. and Edenberg, H.J. (2005). Alternative splicing and promoter use in the human *GABRA2* gene. *Molec. Brain Res.*, **137**, 174–183.

Trucco, E.M., Colder, C.R. and Wieczorek, W.F. (2011). Vulnerability to peer influence: a moderated mediation study of early adolescent alcohol use initiation. *Addict. Behav.*, **36**, 729–736.

Trucco, E.M., Colder, C.R., Wieczorek, W.F., Lengua, L.J. and Hawk, L.W., Jr. (2014a). Early adolescent alcohol use in context: how neighborhoods, parents, and peers impact youth. *Develop. Psychopathol.*, **26**, 425–436.

Trucco, E.M., Villafuerte, S., Burmeister, M. and Zucker, R.A. (2014b). Reconceptualizing G × E interactions in adolescence: the role of *GABRA2* variants as susceptibility factors to positive peer influence. Society for Research on Adolescence. Austin, TX.

Trucco, E.M., Villafuerte, S., Heitzeg, M.M., Burmeister, M. and Zucker, R.A. (2014c). Rule breaking mediates the developmental association between GABRA2 and adolescent substance use. *J. Child Psychol. Psych.*, 55, 1372–1379.

Trucco, E.M., Villafuerte, S., Heitzeg, M.M., Burmeister, M. and Zucker, R.A. (in press). Susceptibility effects of GABRA2 variants and parental monitoring on externalizing behavior trajectories: Risk and Protection are both conveyed by the minor allele. *Development and Psychopathology*.

Uhart, M., Weerts, E.M., McCaul, M.E., *et al.* (2013). GABRA2 markers moderate the subjective effects of alcohol. *Addict. Biol.*, 18, 357–369.

Villafuerte, S., Heitzeg, M.M., Foley, S., *et al.* (2012). Impulsiveness and insula activation during reward anticipation are associated with genetic variants in *GABRA2* in a family sample enriched for alcoholism. *Molec. Psych.*, **17**, 511–519.

Villafuerte, S., Strumba, V., Stoltenberg, S.F., Zucker, R.A. and Burmeister, M. (2013). Impulsiveness mediates the association between *GABRA2* SNPs and lifetime alcohol problems. *Genes Brain Behav.*, **12**, 525–531.

Villafuerte, S., Trucco, E.M., Heitzeg, M.M., Burmeister, M. and Zucker, R.A. (2014). Genetic variation in GABRA2 moderates peer influence on externalizing behavior in adolescents. *Brain and Behavior*, **4**, 833–840.

Way, B.M. and Taylor, S.E. (2010). Social influences on health: is serotonin a critical mediator? *Psychosom. Med.*, **72**, 107–112.

Young, S.E., Stallings, M.C., Corley, R.P., Krauter, K.S. and Hewitt, J.K. (2000). Genetic and environmental influences on behavioral disinhibition. *Am. J. Med. Genet.*, **96**, 684–695.

Zintzaras, E. (2012). Gamma-aminobutyric acid A receptor, alpha-2 (*GABRA2*) variants as individual markers for alcoholism: a meta-analysis. *Psych. Genet.*, **22**, 189–196.

Zucker, R.A. (1991). Scaling the developmental momentum of alcoholic process via the lifetime alcohol problems score. *Alcohol Alcohol.*, **1**, 505–510.

Zucker, R.A., Heitzeg, M.M. and Nigg, J.T. (2011). Parsing the undercontrol/disinhibition pathway to substance use disorders: a multilevel developmental problem. *Child Develop. Perspect.*, **5**, 248–255.

Zucker, R.A., Davies, W.H., Kincaid, S.B., Fitzgerald, H.E. and Reider, E.E. (1997). Conceptualizing and scaling the developmental structure of behavior disorder: the lifetime alcohol problems score as an example. *Develop. Psychopathol.*, **9**, 453–471.

Zuckermann, M. (1999). *Diathesis–Stress Model. Vulnerability to Psychopathology: A Biosocial Model*. American Psychological Association, Washington, DC.

12 Copy number variation in monozygous twins

Erwin Brosens, K. G. Snoek, D. Veenma, H. Eussen, D. Tibboel,
and A. de Klein

12.1. Introduction

12.1.1. Microarray technology in GWAS and copy number variations

Genome-wide association studies (GWAS) are used to associate human disease and traits with a certain chromosomal locus or loci. This is done by whole-genome genotyping of many cases and controls using so-called tag-SNP probes containing a relative common SNP associated with a region in linkage disequilibrium (LD). If a certain tag-SNP has a statistically significant higher frequency in cases compared to controls, the studied trait or disease is said to be associated with that specific locus. This method has proven to be successful in many occasions (Hindorff *et al.*, 2009; Manolio 2013) and has increased our knowledge of disease etiology and human development significantly. Genotyping has mainly been done using microarrays with up to several million allele-specific oligo-probes. These microarrays contain information on the relative amount of DNA at a given locus. Using both genotype and quantitative information, these SNP-based GWAS studies can be extended to include DNA gains or losses. These segmental variations in DNA copy number can arise *de novo* or be inherited in a Mendelian matter. Many of the characteristics of other types of genomic variation are shared: they can be ancestry-specific, are driven by selection pressure, and can influence the expression of genes by altering their copy number or affecting gene regulatory regions (Zhang *et al.*, 2009; Schlattl *et al.*, 2011). Losses can result in haploinsufficiency of one or several genes or truncated proteins, while gains can increase gene expression or can also lead to altered protein structure, reduced protein levels or function.

12.1.2. Copy number variation; common, rare, and *de novo*

Many gains and losses are rather common and most likely represent the normal population variance (Zhang *et al.*, 2009; Stankiewicz and Lupski, 2010). These

Genome-Wide Association Studies: From Polymorphism to Personalized Medicine, ed.
K. Appasani. Published by Cambridge University Press.

recurrent DNA variations have an allele frequency of over 1% and are called *common* copy number polymorphisms (CNPs). CNPs account for a significant proportion of the healthy human genome (Iafrate *et al.*, 2004; Sebat *et al.*, 2004). They often arise after non-allelic homologous recombination of misaligned DNA segments due to the presence of low copy repeats (Stankiewicz and Lupski, 2010). In general, common copy number variations (CNVs) are not associated with severe congenital anomalies, but could influence human traits such as height (van Duyvenvoorde *et al.*, 2014), or aging (Iakoubov *et al.*, 2013). Also, they could raise disease susceptibility for a limited number of complex diseases as age-related macular degeneration (Cantsilieris *et al.*, 2012), Crohn's disease (McCarroll *et al.*, 2008), obesity (Wheeler *et al.*, 2013), diabetes, and rheumatoid arthritis (Craddock *et al.*, 2010). If recurrent gains and losses have an allele frequency under < 1%, they are called *rare* copy number variations (CNVs) (McCarroll, 2008). These CNV are proposed to arise after replication errors, e.g. non-homologs end-joining, fork-stalling and template-switching, or micro homology-mediated break-induced replication (Stankiewicz and Lupski, 2010). These rare CNVs can contribute to a wide range of diseases, for instance in congenital heart disease (Hitz *et al.*, 2012; Soemedi *et al.*, 2012). Interpretation of the significance of a rare CNV can sometimes be difficult because CNV can be ancestry-specific (Oei *et al.*, 2014). Common CNPs and rare CNVs can be used in GWAS (McCarroll and Altshuler, 2007). However, it has to be expected that many associated loci are likely to have been previously detected by tag-SNPs from "regular" GWAS studies (Craddock *et al.*, 2010). In contrast, often large (> 500 kb) *de novo* CNV, affecting many genes and regulatory sites, are related to pathogenic conditions. For instance, in congenital anomalies as congenital diaphragmatic hernia (Scott *et al.*, 2007; Wat *et al.*, 2011), or Soto's syndrome (Visser *et al.*, 2010), tumor genesis (Stadler *et al.*, 2012), autism (Sebat *et al.*, 2007), and intellectual disability (Hehir-Kwa *et al.*, 2011). It can be difficult to pinpoint the causal gene in these large *de novo* CNVs because many sites are affected. However, overlapping CNVs in similar patients can result in a reduction of the "region of interest." Summarizing, CNPs are usually associated with traits and not to severe early-onset diseases. Rare inherited CNVs can be associated with traits, common and rare diseases with differences in timing and penetrance in individuals. Finally, large *de novo* CNV can be causal in both early and late-onset diseases, congenital anomalies, and tumorgenesis (Figure 12.1).

DNA gains and losses affect large parts of the otherwise diploid human genome. Variation recurring with a population frequency over 1% are called common CNPs and usually have little impact on health, but could influence human traits. CNVs have an allele frequency under <1 and could raise disease susceptibility for a limited number of sporadic or common complex diseases. Congenital anomalies are mostly associated with *de novo* CNV, although again, rare CNV could contribute often in a multifactorial disease model.

12.1.3. The twinning process

The prevalence of spontaneous multiple offspring in a single gravidity, twinning, is about 1 in 40 pregnancies, although large regional differences are present (Hall,

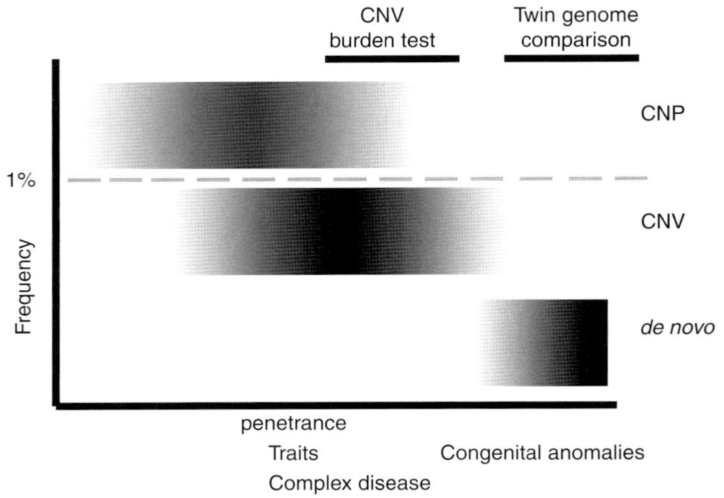

Figure 12.1. Copy number variation frequency and association with traits and their diseases.

2003). The incidence of twin pregnancies has increased in the developed world due to the increasing use of fertility treatments, increased maternal age, weight and height (Russell *et al.*, 2003; Basso *et al.*, 2004; Reddy *et al.*, 2005). More congenital anomalies are observed in twin pregnancies. Twins can originate either from one oocyte (also called monozygous, ~30%) or from multiple oocytes (di- or polyzygotic, ~70%) (Weber and Sebire, 2010). The fetal placental membrane, the chorion, starts to develop around day 3–4 and the amnion forms between day 6 and 8 (Hall, 2003). Depending on time of separation, monozygous twins either have independent amnions, chorions and placentas, or share one or more of these structures (Weber and Sebire, 2010). In almost all dizygotic pregnancies each embryo develops an individual amnion, chorion, and placenta. There are, however, rare incidences of other mechanisms (Souter *et al.*, 2007). Dichorionic diamnionic monozygous (DCDA) twins separate before day 3, have two placentas, and represent about 20–30% of monozygous twins. Monochorionic diamniotic (70–75%) twins are roughly thought to separate between days 5 and 7 after fertilization. In contrast, monochorionic monoamniotic twins (1–2%) have one placenta and separate after day 8, while division at or after day 13 results in conjoined twins (Hall, 2003; Pharoah, 2005). These placentas can be joined or connected via vascular anastomoses, just as this can happen in dizygotic twins (Hall, 2003). Genetic factors, *in vitro* fertilization, and increased concentrations of follicle-stimulating hormone are factors associated with dizygous twinning, while monozygous twinning etiology is less clear. It has been suggested that inner cell mass damage, breaks in the zona pelucida, genetic factors, or skewed X-inactivation could be mechanisms resulting in monozygous twinning (Hall, 2003). Ultrasonographic assessment of chorionicity is optimally performed in the first or early second trimester (Shetty and Smith, 2005; Lee *et al.*, 2006).

12.1.4. Classical twin design; nature versus nurture

Using cohorts of concordant and discordant monozygotic and dizygotic twins, researchers are able to determine the impact of environment and genetic factors in traits and common or complex diseases (van Dongen *et al.*, 2012). Two advantages of twin studies are that twins share an intrauterine environment and are perfect age-matched controls. Moreover, the genetic contribution of a trait or disease can be inferred by comparing the genetic variance with the environmental difference. Because monozygotic twins are (mostly) genetically identical, any phenotypical discordance should be attributed to an environmental exposure *in utero* or later in the life span. Dizygotic twins only share on average half of their DNA – that is, like common siblings – and a greater variance between dizygotic versus monozygotic twins is supportive of an increased genetic contribution. Comparing the DNA of twins is now possible up to the base pair level and allows for the mapping of disease genes in dizygotic twins and intra-twin comparisons in monozygotic twins (Tan *et al.*, 2010).

12.1.5. Associations of CNV and traits or diseases

Enrichment of CNVs in classical case-control studies can result in a statistically significant association to a given phenotype. There is a huge increase in the number of publications of GWAS studies establishing associations of CNV with a specific disease or trait, e.g., in osteoporotic fractures (Oei *et al.*, 2014), colorectal cancer (Fernandez-Rozadilla *et al.*, 2014; Yang *et al.*, 2014), congenital heart disease (Soemedi *et al.*, 2012), alcohol dependence (Lin *et al.*, 2012), and schizophrenia (Bergen *et al.*, 2012). Similarly to these studies, the enrichment of phenotypical concordance in monozygotic twin cohorts compared to their dizygotic counterparts hints to a more genetic background. Moreover, genetic discordance in phenotypical discordant monozygotic twins can give strong clues about genetic background (Table 12.1). Specific CNVs more prevalent in affected monozygous twins and their virtual absence in unaffected dizygous twins could identify causal CNVs in traits or diseases. CNV-twin studies can be either additive to "regular" GWAS or "classical" twin studies, or used as an independent study protocol. Therefore, monozygotic twins, either concordant or discordant for a given phenotype or trait, can be very helpful in these CNV association studies.

Surprisingly, CNV association studies in twins are even rarer than regular CNV-GWAS studies. Few studies focus on single genes or locus, e.g. *CES1* gene duplication with susceptibility to type 2 diabetes (Friedrichsen *et al.*, 2013). In this study, mRNA expression levels and *CES1* gene copy number of 295 monozygous and 170 dizygous twins discordant for obesity were compared. A positive association was found between *CES1* mRNA levels and body mass index, insulin resistance, insulin levels, fasting glucose levels, and triglyceride levels. Moreover, gene duplication was associated with insulin sensitivity, and glucose tolerance, and negatively associated with insulin resistance. Interestingly, *CES1* mRNA levels were not associated with *CES1* copy number; most likely other factors also contribute to the associations found in this study. In a genome-wide CNV association study

Table 12.1. Copy number alterations in traits and disease.

	CNP-inherited events	CNV-rare or private, inherited from affected parent	CNV-rare or private, inherited from unaffected parents	CNV-rare or private, *de novo*
Population frequency	High	Low or even absent	Low or even absent	Low or even absent
Disease or trait	May be associated with common late-onset diseases (susceptibility) and traits	May be associated in rare diseases if relevant genes are affected	Possible modifying or predisposing factors in rare diseases	Likely causal in rare diseases if relevant genes are affected
In case-control studies	Enrichment in case-control studies	Associated with disease	Almost absent in controls and enriched in cases; absolute frequency is low	Virtually absent in controls and enriched in cases; absolute frequency is low
MZ twin phenotype is concordant	Associated if parent with CNP has trait or disease (susceptibility) is detected in case-control studies	Associated if monozygous twins are both affected	Possible modifying or predisposing factors in rare diseases	Likely causal in rare diseases if relevant genes are affected
MZ twin phenotype is discordant	CNP not associated with trait or disease (susceptibility) and traits	Possible role in multifactorial etiology with reduced penetrance[a]	Possible role in multifactorial etiology with reduced penetrance if enriched in cases[a]	Possible role in multifactorial etiology with reduced penetrance[a]

CNP; copy number polymorphism, CNV; copy number variation;
[a] additional functional studies are necessary to prove or exclude causality.Partly adapted from McCarroll (2008).

(Grayson *et al.*, 2010), the copy number profiles of 20 sporadic patients with type 1 diabetes was compared with 20 healthy controls. The resulting 39 CNVs were subsequent validated in a discordant monozygous twin pair cohort (n = 10). Five of these CNV, enriched in the twin cohort, pointed to susceptibility loci for type 1 diabetes. Nag and coworkers studied intraocular pressure in 992 twins enlisted in the UK twin registry and replicated their findings in two independent twin cohorts (n = 467 and n = 1620). They found an association with a CNV on chromosome 5q21 (Nag *et al.*, 2013). With a CNV burden test, associations were obtained between specific CNV and attention deficit disorder. Although Ehli and coworkers detected two *de novo* CNV, these were concordant in the monozygous twins (Ehli *et al.*, 2012). However, CNV burden in patients was significantly higher and the affected individuals had larger, more often rare and *de novo*, CNVs compared to unaffected individuals. These rare CNVs could contribute to the attention deficit disorder. This is in line with the "common disease, rare variant" hypothesis or rare allele model, in which recently emerged (in the population) rare alleles have a large effect on disease etiology. In this model, penetrance may be influenced by other factors (Gibson, 2011; Nelson *et al.*, 2012; Tennessen *et al.*, 2012).

12.1.6. Identical twins; exceptions to the dogma

There are some exceptions to consider before interpreting twin study results. First of all, the intrauterine environmental conditions are not always equal. For example, twin-to-twin transfusion syndrome (TTTS), one of the most serious complications of monochorionic multiple gestations, can occur. TTTS develops as discordant loss of arteriovenous anastomoses resulting in asymmetrical flow resistance (Sebire *et al.*, 2001). In monozygotic twins often the placenta is shared, blood including nutrients, oxygen, CO_2, metabolites, and stem cells can be transferred from one twin to the other via placental anastomoses due to increasing hydrostatic and osmotic forces; the resulting disproportional blood supply causes developmental delay and decreased urinary output in the "donor" twin (and oligohydramnion), and increased urinary output and resulting polyhydramnios in the "recipient" twin. The increased blood flow can result in cardiac anomalies in the later twin. The unequal access to maternal blood, transfusion and waste of nutrients, and oligohydramnios and polyhydramnios result in differences in environmental exposure. TTTS develops in 10–15% of monochorionic twins and is associated with several congenital anomalies. For instance, TTTS and undiagnosed loss of a twin fetus in an early stage of pregnancy could be a substantial factor contributing to congenital anomalies as VACTERL association or gastrointestinal atresia (Pharoah, 2005; Pharoah *et al.*, 2009). Second, DNA changes, mutations, or CNVs can arise *de novo* at any time during the human life span. Somatic DNA variations, occurring early in development or later in life, can be either absent or present in a different frequency in monozygotic twins depending on the timing of egg division and mutation (Figure 12.2). Clear examples are those CNV that are involved in tumorgenesis (Bruder *et al.*, 2008), although these DNA changes can occur much earlier and be responsible for less dramatic somatic differences between newborns (Essaoui *et al.*, 2013). It has even been proposed that DNA mutations cause cells in the blastocyst to recognize each other as foreign, resulting in separation of these cells and causing the twinning process (Hall, 2003).

12.1.7. Current twin study design and discordant twin studies

Improvements in genetic and epidemiological insights enable researchers to go beyond the classical twin design. The genomes of parents, mono-, and dizygous twins are characterized and very large twin registries enable us to follow the phenotype of interest over time. Several experimental set-ups are now possible. For instance, the DNA of phenotypical discordant monozygotic twins can now be compared and any discrepancy could point to disease-causing mutations. Subsequent screening of large cohorts of cases and controls could confirm the impact of this mutation on disease etiology. In a classical twin design the

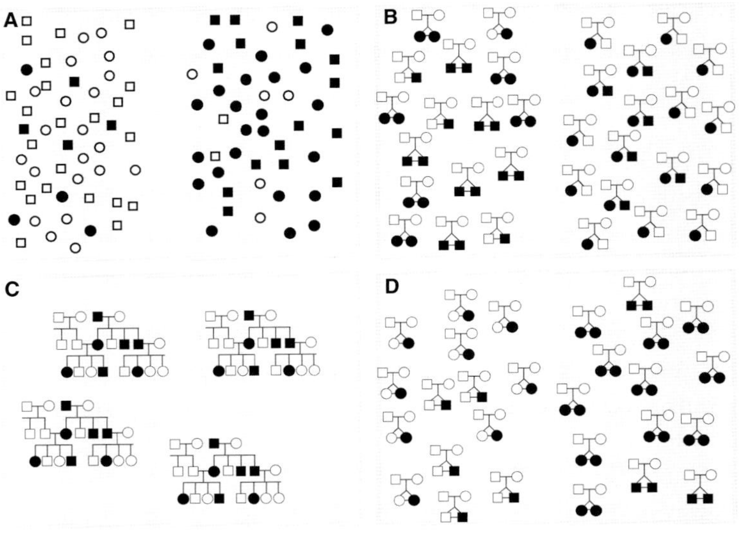

Figure 12.2. Examples of experimental set-ups to determine the genetic basis or contribution of a disease or trait. (A) With GWAS in sporadic cases enrichment of a DNA variant in cases (right) and a significant over-representation of a trait or disease in these cases (▪ and •) compared to controls (left, ▫ and ◦) points to the genetic contribution of a specific locus. (B) If the concordance rate in monozygotic twins is higher compared to dizygous twins, it is likely that genetic factors contribute to the trait or disease. (C) A trait or disease can be hereditary. Pedigree and linkage analysis can pinpoint the inheritance pattern and locus of the contributing or causal genetic factor. (D) Comparing the genomes of monozygotic twins can aid in finding the causal genetic factor of a trait or disease. In discordant monozygous twins (left), finding a discordant genetic variant would provide first evidence of this causal factor. Finding *de novo* variants in concordant monozygous twins with healthy parents could also provide this evidence. Finding variants that disturb protein function, structure, or expression level in several independent families or twin–parent combinations strengthens this evidence for a causal gene; subsequent functional studies would prove causality.

concordance rate of monozygotic and dizygotic twins is compared. Another approach could be to compare CNV burden of affected concordant monozygous twins and compare this to cohorts of affected concordant dizygotic twins (Figure 12.2). If a specific type of CNV is more frequent in the monozygous cohort, it is likely to be associated with the trait or disease studied. An essential aspect of twin studies, which is independent of study design, is the need for proper zygosity determination.

12.2. Experimental procedures

12.2.1. Determination of zygosity

Zygosity can be determined in a number of ways, with variable accuracy in different study populations: for example, blood group, gender, chorionicity, dermatoglyphics, and phenotypical appearances. The selection of same-sex pairs combined with questionnaires could be accurate enough in adult twin studies.

Although relatively reliable in adults, zygosity typing based on gender, chronicity, and appearances alone is not so reliable in newborns. For instance, monochorionicity does not always mean that twins are monozygous (Smeets *et al.*, 2013), same-sex pairs are not always monozygous (Veenma *et al.*, 2012), nor does having an opposite gender confirm dizygosity (Razzaghian *et al.*, 2013). A DNA-based test gives an almost 100% certainty; for eample, comparing the length of several polymorphic tetra-nucleotide short tandem repeat loci, using several SNPs (not in LD), or genome-wide SNP-arrays (Bruder *et al.*, 2008; Machin, 2009; Veenma *et al.*, 2012).

12.2.2. Copy number variation profiling

Determining the genome-wide segmental copy number can be done with a DNA-array using single nucleotide polymorphism (SNP)-based probes or by using oligonucleotides in a comparative genomic hybridization (CGH) experiment. SNP-based microarrays have the advantage that they rely on two types of information: allelic, and segmental copy number state. Therefore, these kinds of arrays are less influenced by DNA input variations. Most laboratories use SNP-arrays for genotyping and GWAS, homozygosity mapping, and CNV analysis. Using these types of arrays there is also no need for zygosity testing in twin births, because zygosity and gender can easily be determined within the genotyping analysis. CNV size and number are dependent on the platform used, the amount and spacing of probes, analysis settings, and CNV calling algorithms and software (Haraksingh *et al.*, 2011; Pinto *et al.*, 2011; Tucker *et al.*, 2011). Throughout the years, the resolution to distinguish CNV from background noise has increased significantly: current microarray technologies allow detecting of genomic imbalances down to only a few kb in size.

12.2.3. Confirmation of results with a second technique

It is always advisable to confirm copy number events of interest with a second technique, preferably as dissimilar to the initial technique as possible. For instance, using fluorescence *in situ* hybridization (FISH) and counting signals on metaphase slides, melting curve analysis or quantitative polymerase chain reaction (qPCR). Ideally, the effect of the discordant chromosomal aberration is measured, for example, with RNA- or protein expression-based studies in relevant tissues. Validation is needed for several reasons: (1) the determination of CNV using microarray is a relative new technique, and each platform has its own sensitivity and specificity; (2) with microarray, the *relative* copy number is determined, that is, compared to one or more reference samples or to a large reference cohort; (3) moreover, it determines the average amount of DNA at each segment, gains and losses at the same locus result in a diploid copy number; (4) the microarray gives limited information on the location or orientation of a given CNV; (5) also, technical variation can influence experimental outcome; for example, the microarray relies on amplification of DNA and measurements using (fluorescence) intensity.

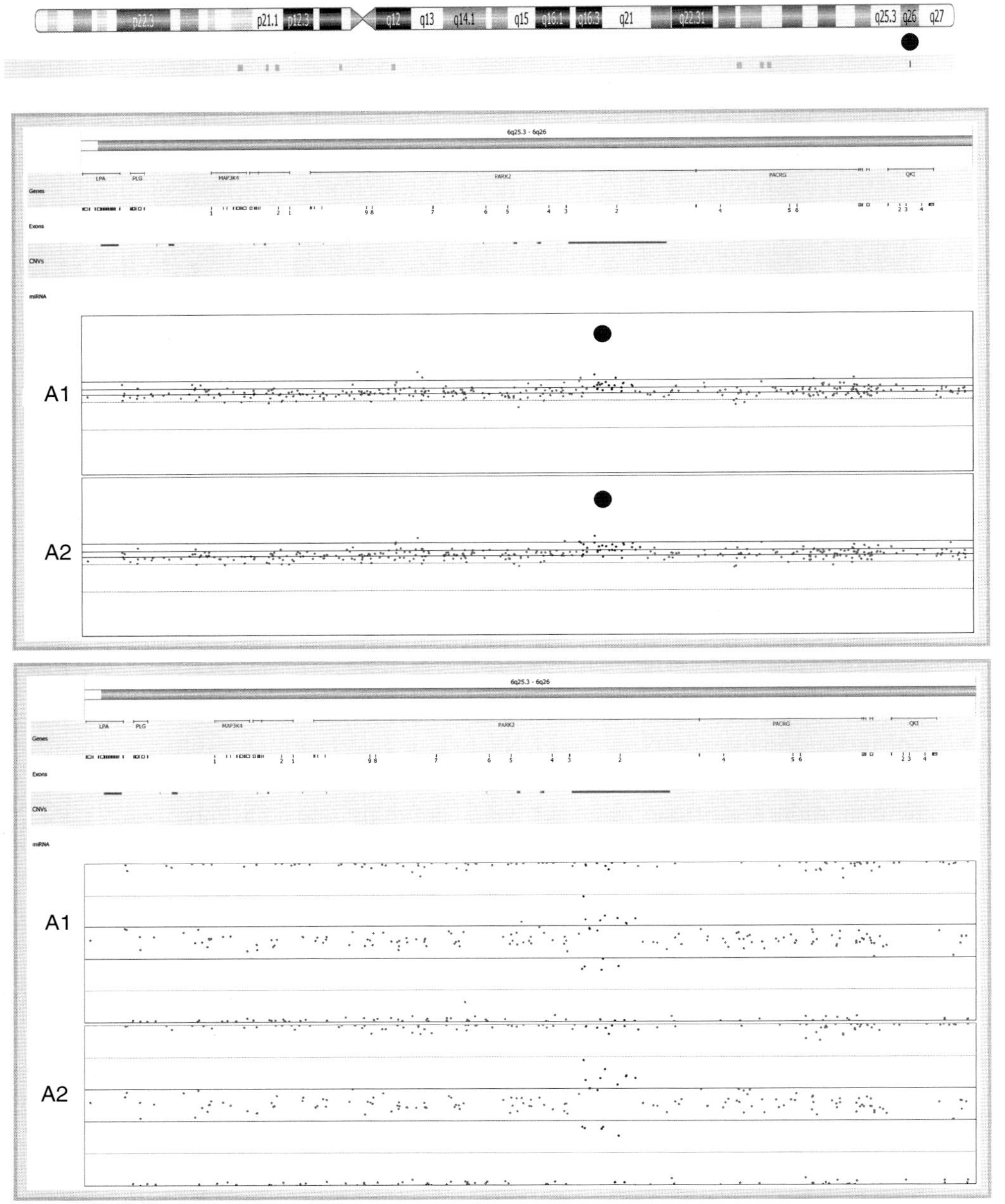

Figure 12.3. Two examples of concordant CNVns in discordant twin-pairs. (A1–A2) On the left, the SNP array results, visualized with Nexus Copy Number v7 (Biodiscovery Inc., Hawthorne, CA, USA), of a monozygotic twin pair discordant for esophageal atresia. Both the affected (upper panel A1 and lower panel A1) and healthy twin (upper panel A2 and lower panel A2) have a 201 kb gain affecting exon 2 of the *PARK2* gene. In the upper panel the $\log_2$ R ratio, representing the copy number status, shows a rise (blue dots) of several probes. This rise is accompanied by a shift in allelic frequency (B-allele frequency, lower panel) indicative for the over-representation of one allele over the other. The combination of rise in $\log_2$ R ratio and shift in B-allele frequency of several allele-specific SNP probes is indicative for a gain in DNA quantity of that specific locus. Both twins have exactly the same gain and this gain was inherited maternally, which was confirmed with qPCR. (B1–B2) On the right, the SNP array results of a monozygous twin pair discordant for congenital diaphragmatic hernia. The affected twin (upper panel B1 and lower panel B1) and unaffected sib (upper panel B2 and lower panel B2) both

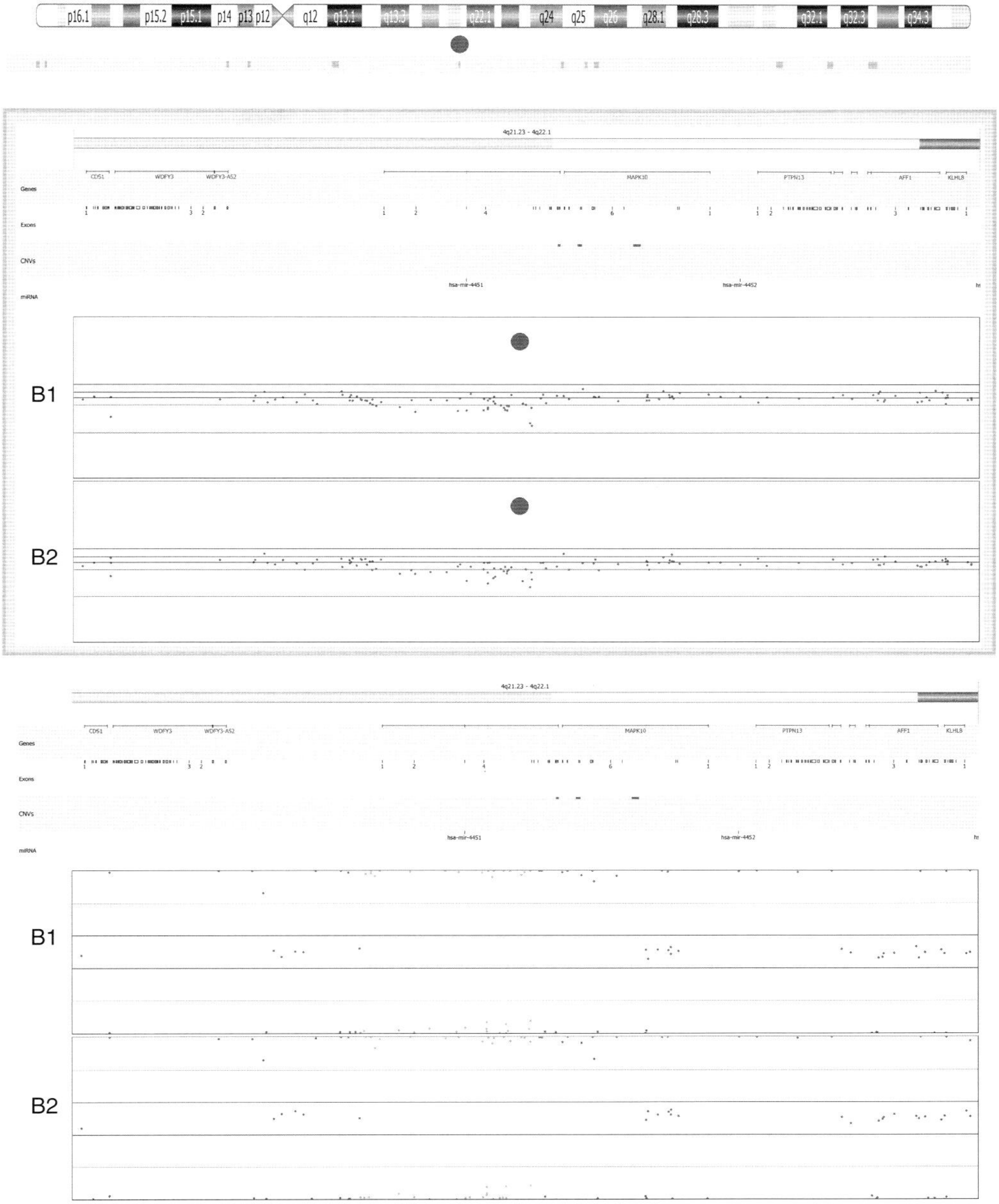

Caption for Figure 12.3. (cont.)

have a 474 kb deletion deleting several exons of the *ARHGAP24* gene. The drop in $\log_2$ R ratio (upper panel, red dots) and loss of one allele type (yellow dots, lower panel) are indicative of a loss of one copy of DNA at that locus. qPCR confirmed that both twins had the same deletion and that this deletion was inherited maternally (Veenma *et al.*, 2012). A black and white version of this figure will appear in some formats. For the color version, please refer to the plate section.

12.3. Considerations in monozygous twins

In general, monozygous twins with a concordant pathogenic genetic aberration have a concordant phenotype. The presence of a newly associated pathogenic CNV in both twins strengthens the implication of the CNV pathogenicity. For instance, two monozygous twin boys with autism both had the newly identified pathogenic 16p11 micro duplication (Tabet *et al.*, 2012). Surprisingly, monozygous twins with concordant chromosomal anomalies, pathogenic CNV, or mutations can also have a discordant phenotype (Figure 12.3). Even monozygous twins with known genetic syndromes can have a highly variable phenotype (Zwijnenburg *et al.*, 2010). For instance, the phenotypical spectrum of the 22q11.2 deletion syndrome (Yamagishi *et al.*, 1998; Vincent *et al.*, 1999) is extremely variable. One of the proposed explanations for the discordance is an undetected difference in deletion size (Halder *et al.*, 2012). Others could be epigenetic differences, differences in blood perfusion, somatic changes, single nucleotide changes and other genetic abnormalities below the detection threshold on the other allele. Phenotypical discordant monozygous twin studies are an excellent model for studying the phenotypical spectrum of known genetic diseases (Zwijnenburg *et al.*, 2010). Critical to consider in some, often late-onset, diseases is that initial phenotypical discordant twins can become concordant – for example, in familial amyloidotic polyneuropathy type 1 (Munar-Ques *et al.*, 1999), schizophrenia (O'Reilly *et al.*, 2013), or spinocerebellar ataxia (O'Reilly *et al.*, 2013). Various factors could contribute to these age of onset differences. For instance, environmental components were suspected to modulate the penetrance of a concordant repeat expansion in a monozygotic twin pair with Huntington's disease (Friedman *et al.*, 2005).

12.3.1. Timing matters; twin separation and mutation

Phenotypical discordance between monozygotic twins at birth can have several reasons, namely, differences in epigenetic modifications, environmental exposure, stochastic factors, but also genetic differences. Here we would like to discuss these DNA differences. This sounds a bit counterintuitive, because by definition the DNA of monozygous twins is the same because they originate from a single zygote. However, there is increasing evidence that there are exceptions and several studies have undertaken the endeavor of comparing their genomes. The rationale behind these types of studies is that discordant somatic mutations arisen during development or life span are responsible for the phenotypical discordance. These discordant twin studies are performed at different resolutions. Karyotyping (see Table 12.2) is able to detect numerical and large structural chromosomal anomalies. CNV analysis (see Table 12.3) is able to detect DNA gains and losses around the gene-level resolution. Finally, whole-genome next-generation sequencing (NGS) which can not only measure SNPs and small DNA gains and losses called Insertion/Deletions (InDel), but also CNV at the exon level, inversions, and translocations. Moreover, this genome-wide screening is not dependent on platform design as is the case with array technology. Here, we will discuss

Table 12.2. Examples of hetero-karyotypic twinning in discordant monozygotic twins.

Genetic abnormality	Location	Tested for monozygosity	Technique	Confirmation method	Discordant phenotype	Ref.
Trisomy	13	Short tandem repeat profiling with seven markers and one gender marker	Quantitative PCR	Karyotyping	NA; radiofrequency ablation of affected twin at 18 weeks of gestation	Dixit *et al.* (2012)
Trisomy	18	Chorionicity (MCDA) and short tandem repeat	Karyotyping after amniocentesis of both sacs	Karyotyping of fibroblasts and chorionic villi taken from different sites of placenta	MCA, 47,XY,+18 versus healthy twin with low-grade (5%) +18 mosaicism in fibroblasts	Reuss *et al.* (2011)
Trisomy	21	Chorionicity and gender	MSS in first trimester and nuchal translucency scan	Karyotyping in one case; others not described	21 cases discordant for Down's syndrome	Egan *et al.* (2014)
Trisomy	21	Chorionicity and short tandem repeat profiling with 10 markers	Karyotyping amniotic fluid cells	Short tandem repeat with seven STR markers on chromosome 21	Placenta was mosaic trisomy 21; 46,XY sib spontaneous aborted. Termination of pregnancy of affected fetus. Twins discordant for Down's syndrome	Nieuwint *et al.* (1999)
Monosomy	X	Chorionicity and short tandem repeat profiling with 11 markers	Karyotyping amniotic fluid cells	Affected twin 45,X [24] and healthy twin low-grade mosaic 45,X [1]/46,XY [29] in fibroblast cultures	Termination of pregnancy; discordant for Turner's syndrome; placenta was mosaic 45,X/46,XY	Nieuwint *et al.* (1999)
Monosomy	X	Chorionicity	Karyotyping and qPCR	FISH; karyotyping of fibroblasts	45,X in one and 46,X, del(X) (p11.1)	Gentilin *et al.* (2008)
Mosaic XY	X, Y	SNP-array	SNP-array	FISH	45,X/46,XX in one and 45,X/46,XX, 46, XY in the other twin	Razzaghian *et al.* (2013)

MSS, maternal serum screening: β-human chorionic gonadotropin (β-HCG) and pregnancy-associated plasma protein A levels in first trimester screening and β-HCG, unconjugated oestriol (μE_3, alpha-fetoprotein (AFP) and inhibin-A levels in the second trimester screening.

Table 12.3. Examples of copy number variation measurements in discordant monozygotic twins.

Genetic abnormality	Location	Tested for monozygosity	Technique	Confirmation method	Discordance	Reference
CNV difference detected						
CNV	12p13; Xp11 and Xq28	Same sex pairs; determination of zygosity not described	Array-CGH	qPCR	CHD discordance	Breckpot *et al.* (2012)
Mosaic CNV (mosaic; deletion)	11q and 4q	SNP-test	BAC and SNP-array	Melting curve analysis and PCR based array	CLL discordance	Bruder *et al.* (2008)[a]
CNV	2q	SNP-array	BAC and SNP-array	Melting curve analysis and PCR based array	Both sibs were healthy	Bruder *et al.* (2008)[a]
Mosaic CNV (mosaic; deletion)	18q21.2qter	STR-profiling with 8 markers	Array-CGH	FISH	Discordant for cleft lip and palate phenotype; 18q21-qter mosaicism in both twins with different frequencies	Essaoui *et al.* (2013)
CNV (mosaic; deletion)	2p25.3	STR-profiling with 25markers	Array-CGH	FISH	Mosaicism; del(2)(p25.3)/dup92)(p25.3) and normal cells (1:1:1) in mildly affected sib; only deleted cells in severely affected sib	Rio *et al.* (2013)
CNV	Multiple	SNP-array	SNP-array	qPCR*	Discordant for schizophrenia	Maiti *et al.* (2011)
CNV	Multiple	SNP-array	SNP-array	–[b]	Discordant for amyotrophic lateral sclerosis	Pamphlett and Morahan (2011)
CNV (loss-*de novo*)	19p13.3	SNP-array	Array-CGH	Custom high-density CGH-array	Discordant for multiple system atrophy	Sasaki *et al.* (2011)

CNV (loss-*de novo*)	22q11	SNP-array?	SNP-array; FISH	–	Discordant 22q11 phenotype deletion size	Halder *et al.* (2012)
No CNV difference detected						
CNV	NA	STR-profiling with 16 markers and SNP-array	SNP-array	qPCR	Discordant for urorectal malformations	Baudisch *et al.* (2013)
CNV	NA	–	BAC-array	–[b]	Discordant for cono-truncal malformations	Jensen *et al.* (2012)
CNV	NA	SNP-test	Array-CGH	qPCR	Discordant for schizophrenia or bipolar disorder	Bloom *et al.* (2013)
CNV	22q11	Red cell antigens; STR-profiling	FISH	–	Twin A: anal atresia, tetralogy of Fallot, pulmonary atresia and characteristic facial features; twin B: only characteristic facial features	Yamagishi *et al.* (1998)
CNV	NA	SNP-genotyping	SNP and CGH-array	DNA-sequencing	Discordant for cleft lip and/or palate	Kimani *et al.* (2009)
CNV	NA	SNP-array	SNP-array	qPCR, DNA-sequencing	Discordant for schizophrenia	Ono *et al.* (2010)
CNV	NA	STR-profiling with 16 markers and SNP-array	SNP-array; NGS-Exome CN	qPCR, FISH	Discordant for CDH or EA/TEF	Veenma *et al.* (2012)

CNV, copy number variation; STR, short tandem repeat; CGH, comparative genomic hybridization; FISH, fluorescence *in situ* hybridization; CDH, congenital diaphragmatic hernia; EA, esophageal atresia; CHD, congenital heart defect; SNP, single nucleotide polymorphism.

[a] Bruder *et al.* (2008) describe several more somatic differences detected with both 32 K BAC-array and 300 K SNP-array. However, these were not described to be validated with a second technique.

[b] No validation experiments addressed in article.

two types of copy number discordance: hetero-karyotypical discordance, and copy number variation.

Mutations or non-disjunction during meiosis or early mitosis can lead to either complete concordant mutated genomic DNA or, via unequal distribution and subsequent rescue mechanisms, to differences in somatic mosaicism in monozygotic twins. Post-zygotic DNA mutations will lead to somatic mosaicism if the cells survive the damage and the mutations are not repaired. These mutations can either be single base-pair changes, deletions and insertions of several bases, or CNVs ranging from dozens of base pairs to gains and losses of several megabases in length. Depending on the developmental stage and cell types affected, CNV can be present throughout the body, or be present at certain frequencies in specific tissues. If germ cells are affected, the mutation can be transmitted to subsequent generations. Differences in timing of separation and mutation result in differences in genetic manifestations; depending on the number of cells with and without the mutation and their distribution across the twins, tissue-specific somatic mosaicism can vary accordingly. Therefore, it is advisable to test different tissues for the presence of a genetic aberration; for example, to search for placental mosaicism, or the presence of low-grade mosaicism in dermal fibroblasts. A source of false-positive results when testing for genetic defects in blood or saliva immediately after birth are those differences caused by transfer of lymphocytes via twin-to-twin transfusion. After several weeks, each twin has their "own" lymphocytes in its circulation, although transfer of hematopoietic stem cells cannot be excluded (Erlich, 2011).

12.3.2. Germline non-disjunction, aneuploidy, and somatic mosaicsm

For example, karyotyping of cultured cells derived after amniocentesis of two monochorionic diamniotic twin boys revealed discordance for trisomy 18. Both twins had trisomy 18 mosaicism of blood and buccal smear samples after birth. However, the placenta turned out to be mosaic for trisomy 18, indicative for a mosaicism in the early zygote. Further testing revealed only a low-grade (5%) mosaicism in skin fibroblasts in one of the sibs (Reuss *et al.*, 2011). Therefore, most likely non-disjunction during meiosis II followed by an unequal distribution of normal/affected cells to the twin sibs led to this discordant karyotype. Analysis of the placenta of monochorionic diamnionic (MCDA) twins, in which one had a non-mosaic deletion of the short arm of chromosome X and the other twin had non-mosaic Turner syndrome, revealed a 45,X/46,X, del(X)(p11.1) karyotype in the placenta (Gentilin *et al.*, 2008). Loss of the p-arm of the X-chromosome has led to complete loss of the aberrant X chromosome during early mitotic events.

Aneuploidy prior to conception –that is, in the germ cells – can result in hetero-karyotypical discordance between monozygotic twins. During meiosis or early post-zygotic mitosis, non-disjunction of sister-chromatids (meiosis II or mitosis) or homologous chromosomes (meiosis I) can lead to an unequal distribution of chromosomes; the resulting cells have either 0 or 2 copies (meiosis I and II non-disjunction), or 1 and 3 copies (mitotic non-disjunction error). Depending on the timing of the twin separation and division of cell lineages, the monozygous twins can be completely discordant for the aneuploidy, or have a certain discordant somatic mosaicism. For example, karyotypes from blood derived from an MCDA twin pregnancy hinted at mosaicism for trisomy 21 in both sibs. However, cheek and skin biopsies revealed two distinct karyotypes: 46,XY in the healthy sib and 47,XY,+21 in the affected sib (Egan *et al.*, 2014). This a clear example of the effect of TTTS responsible for the admixture of 47,XY,+21 cells of the affected to the healthy twin. Also, because both twins were mosaic in blood, there was blood flow from the healthy twin to the affected twin, effectively diluting the 47, XY,+21 cells with 46,XY cells. Because the trisomy was present in skin and blood, non-disjunction during meiosis is a plausible cause of this discordant aneuploidy.

12.3.3. Genetic discordance; somatic mosaicism and *de novo* CNV

Another source of genetic discordance are *de novo* CNVs. Such CNV could arise during meiosis in the parental germ cells, during early zygotic mitotic divisions, or later in life. Therefore, it is recommended to compare the CNV profiles of twins to that of their parents and siblings in addition to the comparison with the traditional cohorts of ancestry-matched affected and non-affected individuals. The *de novo* CNV rate per generation is much higher than the mutation rate; estimates for CNV are 100–10,000 times higher than the single base-pair mutation rate per generation; that is, ~ 2×10^{-8} versus ~$1–2 \times 10^{-6}$ to 10^{-4} (Zhang *et al.*, 2009). Ehli *et al.* describe 26 *de novo* CNV, measured with SNP microarray, in 25 attention deficient concordant and discordant monozygous twin pairs (Ehli *et al.*, 2012). Some of these *de novo* CNVs were recurrent; there were 18 unique targets for a validation experiment. Quantitative PCR primers could be developed on 11 of these target regions. Two of 17 tested *de novo* CNVs could be validated with qPCR and one could not be proven or excluded; that is, 13 seemingly *de novo* CNV were in fact inherited from one of the parents. One of the *de novo* CNV that was present was duplication in chromosomal band 15q11. This duplication was present in a male *concordant unaffected* twin-pair and was a pre-twinning event. The other, a post-twinning, *de novo* CNV was a large 1.3 Mb deletion located on band 4q35 and only observed in the oldest sib of a male concordant affected twin pair. The genotype–phenotype correlation in both instances is difficult because both unaffected twins had a *de novo* duplication and only one of the affected twins had a *de novo* deletion. Somatic mosaicism could not be excluded or confirmed because only buccal swab-derived DNA was available.

Congenital anomalies are a major contributor to prenatal and postnatal mortality and childhood morbidity. Although some of these anomalies can be completely explained by chromosomal anomalies, other genetic defects, teratogens, or malnutrition, the majority of patients still have an unknown disease etiology. Several explanations have been put forward for these unknown etiologies; for example, as-yet undiscovered genetic or epigenetic changes, environmental or nutritional factors, disturbances in biological processes, mechanical explanations or otherwise, and the lack of systematic data collection of the affected child and the parents. Twin studies can help in assessing the contribution of environmental and genetic factors in many of these congenital anomalies. Several studies specifically searched for differences in DNA copy number between discordant monozygous twins with congenital anomalies.

Breckpot *et al.* compared the copy number profiles of six monozygous twin pairs discordant for congenital heart defects. In female twin pair 1, a 186 kb large duplication was identified in the affected sib. In male pair 3, two small 2 kb losses were detected at chromosomal loci 1p33 and 6p25 in the unaffected sib (all false-positive differences). In male twin pair 6, three confirmed and three false-positive CNVs were detected in the affected male sib: the 4 kb gain on chromosomal 12p13, a gain on Xp11.23, and a gain on Xq28 were confirmed with qPCR (Breckpot *et al.*, 2012). In a monozygous twin pair discordant for cleft lip and palate and growth restriction, array-CGH on DNA derived from uncultured amniocytes revealed a terminal deletion of chromosome 18q in the affected sib of a MCDA female twin pair. Subsequent FISH confirmation on cultured cells confirmed a non-mosaic 46,XX,del(18)(q21qter) karyotype in the affected twin, while all cells from the healthy sib were normal. However, fetal blood, skin, tendons, and lung biopsies were mosaic for the chromosome 18q deletion in varying frequencies in both twins (Essaoui *et al.*, 2013). Both fetal blood samples had a 25% mosaicism, while skin, tendon, and lung were low-grade mosaic in one twin and high-grade mosaic in the other. Unequal distribution of affected cells prior to twin separation had resulted in tissue-specific differences in mosaic frequency. Twin-to-twin transfusion resulted in a mixture of normal and affected cells in both twins; it diluted the 18q21 deletion with normal cells in one and transferred affected cells to the other twin.

An intriguing example of somatic copy number variation mosaicism was published recently. Rio and coworkers (2013) described two DCDA monozygous twin sisters in which one girl has facial dysmorphisms, and a delay in fine motor skills, language skills, and mental development. Her sister did not have facial dysmorphisms and only has delayed speech and a unilateral single palmar crease. In the more severely affected twin, a *de novo* non-mosaic deletion in chromosomal band 2p23 was detected with array-CGH and confirmed with locus-specific FISH, while in the mildly affected twin, normal cells, 2p25 deletion and 2p25 duplication cells were detected using FISH only. Array-CGH in the latter was normal because it is an average of all cells, the 2p25 duplicated balanced the signal of the 2p25 deleted cells to a seemingly 2n state. Post-zygotic non-allelic recombination has led to these three distinct cell populations: normal 46,XX cells, 46,XX dup(2)(p25.3), and 46,XX, del(2)(p25.3) cells. This post-zygotic event had to be prior to separation, making the drastic differential distribution of three cell lineages possible.

Although several studies report CNV differences in monozygotic twins discordant for specific congenital anomalies (Breckpot *et al.* 2012; Essaoui *et al.*, 2013; Rio *et al.*, 2013), others have not found these differences (Kimani *et al.* 2009; Veenma *et al.* 2012; Baudisch *et al.*, 2013). In our study, the DNA of phenotypical discordant twins for esophageal atresia (seven pairs) and congenital diaphragmatic hernia (four pairs) were compared using SNP-array. Genotyping differences were validated with Sanger sequencing and CNVs confirmed with qPCR. No SNP or (mosaic) CNV differences were present in these twins (Veenma *et al.*, 2012). Several concordant CNPs and CNVs were observed. Although no shared rare inherited CNV were observed in this limited number of patients, a few inherited CNV could be modulators or predisposing factors which, in combination with other factors, could have relevance to the disease phenotype. For instance, the duplication observed in both sibs of an esophageal atresia discordant monozygotic twin pair affects the *PARK2* gene; this gene is also duplicated in patients with low-grade dysplasia of the esophagus (Gu *et al.*, 2010). The deletion of (part of) the *ARHGAP24* gene in both the affected and unaffected members of a monozygous twin pair discordant for congenital diaphragmatic hernia by disturbing normal angiogenesis was seen (Veenma *et al.*, 2012). Baudisch and coworkers performed a similar experiment in four monozygous twins discordant for urorectal malformations. They also did not find any CNV differences between twins or *de novo* CNV in both twins (Baudisch *et al.*, 2013).

12.3.4. Post-twinning mutation and tissue-specific somatic mosaicism

Somatic changes can be responsible for slight genomic differences between monozygotic twins, but may also result in differences within different body tissues of the same individual. For instance, Liang *et al.* analyzed the genomes of clones from mouse embryonic stem cells and measured differences in copy number at several loci in those clones. These CNV arose during mitotic cell divisions. If similar events take place in human stem cells we could have many somatic CNV mosaicism in our tissues (Liang *et al.*, 2008). There can be accumulation of somatic CNV in the adult with an increased incidence over time. In a large cohort of monozygotic twins, Forsberg could detect mosaicism for megabase size CNV in 3.4% of individuals of 60 years and older, while these were absent in the younger study cohort (33–55 years) (Forsberg *et al.*, 2012). In five of these monozygous twin pairs, the CNV profiles were determined over time; a 5q deletion increased in frequency in female MZ sib with dysthymic disorder, cholelithiasis, and cerebral infarction. In other, unrelated, patients with several forms of malignancies, specific CNV accumulation was present. In fact, in two of these patients the CNV were indicative of, but at that time not yet diagnosed, myelodysplastic syndrome. In an apparently healthy individual, Piotrowski and coworkers described somatic CNV differences between cardiac and cerebellar cortex tissue-derived DNA (Piotrowski *et al.*, 2008). In fact, human neuronal cells can differ greatly in DNA content (Fischer *et al.*, 2012) and discordant CNVs are measured in single cell assays (McConnell *et al.*, 2013). Accumulation of these genomic imbalances could be an underlying mechanism of late-onset neuropathies (Fischer *et al.*, 2012; McConnell *et al.*, 2013). Perhaps it is advisable to screen individuals not only several times during their life, but also to evaluate multiple tissues.

CNVs that arise after the separation of the monozygous twins are completely discordant between those twins. After zygosity testing, Bruder *et al.* compared the genome of 9 Parkinson or Parkinson-like discordant monozygotic twins and 10 healthy monozygotic twins (Bruder *et al.*, 2008). They found (and confirmed with additional techniques) three somatic intra-twin differences. In an 80-year-old male twin pair, in the sib with Parkinson's disease, an approximately 20% mosaicism was detected for two large deletions: one on chromosome 11q and the other on chromosome 4q. This aberration is often seen in patients with chronic lymphocytic leukemia and this patient with Parkinson's 's disease was previously diagnosed with this condition. In addition to these tumorigenesis-related CNVs, another discordant CNV was detected in a concordant healthy 60-year-old female monozygous twin pair. This CNV, a 70–80% mosaic deletion of chromosome 2p23, was only present in one of the sisters and was confirmed with high-resolution melting curve analysis and pyro-sequencing.

Sasaki and colleagues compared the CNV profile of a 67-year-old monozygotic twin pair (Sasaki *et al.*, 2011). The affected sib was diagnosed with multiple system atrophy (MSA) at the age of 57 years; atrophy of the brainstem and cerebellum was confirmed by MRI. His brother did not have MSA symptoms. Three discordant CNVs, present in the affected twin only, were identified with array-CGH: loss in chromosomal bands 2p25, loss on 4q35, and a loss on 19p13. Two independent array experiments (SNP-array and a custom CGH-array) confirmed the presence of the 19p13 loss in 10 additional MSA patients. This loss was absent in 125 controls. These authors successfully used a twin-CNV discordance experiment to find a possible disease-associated CNV and confirmed the pathogenicity of this CNV with a burden test in a case control study.

CNV analysis in concordant and discordant monozygotic twins can be a valuable asset in determining the presence and identity of contributing genetic factors to specific traits and diseases. Caution has to be taken in the interpretation of these twin-study results, because environmental and genetic components are not always exactly identical in monozygous twins. Moreover, twin pregnancy may not be representative of a singleton pregnancy due to atypical growth *in utero* (Hall, 2003). TTTS can cause intrauterine environmental differences; somatic mutations can lead to genetic mosaicism in different tissues within one individual and between twins. It is advisable to screen for genetic abnormalities at several time-points during a person's life span and in different tissues. The accumulation of somatic changes could be responsible for a significant proportion of human disease. Moreover, determining the zygosity status in all pregnancies is advisable. Undiagnosed "vanishing" twins and TTTS could cause congenital malformations, but MZ twin pregnancies could also result in masking genetic aberrations in blood cells. Comparing different tissues, if possible, could be crucial in determining genetic aberrations in congenital anomalies. Comparing the genomes of large cohorts of twins and their parents may reveal not only *de novo* genetic changes, but perhaps also shed light on predisposing factors modulating penetrance in traits and diseases in combination with environmental or other factors.

12.4. Future prospects

12.4.1. Epigenetic studies in discordant monozygotic twins

Epigenetic mechanisms such as X-inactivation, DNA-methylation, histone modifications, or other mechanisms influence chromatin structure, maintenance, or remodeling and thereby influence the expression of genes. These epigenetic signatures can be heritable, vary over time, and are responsive to environmental cues (Bonasio *et al.*, 2010). For example, DNA methylation differences can develop over time in monozygous twins, possibly caused by environmental influences (Wong *et al.*, 2010). Although these epigenetic signatures in monozygous twins are mostly very similar, they can differ between monozygous twins. These differences in epigenetic signature have been observed in discordant monozygous twin pairs – for example, in systemic lupus erythematosus (Javierre *et al.*, 2010), RETT syndrome (cerebroatrophic hyperammonemia; Miyake *et al.*, 2013), and in Beckwith–Wiedeman syndrome (Weksberg *et al.*, 2002). It has been proposed that epigenetic changes could be a major contributor to monozygotic twin discordance (Czyz *et al.*, 2012), and studying these epigenetic differences in monozygotic twins could bridge the gap between the environment and our genome (Zhang *et al.*, 2012). Large-scale studies in discordant monozygous twins could be a successful strategy to identify epigenetic variation responsible for complex traits and diseases (Bell and Spector, 2011). Technically, epigenetic profiling is, like copy number measurements, in transition from single gene and genome-wide array-based techniques to high-throughput techniques based on NGS (Huss, 2010). So many more examples of discordant methylation patterns will follow in the near future.

12.4.2. NGS-based copy number profiling and GWAS

Using whole-genome NGS data, both copy number and nucleotide sequence can be determined. GWAS on both SNP and CNV level could be performed on these data, and not the *associated* locus but the *responsible* genetic variant could be identified. While variant calling on nucleotide resolution has seen tremendous progress, InDel and CNV calling are still in their infancy. Current NGS-based methods use read depth, the number of unique reads that map to a given locus in the genome, to determine the copy number status at that locus. Comparing the read counts of individual reference samples (paired analysis) or using a normalized cohort of reference samples and comparing this with patient samples will give a fairly good estimate of the relative copy number. This information can be combined with allelic frequency data obtained from the variant files, much like it is with SNP-array analysis. Several tools exist, each with its advantages and disadvantages, such as reliability to accurately detect relative gains or losses (de Ligt *et al.*, 2013; Duan *et al.*, 2013; Guo *et al.*, 2013). Challenges in accurately detecting copy number from sequencing data are aligning sequence reads in repeat regions, and issues that arise from capturing segments of the DNA during exome-NGS (Teo *et al.*, 2012). It is likely that SNP-array-based GWAS studies will be replaced by NGS-based methods in the upcoming years. Although GWAS have proven their value in medical research, publications in CNV-based association studies in twins are relatively scarce. Burden studies in monozygous twins, both concordant and discordant for a given phenotype, could prove to be a valuable addition to the phenotype–genotype toolset. Tissue-specific whole-genome and epigenome profiling in large cohorts of concordant and discordant monozygous twins could be a promising key to unraveling the impact of gene, environment, and gene–environment interactions.

REFERENCES

Basso, O., Nohr, E.A., Christensen, K. and Olsen, J. (2004). Risk of twinning as a function of maternal height and body mass index. *J. Am. Med. Assoc.*, **291**(13), 1564–1566.

Baudisch, F., Draaken, M., Bartels, E., *et al.* (2013). CNV analysis in monozygotic twin pairs discordant for urorectal malformations. *Twin Res. Hum. Genet.*, **16**(4), 802–807.

Bell, J.T. and Spector, T.D. (2011). A twin approach to unraveling epigenetics. *Trends Genet.*, **27**(3), 116–125.

Bergen, S.E., O'Dushlaine, C.T., Ripke, S., *et al.* (2012). Genome-wide association study in a Swedish population yields support for greater CNV and MHC involvement in schizophrenia compared with bipolar disorder. *Molec. Psych.*, **17**(9), 880–886.

Bloom, R.J., Kahler, A.K., Collins, A.L., *et al.* (2013). Comprehensive analysis of copy number variation in monozygotic twins discordant for bipolar disorder or schizophrenia. *Schizophr. Res.*, **146**(1–3), 289–290.

Bonasio, R., Tu, S. and Reinberg, D. (2010). Molecular signals of epigenetic states. *Science*, **330**(6004), 612–616.

Breckpot, J., Thienpont, B., Gewillig, M., *et al.* (2012). Differences in copy number variation between discordant monozygotic twins as a model for exploring chromosomal mosaicism in congenital heart defects. *Molec. Syndromol.*, **2**(2), 81–87.

Bruder, C.E., Piotrowski, A., Gijsbers, A.A., *et al.* (2008). Phenotypically concordant and discordant monozygotic twins display different DNA copy-number-variation profiles. *Am. J. Hum. Genet.*, **82**(3), 763–771.

Cantsilieris, S., White, S.J., Richardson, A.J., Guymer, R.H. and Baird, P.N. (2012). Comprehensive analysis of copy number variation of genes at chromosome 1 and 10 loci associated with late age related macular degeneration. *PLoS ONE*, **7**(4), e35255.

Craddock, N., Hurles, M.E., Cardin, N., *et al.* (2010). Genome-wide association study of CNVs in 16,000 cases of eight common diseases and 3,000 shared controls. *Nature*, **464**(7289), 713–720.

Czyz, W., Morahan, J.M., Ebers, G.C. and Ramagopalan, S.V. (2012). Genetic, environmental and stochastic factors in monozygotic twin discordance with a focus on epigenetic differences. *BMC Med*, **10**, 93.

de Ligt, J., Boone, P.M., Pfundt, R., *et al.* (2013). Detection of clinically relevant copy number variants with whole-exome sequencing. *Hum. Mutat.*, **34**(10), 1439–1448.

Dixit, A., Tanteles, G., Ocraft, K., McEwan, A. and Sarkar, A. (2012). Monozygotic twins discordant for trisomy 13: counselling and management issues. *J. Perinatol.*, **32**(8), 639–641.

Duan, J., Zhang, J.G., Deng, H.W. and Wang, Y.P. (2013). Comparative studies of copy number variation detection methods for next-generation sequencing technologies. *PLoS ONE*, **8**(3), e59128.

Egan, E., Reidy, K., O'Brien, L., Erwin, R. and Umstad, M. (2014). The outcome of twin pregnancies discordant for trisomy 21. *Twin Res. Hum. Genet.*, **17**(1), 38–44.

Ehli, E.A., Abdellaoui, A., Hu, Y., *et al.* (2012). *De novo* and inherited CNVs in MZ twin pairs selected for discordance and concordance on Attention Problems. *Eur. J. Hum. Genet.*, **20**(10), 1037–1043.

Erlich, Y. (2011). Blood ties: chimerism can mask twin discordance in high-throughput sequencing. *Twin Res. Hum. Genet.*, **14**(2), 137–143.

Essaoui, M., Nizon, M., Beaujard, M.P., *et al.* (2013). Monozygotic twins discordant for 18q21.2qter deletion detected by array CGH in amniotic fluid. *Eur. J. Med. Genet.*, **56**(9), 502–505.

Fernandez-Rozadilla, C., Cazier, J.B., Tomlinson, I., *et al.* (2014). A genome-wide association study on copy-number variation identifies a 11q11 loss as a candidate susceptibility variant for colorectal cancer. *Hum. Genet.*, **133**(5), 525–534.

Fischer, H.G., Morawski, M., Bruckner, M.K., *et al.* (2012). Changes in neuronal DNA content variation in the human brain during aging. *Aging Cell*, **11**(4), 628–633.

Forsberg, L.A., Rasi, C., Razzaghian, H.R., *et al.* (2012). Age-related somatic structural changes in the nuclear genome of human blood cells. *Am. J. Hum. Genet.*, **90**(2), 217–228.

Friedman, J.H., Trieschmann, M.E., Myers, R.H. and Fernandez, H.H. (2005). Monozygotic twins discordant for Huntington disease after 7 years. *Arch. Neurol.*, **62**(6), 995–997.

Friedrichsen, M., Poulsen, P., Wojtaszewski, J., *et al.* (2013). Carboxylesterase 1 gene duplication and mRNA expression in adipose tissue are linked to obesity and metabolic function. *PLoS ONE*, **8**(2), e56861.

Gentilin, B., Guerneri, S., Bianchi, V., *et al.* (2008). Discordant prenatal phenotype and karyotype of monozygotic twins characterized by the unequal distribution of two cell lines investigated by different methods: a review. *Twin Res. Hum. Genet.*, **11**(3), 352–356.

Gibson, G. (2011). Rare and common variants: twenty arguments. *Nature Rev. Genet.*, **13**(2), 135–145.

Grayson, B.L., Smith, M.E., Thomas, J.W., *et al.* (2010). Genome-wide analysis of copy number variation in type 1 diabetes. *PLoS ONE*, **5**(11), e15393.

Gu, J., Ajani, J.A., Hawk, E.T., *et al.* (2010). Genome-wide catalogue of chromosomal aberrations in barrett's esophagus and esophageal adenocarcinoma: a high-density single nucleotide polymorphism array analysis. *Cancer Prev. Res. (Phila.)*, **3**(9), 1176–1186.

Guo, Y., Sheng, Q., Samuels, D.C., *et al.* (2013). Comparative study of exome copy number variation estimation tools using array comparative genomic hybridization as control. *Biomed. Res. Int.*, **2013**, 915636.

Halder, A., Jain, M., Chaudhary, I. and Varma, B. (2012). Chromosome 22q11.2 microdeletion in monozygotic twins with discordant phenotype and deletion size. *Molec. Cytogenet.*, **5**(1), 13.

Hall, J.G. (2003). Twinning. *Lancet*, **362**(9385), 735–743.

Haraksingh, R.R., Abyzov, A., Gerstein, M., Urban, A.E.and Snyder, M. (2011). Genome-wide mapping of copy number variation in humans: comparative analysis of high resolution array platforms. *PLoS ONE*, **6**(11), e27859.

Hehir-Kwa, J.Y., Rodriguez-Santiago, B., Vissers, L.E., *et al.* (2011). *De novo* copy number variants associated with intellectual disability have a paternal origin and age bias. *J. Med. Genet.*, **48**(11), 776–778.

Hindorff, L.A., Sethupathy, P., Junkins, H.A., *et al.* (2009). Potential etiologic and functional implications of genome-wide association loci for human diseases and traits. *Proc. Natl Acad. Sci. USA*, **106**(23), 9362–9367.

Hitz, M.P., Lemieux-Perreault, L.P., Marshall, C., *et al.* (2012). Rare copy number variants contribute to congenital left-sided heart disease. *PLoS Genet.*, **8**(9), e1002903.

Huss, M. (2010). Introduction into the analysis of high-throughput-sequencing based epigenome data. *Brief Bioinform.*, **11**(5), 512–523.

Iafrate, A.J., Feuk, L., Rivera, M.N., *et al.* (2004). Detection of large-scale variation in the human genome. *Nature Genet.*, **36**(9), 949–951.

Iakoubov, L., Mossakowska, M., Szwed, M., *et al.* (2013). A common copy number variation (CNV) polymorphism in the *CNTNAP4* gene: association with aging in females. *PLoS ONE*, **8**(11), e79790.

Javierre, B.M., Fernandez, A.F., Richter, J., *et al.* (2010). Changes in the pattern of DNA methylation associate with twin discordance in systemic lupus erythematosus. *Genome Res.*, **20**(2), 170–179.

Jensen, H., Kjeldsen, E. and Hjortdal, V.E. (2012). Could submicroscopical chromosomal imbalances cause cono-truncal malformations in twins? *Congenit. Heart Dis.*, **7**(2), 170–177.

Kimani, J.W., Yoshiura, K., Shi, M., *et al.* (2009). Search for genomic alterations in monozygotic twins discordant for cleft lip and/or palate. *Twin Res. Hum. Genet.*, **12**(5), 462–468.

Lee, Y.M., Cleary-Goldman, J., Thaker, H.M. and Simpson, L. L. (2006). Antenatal sonographic prediction of twin chorionicity. *Am. J. Obstet. Gynecol.*, **195**(3), 863–867.

Liang, Q., Conte, N., Skarnes, W.C. and Bradley, A. (2008). Extensive genomic copy number variation in embryonic stem cells. *Proc. Natl Acad. Sci. USA*, **105**(45), 17453–17456.

Lin, P., Hartz, S.M., Wang, J.C., *et al.* (2012). Copy number variations in 6q14.1 and 5q13.2 are associated with alcohol dependence. *Alcohol. Clin. Exp. Res.*, **36**(9), 1512–1518.

Machin, G. (2009). Non-identical monozygotic twins, intermediate twin types, zygosity testing, and the non-random nature of monozygotic twinning: a review. *Am. J. Med. Genet. C Semin. Med. Genet.*, **151c**(2), 110–127.

Maiti, S., Kumar, K.H., Castellani, C.A., O'Reilly, R.and Singh, S.M. (2011). Ontogenetic *de novo* copy number variations (CNVs) as a source of genetic individuality: studies on two families with MZD twins for schizophrenia. *PLoS ONE*, **6**(3), e17125.

Manolio, T.A. (2013). Bringing genome-wide association findings into clinical use. *Nature Rev. Genet.*, **14**(8), 549–558.

McCarroll, S.A. (2008). Extending genome-wide association studies to copy-number variation. *Hum. Molec. Genet.*, **17**(R2), R135–142.

McCarroll, S.A. and Altshuler, D.M. (2007). Copy-number variation and association studies of human disease. *Nature Genet.*, **39**(7 Suppl), S37–42.

McCarroll, S.A., Huett, A., Kuballa, P., *et al.* (2008). Deletion polymorphism upstream of IRGM associated with altered IRGM expression and Crohn's disease. *Nature Genet.*, **40**(9), 1107–1112.

McConnell, M.J., Lindberg, M.R., Brennand, K.J., *et al.* (2013). Mosaic copy number variation in human neurons. *Science*, **342**(6158), 632–637.

Miyake, K., Yang, C., Minakuchi, Y., *et al.* (2013). Comparison of genomic and epigenomic expression in monozygotic twins discordant for Rett syndrome. *PLoS ONE*, **8**(6), e66729.

Munar-Ques, M., Pedrosa, J.L., Coelho, T., *et al.* (1999). Two pairs of proven monozygotic twins discordant for familial amyloid neuropathy (FAP) TTR Met 30. *J. Med. Genet.*, **36**(8), 629–632.

Nag, A., Venturini, C., Hysi, P.G., *et al.* (2013). Copy number variation at chromosome 5q21.2 is associated with intraocular pressure. *Invest. Ophthalmol. Vis. Sci.*, **54**(5), 3607–3612.

Nelson, M.R., Wegmann, D., Ehm, M.G., *et al.* (2012). An abundance of rare functional variants in 202 drug target genes sequenced in 14,002 people. *Science*, **337**(6090), 100–104.

Nieuwint, A., Van Zalen-Sprock, R., Hummel, P., *et al.* (1999). "Identical" twins with discordant karyotypes. *Prenat. Diagn.*, **19**(1), 72–76.

O'Reilly, R., Torrey, E.F., Rao, J. and Singh, S. (2013). Monozygotic twins with early-onset schizophrenia and late-onset bipolar disorder: a case report. *J. Med. Case Rep.*, **7**(1), 134.

Oei, L., Hsu, Y.H., Styrkarsdottir, U., *et al.* (2014). A genome-wide copy number association study of osteoporotic fractures points to the 6p25.1 locus. *J. Med. Genet.*, 51(2), 122–131.

Ono, S., Imamura, A., Tasaki, S., *et al.* (2010). Failure to confirm CNVs as of aetiological significance in twin pairs discordant for schizophrenia. *Twin Res. Hum. Genet.*, **13**(5), 455–460.

Pamphlett, R. and Morahan, J.M. (2011). Copy number imbalances in blood and hair in monozygotic twins discordant for amyotrophic lateral sclerosis. *J. Clin. Neurosci.*, **18**(9), 1231–1234.

Pharoah, P.O. (2005). Causal hypothesis for some congenital anomalies. *Twin Res. Hum. Genet.*, **8**(6), 543–550.

Pharoah, P.O., Glinianaia, S.V. and Rankin, J. (2009). Congenital anomalies in multiple births after early loss of a conceptus. *Hum. Reprod.*, **24**(3), 726–731.

Pinto, D., Darvishi, K., Shi, X., *et al.* (2011). Comprehensive assessment of array-based platforms and calling algorithms for detection of copy number variants. *Nature Biotechnol.*, **29**(6), 512–520.

Piotrowski, A., Bruder, C.E., Andersson, R., *et al.* (2008). Somatic mosaicism for copy number variation in differentiated human tissues. *Hum. Mutat.*, **29**(9), 1118–1124.

Razzaghian, H.R., Forsberg, L.A., Prakash, K.R., *et al.* (2013). Post-zygotic and inter-individual structural genetic variation in a presumptive enhancer element of the locus between the *IL10Rbeta* and *IFNAR1* genes. *PLoS ONE*, **8**(9), e67752.

Reddy, U.M., Branum, A.M. and Klebanoff, M.A. (2005). Relationship of maternal body mass index and height to twinning. *Obstet. Gynecol.*, **105**(3), 593–597.

Reuss, A., Gerlach, H., Bedow, W., *et al.* (2011). Monozygotic twins discordant for trisomy 18. *Ultrasound Obstet. Gynecol.*, **38**(6), 727–728.

Rio, M., Royer, G., Gobin, S., *et al.* (2013). Monozygotic twins discordant for submicroscopic chromosomal anomalies in 2p25.3 region detected by array CGH. *Clin. Genet.*, **84**(1), 31–36.

Russell, R.B., Petrini, J.R., Damus, K., Mattison, D.R.and Schwarz, R.H. (2003). The changing epidemiology of multiple births in the United States. *Obstet. Gynecol.*, **101**(1), 129–135.

Sasaki, H., Emi, M., Iijima, H., *et al.* (2011). Copy number loss of (src homology 2 domain containing)-transforming protein 2 (*SHC2*) gene: discordant loss in monozygotic twins and frequent loss in patients with multiple system atrophy. *Molec. Brain*, **4**, 24.

Schlattl, A., Anders, S., Waszak, S.M., Huber, W. and Korbel, J.O. (2011). Relating CNVs to transcriptome data at fine resolution: assessment of the effect of variant size, type, and overlap with functional regions. *Genome Res.*, **21**(12), 2004–2013.

Scott, D.A., Klaassens, M., Holder, A.M., *et al.* (2007). Genome-wide oligonucleotide-based array comparative genome hybridization analysis of non-isolated congenital diaphragmatic hernia. *Hum. Molec. Genet.*, **16**(4), 424–430.

Sebat, J., Lakshmi, B., Troge, J., *et al.* (2004). Large-scale copy number polymorphism in the human genome. *Science*, **305**(5683), 525–528.

Sebat, J., Lakshmi, B., Malhotra, D., *et al.* (2007). Strong association of *de novo* copy number mutations with autism. *Science*, **316**(5823), 445–449.

Sebire, N.J., Talbert, D. and Fisk, N.M. (2001). Twin-to-twin transfusion syndrome results from dynamic asymmetrical reduction in placental anastomoses: a hypothesis. *Placenta*, **22**(5), 383–391.

Shetty, A. and Smith, A.P. (2005). The sonographic diagnosis of chorionicity. *Prenat. Diagn.*, **25**(9), 735–739.

Smeets, D., van Vugt, J.M., Gomes, I., *et al.* (2013). Monochorionic dizygous twins presenting with blood chimerism and discordant sex. *Twin Res. Hum. Genet.*, **16**(4), 799–801.

Soemedi, R., Wilson, I.J., Bentham, J., *et al.* (2012). Contribution of global rare copy-number variants to the risk of sporadic congenital heart disease. *Am. J. Hum. Genet.*, **91**(3), 489–501.

Souter, V.L., Parisi, M.A., Nyholt, D.R., *et al.* (2007). A case of true hermaphroditism reveals an unusual mechanism of twinning. *Hum. Genet.*, **121**(2), 179–185.

Stadler, Z.K., Esposito, D., Shah, S., *et al.* (2012). Rare *de novo* germline copy-number variation in testicular cancer. *Am. J. Hum. Genet.*, **91**(2), 379–383.

Stankiewicz, P. and Lupski, J.R. (2010). Structural variation in the human genome and its role in disease. *Annu. Rev. Med.*, **61**, 437–455.

Tabet, A.C., Pilorge, M., Delorme, R., *et al.* (2012). Autism multiplex family with 16p11.2p12.2 microduplication syndrome in monozygotic twins and distal 16p11.2 deletion in their brother. *Eur. J. Hum. Genet.*, **20**(5), 540–546.

Tan, Q., Ohm Kyvik, K., Kruse, T.A.and Christensen, K. (2010). Dissecting complex phenotypes using the genomics of twins. *Funct. Integr. Genomics*, **10**(3), 321–327.

Tennessen, J.A., Bigham, A.W., O'Connor, T.D., *et al.* (2012). Evolution and functional impact of rare coding variation from deep sequencing of human exomes. *Science*, **337**(6090), 64–69.

Teo, S.M., Pawitan, Y., Ku, C. S., Chia, K.S. and Salim, A. (2012). Statistical challenges associated with detecting copy number variations with next-generation sequencing. *Bioinformatics*, **28**(21), 2711–2718.

Tucker, T., Montpetit, A., Chai, D., *et al.* (2011). Comparison of genome-wide array genomic hybridization platforms for the detection of copy number variants in idiopathic mental retardation. *BMC Med. Genomics.*, **4**, 25.

van Dongen, J., Slagboom, P.E., Draisma, H.H., Martin, N.G. and Boomsma, D.I. (2012). The continuing value of twin studies in the omics era. *Nature Rev. Genet.*, **13**(9), 640–653.

van Duyvenvoorde, H.A., Lui, J.C., Kant, S.G., *et al.* (2014). Copy number variants in patients with short stature. *Eur. J. Hum. Genet.*, 22(5), 602–609.

Veenma, D., Brosens, E., de Jong, E., *et al.* (2012). Copy number detection in discordant monozygotic twins of Congenital Diaphragmatic Hernia (CDH) and Esophageal Atresia (EA) cohorts. *Eur. J. Hum. Genet.*, **20**(3), 298–304.

Vincent, M.C., Heitz, F., Tricoire, J., *et al.* (1999). 22q11 deletion in DGS/VCFS monozygotic twins with discordant phenotypes. *Genet. Couns.*, **10**(1), 43–49.

Visser, R., Gijsbers, A., Ruivenkamp, C., *et al.* (2010). Genome-wide SNP array analysis in patients with features of sotos syndrome. *Horm. Res. Paediatr.*, **73**(4), 265–274.

Wat, M.J., Veenma, D., Hogue, J., *et al.* (2011). Genomic alterations that contribute to the development of isolated and non-isolated congenital diaphragmatic hernia. *J. Med. Genet.*, **48**(5), 299–307.

Weber, M.A. and Sebire, N.J. (2010). Genetics and developmental pathology of twinning. *Semin. Fetal Neonatal Med.*, **15**(6), 313–318.

Weksberg, R., Shuman, C., Caluseriu, O., *et al.* (2002). Discordant *KCNQ1OT1* imprinting in sets of monozygotic twins discordant for Beckwith–Wiedemann syndrome. *Hum. Molec. Genet.*, **11**(11), 1317–1325.

Wheeler, E., Huang, N., Bochukova, E.G., *et al.* (2013). Genome-wide SNP and CNV analysis identifies common and low-frequency variants associated with severe early-onset obesity. *Nature Genet.*, **45**(5), 513–517.

Wong, C.C., Caspi, A., Williams, B., *et al.* (2010). A longitudinal study of epigenetic variation in twins. *Epigenetics*, **5**(6), 516–526.

Yamagishi, H., Ishii, C., Maeda, J., *et al.* (1998). Phenotypic discordance in monozygotic twins with 22q11.2 deletion. *Am. J. Med. Genet.*, **78**(4), 319–321.

Yang, R., Chen, B., Pfutze, K., *et al.* (2014). Genome-wide analysis associates familial colorectal cancer with increases in copy number variations and a rare structural variation at 12p12.3. *Carcinogenesis*, 35(2), 315–323.

Zhang, D., Li, S., Tan, Q. and Pang, Z. (2012). Twin-based DNA methylation analysis takes the center stage of studies of human complex diseases. *J. Genet. Genom.*, **39**(11), 581–586.

Zhang, F., Gu, W., Hurles, M.E. and Lupski, J.R. (2009). Copy number variation in human health, disease, and evolution. *Annu. Rev. Genom. Hum. Genet.*, **10**, 451–481.

Zwijnenburg, P.J., Meijers-Heijboer, H. and Boomsma, D.I. (2010). Identical but not the same: the value of discordant monozygotic twins in genetic research. *Am. J. Med. Genet. B Neuropsychiatr. Genet.*, **153b**(6), 1134–1149.

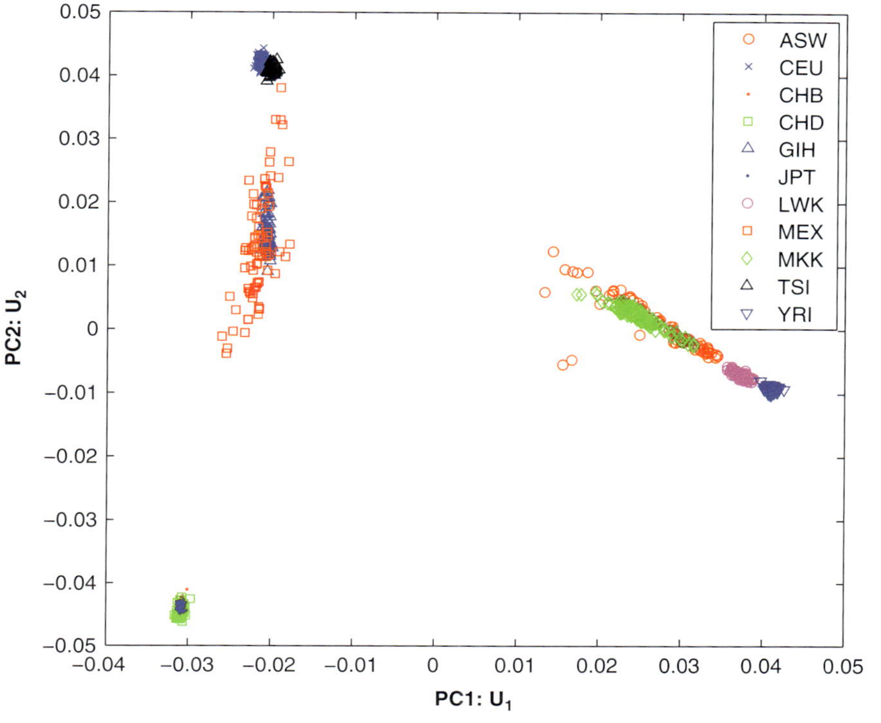

Figure 3.1. PCA analysis of HapMap data. Genotypes of 1184 individuals from 11 sites around the world are available at the HapMap website. Non-founders are removed from PCA analysis, and 988 individuals remain. ASW, African ancestry in Southwest USA; CEU, Utah residents with Northern and Western European ancestry from the CEPH collection; CHB, Han Chinese in Beijing, China; CHD, Chinese in Metropolitan Denver, Colorado; GIH, Gujarati Indians in Houston, Texas; JPT, Japanese in Tokyo, Japan; LWK, Luhya in Webuye, Kenya; MXL, Mexican ancestry in Los Angeles, California; MKK, Maasai in Kinyawa, Kenya; TSI, Toscani in Italia; YRI, Yoruba in Ibadan, Nigeria.

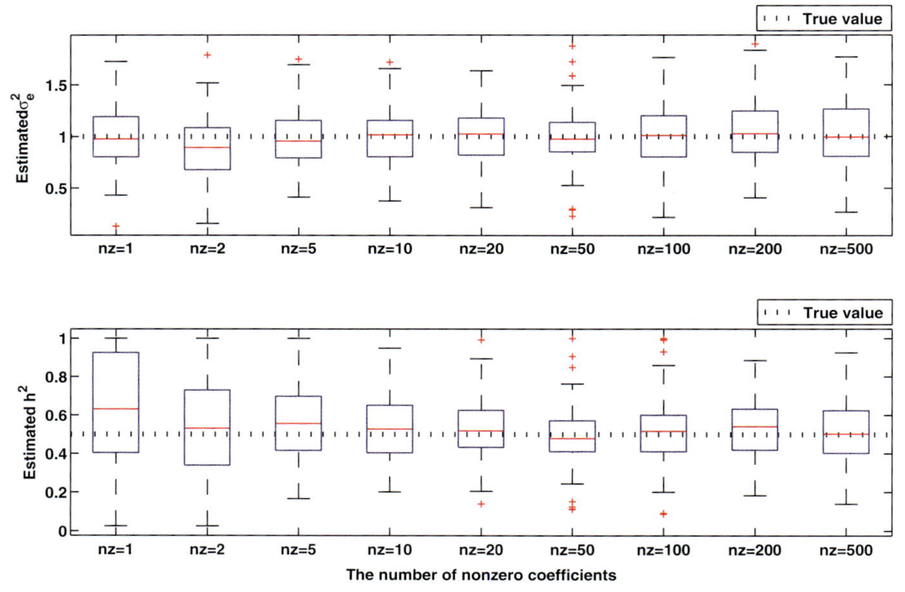

Figure 3.3. Performance of LMM for variance component and heritability estimation. Upper panel: the estimation of σ_e^2 under different *nz*. Lower panel: heritability estimation (h_g^2) under different *nz*. The results were summarized on 100 replications. We can see that LMM works well for heritability estimation under different *nz*. Although bias estimation of h_g^2 has been observed when $nz = 1$, it will disappear as the sample size tends to infinity. Here the sample size is $n = 200$.

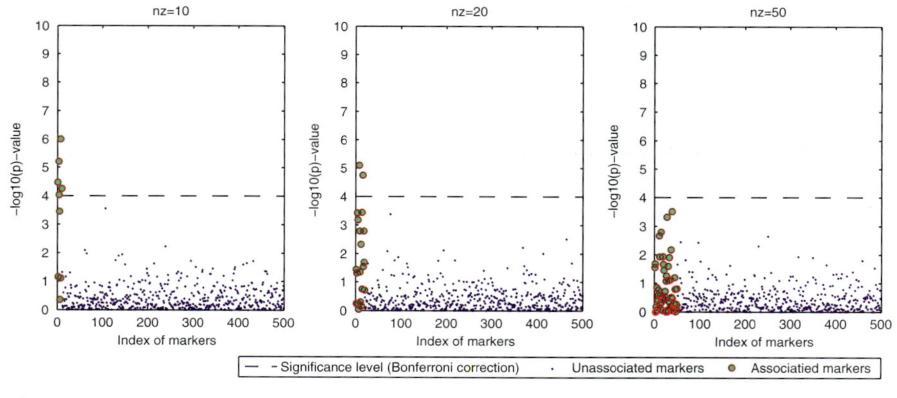

Figure 3.4. Illustration of association mapping ($n = 200$, $M = 500$) for different *nz*. The results from left to right are corresponding to $nz = 10$, 20, and 50, respectively.

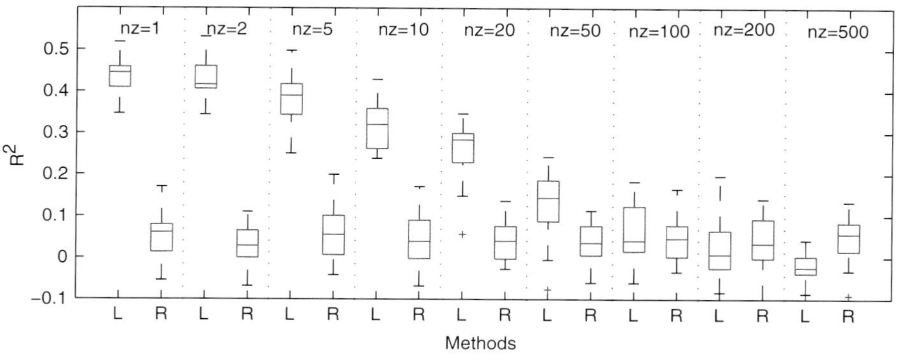

Figure 3.5. Comparison of lasso (L) and ridge (R) regression for different nonzero coefficients ($nz =$ 1, 2, 5, 10, 20, 50, 100, 200, 500). The results were summarized based on 20 replications.

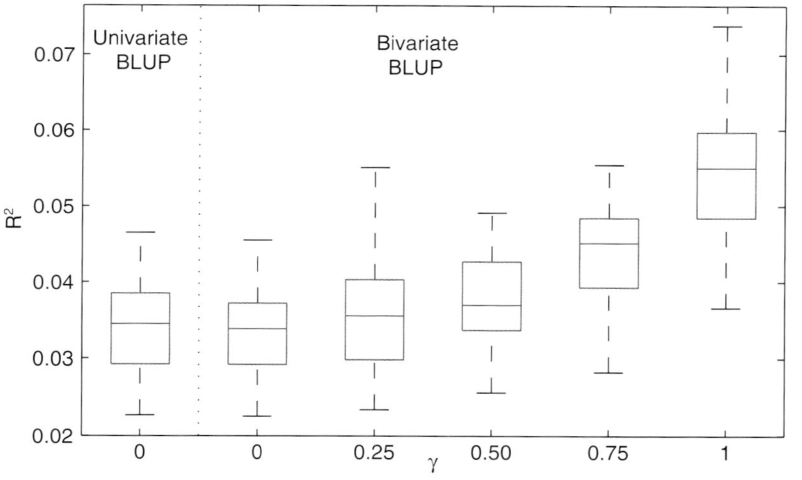

Figure 3.6. The comparison of prediction accuracies of bivariate BLUP and univariate BLUP. Sample size for each phenotype is 2000. Total number of SNPs is 20,000. Number of causal SNPs is 2000 for each phenotype. The parameter γ represents the proportion of shared causal SNPs. As γ increases, bivariate BLUP achieves better prediction accuracy than univariate BLUP.

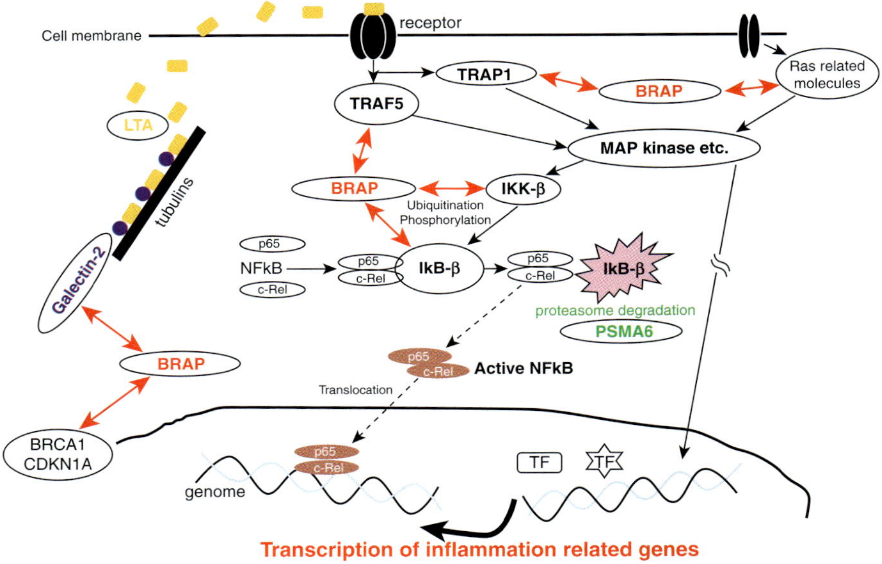

Figure 6.2. MI-related inflammatory cascade. Red arrows indicate direct interaction of BRAP. TRAF5, TNF receptor associated factor 5; TRAP1, TNF receptor-associated protein 1; IκK-β, I kappa kinase β; IκB-β; I kappa B-β; BRCA1, breast cancer 1; CDKN1A, cyclin-dependent kinase inhibitor 1A; TF, transcription factor.

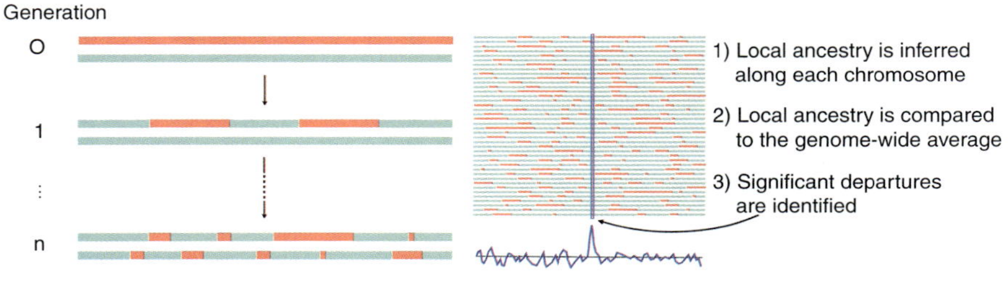

Figure 7.1. Illustration of IBD chromosomal segments in *n* generations following admixture (left) and strategy for mapping of a disease locus by admixture linkage disequilibrium (right).

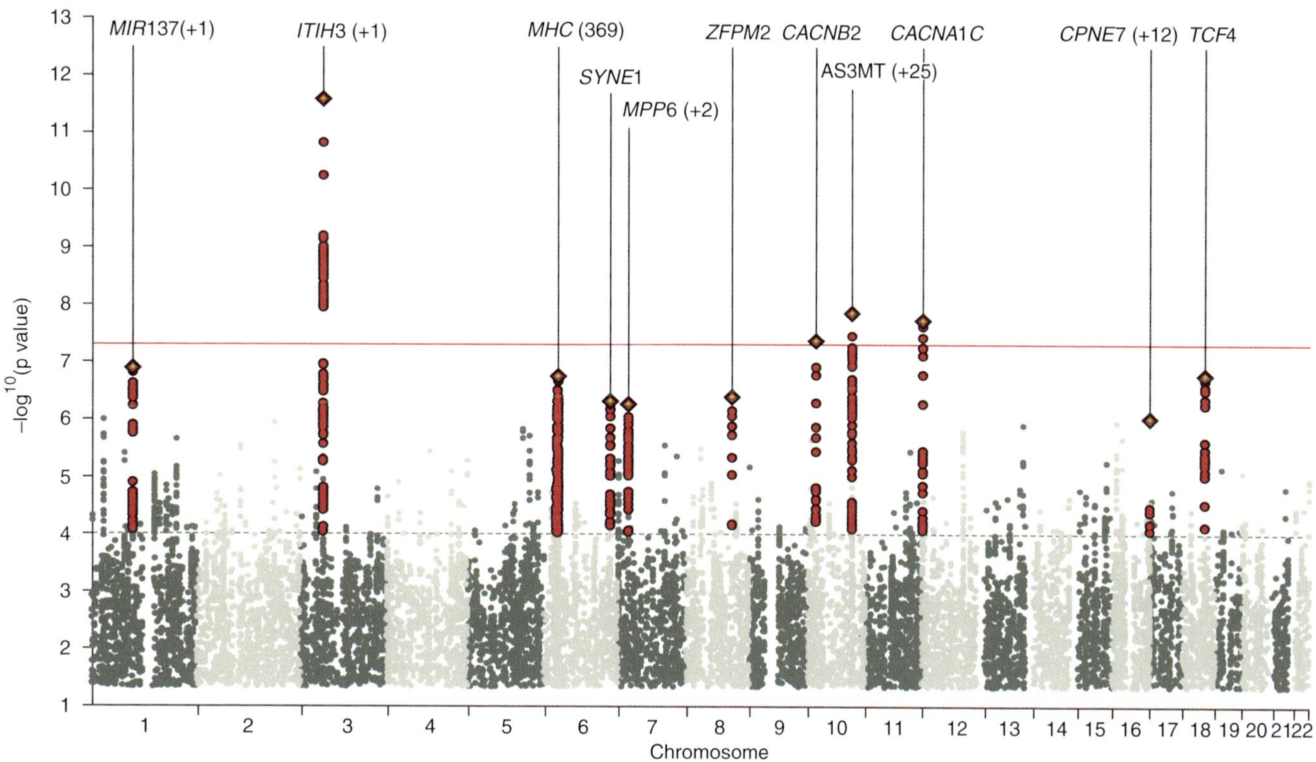

Figure 8.1. Manhattan plot of cross-disorder meta-analysis. Horizontal line represents genome-wide significance threshold ($p < 5 \times 10^{-8}$) (from Smoller *et al.*, 2013).

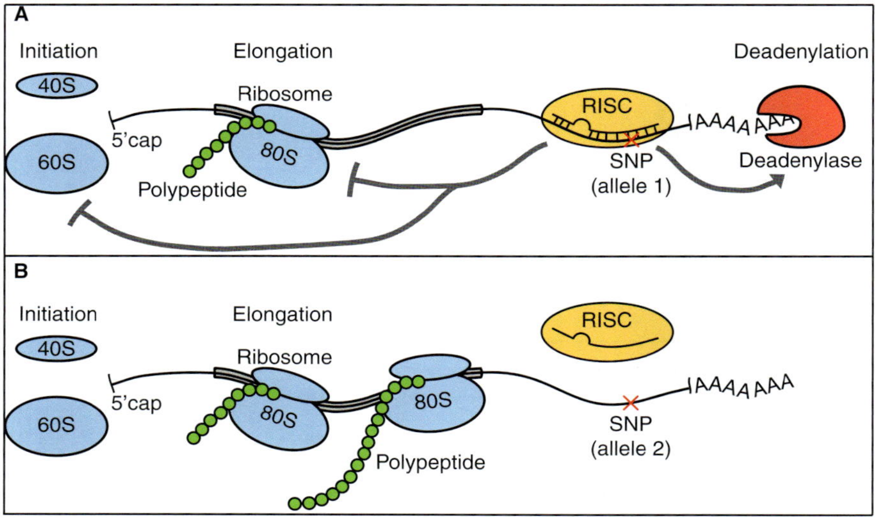

Figure 10.1. SNPs in miRNA target sites. An mRNA depicted by a coding region (in gray), its 5′ and 3′ UTRs, a 5′ cap and a polyA tail, is being translated by the ribosome (depicted by the 40S and 60S subunits and the 80S complex) into a protein (shown by the polypeptide chain in green). Panel (A) shows an mRNA with the allele 1 at a SNP located in a miRNA target site in the 3′ UTR of the mRNA. The miRNA can bind to the target site with this allele, guiding the RISC complex to the target mRNA, which results in gene downregulation by translation inhibition at initiation or elongation of the polypeptide, or by activating deadenylation of the mRNA, through deadenylase, destabilizing the transcript. Panel (B) shows an mRNA with the alternative allele at the SNP, which disrupts the affinity between the miRNA and the mRNA, resulting in the absence of miRNA silencing and a higher gene expression compared to the other allele.

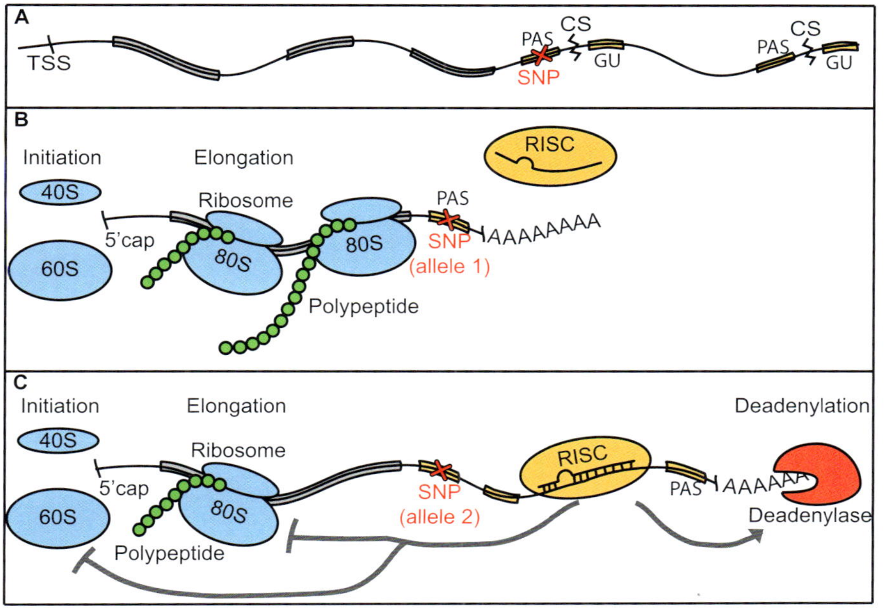

Figure 10.2. SNPs in polyadenylation signals. Panel (A) shows a gene on a DNA strand, including its transcription start site (TSS), its 5′ UTR, its coding region (in gray) divided by two introns, and its 3′ UTR containing two polyadenylation cleavage sites (CS) together with their respective PAS and GU-rich region. A SNP is located in the proximal PAS. Panel (B) shows the transcribed mRNA of the gene in panel A, with allele 1 at the SNP, which increases affinity of the polyadenylation machinery to the proximal PAS, resulting in cleavage at the proximal polyadenylation site, a shorter 3′ UTR, and loss of potential downstream miRNA binding sites. Panel (C) shows the transcript with allele 2 at the SNP, which disrupts the proximal PAS, and results in the usage of the distal one. The longer 3′ UTR contains miRNA target sites that can trigger gene downregulation by miRNAs.

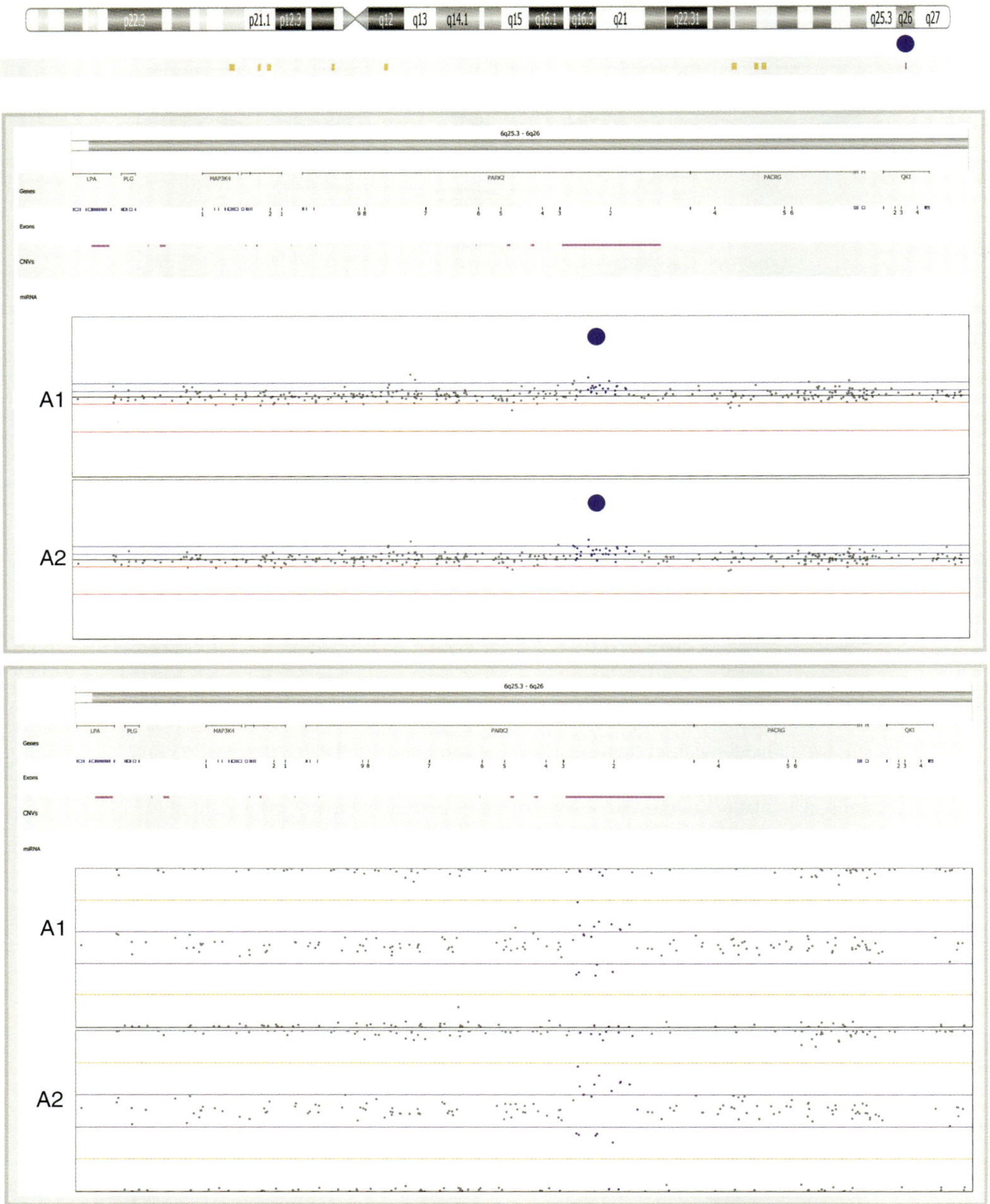

Figure 12.3. Two examples of concordant CNVns in discordant twin-pairs. (A1–A2) On the left, the SNP array results, visualized with Nexus Copy Number v7 (Biodiscovery Inc., Hawthorne, CA, USA), of a monozygotic twin pair discordant for esophageal atresia. Both the affected (upper panel A1 and lower panel A1) and healthy twin (upper panel A2 and lower panel A2) have a 201 kb gain affecting exon 2 of the *PARK2* gene. In the upper panel the $\log_2$ R ratio, representing the copy number status, shows a rise (blue dots) of several probes. This rise is accompanied by a shift in allelic frequency (B-allele frequency, lower panel) indicative for the over-representation of one allele over the other. The combination of rise in $\log_2$ R ratio and shift in B-allele frequency of several allele-specific SNP probes is indicative for a gain in DNA quantity of that specific locus. Both twins have exactly the same gain and this gain was inherited maternally, which was confirmed with qPCR. (B1–B2) On the right, the SNP array results of a monozygous twin pair discordant for congenital diaphragmatic hernia. The affected twin (upper panel B1 and lower panel B1) and unaffected sib (upper panel B2 and lower panel B2) both have a 474 kb deletion deleting several exons of the *ARHGAP24* gene. The drop in $\log_2$ R ratio (upper panel, red dots) and loss of one allele type (yellow dots, lower panel) are indicative of a loss of one copy of DNA at that locus. qPCR confirmed that both twins had the same deletion and that this deletion was inherited maternally (Veenma *et al.*, 2012).

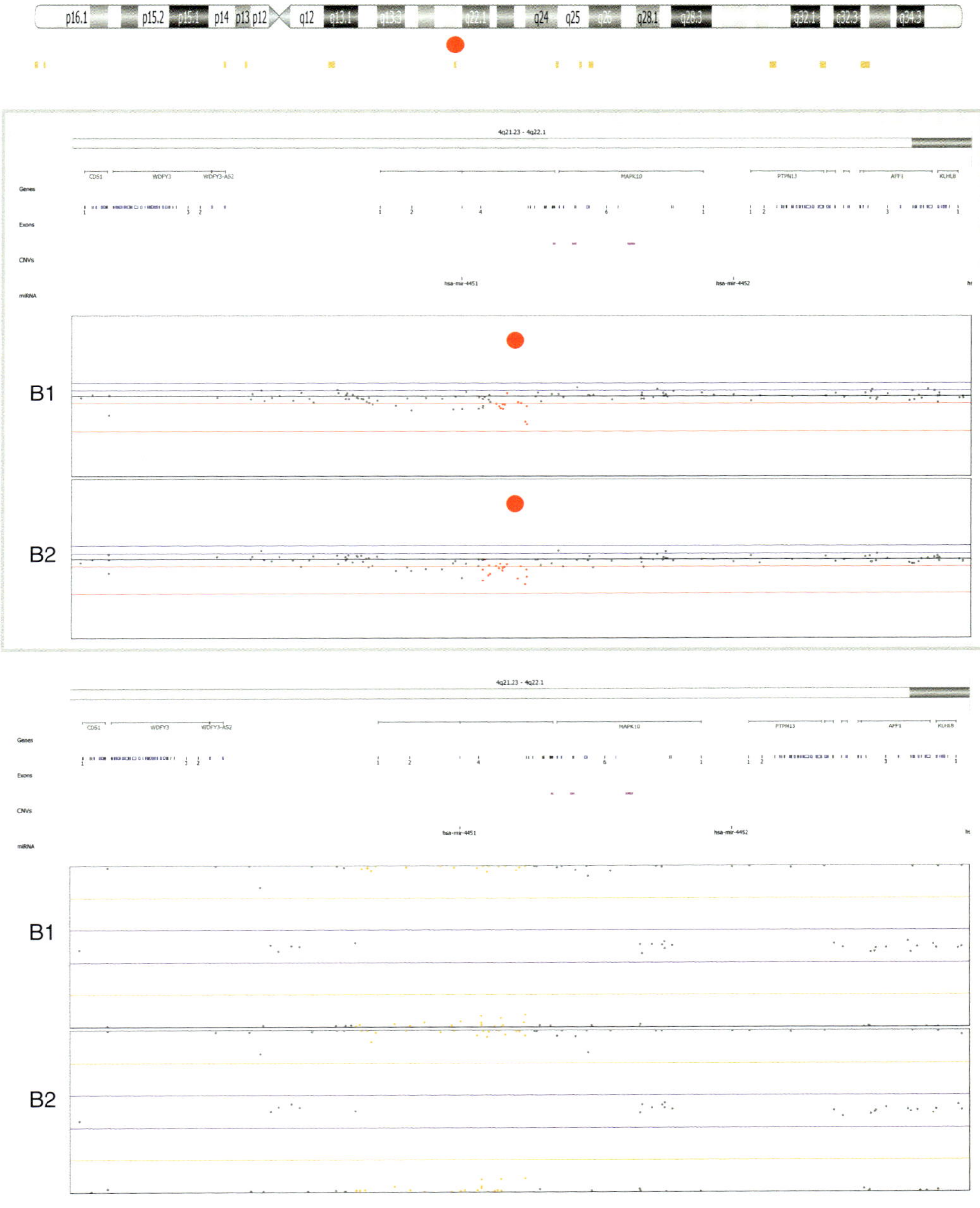

Figure 12.3. (cont.)

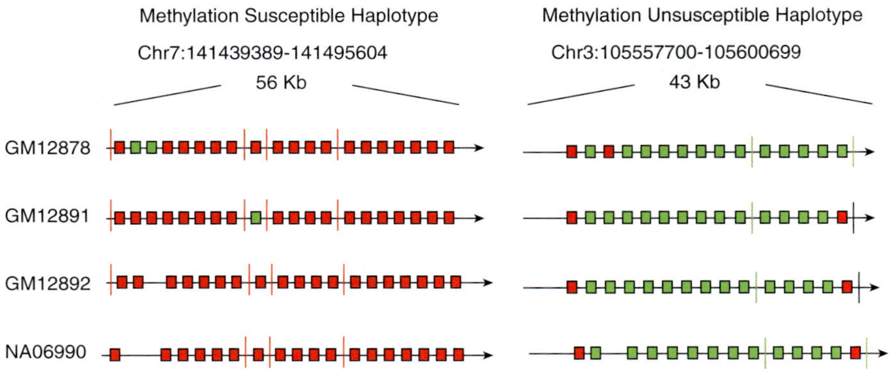

Figure 13.5. Examples of association between CGS haplotype and DNA methylation. Two examples of regions with methylation-susceptible haplotype and methylation-unsusceptible haplotype in GM12878, GM12891, GM12892, and NA06990 were displayed. On the left, the LD region occupies 56 kb in length and contains four CGS-Cs (red dash), hypermethylation regions (red square), and hypomethylation regions (green square). On the right, the LD region occupies 43 kb in length and contains two CGS-Ds (green dash), hypermethylation regions (red square), and hypomethylation regions (green squre). Each square represents about 50 bp and each dash represents 1 bp. Abbreviations: CGS, CpG-related SNP; CGS-C, CpG-related SNP with the allele to create CpG dinucleotides; CGS-D, CpG-related SNP with the allele to disrupt CpG dinucleotides.

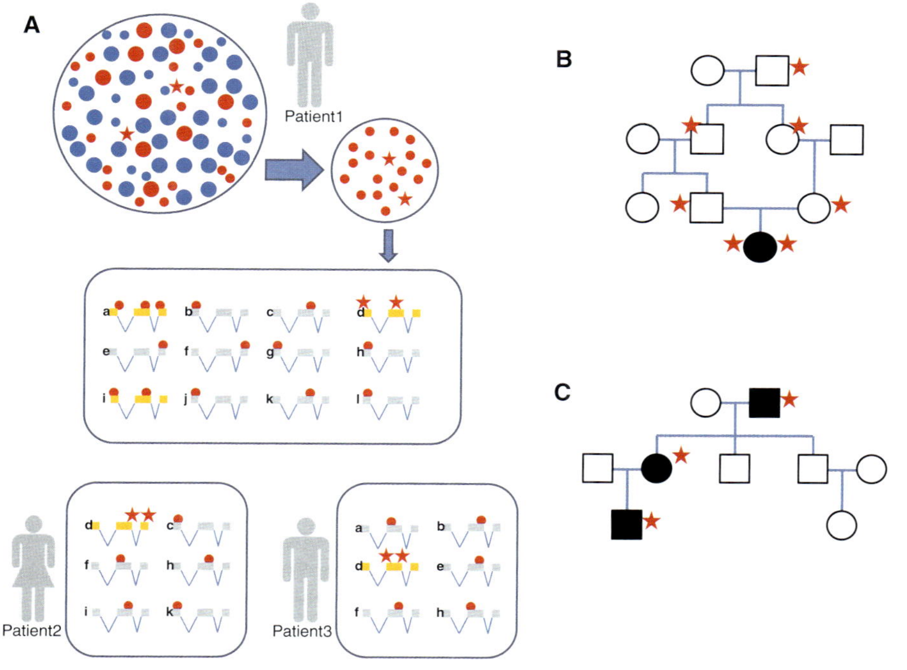

Figure 15.1. Three approaches to the study of rare disorders by whole-exome sequencing. (A) Every individual's genome harbors thousands of functional (red) and non-functional variants (blue). Only rare functional variants (small circles) are considered as candidate causal variants, while common (big circles) and non-functional variants are discarded. Only two of the variants are the causal mutations for the disease (stars). The list of candidate genes (orange) includes all *loci* with two or more functional variants for which the patient is a compound heterozygote. The comparison of candidate genes list across different patients allows detection of the causal gene. (B) In the case of inbreeding, most probably the causal variant will have been inherited twice from the same relative and found in homozygosis in the patient. (C) In a linkage analysis, the causal variant is expected to be present in all the patients from the same family and not present in the unaffected relatives of the pedigree.

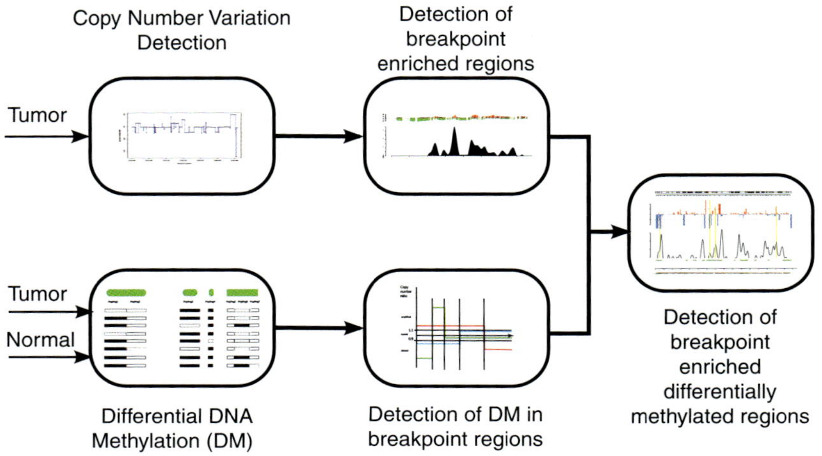

Figure 17.1. Analysis method in order to find DMRs associated with BERs. ROMA genome-wide copy number profiles from breast tumors were combined to partition the genome into variable intervals of stable copy number state in which we estimate DNA methylation levels using MOMA measurements from tumor and normal samples. A Hotelling's T^2 test is performed to identify significant DMRs. On the other track, the locations of BERs are obtained from the ROMA profiles and the list of both significant DMRs and BERs are further evaluated for statistical association (Figure 17.3).

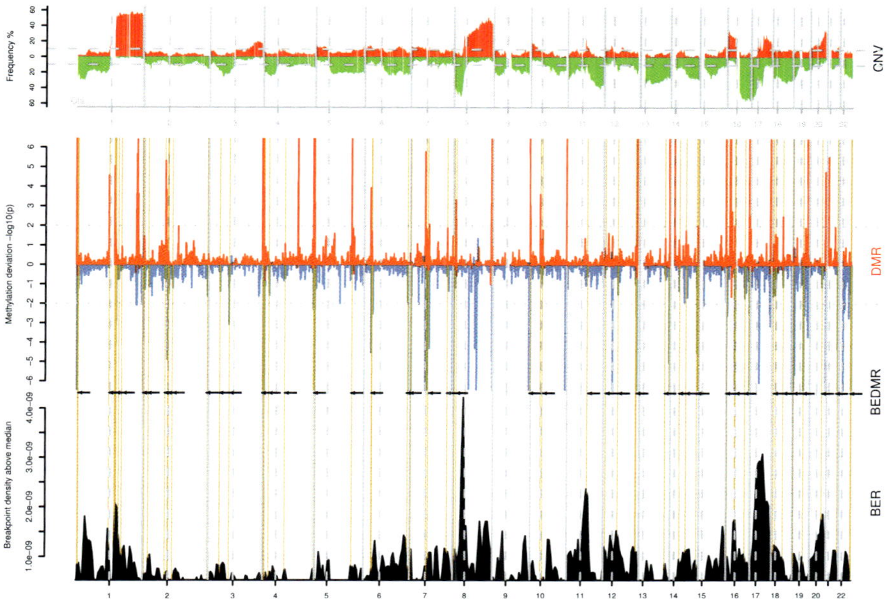

Figure 17.2. Significant DMRs tend to co-localize with breakpoint-enriched regions. The copy number profile of all 108 breast tumors is shown on the top track (CNV). The middle track (methylation) shows the amplitude of the DNA methylation level change compared to normal across genome. Hypomethylated regions are assigned a negative score, defined as $\log_{10}(p)$, while hypermethylated regions take a score equal to $-\log(p)$. Significant DMRs are marked by peaks with a score greater than ±2. The bottom track (breakpoint density) shows the locations of BERs. Breakpoint-enriched DMRs, i.e., DMRs occurring in the vicinity of a BER, are marked by vertical yellow lines and black arrows. Figure reproduced from Tang *et al.* (2012).

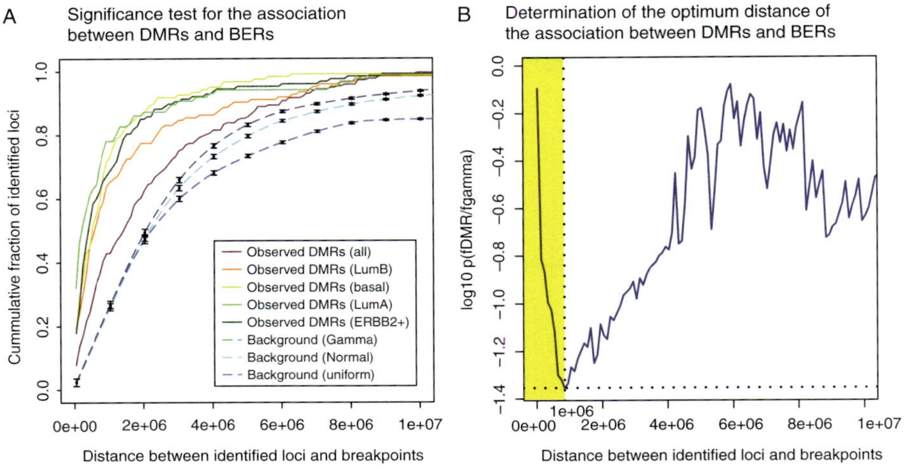

Figure 17.3. Differentially methylated regions co-localize with breakpoint-enriched regions. (A) DMRs tend to be more proximal to BER than expected. (B) The most significant distance of the association between DMRs and BERs occurs at a distance of 1 Mb (shown in yellow).

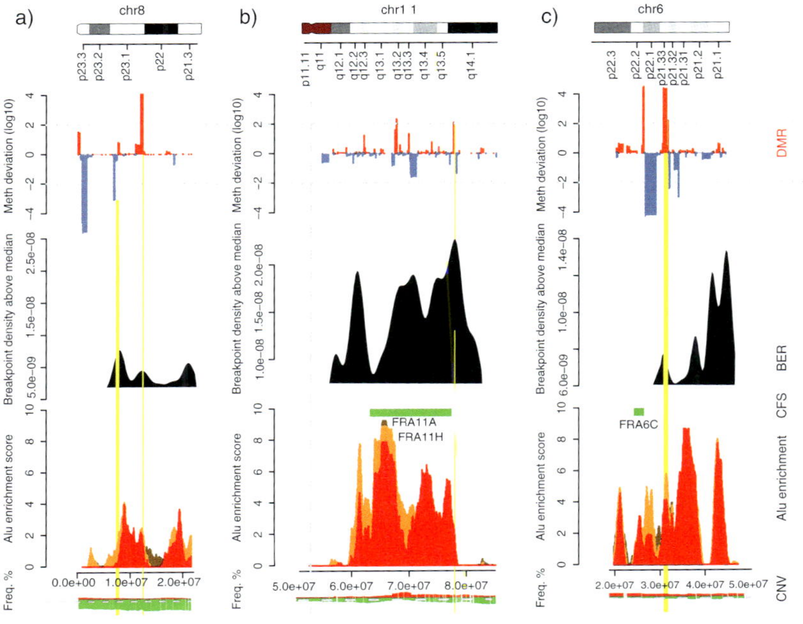

Figure 17.4. Localization of BEDMRs in the genome. BEDMRs tend to occur in genomic contexts. For example A: 8p23.1 (*DUB3*); B: 11q14.1 (*GAB2*); C: 6p21.33 (*DDR1*). These regions contain important cancer-related genes and can be both deleted and hypermethylated or amplified and demethylated.

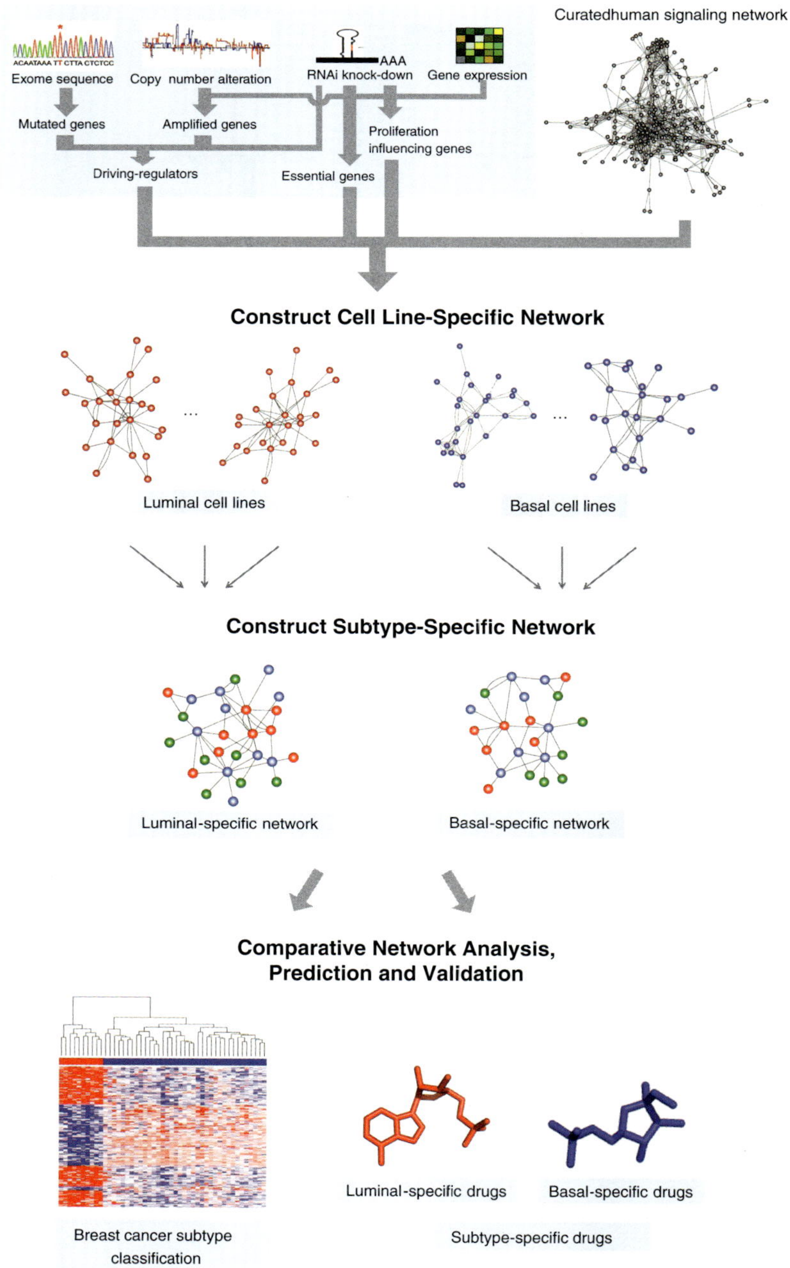

Figure 18.1. Integrated networks for breast cancer cell survival and proliferation framework. The data of genome sequencing, genome-wide RNAi screening, copy number variations, and gene expression profiles of individual lines were used for constructing an integrated network for each individual cell line. Cell-line specific networks across each of the breast cancer subtypes were used for constructing subtype-specific networks for cancer cell survival and proliferation. Comparative and differential analysis of the subtype-specific networks allowed us to predict subtype-specific treatments and significantly classify breast tumor samples.

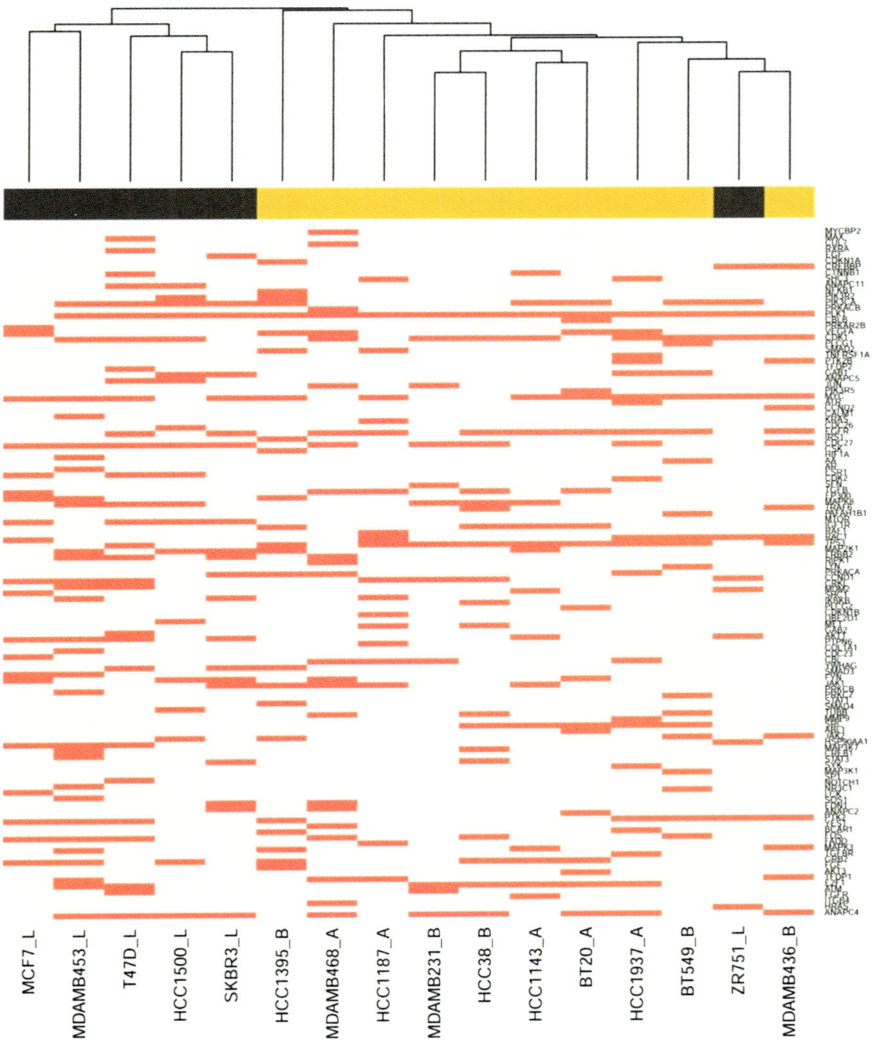

Figure 18.2. Subtype-specific survival signaling networks. Subtype-specific survival signaling networks for basal A (A), basal B (B), and luminal (C) subtypes. Nodes represent genes while links represent regulation (directed links) or interaction (neutral links) between genes. A node is represented by a pie chart that shows each gene's distribution as an essential gene (red), a driving-regulator (blue), or a proliferation-influencing gene (cream) in its subtype. The background color behind the clusters represents a cluster's function in relation to one of the cancer hallmarks: apoptosis (pink), cell proliferation (green), and cell cycle (blue). Cytoscape (Saito *et al.*, 2012) was used to present and visualize the networks.

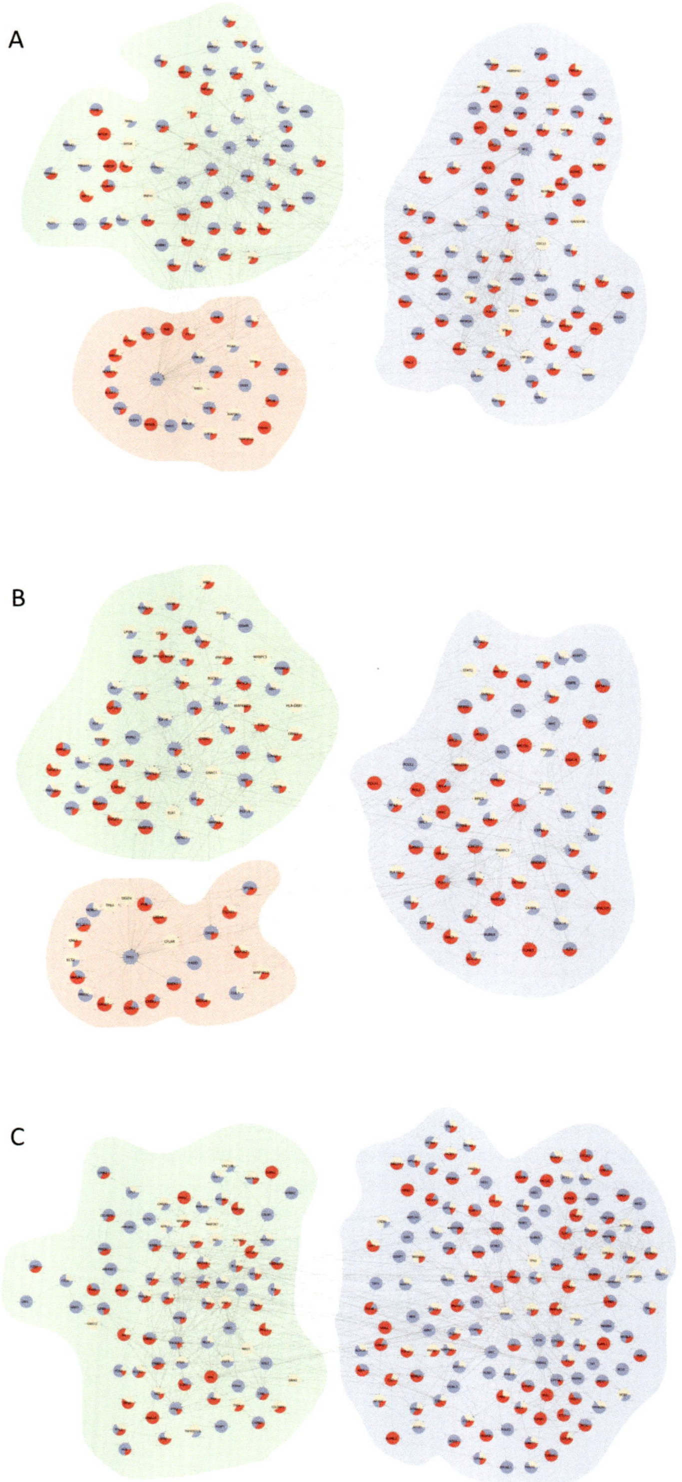

Figure 18.3. Clustering of 402 breast tumor samples using the hubs from subtype-specific networks. Hierarchical clustering of the 402 breast tumor samples using the differential hubs from the subtype-specific networks of luminal and basal subtypes. In the heatmap, red represents mutated genes or amplified genes that are among the top 50% of the expressed genes for tumor samples; otherwise, it appears in beige. On the side bar, gray and yellow represent luminal and basal cell lines, respectively.

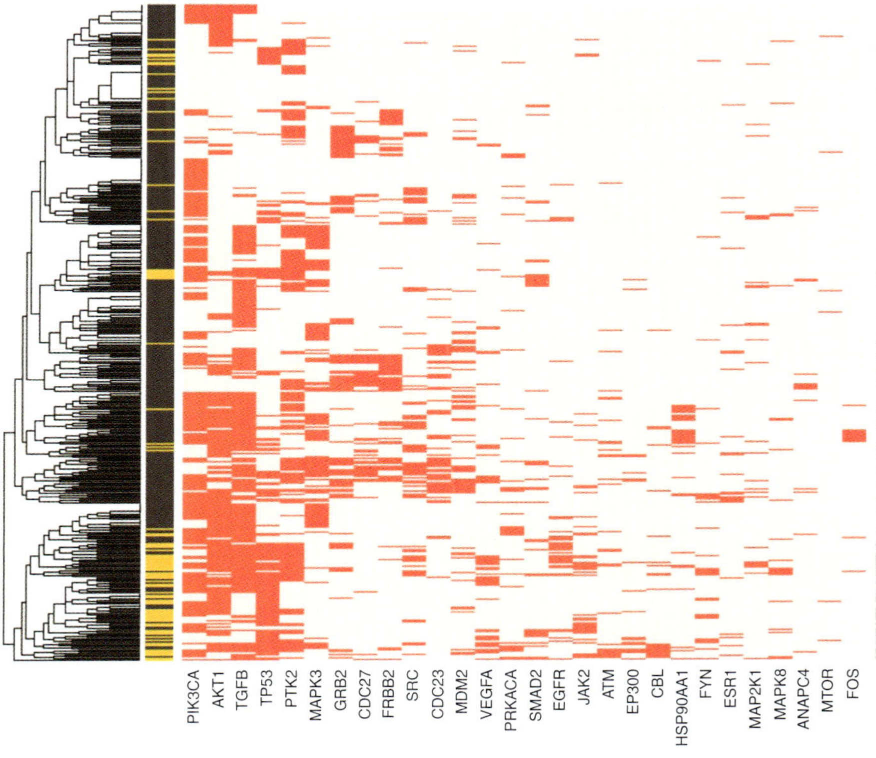

Figure 18.4. Clustering of 402 breast tumor samples using the hubs from subtype-specific networks. Hierarchical clustering using the differential hubs from the subtype-specific networks of luminal and basal subtypes. In the heatmap, for a given sample, red represents mutated genes or amplified genes that are among the top 50% of the expressed genes for tumor samples; otherwise, it appears in beige. On the side bar, gray and yellow represent luminal and basal cell lines, respectively.

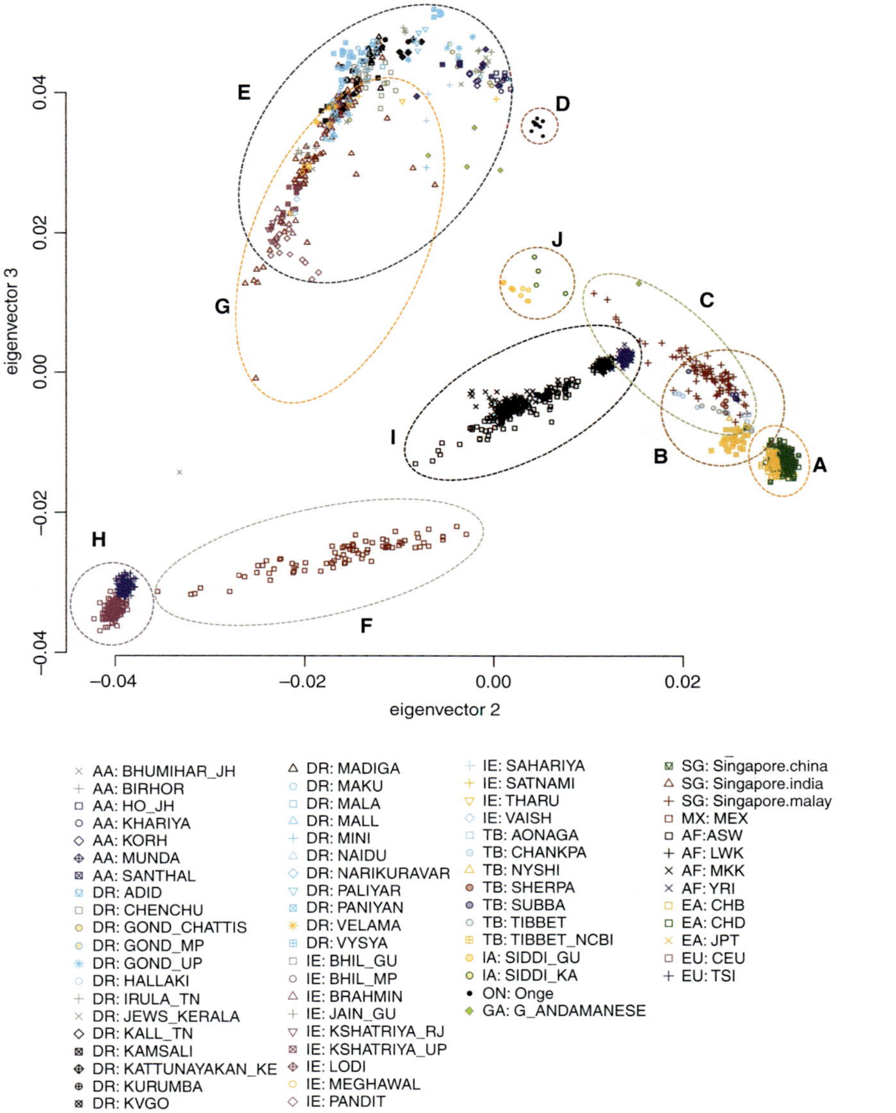

Figure 21.1. Principal component analysis (PCA) of the Indian population with HapMap samples revealed that the genetic architecture of the Indian population is different and the gradient of genetic relatedness with Europeans and the Onge population. Individuals from the same population making a tight cluster suggests endogamy and points toward founder events in the majority of the population. Cluster A: HapMap East Asians (JPT, CHB, CHS) and Singaporean Chinese; cluster B: Tibeto-Burmans of India; cluster C: Singaporean Malay; cluster D: Onge; cluster E: Indo-Europeans, Dravidians and Austroasiatic; cluster F: HapMap Mexican population (MXL); cluster G: Singaporean Indians; cluster H: HapMap European population (CEU and TSI); cluster I: HapMap African population (MKK, YRI and ASW); cluster J: Indo-Africans. (AA, Austroasiatic; DR, Dravidians; IE, Indo-Europeans; TB, Tibeto-Burmans; IN, Gujarati Indians of Houston; EA, East Asians; AF, Africans; EU: Europeans.)

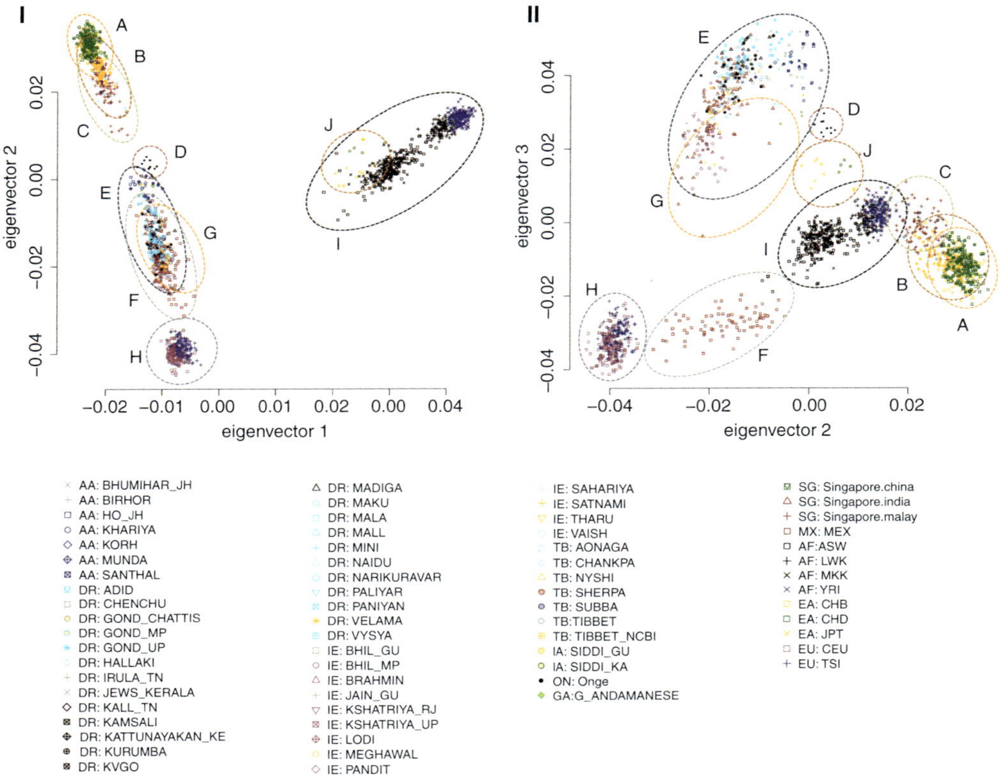

Figure 21.2. Principle component analysis (PCA) of different ethnic populations, using pharmacogenomically important SNPs. (I) On eigenvector 1 the Indian population splits out into two clusters; one consists of IA population while the rest of the population is in the second cluster. On eigenvector 2, Tibeto-Burmans split from the second cluster. (II) Mexican and Singaporean Indians which are clustering on both eigenvectors 1 and 2 with the Indian population split on eigenvector 3, which reveals that Singaporean Indians cannot be used as representative of Indians. Notations are the same as in Figure 21.1.

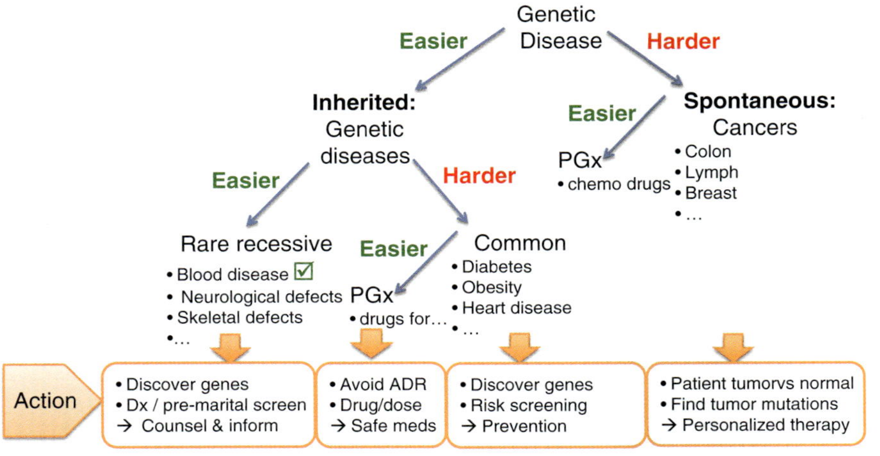

Figure 22.1. Major applications of genomics in healthcare. Shown are the major disease areas from a genetic perspective, with an indication of which are easier or harder to address in population-scale programs to solve disease. Also shown are the major actions that can be taken in each case. PGx, pharmaco-genomics. The most tractable areas of genomic medicine and in diagnosis or carrier screening for rare disease, and PGx for proper drug selection relevant to common diseases and cancer.

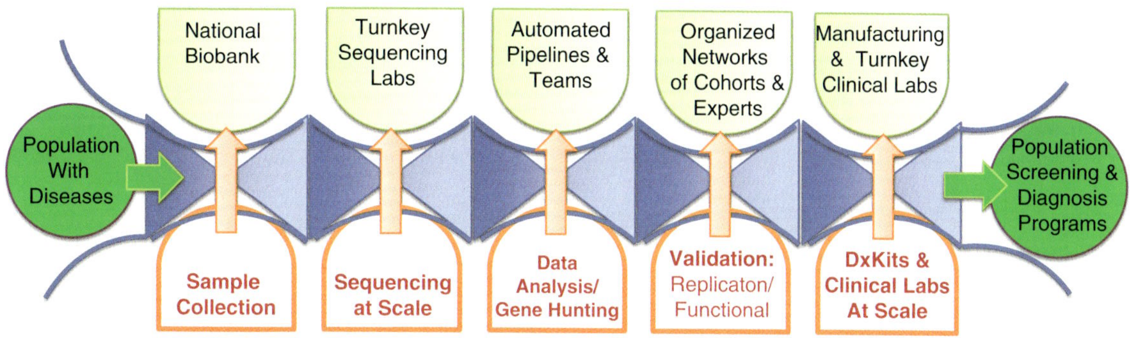

Figure 22.2. How to solve disease on a whole-population scale: process and critical bottlenecks. The major steps of such programs are shown, starting from a population with disease, and ending with that population having personalized screening and diagnosis programs. The lower track shows the functional activities, and the upper track shows the solutions that support these. At each stage there is a major bottleneck of obtaining the needed functionality in a coherent way to support the goals of the project. Both discovery and clinical tracks share similar bottlenecks at this level of description.

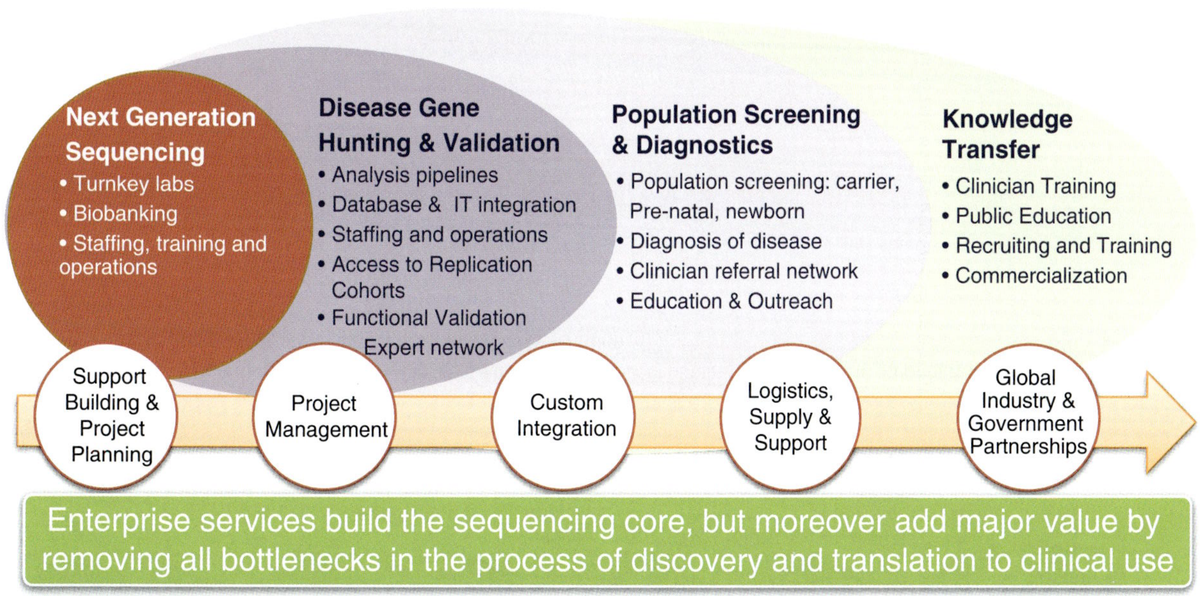

Figure 22.3. Enterprise genomics solutions for solving disease at the population scale. The range of services and solutions that are ideally provided by an enterprise end-to-end support provider. Establishing biobanking and next-generation sequencing facilities is the core, at left, but much more must be done, as indicated by the expanding layers of support activities, to fill all gaps and remove all bottlenecks in the process. At the two extremes of the end-to-end process, at the start there is help building support among sponsors, and at the end there is assistance commercializing the results of the program.

13 Haplotypes of CpG-related SNPs and associations with DNA methylation patterns

Yiyi Ma, Caren E. Smith, Yu-Chi Lee, Laurence D. Parnell, Chao-Qiang Lai and José M. Ordovás

13.1. Introduction

Many technological advances have greatly facilitated the cataloging of millions of single nucleotide polymorphisms (SNPs). In spite of these, an understanding of the functional implications of single base pair changes remains limited. For example, genome projects including the Human Genome Project, HapMap project (International HapMap Consortium, 2003), and 1000 Genomes project have sequenced approximately 6 giga-base pairs of DNA nucleotides to identify 15 million SNPs (1000 Genomes Project Consortium *et al.*, 2010; Clarke *et al.*, 2012). Accompanying the discovery of novel SNPs is growing recognition of associations between SNPs and human diseases, many through genome-wide association studies (GWAS). However, despite this vast growth in genome databases, mechanistic understanding of how SNPs alter phenotypes is largely unknown.

Recent findings have suggested that links between epigenetic status and genetic variants may underlie the functionality of SNPs. Of the major types of epigenetic processes (Portela and Esteller, 2010), DNA methylation has been most frequently linked to human diseases, including cardiovascular diseases (Ordovas and Smith, 2010), diabetes (Hidalgo *et al.*, 2014), obesity (Milagro *et al.*, 2012), dyslipidemia (Guay *et al.*, 2013), and cancer (Weisenberger 2014). Recently, extensive studies have identified the genetic contribution to DNA methylation. Genetic manipulation in mouse stem cells demonstrated that local sequences regulate DNA methylation in both a necessary and sufficient manner (Lienert *et al.*, 2011). SNPs are commonly occurring and well-documented examples of DNA sequence variation, and SNP regulation of DNA methylation is widespread in humans (Bell *et al.*, 2010; Schalkwyk *et al.*, 2010; Shoemaker *et al.*, 2010; Zhang *et al.*, 2010; Gertz *et al.*, 2011; Qu *et al.*, 2012). For example, 80% of the variation in DNA methylation can be predicted by overall genotype of 40 sequence-dependent autosomal regions (Gertz *et al.*, 2011). Additionally, sequence-based regulation of DNA methylation

Genome-Wide Association Studies: From Polymorphism to Personalized Medicine, ed. K. Appasani. Published by Cambridge University Press.

is flexible, occurring either in *cis* or in *trans* (Schalkwyk *et al.*, 2010). Further, methylation-related SNPs have been shown to affect gene expression (Schalkwyk *et al.*, 2010), alternative splicing (Bell *et al.*, 2010), and binding by certain transcription factors (Bell *et al.*, 2010; Paliwal *et al.*, 2013), all of which are potential mechanisms by which SNPs may alter phenotypes.

CpG-related SNPs (CGSs) constitute a group of SNPs with a particular relationship to DNA methylation. By definition, CGSs referred to those SNPs which can change the formation of CpG dinucleotides, which has been established as the primary target site of DNA methylation. For example, studies with human differentiated cells found that 99.98% of the methylation occurs on CpG dinucleotides (Lister *et al.*, 2009). CGSs have been found to contribute a significant fraction (38–88%) of allele-specific methylation regions in the human genome (Shoemaker *et al.*, 2010; Gertz *et al.*, 2011). Also, over 80% of CGSs were shown to play a regulation role in DNA methylation (Zhi *et al.*, 2013).

However, almost all current research has focused on the effects on DNA methylation by a single SNP, rather than the combination of multiple SNPs within one region. With the growth of knowledge in genetics, the concept of regions with multiple variants outweighs the concept of a single variant in terms of functional significance. In addition, SNPs in close proximity tend to be in linkage disequilibrium (LD) because of their less chance for recombination compared to those SNPs far away from each other. As nearby SNPs often are linked from the perspective of variant alleles, the methylation status of adjacent CpG sites also tends to be similar (Shoemaker *et al.*, 2010). In this case, we hypothesize that CGSs affect DNA methylation patterns in the form of haplotype.

To test our hypothesis, we conducted a pilot study on a genome-wide scale to explore the potential relationships between DNA methylation and CGSs in three perspectives: (1) whether the genotype of a CGS is correlated with the DNA methylation level of that CpG's dinucleotides; (2) whether two CGSs in high LD tend to carry the same type of allele, either creating or disrupting the CpG site; and (3) whether the haplotype consisting of multiple CGSs in high LD is associated with the DNA methylation pattern of that LD region (Table 13.1).

13.2. Materials and methods

13.2.1. Definition of CGS, CGS-C and CGS-D

In this study, CGSs are defined as those SNPs which could change the formation of CpG dinucleotides. Each CGS has two types of alleles, which can either create (CGS-C) or disrupt (CGS-D) CpG dinucleotides. For example, for a SNP with adenosine/cytosine (A/C) allele substitution, if the nucleotide on the 3′ of this SNP is guanosine (G), then this SNP can be defined as CGS because the allele cytosine (C) of this SNP and the 3′ guanosine (G) can form a CpG dinucleotides site. In this case, allele cytosine (C) is the CGS-C allele that will create while allele adenosine (A) is the CGS-D allele that will disrupt CpG dinucleotides.

Table 13.1. Hypotheses of this study.

N	SNP	DNA methylation of	Hypothesis
1	Single CGS	Single nucleotide	Genotype of single CGS affects DNA methylation status at that site
2	Two CGSs in high LD	–	CGSs in high LD tend to carry the same type of allele; i.e., CGS-Ci and CGS-Cp, or CGS-Di and CGS-Dp
3	Multiple CGSs in high LD	Thousands of nucleotides	Haplotype of multiple CGSs in high LD is associated with DNA methylation pattern of that LD region

CGS, CpG-related SNP; LD, linkage disequilibrium; CGS-Ci, index CGS with the allele to create CpG dinucleotides; CGS-Cp, proxy CGS with the allele to create CpG dinucleotides; CGS-Di, index CGS with the allele to disrupt CpG dinucleotides; CGS-Dp, proxy CGS with the allele to disrupt CpG dinucleotides.

Considering the possibility of methylation when the CpG site was present, the expected probabilities of CGS-C allele and CGS-D allele were differently defined in this study, depending on the level of data available, on the level of allele or genotype. If the data were on the level of allele, the expected probabilities of each allele would be accurately estimated as 50% (or 1/2), assuming an equal mutation rate across all SNPs. However, if the data available were on the level of genotype, we defined the heterozygotes as CGS-C rather than CGS-D, because one allele of the CpG site has the potential to be methylated, which resulted in the overestimation of the probability of CGS-C allele and the underestimation of the probability of CGS-D allele. By this definition, thus, there are 67% (or 2/3) CGSs that have the CGS-C allele and 33% (or 1/3) with the CGS-D allele, assuming an equal mutation rate across all SNPs. Taking the above SNP with A/C allele substitution as one example, individuals with a genotype of CC or AC were grouped altogether into the CGS-C allele, while those with a genotype of AA were grouped into the CGS-D allele.

13.2.2. Retrieval of CGSs

As a first step, all types of variants were downloaded from the SNP135 database from UCSC human genome browser (Kent *et al.*, 2002). Because our interest was limited to SNPs, we extracted just those variants with class defined as "single" and a difference between the starting and ending coordinates of 1. Furthermore, we excluded ambiguous SNPs which had multiple different coordinates in one database. Then, the bioinformatics tool Galaxy (Goecks *et al.*, 2010) was used to fetch human hg19 sequence data provided by the UCSC genome browser in order to retrieve one adjacent nucleotide on the 5′ and 3′ ends of each SNP. From this data set, all CGSs on the genome-wide scale were extracted. In addition, a subgroup of extracted CGSs were further determined, which are located within 3kb up- and downstream of both the transcription start site (TSS) of the annotated gene, identified by the NCBI Reference Sequence (RefSeq) project from the "refGene" database (as of February 2, 2012; Pruitt *et al.*, 2007), and CpG islands from the "cpgIslandExt" database (Gardiner-Garden and Frommer, 1987), both of which were downloaded from the UCSC genome browser. An R script was created to implement all these procedures.

13.2.3. Data sources for methylation, phased allele, and genotypes

Three sources of methylation data were used in this study. One provides the methylation values by a single nucleotide based on the method of reduced representation bisulfite sequencing (RRBS; Meissner *et al.*, 2008), called Methyl-RRBS. The other two data sets, Methyl-seq and Methyl-450, provide the methylation levels for 50 nucleotides on average. Methyl-seq methodology combines DNA digestion by HpaII and MspI with the Illumina DNA sequencing platform (Brunner *et al.*, 2009). Methyl-450 measures DNA methylation by bisulfite conversion and then is assayed with methylated and unmethylated probes on Illumina Infinium Human Methylation 450 Bead Array platform. A detailed description of the methodology can be found on the UCSC website. Within the above-described data, one site in the Methyl-RRBS database equaled one nucleotide, while one site within the Methl-seq and Methyl-450 databases represented on average approximately 50 nucleotides, which have been taken into consideration for the calculation of hypermethylated nucleotides for each respective region as described below. All three databases were downloaded from the Encyclopedia of DNA Elements (ENCODE) project (ENCODE Project Consortium, 2011) through the UCSC genome browser. According to UCSC genome browser criteria, hypermethylated sites are those with ≥50% of reads showing methylation at the given position in the genome for Methyl-RRBS, a score of 1000 for Methyl-seq, or a score ≥600 for Methyl-450.

We downloaded genotype information for three individuals of European ancestry: GM12878, GM12891, and GM12892, whose genomes have been made publicly available via the 1000 Genomes Project high-coverage pilot study (1000 Genomes Project Consortium *et al.*, 2010). These data were accessed from the UCSC genome browser. Genotype information for a fourth individual, NA06990, was downloaded from NCBI. In order to get the exact haplotype information, we downloaded phase III phased allele data (released in February 2009) of Utah residents with Northern and Western European Ancestry (CEU) in HapMap project. Their corresponding genotype data were also downloaded, which were released in August 2010 with release number 28. There were 113 individuals in the phased allele data set, while 174 individuals were available in the unphased genotype data set (International HapMap Consortium, 2003; International HapMap 3 Consortium *et al.*, 2010).

13.2.4. LD region development

Each CGS within 3 kb distance from TSS and CpG islands was defined as an index SNP. Proxy SNPs for each index SNP were obtained using SNAP software (Johnson *et al.*, 2008) with the following searching parameters: (1) the SNP data set was the 1000 Genomes Pilot 1; (2) the population panel was CEU; (3) the r^2 threshold was 1.0; and (4) the distance limit was 500 kb. Those CGS pairs with the same index CGS were grouped and the LD region (block) was defined as the region starting from the CGS at the most 5′ side and proceeding to the one on the most 3′ side. Any duplicated LD regions were removed.

13.2.5. Statistical analysis

Both data of phased allele and unphased genotype were used to test the haplotype defined by two highly linked CGSs: (1) both CGSs carry the alleles to create CpG dinucleotides; that is, index CGS-C with proxy CGS-C (CGS-Ci and CGS-Cp); (2) both CGSs carry the alleles to disrupt CpG dinucleotides; that is, index CGS-D with proxy CGS-D (CGS-Di and CGS-Dp); (3) the index CGS carries the allele to create while the proxy CGS carries the allele to disrupt CpG dinucleotides (CGS-Ci and CGS-Dp); and (4) the index CGS carries the allele to disrupt while the proxy CGS carries the allele to create CpG dinucleotides (CGS-Di and CGS-Cp). One-way Chi-square analysis was applied to test whether the observed probabilities of these four haplotypes were different from their corresponding expected probabilities, all of which should be equaled to 25% (or 1/4) because the expected probabilities of CGS-C and CGS-D are both 50% (or 1/2), based on the assumption of independence. Considering the fact that gametic phase information was not available for DNA methylation data used by this study, and that there were limited amount of individuals with both phased allele data and DNA methylation to analyze the association between haplotype and DNA methylation, an indirect estimation of haplotype based on unphased genotype information was further studied. Heterozygotes were determined to be included in the analysis because of statistical power, and they were further categorized into the CGS-C group not CGS-D group. In this case, the probability of CGS-C allele was overestimated and their expected values were 67% (or 2/3), higher than the accurate estimation of 50% (or 1/2), while the probability of CGS-D allele was underestimated with expected values of 33% (or 1/3), lower than the accurate estimation of 50% (or 1/2). The corresponding expected probabilities were 44.4% (or 4/9), 11.1% (or 1/9), 22.2% (or 2/9), and 22.2% (or 2/9) for haplotype CGS-Ci and CGS-Cp, CGS-Di and CGS-Dp, CGS-Ci and CGS-Dp, and CGS-Di and CGS-Cp, respectively. In order to further know to which haplotype two CGSs tend to belong, the observed probabilities of the haplotype when both CGSs are with the same type of allele was compared to the random probability assuming that two CGSs show no interdependence, which was 50% based on right estimation and 56% based on both overestimation of CGS-C and underestimation of CGS-D.

To conduct the association with DNA methylation in a given region, the haplotype of multiple CGSs, derived with genotype data, within one LD region was analyzed as both categorical and continuous variables. For the categorical analysis, LD regions were dichotomized into two groups based on whether or not there were more than 50% CGSs with CGS-C allele. Analysis of variance (ANOVA) was used to test whether those LD regions with >50% CGS-Cs tend to have more hypermethylated nucleotides compared to those LD regions with ≤50% CGS-Cs. In the continuous analysis, the exposure variable is the number of CGSs with CGS-C allele within one LD region. The dependent variable is the sum of the nucleotides with hypermethylation from Methyl-seq, Methyl-450, and Methyl-RRBS data sets within that LD region. Pearson correlation and linear regression were applied to analyze whether the haplotype of multiple CGSs, defined as the number of CGS-Cs, within one LD region was related to the number of hypermethylated nucleotides in that region. In order to reduce potential bias, an adjusted number of hypermethylated nucleotides and an adjusted number of CGSs in one LD region were calculated by dividing the length of the LD region from these measurements, and then multiplied by 1 Mb. Because the total number of CGSs and CpG sites with methylation measurements in one LD region was determined by the technological resolution of genotyping platform and methylation chip, these two parameters were further adjusted in the regression model. All data were analyzed using SAS (version 9.2 for Windows; SAS Institute, Inc. Cary, NC, USA). A two-tailed P-value of <0.05 was considered statistically significant.

13.3. Results and discussion

13.3.1. Distribution of CGSs

There were 12,023,433 CGSs from a total 54,212,080 variants within the SNP135 data set after excluding those polymorphisms which were not SNPs (6,091,064, typically insertions and deletions), those ambiguous variants (4,178,554, variants with different coordinates in the database), and those SNPs that do not change CpG dinucleotides (31,919,029). Among these CGSs, 602,700 were close to the TSS of annotated genes as identified by the NCBI RefSeq project and 359,151 were also close to CpG islands.

13.3.2. Genotype of single CGS and its methylation

There were approximately 1000 sites across the human genome with both genotype data and methylation measurements at the level of the single nucleotide as assessed by the method of Methyl-RRBS (1184 in GM12878, 1076 in GM12891, 956 in GM12892, and 701 in NA06990) (Figure 13.1). The means of DNA methylation level for CGS-C sites were above 40% (43.94% in GM12878, 47.52% in GM12891, 46.52% in GM12892, and 51.68% in NA06990), while those means were less than 3% for CGS-D sites (0.96% in GM12878, 2.33% in GM12891, 2.31% in GM12892, and 1.02% in NA06990). The variation of DNA methylation level for CGS-C sites was greater than those for CGS-D sites (35.29% vs. 8.47% in GM12878, 38.47% vs. 12.6% in GM12891, 37.44% vs. 10.87% in GM12892, and 37.05% vs. 6.48% in NA06990). Less than 10% of the CGS-D sites were methylated (7% in GM12878, 6% in GM12891, 10% in GM12892, and 5% in NA06990). However, the majority of CGS-C sites (approximately 80%) had a methylation level greater than 0 (82% in GM12878, 76% in GM12891, 79% in GM12892, and 85% in NA06990).

This observation is highly consistent with the findings that most DNA methylation occurs on CpG dinucleotides in humans (Lister *et al.*, 2009). Interestingly, our findings further indicate that methylation is prevalent when CpG dinucleotides were present. However, approximately 60% of CpG sites in Figure 13.1 were neither completely methylated nor completely unmethylated. This is in line with the widespread existence of allelic skewing of DNA methylation in the human genome, a phenomenon in which the allele-specific methylation is

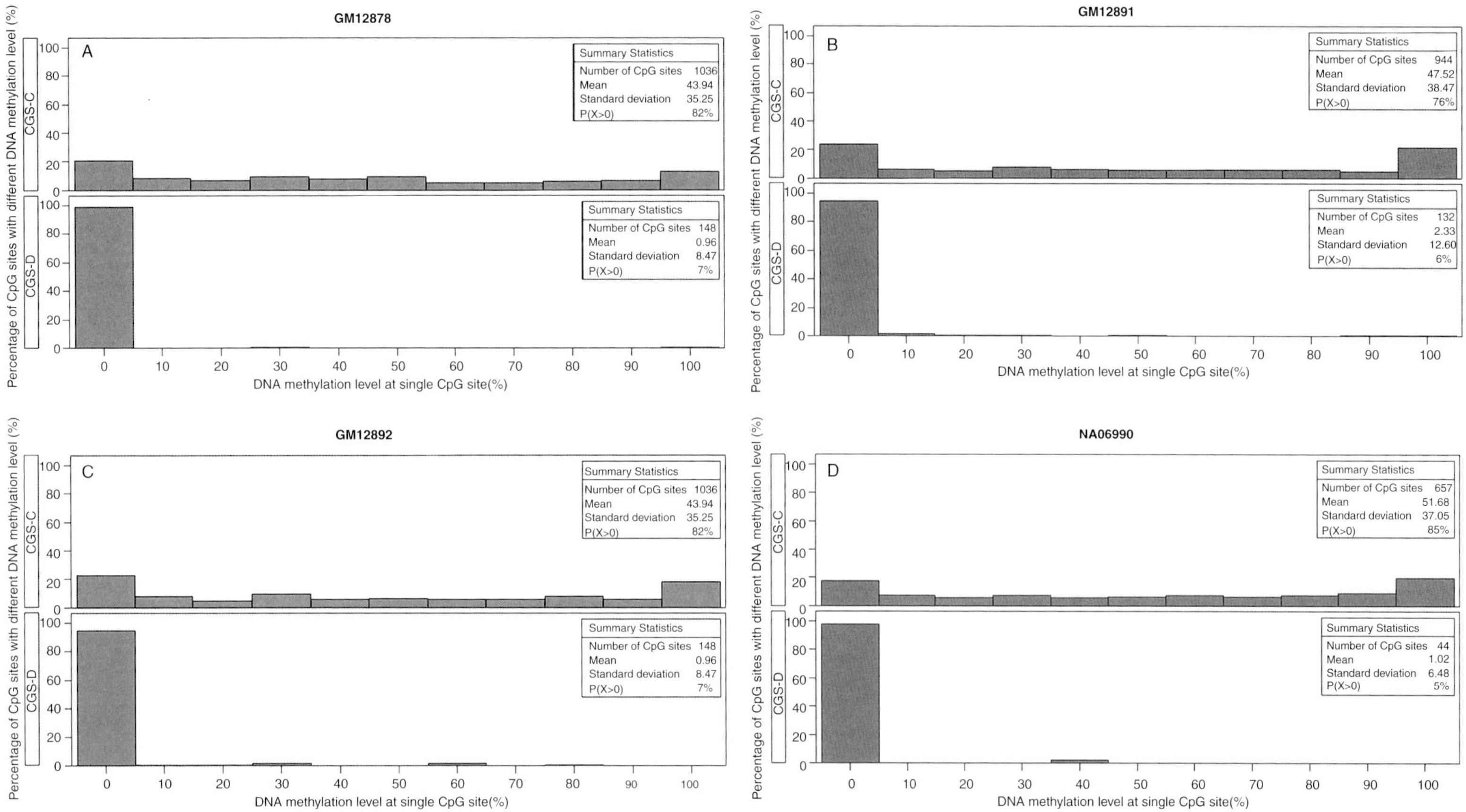

Figure 13.1. Methylation status of CGSs with different types of alleles. Histograms of methylation status for CGS-C sites (upper panel) and CGS-D sites (lower panel) in human blood lymphocyte cell lines. Data from GM12878 (A), GM12891 (B), GM12892 (C), NA06990 (D) were shown. The summary statistics of methylation status, including number of CpG sites, mean, standard deviation, and percentage of CpG sites with methylation level greater than 0 ($P(x > 0)$), were presented on the right upper corner of each panel. Abbreviation: CGS, CpG-related SNPs; CGS-C, CpG-related SNP with the allele to create CpG dinucleotides; CGS-D, CpG-related SNP with the allele to disrupt CpG dinucleotides; $P(x > 0)$, percentage of CpG sites with methylation level greater than 0.

relatively subtle rather than clear-cut (Schalkwyk *et al.*, 2010). Also, 7.6% of CpG islands in polyclonal or monoclonal cell lines were found to be predominant with CpG dinucleotides with intermediate levels of methylation (25–75%) (Deng *et al.*, 2009). It has been hypothesized that the usual small effect sizes and failure of replications in genetic association studies were because of the existence of these CpG sites with intermediate methylation levels (Schalkwyk *et al.*, 2010).

13.3.3. Haplotype of two highly linked CGSs

The phased allele data provided direct test for the haplotype of two highly linked CGSs. In most CEU individuals, the observed probabilities of four haplotypes were significantly different from their expected ones with assumption of independency ($P < 0.05$) (Figure 13.2A). Furthermore, the probability of the condition when both CGSs with CGS-C allele or CGS-D allele is significantly higher than the random probability of 50% ($P < 0.001$) (Figure 13.2B). Figure 13.2A and B show the results of the analysis with transmitted allele and the analysis with untransmitted allele yielded similar results (data not shown).

Consistent with the results with phased allele data, tests with unphased genotype information also showed not only the significant differences between the observed and expected probabilities of four haplotypes ($P < 0.05$) (Figure 13.2C), but also a higher than expected probability of haplotypes of CGS-Ci and CGS-Cp and CGS-Di and CGS-Dp ($P < 0.0001$) (Figure 13.2D), although the corresponding expected probability was overestimated from the original 50% up towards 56%. Although the probability of CGS-C was overestimated by genotype data, the similarity of the significant findings from unphased genotype data to those from phased allele data ameliorated the effect of the overestimation and validated the usage of genotype information to derive haplotype structure.

Both tests indicate that two CGSs tend to contain the same type of allele (e.g., either both create or both disrupt the formation of CpG dinucleotides). The potential mechanism for the observed haplotype may be related to the similar selection pressure and mutation rate introduced by correlated DNA methylation levels among closely located CpG dinucleotides. Methylated cytosines at CpG dinucleotides can be mutated to thymines (T) through deamination (Lindahl and Nyberg, 1972; Coulondre *et al.*, 1978; Cooper and Krawczak, 1989; Sved and Bird, 1990). If the neighboring CpG dinucleotides were methylated, then the incidence of SNP significantly increased by ~50% within the region within 10 bp of a CpG site (Qu *et al.*, 2012). Additionally, CpG sites closely located to each other tend to have similar methylation levels. The correlation coefficients (R^2) for the methylation levels across 30–48% of the CpG sites were greater than 0.3 (Shoemaker *et al.*, 2010). Also, the methylation state of CpG sites located in proximal promoter regions were found to be highly correlated (Guay *et al.*, 2012). In this case, it is plausible for two highly linked CGSs to contain the same type of allele: either both will create, or both will disrupt CpG dinucleotides.

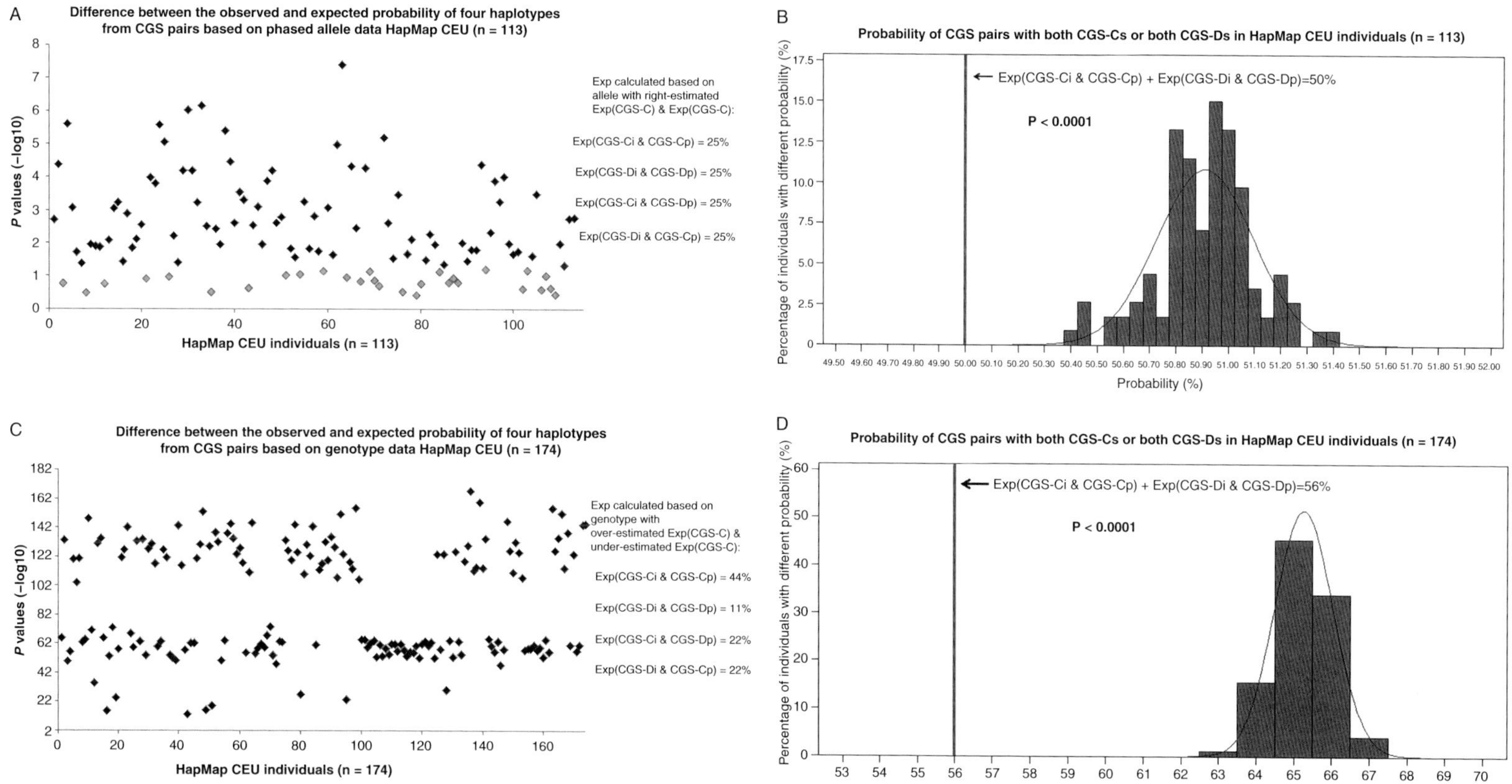

Figure 13.2. Haplotype of two CGSs with high LD in HapMap CEU. Phased allele data ($n = 113$) (panel A and B) and unphased genotype data ($n = 174$) (panel C and D) were used to measure the haplotype of two linked CGSs. (A,C) Scatter plot of all P values (–log10) from Chi-square tests of the difference between the observed and expected probability of four haplotypes derived from CGS pairs in each HapMap CEU individual, shown as diamond (black for $P < 0.05$ and gray for $P \geq 0.05$). (B,D) Histogram of the distribution of the probability of the haplotypes when both CGSs are with the same type of alleles, both with CGS-Cs or both with CGS-Ds. Vertical lines represent the expected probability, which was 50% for the estimation based on phased allele and 56% for the estimation based on unphased genotype. Abbreviations: CGS, CpG-related SNP; LD, linkage disequilibrium; CEU, Utah residents with ancestry from northern and western Europe; CGS-C, CpG-related SNP with the allele to create CpG dinucleotides; CGS-D, CpG-related SNP with the allele to disrupt CpG dinucleotides; Exp, expected probability; CGS-Ci, index CGS-C; CGS-Cp, proxy CGS-C; CGS-Di, index CGS-D; CGS-Dp, proxy CGS-D.

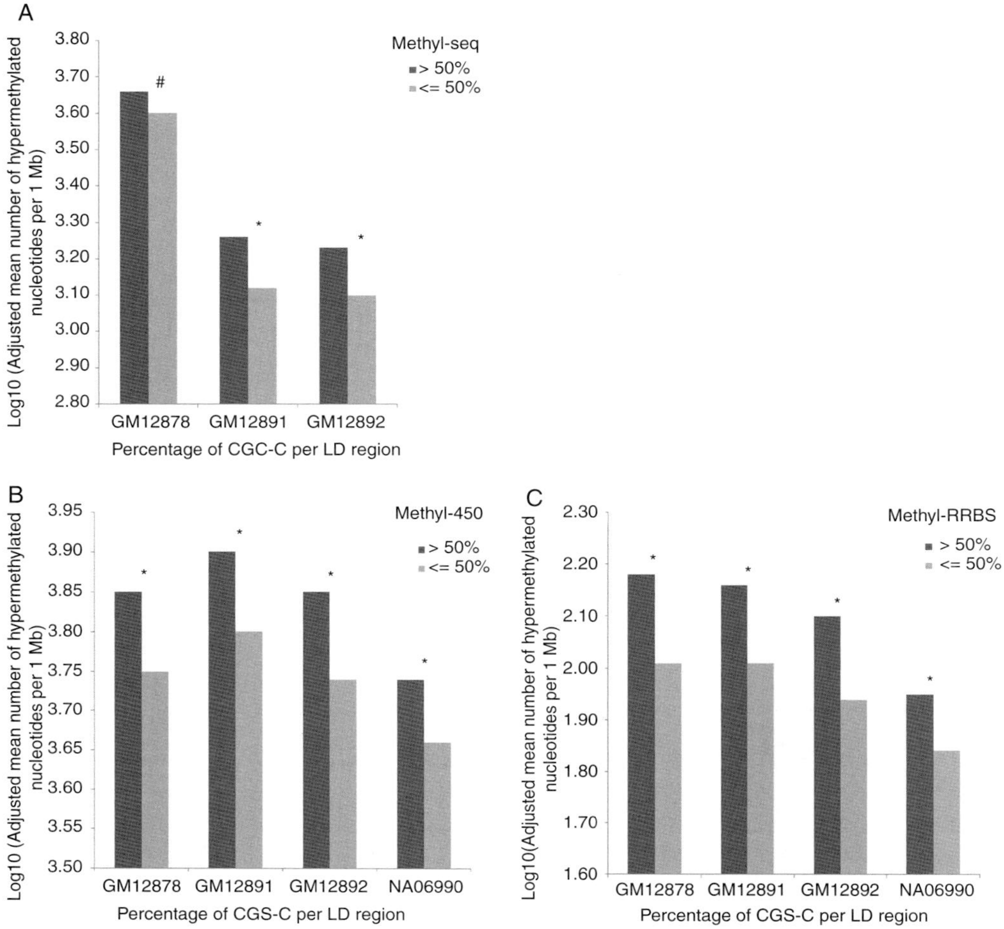

Figure 13.3. Categorical analysis of association between CGS haplotype and DNA methylation. Adjusted mean number of hypermethylated nucleotides per 1 Mb region were compared between those LD regions with >50% CGS-C (dark gray) and those with ≤50% CGS-C (light gray), with different techniques to measure DNA methylation: Methyl-seq (A), Methyl-450 (B), and Methyl-RRBS (C). The model was adjusted for total number of proxy CGSs and total number of nucleotides with methylation measurement ($^{\#}P < 0.05$ and $^{*}P < 0.01$). Abbreviations: LD, linkage disequilibrium; CGS-C, CpG-related SNP with the allele to create CpG dinucleotides.

13.3.4. Haplotype of multiple CGSs and methylation

According to categorical analysis (Figure 13.3), those LD regions with more than 50% CGS-Cs have more hypermethylated nucleotides than those regions with less than 50% CGS-Cs ($P < 0.05$). This pattern is highly consistent across not only four individuals (GM12878, GM12891, GM12892, and NA06990), but also across three different methods to measure DNA methylation pattern (Methyl-seq, Methyl-450, and Methyl-RRBS).

Based on continuous analysis (Figure 13.4), there was a positive relationship between the adjusted number of CGS-Cs and the adjusted number of nucleotides with hypermethylation per 1 Mb region across the entire genome. The correlation

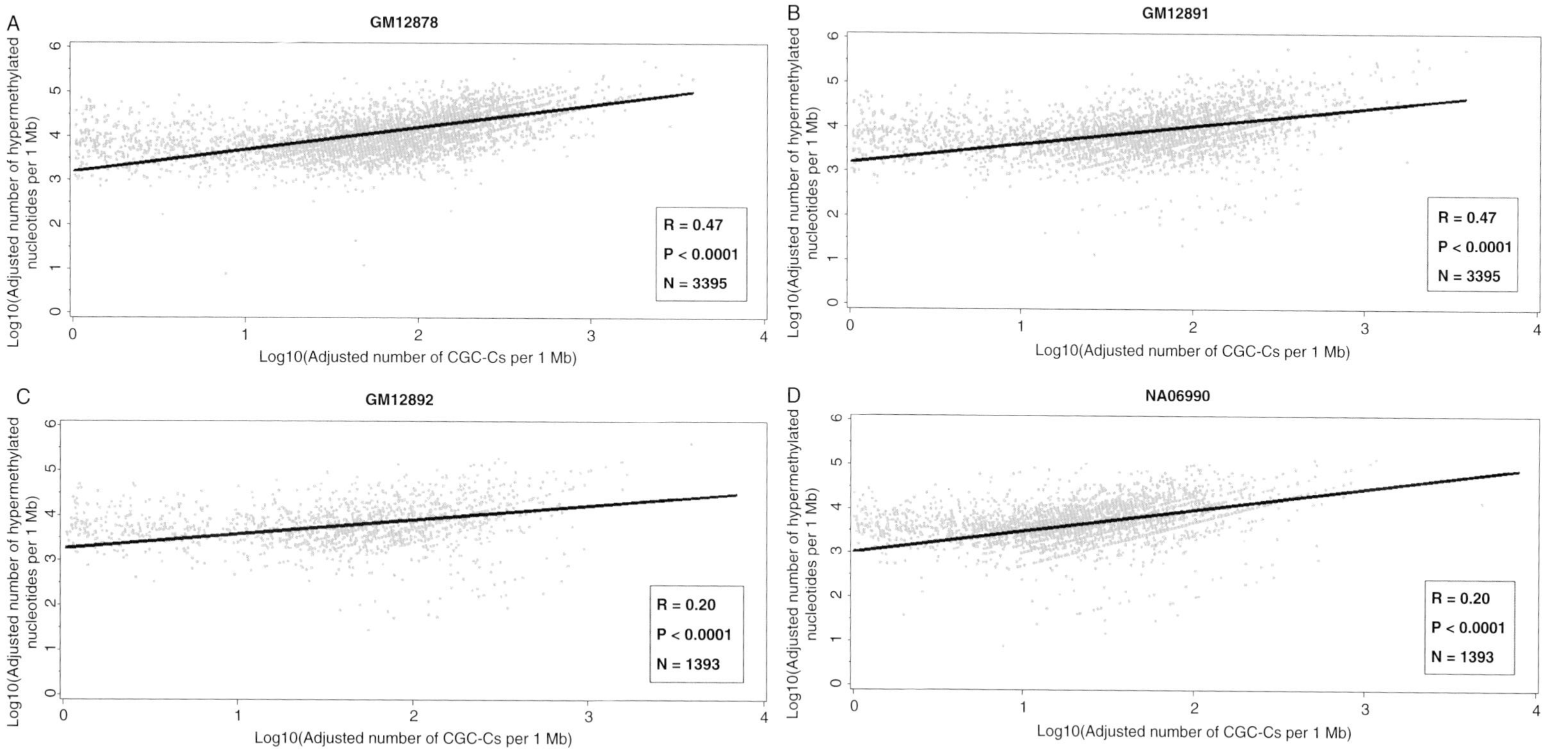

Figure 13.4. Continuous analysis of association between CGS haplotype and DNA methylation. Correlation between adjusted number of CGS-Cs and adjusted number of hypermethylated nucleotides per 1 Mb in GM12878 (A), GM12891 (B), GM12892 (C), and NA06990 (D) are presented. A gray dot represents each LD region. The black line is the predicted correlation line adjusting for total number of proxy CGSs and total number of nucleotides with methylation measurement. *R* and *P* represent the Pearson correlation coefficient and the corresponding *P* value for its significance test, respectively. *N* represents the total number of LD regions in each individual. Abbreviation: CGS-C, CpG related SNP with the allele to create CpG dinucleotides; LD, linkage disequilibrium.

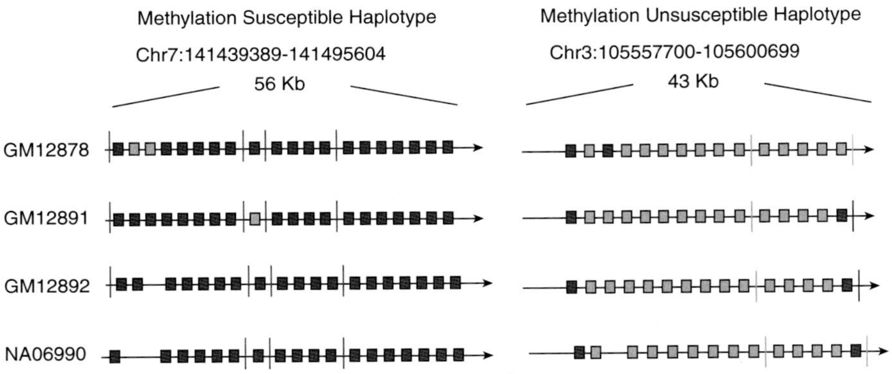

Figure 13.5. Examples of association between CGS haplotype and DNA methylation. Two examples of regions with methylation-susceptible haplotype and methylation-unsusceptible haplotype in GM12878, GM12891, GM12892, and NA06990 were displayed. On the left, the LD region occupies 56 kb in length and contains four CGS-Cs (red dash), hypermethylation regions (red square), and hypomethylation regions (green square). On the right, the LD region occupies 43 kb in length and contains two CGS-Ds (green dash), hypermethylation regions (red square), and hypomethylation regions (green squre). Each square represents about 50 bp and each dash represents 1 bp. Abbreviations: CGS, CpG-related SNP; CGS-C, CpG-related SNP with the allele to create CpG dinucleotides; CGS-D, CpG-related SNP with the allele to disrupt CpG dinucleotides. A black and white version of this figure will appear in some formats. For the color version, please refer to the plate section.

coefficients range from 0.2 to 0.5 and all of these correlations are statistically significant ($P < 0.0001$) across four individuals (GM12878, GM12891, GM12892, and NA06990).

Two example LD regions for the association between haplotype of multiple highly linked CGSs and DNA methylation pattern within one LD region are depicted in Figure 13.5. The LD region with methylation-susceptible haplotype is shown on the left side of the figure, in which all four CGSs are CGS-Cs. In this region, about 1000 nucleotides have measurements for methylation and most of these nucleotides show hypermethylation. In contrast, the right side of the figure shows the LD region with methylation-resistant haplotype, consisting of two CGS-Ds and most nucleotides in this LD region exhibit a hypomethylation pattern.

Our finding is consistent with previous studies showing that CpG density within the methylation-determining regions is one important factor for regulation of DNA methylation (Lienert *et al.*, 2011). However, the direction of the effect of CpG density on methylation based on our observations is not the same as that previous study. Lienert, *et al.* found a significant negative correlation ($r = -0.49$) between the hypermethylation level and the number of CpG sites within methylation-determining regions of length less than 1 kb. In contrast, our study found that the more CpG sites introduced by CGS-Cs in one LD region, the greater the number of hypermethylated nucleotides. A lack of measurement of exact CpG density in our study may account for this discrepancy. All the observations in our study were based on those nucleotides determined by the chip for methylation and genotyping, so not all the CpG sites and CGSs within the region of interest were measured. In this case, although the adjustment by length of each LD region was taken into account, the exact value of CpG density is still unknown. In addition, differences in the length of region of interest may explain, in part, this conflict. The regions of interest in the previous study are fixed with size less than 1 kb, while the length of regions in our study was flexible depending on the distance that encompassed those highly linked CGSs, which spans from 100 bp to 1 Mb. Discrepancies may also be related to species differences, since the previous study was based on mice data and our findings were derived from humans. Nonetheless, the positive relationship between haplotype and DNA methylation is consistent with our previous finding that 80% of the CGSs can be methylated when these SNPs contain the allele to create CpG dinucleotides, indicating the susceptibility to methylation of CGSs.

13.3.5. Strength and limitation

An important strength of this study is the replication not only across multiple individuals but also across different experimental methodologies. The consistency of the replication validates the findings of our study. Additionally, this study was performed on a genome-wide scale, providing a more systematic and comprehensive view of the whole genome. Finally, this study is based on publically available data sets, whose quality has been vetted by numerous groups conducting a myriad of other analyses. Although some experimental bias may remain, replication across different methodologies negates these biases to a significant degree.

This study does have some limitations. All findings are from B lymphocytes, so the identified relationship between CGSs and DNA methylation might not be generalized to other tissue types. We are limited by a relatively small sample size of individuals for which both genotype and DNA methylation data were available. In addition, all DNA methylation data sets were derived from cell lines rather than primary cells, which are more directly applicable to any inferred mechanistic consequences and which indeed may have different methylation patterns than those observed here. Also, the methylation data sets available did not allow us to analyze the differences in gametic phases.

13.4. Concluding remarks

To our knowledge, this is the first study to focus specifically on CpG-related SNPs (CGSs) and their relationship with DNA methylation through genome-wide scale and integrated bioinformatics analysis of publicly available data sets. Our study is one of the few to examine the hypothesis that CGSs are capable of altering the formation of CpG dinucleotides, the target site for DNA methylation. We found that approximately 80% of CGSs were methylated when they carry the allele to create CpG dinucleotides. In addition, when two CGSs are in high LD, they tend to act in a coordinate fashion, meaning that if the allele of one CGS creates the CpG site, then the allele of another CGS in high LD also tends to create another CpG site. This finding is highly consistent in all 113 individuals with phased allele data and 174 individuals with unphased genotype data of the HapMap CEU population.

Finally, the haplotype of one LD block, consisting of multiple CGSs close to both CpG islands and promoter regions, is correlated with the DNA methylation patterns according to both categorical and continuous analyses. The impact of this study resides not only in providing a candidate functional mechanism to link SNPs and DNA methylation, but also in its potential contributions to personalized medicine that relies on knowledge of functional genomic regions.

Acknowledgments

We thank all the investigators and participants for The International HapMap Consortium, ENCODE project, and 1000 Genome Project, and whole genetic community to facilitate scientific researches and findings by making these data publically available. Mention of trade names or commercial products in this publication is solely for the purpose of providing specific information and does not imply recommendation or endorsement by the US Department of Agriculture. The USDA is an equal opportunity provider and employer. This material is based upon work supported by the US Department of Agriculture, under agreement No. 58-1950-0-014. Any opinions, findings, conclusion, or recommendations expressed in this publication are those of the authors and do not necessarily reflect the view of the US Department of Agriculture.

REFERENCES

1000 Genomes Project Consortium, Abecasis, G.R, Altshuler, D., *et al.* (2010). A map of human genome variation from population-scale sequencing. *Nature*, **467**(7319), 1061–1073.

Bell, C.G., Finer, S., Lindgren, C.M., *et al.* (2010). Integrated genetic and epigenetic analysis identifies haplotype-specific methylation in the FTO type 2 diabetes and obesity susceptibility locus. *PLoS ONE*, **5**(11), e14040.

Brunner, A.L., Johnson, D.S., Kim, S.W., *et al.* (2009). Distinct DNA methylation patterns characterize differentiated human embryonic stem cells and developing human fetal liver. *Genome Res.*, **19**(6), 1044–1056.

Clarke, L., Zheng-Bradley, X., Smith, R., *et al.* (2012). The 1000 Genomes Project: data management and community access. *Nature Meth.*, **9**(5), 459–462.

Cooper, D.N. and Krawczak, M. (1989). Cytosine methylation and the fate of CpG dinucleotides in vertebrate genomes. *Hum. Genet.*, **83**(2), 181–188.

Coulondre, C., Miller, J.H., Farabaugh, P.J., *et al.* (1978). Molecular basis of base substitution hotspots in *Escherichia coli*. *Nature*, **274**(5673), 775–780.

Deng, J., Shoemaker, R., Xie, B., *et al.* (2009). Targeted bisulfite sequencing reveals changes in DNA methylation associated with nuclear reprogramming. *Nature Biotechnol.*, **27**(4), 353–360.

ENCODE Project Consortium. (2011). A user's guide to the encyclopedia of DNA elements (ENCODE). *PLoS Biol.*, **9**(4), e1001046.

Gardiner-Garden, M. and Frommer, M. (1987). CpG islands in vertebrate genomes. *J. Molec. Biol.*, **196**(2), 261–282.

Gertz, J., Varley, K.E., Reddy, T.E., *et al.* (2011). Analysis of DNA methylation in a three-generation family reveals widespread genetic influence on epigenetic regulation. *PLoS Genet.*, **7**(8), e1002228.

Goecks, J., Nekrutenko, A., Taylor, J., *et al.* (2010). Galaxy: a comprehensive approach for supporting accessible, reproducible, and transparent computational research in the life sciences. *Genome Biol.*, **11**(8), R86.

Guay, S.P., Brisson, D., Munger, J., *et al.* (2012). *ABCA1* gene promoter DNA methylation is associated with HDL particle profile and coronary artery disease in familial hypercholesterolemia. *Epigenetics*, **7**(5), 464–472.

Guay, S.P., Brisson, D., Lamarche, B., *et al.* (2013). DNA methylation variations at CETP and LPL gene promoter loci: new molecular biomarkers associated with blood lipid profile variability. *Atherosclerosis*, **228**(2), 413–420.

Hidalgo, B., Irvin, M.R., Sha, J., *et al.* (2014). Epigenome-wide association study of fasting measures of glucose, insulin, and HOMA-IR in GOLDN. *Diabetes*, 63(2): 801–807.

International HapMap Consortium. (2003). The International HapMap Project. *Nature*, **426** (6968), 789–796.

International HapMap 3 Consortium, Altshuler, D.M., Gibbs, R.A., *et al.* (2010). Integrating common and rare genetic variation in diverse human populations. *Nature*, **467**(7311), 52–58.

Johnson, A.D., Handsaker, R.E., Pulit, S.L., *et al.* (2008). SNAP: a web-based tool for identification and annotation of proxy SNPs using HapMap. *Bioinformatics*, **24**(24), 2938–2939.

Kent, W.J., Sugnet, C.W., Furey, T.S., *et al.* (2002). The human genome browser at UCSC. *Genome Res.*, **12**(6), 996–1006.

Lienert, F., Wirbelauer, C., Som, I., *et al.* (2011). Identification of genetic elements that autonomously determine DNA methylation states. *Nature Genet.*, **43**(11), 1091–1097.

Lindahl, T. and Nyberg, B. (1972). Rate of depurination of native deoxyribonucleic acid. *Biochemistry*, **11**(19), 3610–3618.

Lister, R., Pelizzola, M., Dowen, R.H., *et al.* (2009). Human DNA methylomes at base resolution show widespread epigenomic differences. *Nature*, **462**(7271), 315–322.

Meissner, A., Mikkelsen, T.S., Gu, H., *et al.* (2008). Genome-scale DNA methylation maps of pluripotent and differentiated cells. *Nature*, **454**(7205), 766–770.

Milagro, F.I., Gomez-Abellan, P., Campion, J., *et al.* (2012). *CLOCK, PER2* and *BMAL1* DNA methylation: association with obesity and metabolic syndrome characteristics and monounsaturated fat intake. *Chronobiol. Int.*, **29**(9), 1180–1194.

Ordovas, J.M. and Smith, C.E. (2010). Epigenetics and cardiovascular disease. *Nature Rev. Cardiol.*, **7**(9), 510–519.

Paliwal, A., Temkin, A.M., Kerkel, K., *et al.* (2013). Comparative anatomy of chromosomal domains with imprinted and non-imprinted allele-specific DNA methylation. *PLoS Genet.*, **9**(8), e1003622.

Portela, A. and Esteller, M. (2010). Epigenetic modifications and human disease. *Nature Biotechnol.*, **28**(10), 1057–1068.

Pruitt, K.D., Tatusova, T. and Maglott, D.R. (2007). NCBI reference sequences (RefSeq): a curated non-redundant sequence database of genomes, transcripts and proteins. *Nucleic Acids Res.*, **35** (Database issue), D61–65.

Qu, W., Hashimoto, S., Shimada, A., *et al.* (2012). Genome-wide genetic variations are highly correlated with proximal DNA methylation patterns. *Genome Res.*, **22**(8), 1419–1425.

Schalkwyk, L.C., Meaburn, E.L., Smith, R., *et al.* (2010). Allelic skewing of DNA methylation is widespread across the genome. *Am. J. Hum. Genet.*, **86**(2), 196–212.

Shoemaker, R., Deng, J., Wang, W., *et al.* (2010). Allele-specific methylation is prevalent and is contributed by CpG-SNPs in the human genome. *Genome Res.*, **20**(7), 883–889.

Sved, J. and Bird, A. (1990). The expected equilibrium of the CpG dinucleotide in vertebrate genomes under a mutation model. *Proc. Natl Acad. Sci. USA*, **87**(12), 4692–4696.

Weisenberger, D.J. (2014). Characterizing DNA methylation alterations from The Cancer Genome Atlas. *J. Clin. Invest.*, **124**(1), 17–23.

Zhang, D., Cheng, L., Badner, J.A., *et al.* (2010). Genetic control of individual differences in gene-specific methylation in human brain. *Am. J. Hum. Genet.*, **86**(3), 411–419.

Zhi, D., Aslibekyan, S., Irvin, M.R., *et al.* (2013). SNPs located at CpG sites modulate genome–epigenome interaction. *Epigenetics*, **8**(8), 802–806

14 eQTL mapping

Mengjie Chen, Can Yang, Cong Li and Hongyu Zhao

14.1. Introduction

With an influx of successful genome-wide association studies to identify genetic variations associated with complex diseases, an unprecedented wealth of knowledge has been accumulated for SNP–phenotype associations (McCarthy *et al.*, 2008; Witte 2010; Manolio 2013). However, many SNP–disease associations do not lend themselves to molecular interpretations, because many of the identified loci are located outside of the coding regions. Even when a gene can be inferred to be causal, there is often a significant gap towards the understanding of the underlying molecular mechanisms (Schadt *et al.*, 2005; McCarthy *et al.*, 2008). Genome-wide eQTL mapping has been one effective approach to bridge this gap (Mackay *et al.*, 2009). In eQTL studies, gene expression levels measured by high-throughput technologies, such as microarrays and RNA-Seq, are treated as quantitative traits. Marker genotypes are also collected from the same set of individuals, and statistical analyses are performed to detect associations between markers and expression traits. By simultaneously capturing many regulatory interactions, eQTLs offer valuable insights on the genetic architecture of expression regulation (Rockman and Kruglyak 2006). The ultimate goal of eQTL studies is to elucidate how genetic variations affect phenotypes by using gene expression levels as intermediate molecular phenotypes (Nica and Dermitzakis 2008). In this chapter, we provide an overview of the eQTL analysis workflow (Figure 14.1), introduce publicly available tools for analysis, and further discuss challenges and issues.

14.2. Data pre-processing

Genome-wide eQTL mapping considers high-density SNP genotype data and gene expression data from the same individuals in a segregating population. Both require appropriate pre-processing as described below for subsequent analysis.

Genome-Wide Association Studies: From Polymorphism to Personalized Medicine, ed. K. Appasani. Published by Cambridge University Press.

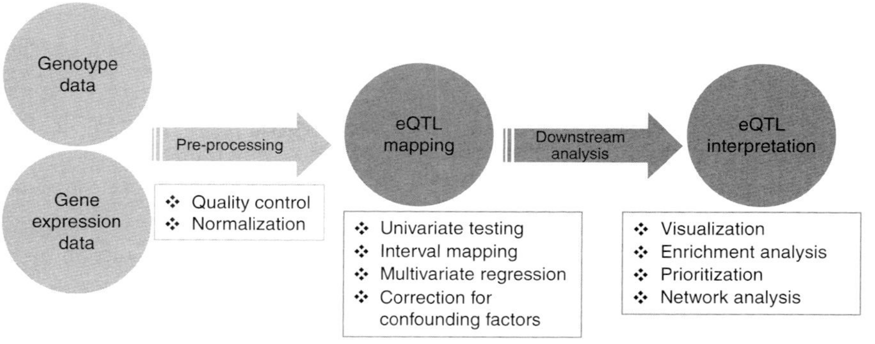

Figure 14.1. The workflow of eQTL data analysis.

14.2.1. Genotype data

Three quality control (QC) criteria are often used in the pre-processing of the genotype data. (1) Missing rate: individuals with a large proportion of missing SNP genotypes (e.g., 10%) should be excluded because the DNA samples of those individuals may be of poor quality. SNPs with a large missing rate (e.g., 5%) should also be filtered out. (2) Hardy–Weinberg Equilibrium (HWE): statistically significant deviations from HWE often result from genotyping errors. Therefore, SNPs that fail an exact HWE test (e.g., a *P*-value less than 0.001) should be filtered out. The criterion does not apply to haploid organisms, such as yeast. (3) Minor allele frequency (MAF): SNPs with low MAF (e.g., 0.05) are sometimes filtered out because of the insufficient statistical power for studies with a relatively small sample size and potentially higher genotype calling error. All the QC measures mentioned above can be accomplished by PLINK (Purcell *et al.*, 2007) for human studies and by R package qtl for animal studies (Broman *et al.*, 2003). After the QC, the remaining missing genotype data need to be carefully handled. For study samples with a reference population panel, missing genotype data may be imputed using a phasing algorithm such as BEAGLE (Browning and Browning, 2007), or IMPUTE2 (Delaneau *et al.*, 2012). For experimental crosses, missing genotype data may be imputed using the hidden Markov model approach implemented in R package qtl (function *fill.geno*). For human studies, to avoid false positives due to population stratification, principal component analysis (PCA) is often used to detect outlier samples with different ancestries from the other samples. EIGENSTRAT (Price *et al.*, 2006) is a popular program that implements such functionality. Individuals found to be outliers based on PCA analysis are often excluded from the subsequent analysis.

14.2.2. Expression data

14.2.2.1. Microarray data

Along with the intrinsic biological variability, microarray data are complicated with non-biological or technical variability from many sources: sample preparation, hybridization, dye affinity, and other factors. Thus it is important to remove or reduce these systematic biases that can substantially affect the results of downstream analysis. This step is called normalization. Many normalization procedures have been proposed based on various assumptions. Among them, quantile normalization is commonly used due to its robustness and simplicity (Bolstad et al., 2003). After quantile normalization, the empirical distribution of the intensities for each array is the same. This can be accomplished by first ranking the intensities in each array in increasing order, followed by substituting each observed value by the corresponding quantile in the reference distribution, which is often derived by the average gene expression level across all the arrays. Quantile normalization implicitly assumes that the great majority of genes have more or less similar expression levels across arrays. This assumption may fail when a large variation is expected in the expression. The invariant set normalization (Li and Wong, 2001) is a procedure that may overcome this problem. It iteratively selects a reference array such that the rankings of the probes within that array are relatively invariant across different samples. The resulting set of probes is then used for an intensity-dependent normalization. Depending on the platform of gene expression data, the above normalization methods could be performed by Bioconductor packages *limma* (Smyth, 2005), *affy* (Gautier *et al.*, 2004), and *beadarray* (Dunning *et al.*, 2007) for data from Agilent, Affymetrix, and Illumina, respectively.

14.2.2.2. RNA-seq data

Compared to microarrays, RNA-seq data have higher resolution and accuracy and allow the detection and quantification of alternative spliced isoforms. In RNA-seq data pre-processing, raw sequence reads need to be transformed into counts for each transcript through transcriptome reconstruction. There are two types of methods for transcriptome reconstruction: assembly-based methods and mapping-based methods. Assembly-based methods, such as Scripture (Guttman *et al.*, 2010), trans-ABySS (Robertson *et al.*, 2010), and Velvet (Zerbino and Birney, 2008), rely solely on an assembled genome from the sequencing reads to build a transciptome *ab initio*. Mapping-based methods rely on the alignment with a reference genome, which are more accurate and computationally efficient when a well-annotated reference genome is available. Software includes Cufflinks (Trapnell *et al.*, 2010) coupled with TopHat (Kim *et al.*, 2013), ERANGE (Mortazavi *et al.*, 2008), and others. To make counts data comparable among samples, similar to microarray gene expression data, a normalization step must be performed. In the analysis of RNA-seq data, reads per kilobase per million reads mapped (RPKM) and fragments per kilobase of exon per million fragments mapped (FPKM) are widely used units to quantify transcript levels (Dillies *et al.*, 2013). Estimated RPKM/FPKM levels after log transformation can be fed directly into the pipeline for the microarray data. In addition to expression at gene level, RNA-seq data also provides isoform abundance information, which could be used to dissect the genetic basis of differential isofrom usage. Many algorithms have been developed to estimate isoform abundance. A subset of these methods are shown in Table 14.1.

Table 14.1. Methods/software for pre-processing of genotype data and gene expression data.

	Methods/software	Notes
Genotype data	PLINK (Purcell *et al.*, 2007)	Quality control for population data
	BEAGLE (Browning and Browning, 2007), MACH (Li *et al.*, 2010a), IMPUTE2 (Delaneau *et al.*, 2012)	Imputation with reference panels for population data
	R package *GGtools* (Carey, 2013)	Quality control
	R package *R/qtl* (Broman *et al.*, 2003)	Imputation using a hidden Markov model approach
	EIGENSTRAT (Broman *et al.*, 2003)	Principle component analysis
Microarray data	Bioconductor package *limma* (Smyth, 2005)	Normalization methods for Agilent platform
	Bioconductor package *affy* (Gautier *et al.*, 2004)	Normalization methods for Affymetrix platform
	Bioconductor package *beadarray* (Dunning *et al.*, 2007)	Normalization methods for Illumina platform
RNA-seq data	Scripture (Guttman *et al.*, 2010), trans-ABySS (Robertson *et al.*, 2010), Velvet (Zerbino and Birney, 2008)	*De novo* transcriptome assembly
	Cufflinks (Trapnell *et al.*, 2010)	Isoform assembly and quantitation
	TopHat (Kim *et al.*, 2013)	Splice junction mapper
	ERANGE (Mortazavi *et al.*, 2008)	Transcript mapping and abundance estimation
	RESM (Li *et al.*, 2010a), MISO (Katz *et al.*, 2010)	Isoform abundance estimation based on generative models
	POEM (Richard *et al.*, 2010; Salzman *et al.*, 2011)	Isoform abundance estimation based on the Poisson model
	Nsmap (Xia *et al.*, 2011), SLIDE (Li *et al.*, 2011), rQuant (Bohnert and Rätsch, 2010), IsoLasso (Li *et al.*, 2006)	Isoform abundance estimation with penalized regression

Consider an eQTL study with N samples. For each sample, there are genotyping data for J genetic markers and gene expression data for I transcripts. After data pre-processing, we use $\mathbf{X}$ to denote the $N \times J$ genotype matrix and $\mathbf{Y}$ to denote the $N \times I$ genotype matrix. Furthermore, we use n to index samples, j to index markers, and i to index transcripts. The parameter μ, denoting the regression intercept, is dropped for notation convenience in all multiple regression settings.

14.3. eQTL mapping

An eQTL is a genetic locus that regulates the expression levels of certain genes. Thus, eQTL mapping amounts to correlating all genetic markers with the expression levels of all measured genes to identify putative regulatory relationships.

14.3.1. Univariate methods

The univariate methods examine one marker at a time with the expression of one target gene. The simplest method is to regress each expression trait on each marker with a linear model. For a pair of transcript i and marker j, this model can be expressed as:

$$Y_i = \mu_i + \beta_{ji} X_j + \varepsilon_i \tag{14.1}$$

where Y_i is the expression value and X_j is the variable encoding the genotype (0, 1, 2) of the marker under consideration, μ_i is the mean target gene expression over all individuals, β_{ji} is the regression coefficient, and ε_i is the error term with each entry i.i.d. following $N(0, \sigma^2)$. If there is evidence of an eQTL at the locus under consideration, a significant linear relationship between Y_i and X_j will be reflected through a nonzero slope β_{ji}. Thus we could perform a t-test against the null hypothesis $H_0 : \beta_{ji} = 0$ to identify the SNP–expression association.

The likelihood ratio (LR) test is a popular alternative. It compares the fit of two models: the model with the eQTL (the "full" model) and the model without the eQTL (the "reduced" model),

$$Y_i = \mu_i + \varepsilon_i. \tag{14.2}$$

The test statistic of the likelihood ratio test is

$$\mathrm{LR} = N\log\left(\frac{\mathrm{RSS}_r}{\mathrm{RSS}_f}\right) \tag{14.3}$$

where N is the number of individuals, and RSS_r and RSS_f are the residual sums of squares for the reduced and full models, respectively. LR under the null hypothesis follows a Chi-square distribution. Thus, a Chi-square test could be applied to test the significance of the association. Besides the parametric tests, one could use non-parametric tests such as Wilcoxon's rank test to identify the associations, which is more robust to outliers but with a compromised statistical power.

Because millions of tests would be performed in an eQTL analysis, multiple testing adjustment is necessary to control the number of false positives. Popular approaches include Bonferroni correction and false discover rate (FDR)-based correction (Benjamini and Hochberg, 1995). The former is known to be overly conservative. The latter exerts control over the FRD and thus is less conservative. Due to the correlation structures in the genome (linkage disequilibrium, LD) and gene expression levels (pathways), FDR-based corrections are often more powerful in the eQTL setting.

eQTLs could be broadly categorized into *cis*-eQTLs and *trans*-eQTLs based on their distances to the target gene. A *cis*-eQTL is typically within 1 Mb (or less) from the target gene. An eQTL distal to the target gene more then 1 Mb is called a *trans*-eQTL. Thus, the number of tests for *trans* and *cis* association is very unbalanced. As a consequence, the FDR correction should be performed separately for *trans* and *cis* associations (Li *et al.*, 2012b). Given a set of P-values, the adjustment may be performed using the *p.adjust()* function in R, where different correction methods could be chosen by the *method* argument (e.g., *"bonferroni"* or *"fdr"*).

14.3.2. Interval mapping for experimental crosses

Compared to human studies where millions of markers are genotyped, the number of markers studied is much less in experimental crosses. Because of the strong dependency among nearby markers in this setting, researchers have proposed methods to infer the associations on ungenotyped intervals between markers by applying the principles of regression. The approach is called interval mapping (Lander and Botstein, 1989). R package *R/qtl* implements two commonly used methods for interval mapping: the expectation maximization (EM) algorithm (Lander and Botstein, 1989) and Haley–Knott regression (Haley *et al.*, 1994). In the EM algorithm, the same parameterization is used as in model (14.1), but the ungenotyped intervals are treated as missing data. In the Haley–Knott regression, a different parameterization is employed and the overall effect of the eQTL is decomposed into two components: additive effects, and dominant effects. For the expression of the *i*-th gene regressed on the *j*-th ungenotyped marker in the interval:

$$
\begin{aligned}
Y_i &= \mu_i + \alpha X_{j1} + \beta X_{j2} + \varepsilon_i, \\
X_{j1} &= P\left(AA|M\right) - P(aa|M\right), \\
X_{j2} &= P(Aa|M).
\end{aligned}
$$

In the above model, for an ungenotyped marker, X_{j2} is the probability of being heterozygous and X_{j1} is based on the probability of homozygous conditioning on the genotype of the flanking markers M. LR could be used to compare the fitness of the model with the dominant effect and the model without the dominant effect. An ungenotyped marker with a significant LR can be identified as an eQTL. Haley–Knott regression is more computationally efficient than the EM algorithm with similar power when the marker set is relatively dense.

14.3.3. Multivariate regression methods

In the univariate regression, expression pattern of a gene is independently and sequentially examined against each marker. However, it is widely believed that genes from the same biological pathway are often co-regulated and may share a common genetic basis that affects their expression levels. Incorporating the relatedness information of genes into the association analysis would help to boost the statistical power. To address the multi-traits in eQTL mapping, we can formulate the problem as a variable selection problem in a multivariate response regression setting as the following:

$$
\mathbf{Y} = \mathbf{XB} + \mathbf{E} \tag{14.4}
$$

where $\mathbf{B}$ is a $J \times I$ matrix of regression coefficients, $\mathbf{E}$ is an $N \times I$ matrix of Gaussian noises, and the other notations were defined before. The columns of $\mathbf{X}$ and $\mathbf{Y}$ need to be lefted before fitting the above model without an intercept. To simplify the notations, we further assume that $\mathbf{X}$ has been standardized in the following sections.

In the multivariate regression setting, the improvement of association analysis could be achieved through either taking advantage of the structure of the data or incorporating prior biological knowledge or both. The choice of strategy determines how to devise an appropriate objective function over $\mathbf{B}$. In eQTL studies, the number of SNPs J is large (millions) while the number of genes I is small (thousands), and it is likely that only a small number of eQTLs exist. For such a $p>>n$ problem (in our case, $J>>N$), lasso (Tibshirani, 1996) offers a stable, sparse and effective method to identify nonzero elements in $\mathbf{B}$. Mathematically, lasso solves an l_1 penalized least squares problem,

$$\min_{\mathbf{B}} \frac{1}{2}||\mathbf{Y} - \mathbf{XB}||_F^2 + \lambda||\mathbf{B}||_1, \tag{14.5}$$

where $||.||_F$ denotes the Frobenius norm, $||.||_1$ denotes the matrix l_1 norm and λ is a tuning parameter that determines the degree of regularization on $\mathbf{B}$. The standard lasso estimation does not take advantage of the relatedness of genes. It is equivalent to selecting associated markers for each gene separately. To capture shared genetic structure among genes, a group lasso penalty should be used instead, which estimates the coefficients by solving the following problem,

$$\min_{\mathbf{B}} \frac{1}{2}||\mathbf{Y} - \mathbf{XB}||_F^2 + \lambda\sum_j||\beta_j||_2, \tag{14.6}$$

where $||.||_2$ denotes the l_2 norm and β_j is the j-th row of $\mathbf{B}$. This penalty encourages the group-wise sparsity by l_2 norms combined through l_1 norm. Because the l_2 penalty itself does not induce sparsity, when the j-th marker is selected as an associated marker, then all the elements of β_j would take nonzero values. Thus, the group lasso is limited in the eQTL setting because it is not realistic to assume that a small set of markers could affect the expression of all genes. To obtain the sparsity within associated markers, a lasso penalty could be added to 6, then the objective function becomes (Simon *et al.*, 2013):

$$\min_{\mathbf{B}} \frac{1}{2}||\mathbf{Y} - \mathbf{XB}||_F^2 + \lambda_1||\mathbf{B}||_1 + \lambda_2\sum_j||\beta_j||_2, \tag{14.7}$$

where λ_1 and λ_2 are the regularization parameters that are often tuned by cross-validation. The blended regularization can induce within-group sparsity.

To incorporate prior knowledge, Lee *et al.* (2010) further extended the above penalty as an adaptive one. The corresponding objective function takes the following form:

$$\min_{\mathbf{B}} \frac{1}{2}||\mathbf{Y} - \mathbf{XB}||_F^2 + \lambda_1\sum_j\theta_j||\mathbf{B}||_1 + \lambda_2\sum_j\rho_j||\beta_j||_2, \tag{14.8}$$

where θ_j and ρ_j are scaling parameters, which reflect the prior information of the j-th SNP by a weighted sum over a set of features:

$$\theta_j = \sum_t \omega_t f_t^j \text{ and } \rho_j = \sum_t v_t f_t^j \tag{14.9}$$

Here, f_t^j is the t-th feature for the j-th SNP. For example, it can be an indicator for whether the SNP is located at a transcription factor binding site, or it can be the conservation score. Because only the relative contributions of different features are of interest, the authors further added the constraints $\sum_t \omega_t = 1$ and $\sum_t v_t = 1$.

To achieve a framework that enjoys a Bayesian interpretation, the problem was finally defined as finding the MAP estimation of β and simultaneously estimating the feature weights (ω, v) (referred to as adaptive multi-task lasso; Lee *et al.*, 2010):

$$\min_{\mathbf{B},\omega,v} \frac{1}{2}||\mathbf{Y} - \mathbf{XB}||_F^2 + \lambda_1 \sum\nolimits_j \theta_j ||\beta_j||_1 + \lambda_2 \sum\nolimits_j \rho_j ||\beta_j||_2 + \log Z(\theta, \rho), \tag{14.10}$$

where $Z(\theta, \rho)$ is a normalization factor from the conditional probability of B under a multivariate normal assumption of X. Thus, in model (14.10), the weights of associations are learned adaptively from prior information on SNPs and data. Because SNPs perturb modules of genes rather than acting on individual genes, for example, a polymorphism at a transcription factor may regulate the expression of a group of related genes, the same group of authors also proposed to incorporate the gene group information. More specifically, they ran a hierarchical agglomerative clustering analysis on gene expression data first, and then incorporated the output of clustering algorithm to the regression framework using a tree-structured penalty (Kim and Xing, 2012).

Recently, Wang *et al.* (2015) proposed to use the bridge penalty to induce group selection of SNPs. Their method (named GroupRemMap) solves the following objective function:

$$\min_{\mathbf{B}} \frac{1}{2}||\mathbf{Y} - \mathbf{XB}||_F^2 + \lambda_1 \sum\nolimits_i ||C_i\beta_i||_1 + \lambda_2 \sum_g^G \omega_g \left(\sum_{k \in A_g} ||C_k\beta_k||_2 \right)^\gamma, \tag{14.11}$$

where $C = (c_{ij})$ is an indicator matrix capturing prior knowledge whether SNP j affects transcript i, G is the number of groups indexed by g, and $0 < \gamma < 1$. In the bridge penalty term, $(.)^\gamma$ induces group selection for SNPs that belong to same gene/block and $||.||_2$ induces group selection for transcripts functioned on each SNP.

14.3.4. Confounding effects

An important issue in eQTL mapping is that a proportion of variations on gene expression may come from non-genetic effects. As recent studies have illustrated, confounding factors would complicate the eQTL analysis by causing spurious associations.

14.3.4.1. Effects from known factors

Sex, age, population structures and batch effects are factors that could have an effect on the gene expression level (Michaelson *et al.*, 2009). When measured, correct estimation of the variation due to known factors would lead to a more powerful statistical analysis. To correct for the sex-specific effect, sex could be used as an additional variable controlling gene expression in the regression model. The same strategy can be applied to age, population structures, and other factors. Batch effects are commonly observed across batches of microarray experiments despite efforts to normalize microarray data. They result from many sources including array manufacturing, image artifacts, or sample preparation. When the batch information is available, a proportion of the batch effect could be removed using some explicit batch correction methods, such as ComBat (Johnson *et al.*, 2007). Essentially, ComBat is an empirical Bayes framework that directly models parameters that represent the batch effects, and its R package named ComBat is publicly available (Johnson *et al.*, 2007).

14.3.4.2. Effects from hidden factors

One major challenge is that some non-genetic factors are unobserved. For example, Pastinen *et al.* (2006) report that cell culture conditions often have an influence on gene expressions. Substantial within-batchs effect were observed in the Microarray Quality Control study (Gagnon-Bartsch and Speed, 2012). When these hidden factors fail to be taken into account in the analysis, expression heterogeneity (EH) arises, which reduces statistical power and causes spurious associations. Many methods have been proposed to account for hidden confounding factors. Surrogate variable analysis (SVA; Leek *et al.*, 2010) identifies principal components of EH as single vectors, each treated as a fixed effect in the model. The number of principal components is chosen from permutation. Inter-sample correlation emended (ICE; Kang *et al.*, 2008) models the hidden factors using a linear mixed model. The inter-sample correlation structure is directly incorporated as a variance component accounting for random effects. The main difference between SVA and ICE is that SVA accounts for the most significant correlations, while ICE accounts for all correlations. However, ICE is inconsistent with a compromised statistical power. Another linear mixed model called LMM-EH (Listgarten *et al.*, 2010) corrects the inconsistency of the estimated EH covariance matrix. It assumes the following structure:

$$\mathbf{Y} = \mathbf{XB} + \mathbf{U} + \mathbf{E} \tag{14.12}$$

where $\mathbf{U}$ denotes random effects, i.e., each column $U_i \sim N(0, \tau\ \Sigma)$, where τ is a scalar, $\Sigma \in R^{N \times N}$, N is number of samples. LMM-EH first estimates $\hat{\Sigma}$ from the null model without SNPs, i.e., $\mathbf{Y} = \mathbf{U} + \mathbf{E}$. Then it plugs $\hat{\Sigma}$ into (14.12) to estimate parameters $\{\mathbf{B}, \tau, \Sigma\}$ for all gene–SNP pairs (one gene versus one SNP at a time). PANAMA (Fusi *et al.*, 2012) further extends LMM-EH by the following model:

$$\mathbf{Y} = \mathbf{XB} + \mathbf{HW} + \mathbf{E} \tag{14.13}$$

The additional term $\mathbf{H}$ represents the effects of hidden factors. PANAMA assigns independent Gaussian priors for $\mathbf{B}$ and $\mathbf{W}$,

$$P(\mathbf{B}) = \Pi_j N\left(0, \alpha_j^2 \mathbf{I}\right),\ P(\mathbf{W}) = \Pi_k N(0, \gamma_k^2 \mathbf{I}), \tag{14.14}$$

leading to a Bayesian formulation of the problem:

$$P(\mathbf{Y}|\mathbf{H}, \Theta) = \Pi_i N\left(\mathbf{0}, \sum_j \alpha_j^2 X_j X_j^T + \sum_k \gamma_k^2 H_k H_k^T + \sigma_e^2 \mathbf{I}\right), \tag{14.15}$$

where k is the index of hidden factors, $\Theta = \{\{\alpha_j^2\}, \{\gamma_k^2\}, \sigma_e^2\}$. In PANAMA, these parameters are estimated by computational techniques from Gaussian process model optimization.

However, because the optimization problem in PANAMA is not convex, PANAMA may return a local optimum solution. Yang *et al.* (2013) recently introduced an alternative formulation to correct for EH during eQTL mapping. In a unified framework, they integrate low-rank representation and sparse regression (LORS). The former accounts for non-genetic factors and the latter performs eQTL mapping. In addition to model (14.4), $\mathbf{L}$ is introduced to account for the variations caused by hidden factors. Thus, the observed gene expression levels can be decomposed into three parts influenced by genetic factors, non-genetic factors, and random noises as the following,

$$\mathbf{Y} = \mathbf{XB} + \mathbf{L} + \mathbf{E}.$$

In addition to the sparsity assumption on $\mathbf{B}$, we further assume $\mathbf{L}$ is a low-rank matrix, i.e., in all hidden factors, only a few may affect gene expression levels. To estimate the coefficients, the following optimization problem is solved:

$$\min_{\mathbf{B},\mathbf{L}} \frac{1}{2} ||\mathbf{Y} - \mathbf{XB} - \mathbf{L}||_F^2 + \rho||\mathbf{B}||_1 + \lambda||\mathbf{L}||_*.$$

The above problem could be efficiently solved with a guaranteed convergence.

We summarize the methods that account for hidden confounding factors in Table 14.2.

14.4. Interpretation of eQTL data

By simultaneously capturing many regulatory interactions, eQTL data may provide valuable insight to the genetic architecture of gene expression. However, sophisticated methods are needed to exploit the full potential of this approach. The eQTL data could be interpreted either based on the original data or the integration with prior knowledge.

14.4.1. Visualization and enrichment analysis for regulatory hotspots

A visualization of eQTL mapping results such as Figure 14.2 may help to identify genomic regions that are of interest. The *cis*-eQTLs are located along the diagonal, while *trans*-eQTL are located distal to the targets. In addition, there may be additional "bands" in the matrix that are enriched for eQTL. Horizontal bands occasionally exist, but they are usually due to artifacts such as population substructure. Thus, horizontal bands are often indicators of statistical problem with the data. Vertical bands are called *trans*-bands or *trans*-eQTL bands, because they are regulatory hotspots that control many target genes (Kang *et al.*, 2008). The presence of *trans*-bands indicates that the region under consideration contains a polymorphic

Table 14.2. eQTL mapping methods/tools.

Category	Methods	Notes	Availability
Univariate methods	Standard linear regression	Identifying associations by *t*-test	Basic R, Matrix eQTL (Shabalin, 2012)
	Likelihood ratio test	Identifying associations by the model fitness	Basic R
	Wilcoxon's rank test	Non-parametric method	Basic R
Interval mapping	EM algorithm (Lander and Botstein, 1989)	Ungenotyped intervals treated as missing data	R package *R/qtl*
	Haley–Knott regression (Haley *et al.*, 1994)	Two-component effect model	R package *R/qtl*
Multivariate regression	Lasso (Tibshirani, 1996)	Penalized regression with $l_1 l_1$ penalty	R package *LARS*
	Sparse group lasso (Simon *et al.*, 2013)	Penalized regression with $l_1 l_1$ and group penalty	R package *SGL*
	Elastic net (Zou and Hastie, 2005)	Penalized regression with $l_1 l_1$ and $l_2 l_2$ penalty	R package *elasticnet*
	Adaptive multi-task Lasso (Lee *et al.*, 2010)	Penalized regression with data-driven adaptive penalty	NA
	Tree guided group lasso (Kim and Xing, 2012)	Penalized regression with a tree-structured penalty	NA
	GroupRemMap (Wang *et al.*, 2015)	Penalized regression with $l_1 l_1$ and bridge penalty	R package GroupRemMap
Correction for confounding effects	SVA (Leek *et al.*, 2010)	Principal components of EH as fixed effect	R package *SVA*
	ICE (Kang *et al.*, 2008)	Linear mixed model	R package *ICE*
	LMM-EH (Listgarten *et al.*, 2010)	Linear mixed model	Software *eLMM*
	PANAMA (Fusi *et al.*, 2012)	Combination of linear mixed model and factor-based model	Software *PANAMA*
	LORS (Yang *et al.*, 2013)	Low-rank representation and penalized regression	Matlab toolkit

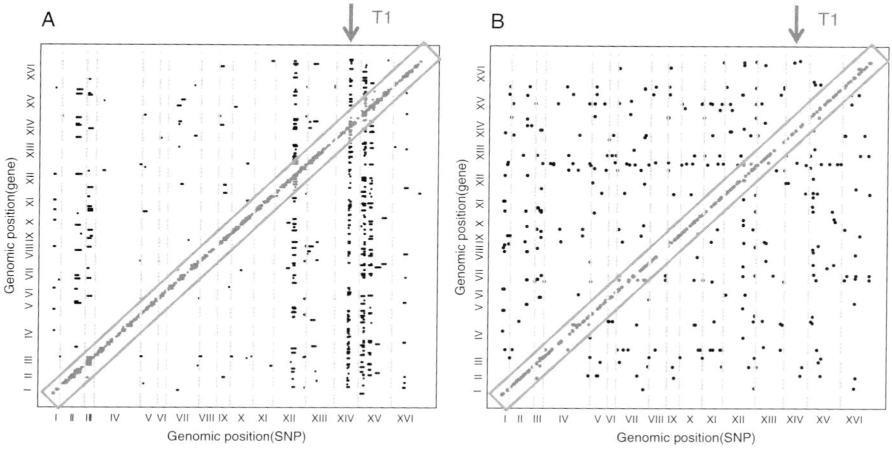

Figure 14.2. Visualization of eQTL results for GSE9376 with marker locations shown in the *x*-axis and expression trait locations shown in the *y*-axis. *cis*-eQTLs are located along the diagonal (shadowed), while *trans*-eQTLs are located distal to the targets. (A) Significant associations identified by standard linear regression. (B) Top 1000 associations identified by LORS based on the absolute value of coefficients. Spurious associations based on standard linear regression have been reduced by LORS. For example, *trans*-band T1 disappears after accounting for non-genetic effects.

transcriptional regulator that influences the expression of several genes. An enrichment analysis could be utilized to help identify co-regulation, that is, the target genes in the *trans*-bands could be checked for the enrichment in terms of of gene ontology (GO; Ashburner *et al.*, 2000) category, pathway, or other functional annotations. Essentially, the enrichment analysis was performed based on a hypergeometric test. Taking the GO category as an example, in a selection of genes, the number of genes with a specific GO annotation follows $P(z|T,S,n) = \frac{\binom{S}{n}\binom{T-S}{n-z}}{\binom{T}{n}}$, where T is total number of genes in the GO database; S is the number of genes in that GO category; n is the number of genes in the *trans*-band; and z is the number of genes within the *trans*-band in that GO category. The P-value can be obtained by calculating tail probability $P(z \geq k)$, which sums over all the probabilities of cases at least as extreme as the observed data. Because multiple GO categories are tested for genes in the *trans*-band, Bonferroni or FDR correction is needed to control the false-positive results. However, spurious *trans*-bands may be artifacts from batch effects or other effects (see an example in the case study). It is essential to account for confounding effects and properly determine the statistical significance of *trans*-bands before the interpretation.

14.4.2. eQTL prioritization

Due to LD, a mapped eQTL often covers multiple genes. To identify the causal gene that is responsible for the observed association in a mapped locus, gene prioritization or refinement techniques are needed. Because genes at a locus are indistinguishable based on eQTL analysis, additional information, such as the physical interaction network, has been incorporated in the eQTL prioritization methods. In eQTL prioritization, the gene priorities are assigned based on a network proximity measure or an information flow measure between a target gene and a set of candidate genes in a physical interaction network. The core idea is that in an eQTL the regulation information will flow from a regulator to the target genes via physical interactions. In eQTL electrical diagrams (eQED; Suthram *et al.*, 2008), the information flow is simulated as an electric current through a wiring diagram of current sources (candidate target genes) and resistors (physical interactions). After solving the circuit for currents, the causal gene is predicted as the one with the maximum current running through it. Although the eQED is methodologically very attractive, it is highly dependent on the completeness and accuracy of the physical interaction networks. If the interaction of a transcription factor and the target gene is unknown, it is unlikely to pinpoint the correct causal gene. Thus, eQED performs best for species with sufficiently dense protein interaction information. Other similar methods model the information flow on physical interaction networks with different techniques, for example, Tu *et al.* (2006) apply a random walk technique and Voevodski *et al.* (2009) use the PageRank algorithm.

Stojmirovi and Yu (2012) point out that gene prioritization may be complicated by the structure of the global network. For example, the result of gene prioritization is highly dependent on the node degree of the gene in the network. Thus, instead of using a global network, some authors suggest to reduce the interaction network to a local neighborhood to reduce the irrelevant interactions and remove the disturbing influence of high degree nodes. In EPSILON (Verbeke *et al.*, 2013), a k-trial shortest path approach is applied to restrict the network that covers the candidate causal genes and a target gene. First of all, all the interaction in the global network will be assigned a cost. Then, an optimal path from each candidate causal gene to the target will be identified using the Dijkstra algorithm. All genes on such paths will be included in the local network. Furthermore, k trials of alternative paths will be performed and further merged to the local network. Finally, the priorities of candidate gene are evaluated based on a distance measure in a locally constructed network. It has been shown in general that using a local network yields significantly better performance compared with using the global network.

We summarize some of the methods for eQTL analysis in Table 14.3.

14.4.3. Network analysis

Through capturing thousands of expression traits and millions of genetic variants simultaneously, eQTL data provide important information about the regulatory relationships among genes. Analogous to Gaussian graphical models used to infer the genetic regulatory network from gene expression data, conditional Gaussian graphical models (cGGM) have been proposed to infer the regulatory network

Table 14.3. Methods/tools to interpret the eQTL results.

Category	Methods	Notes	Availability
Visualization	eQTL viewer (Zou *et al.*, 2007)	Visualizing eQTL mapping results and genome annotation	Web-based service
	Trigger (Chen *et al.*, 2011)	Genome-wide eQTL mapping plot	R package
Enrichment analysis	GOrilla (Eden *et al.*, 2009)	Identifying and visualizing enriched GO terms	Web-based service
	DAVID (Da Wei Huang *et al.*, 2008)	Gene annotation, visualization and integrated discovery	Web-based service
Prioritization	eQED (Suthram *et al.*, 2008)	Prioritizing with a global network using electrical diagrams	Matlab toolkit
	Tu *et al.* (2006)	Prioritizing with a global network using random walk	JAVA program
	Voevodski *et al.* (2009)	Prioritizing with a global network using the PageRank algorithms	Web-based service
	ITM probe (Stojmirovi and Yu, 2009)	Modeling information flow with absorbing or emitting channel	Web-based service
	EPSILON (Verbeke *et al.*, 2013)	Prioritizing with a local network using the shortest path	Matlab toolkit

from eQTL data. By identifying the conditional dependency among a set of genes after removing the effects from genetic variants, the graph inferred from cGGM may better reflect gene regulation at the expression level. More specifically, the problem is formulated as:

$$Y_n = \mathbf{B}^T X_n + E_n, \tag{14.16}$$

where n is the index for individuals, $Y_n = (Y_{n1}, \ldots, Y_{nI})^T$ denote the random vector of gene expression levels for I genes, $X_n = (X_{n1}, \ldots, X_{nJ})^T$ denote the random vector of J genetic variants, and E_n is an $I \times 1$ random vector with mean zero, precision matrix $\boldsymbol{\Omega}$ and covariance matrix $\Sigma = \boldsymbol{\Omega}^{-1}$. Inferring the conditional dependency graph of genes involves simultaneous estimation of the regression coefficient matrix $\mathbf{B}$ and the residual precision matrix , where B reflects the genetic effect and $\boldsymbol{\Omega}$ reflects the dependency structure among genes. Several approaches have been proposed to solve this joint estimation problem, such as a linear approach in Obozinski *et al.* (2011) and a nonlinear method proposed in Li *et al.* (2012a). Yin and Li (2011) proposed a two-stage approach, where a constrained l_1 minimization approach was used to first estimate $\mathbf{B}$, and then precision matrix $\boldsymbol{\Omega}$ was estimated using the regression residual from the the first stage. In their method, z were assumed to follow a multivariate normal distribution. Under the same framework, Cai *et al.* (2013) proposed a method called CAPME, which loosens the multivariate normal assumption.

Let $\overline{Y} = N^{-1}\sum_n Y_n$, $\overline{X} = N^{-1}\sum_n X_n$, $\overline{E} = N^{-1}\sum_n E_n$, $S_{XY} = N^{-1}\sum_n \left(Y_n - \overline{Y}\right)\left(X_n - \overline{X}\right)^T$ and $S_{XX} = N^{-1}\sum_n \left(X_n - \overline{X}\right)\left(X_n - \overline{X}\right)^T$. CAPME estimates $\mathbf{B}$ by solving the optimization problem:

$$\hat{\mathbf{B}} = \underset{\mathrm{B}}{argmin}\{||\mathbf{B}||_1 : ||S_{XY} - \mathbf{B}^T S_{XX}||_\infty \leq \lambda\} \tag{14.17}$$

where λ is a tuning parameter. Here, the estimation of Γ is done by treating the coefficient estimation problem as a multivariate extension of the Danzig selector, which is asymptotically equivalent to the Lasso estimator mentioned in Section 14.3.3. In the second stage, the precision matrix $\boldsymbol{\Omega}$ is estimated by solving the optimization problem:

$$\hat{\boldsymbol{\Omega}}_1 = \underset{\boldsymbol{\Omega}}{argmin}\{||\boldsymbol{\Omega}||_1 : ||I - S_{YY}\boldsymbol{\Omega}||_\infty \leq \tau\} \tag{14.18}$$

where $S_{YY} = N^{-1}\sum_n \left(Y_n - \hat{\mathbf{B}}^T\overline{X}\right)\left(Y_n - \hat{\mathbf{B}}^T\overline{X}\right)^T$ and τ is a tuning parameter. The solution to problem (14.18) may not be symmetrical in general. Thus, the final estimator of $\boldsymbol{\Omega}$, denoted by $\hat{\boldsymbol{\Omega}}$, is $\hat{\boldsymbol{\Omega}}_1$ after symmetrization. In practice, the tuning parameters λ and τ are determined via cross-validation with a grid search.

Chen *et al.* (2015) recently proposed a tuning-free two-stage approach ANTAC (asymptotically normal estimation with thresholding after adjusting covariates) for cGGM inference when the precision matrix is sufficiently sparse. For eQTL data, this approach not only reflects the strength of the regulatory relationships among genes through partial correlations, but also provides P-values to reflect the statistical significance of the detected relationships. Similar to the first stage in CAPME, $\mathbf{B}$ is estimated with l_1 constrained penalty. In the second stage, ANTAC estimates the precision matrix by estimating each element of $\boldsymbol{\Omega}$. Specifically, define a pair of genes i_1 and i_2 in a set $A = \{i_1, i_2\}$, A^c as the complement of this set including all the other genes and use an $N \times I$ matrix $\hat{\mathbf{E}}$ to denote the residues obtained from model (14.16). From each $i \in A$, ANTAC applies a scaled lasso penalization (Sun and Zhang, 2012) to $\hat{E}_i$ against $\hat{E}_{A^c}$ as follows,

$$\{\hat{\delta}_i, \hat{\theta}_{ii}^{1/2}\} = \underset{d,\sigma}{argmin}\left\{\frac{||\hat{E}_i - \hat{E}_{A^c}d||^2}{2N\sigma} + \frac{\sigma}{2} + \lambda\sum_l \frac{||\hat{E}_l||_1}{\sqrt{N}}|d_l|\right\} \tag{14.19}$$

where $\hat{E}_i$ is the i-th column of $\hat{\mathbf{E}}$, $\hat{E}_{A^c}$ is an $N \times (I-2)$ matrix excluding the i_1-th and i_2-th column from $\hat{\mathbf{E}}$, $\hat{\theta}_{ii}$ is the estimated variance of E_i and λ is a tuning parameter. Define the residuals of the above regression by $\hat{\varepsilon}_A = \hat{E}_A - \hat{E}_{A^c}\hat{\delta}$ and $\hat{\Theta}_{A,A} = \hat{\varepsilon}_A^T\hat{\varepsilon}_A/N$, where $\hat{\delta}$ have two columns $\hat{\delta}_{i_1}$ and $\hat{\delta}_{i_2}$, entries in $\boldsymbol{\Omega}_{A,A}$ are estimated by simply inverse $\hat{\Theta}_{A,A}$, i.e., each entry in $\boldsymbol{\Omega}$ is estimated by the corresponding off-diagonal element of $\boldsymbol{\Omega}_{A,A}$.

14.5. A case study

In this section, we use the yeast expression and genotype data set from Brem et al. (2002) (GEO accession number GSE9376) to illustrate the workflow of eQTL analysis. This data set consists of 5493 probes and 2956 genotyped loci measured in 109 segregants derived from a cross between BY, a standard laboratory strain and RM, a wild isolate from a California vineyard. Quality control was done by Fusi *et al.* (2012) and there were no missing data in the downloaded data set.

We applied both standard linear regression and LORS to this data set. An association is defined as *cis* if the location of the SNP and the location of the opening reading frame (ORF) of the gene were within 10 kb, and *trans* otherwise. For standard linear regression, associations are defined as significant if Bonferroni corrected *P*-value less than 0.01 and the multiple testing adjustment was performed for *cis*-acting and *trans*-acting separately. As shown in Figure 14.2, we see that some spurious associations may be caused by confounding effects. LORS detects associations based on nonzero coefficients (**B**) instead of hypothesis testing; thus, the associations are identified based on a cut-off of the absolute value of **B**. In total, LORS outputs about 10,000 nonzero **B** values. In principle, people may be interested in the top signals that are promising for validation experiments. Here we report the top 1000 associations based on the absolute value of **B**. Shown in Figure 14.2, some *trans*-band associations based on standard linear regression have been reduced successfully by LORS, for example *trans*-band T1 disappears after accounting for non-genetic effects. This indicates that non-genetic effects may complicate the eQTL analysis by spurious associations; thus, it is important to account for those effects in the analysis. We further investigated the top 10 regulatory hotspots by GO enrichment analysis. For all these hotspots, associated genes are enriched with at least one GO category. Among these, hotspot 9 is a novel hotspot detected by LORS but not by standard linear regression. Five of the 18 associated genes are functional in response to toxin; they are AAD4, YDL218 W, YLL056C, AAD6, and SPS100. The hot-spot eQTL is *cis*-linked to one of them, AAD4, which apparently explains the detected association.

We also applied ANTAC on a subset of these data. We extracted 54 genes that belong to the yeast mitogen-activated protein kinase (MAPK) signalling pathway from KEGG (Kanehisa and Goto, 2000) and then selected the genetic markers based on simple screening. There are 139 markers that are marginally associated with at least 3 of the 54 genes with an FDR controlled P-value less than or equal to 0.01. To visualize the network, we used a cutoff of 0.01 on the FDR controlled P-values. Forty-one edges were identified and are shown in Figure 14.3. A promising result of this analysis arises from the subnetwork involving *STE2, STE3, STE4, STE12, SWI4, SWI6, SLT2*, and *BCK1*. In budding yeast, five MAPK pathways regulate mating, invasive growth, cell wall integrity, hyperosmolarity responses, and ascospore formation. *STE2, STE3, STE4*, and *STE12* are primary targets of the yeast mating response, while *SWI4, SWI6, SLT2*, and *BCK1* form part of the cell integrity sensing pathway (Hamel *et al.*, 2012).

14.6. Discussion and conclusion

In this chapter, we give an overview of the workflow of eQTL mapping analysis and cover a wide variety of related methods. Much of what we learned from traditional QTL mapping has contributed to addressing the questions posed by

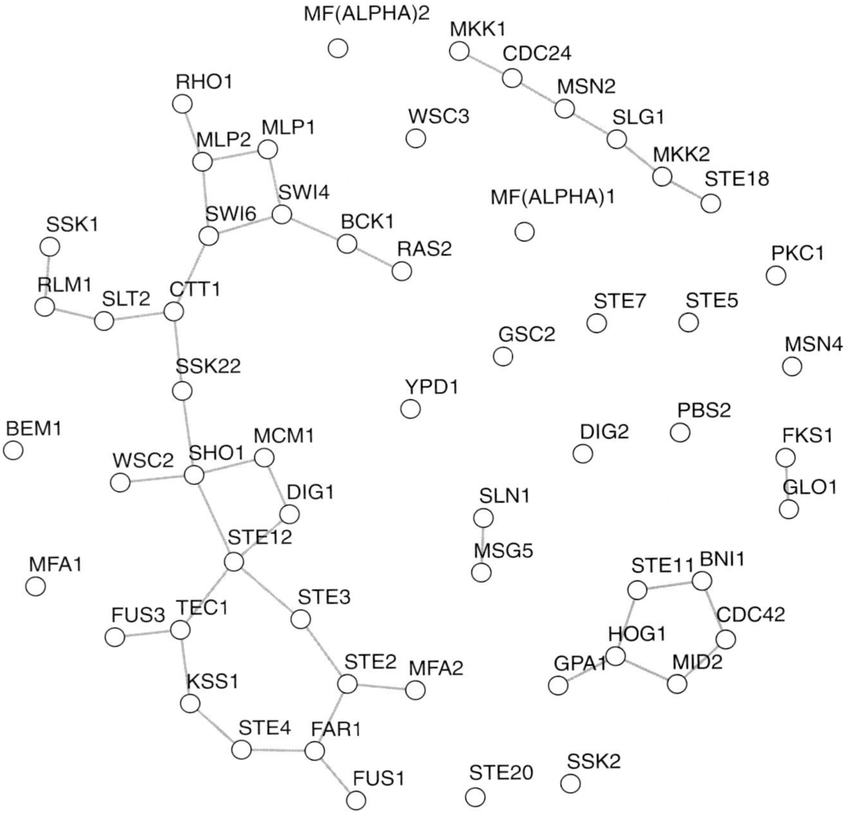

Figure 14.3. Visualization of the network constructed for genes in MAPK pathway from the yeast data by ANTAC.

eQTL mapping. However, compared to traditional QTL analysis, eQTL analysis is challenging, especially in two aspects: (1) identifying "many to many" associations from ultra high-dimensional data is statistically difficult; and (2) unobserved non-genetic confounding factors may cause spurious associations. To address the first problem, many penalized regression techniques mentioned in Section 14.3.3 have been proposed to yield a sparse and interpretable solution. However, how to effectively incorporate the structure of data or prior knowledge in the eQTL setting remains an open and promising problem. To address the second problem, accounting for non-genetic effects in the analysis is of importance. Thus incorporating the hidden factors into the current penalized multivariate regression framework is another promising future direction.

Moreover, due to the complex nature of eQTL data, sophisticated methods are needed to interpret the eQTL mapping results to exploit their full potential. In this chapter, we have introduced some tools and methods for visualization, enrichment analysis, prioritization analysis, and network analysis. Depending on the study design and the availability of external information, these methods may be applied to gain insight into transcriptional regulation.

Finally, we note that here we assume that log-transformed RNA-seq count data follow a normal distribution and then treat transformed data as microarray data in the downstream analysis. However, discrete distributions such as beta-binomial or negative binomial distributions are more appropriate choices to model the RNA-seq count data. It is much more challenging to study multivariate beta-binomial or negative binomial distributions. This is also an interesting direction that deserves further investigation as RNA-seq data will become the more general platform for measuring gene expression levels. We refer interested readers to a review (Sun and Hu, 2013) on this topic.

As next-generation sequencing technologies collect genomic data at an unprecedented high resolution, we anticipate the integrative analysis of variations in transcriptome and sequence will result in unprecedented accuracy in eQTL detection and interpretation, which will bring insights into the molecular pathogenesis of complex traits.

REFERENCES

Ashburner, M., Ball, C.A., Blake, J.A., *et al.* (2000). Gene ontology: tool for the unification of biology. *Nature Genet.*, **25**, 25–29.

Benjamini, Y. and Hochberg, Y. (1995). Controlling the false discovery rate: a practical and powerful approach to multiple testing. *J. R. Statist. Soc. Ser. B (Method.)*, **57**, 289–300.

Bohnert, R. and Rätsch, G. (2010). rQuant.web: a tool for RNA-Seq-based transcript quantitation. *Nucleic Acids Res.*, **38**, W348–W351.

Bolstad, B.M., Irizarry, R.A., Åstrand, M. and Speed, T.P. (2003). A comparison of normalization methods for high density oligonucleotide array data based on variance and bias. *Bioinformatics*, **19**, 185–193.

Brem, R.B., Yvert, G., Clinton, R. and Kruglyak, L. (2002). Genetic dissection of transcriptional regulation in budding yeast. *Science*, **296**, 752–755.

Broman, K.W., Wu, H., Sen, . and Churchill, G.A. (2003). R/QTL: QTL mapping in experimental crosses. *Bioinformatics*, **19**, 889–890.

Browning, S.R. and Browning, B.L. (2007). Rapid and accurate haplotype phasing and missing-data inference for whole-genome association studies by use of localized haplotype clustering. *Am. J. Hum. Genet.*, **81**, 1084–1097.

Cai, T.T., Li, H., Liu, W. and Xie, J. (2013). Covariate-adjusted precision matrix estimation with an application in genetical genomics. *Biometrika*, **100**, 139–156.

Carey, V.J. (2013). GGtools: Genetics of Gene Expression with Bioconductor. R package version 4.6.2.

Chen, L.S., Sangurdekar, D.P. and Storey, J.D. (2011). *trigger: Transcriptional Regulatory Inference from Genetics of Gene ExpRession.* R package version 1.4.0.

Chen, M., Ren, Z., Zhao, H. and Zhou, H. (2015). Asymptotic normal estimation of covariate-adjusted gaussian graphical model. *J. Am. Stat. Ass. Theory Meth.* (in press).

Da Huang, W., Sherman, B.T. and Lempicki, R. A. (2008). Systematic and integrative analysis of large gene lists using DAVID bioinformatics resources. *Nature Protocols*, **4**, 44–57.

Delaneau, O., Zagury, J.-F. and Marchini, J. (2012). Improved whole-chromosome phasing for disease and population genetic studies. *Nature Meth.*, **10**, 5–6.

Dillies, M.-A., Rau, A., Aubert, J., *et al.* (2013). A comprehensive evaluation of normalization methods for illumina high-throughput RNA sequencing data analysis. *Brief. Bioinform.*, 14, 671–683.

Dunning, M.J., Smith, M.L., Ritchie, M.E. and Tavaré, S. (2007). beadarray: R classes and methods for Illumina bead-based data. *Bioinformatics*, **23**, 2183–2184.

Eden, E., Navon, R., Steinfeld, I., Lipson, D. and Yakhini, Z. (2009). GOrilla: a tool for discovery and visualization of enriched GO terms in ranked gene lists. *BMC Bioinform.*, **10**, 48.

Fusi, N., Stegle, O. and Lawrence, N.D. (2012). Joint modelling of confounding factors and prominent genetic regulators provides increased accuracy in genetical genomics studies. *PLoS Comput. Biol.*, **8**, e1002330.

Gagnon-Bartsch, J.A. and Speed, T.P. (2012). Using control genes to correct for unwanted variation in microarray data. *Biostatistics*, **13**, 539–552.

Gautier, L., Cope, L., Bolstad, B.M. and Irizarry, R.A. (2004). affy – analysis of Afymetrix GeneChip data at the probe level. *Bioinformatics*, **20**, 307–315.

Guttman, M., Garber, M., Levin, J.Z., *et al.* (2010). *Ab initio* reconstruction of cell type-specific transcriptomes in mouse reveals the conserved multi-exonic structure of lincrnas. *Nature Biotechnol.*, **28**, 503–510.

Haley, C.S., Knott, S.A. and Elsen, J. (1994). Mapping quantitative trait loci in crosses between outbred lines using least squares. *Genetics*, **136**, 1195–1207.

Hamel, L.-P., Nicole, M.-C., Duplessis, S. and Ellis, B.E. (2012). Mitogen-activated protein kinase signaling in plant-interacting fungi: distinct messages from conserved messengers. *Plant Cell Online*, **24**, 1327–1351.

Johnson, W.E., Li, C. and Rabinovic, A. (2007). Adjusting batch effects in microarray expression data using empirical bayes methods. *Biostatistics*, **8**, 118–127.

Kanehisa, M. and Goto, S. (2000). Kegg: Kyoto Encyclopedia of Genes and Genomes. *Nucleic Acids Res.*, **28**, 27–30.

Kang, H.M., Ye, C. and Eskin, E. (2008). Accurate discovery of expression quantitative trait loci under confounding from spurious and genuine regulatory hotspots. *Genetics*, **180**, 1909–1925.

Katz, Y., Wang, E.T., Airoldi, E.M. and Burge, C.B. (2010). Analysis and design of RNA sequencing experiments for identifying isoform regulation. *Nature Meth.*, **7**, 1009–1015.

Kim, D., Pertea, G., Trapnell, C., *et al.* (2013). TopHat2: accurate alignment of transcriptomes in the presence of insertions, deletions and gene fusions. *Genome Biol.*, **14**, R36.

Kim, S. and Xing, E.P. (2012). Tree-guided group lasso for multi-response regression with structured sparsity, with an application to eQTL mapping. *Ann. Appl. Stat.*, **6**, 1095–1117.

Lander, E.S. and Botstein, D. (1989). Mapping Mendelian factors underlying quantitative traits using RFLP linkage maps. *Genetics*, **121**, 185–199.

Lee, S., Zhu, J. and Xing, E.P. (2010). Adaptive multi-task lasso: with application to eQTL detection. In *Advances in neural information processing systems*, pp. 1306–1314.

Leek, J.T., Scharpf, R.B., Bravo, H.C., *et al.* (2010). Tackling the widespread and critical impact of batch effects in high-throughput data. *Nature Rev. Genet.*, **11**, 733–739.

Li, B., Ruotti, V., Stewart, R.M., Thomson, J.A. and Dewey, C.N. (2010a). RNA-Seq gene expression estimation with read mapping uncertainty. *Bioinformatics*, **26**, 493–500.

Li, B., Chun, H. and Zhao, H. (2012b). Sparse estimation of conditional graphical models with application to gene networks. *J. Am. Statist. Ass.*, **107**, 152–167.

Li, C. and Wong, W.H. (2001). Model-based analysis of oligonucleotide arrays: model validation, design issues and standard error application. *Genome Biol.*, **2**, 1–11.

Li, J.J., Jiang, C.-R., Brown, J.B., Huang, H. and Bickel, P.J. (2011). Sparse linear modeling of next-generation mRNA sequencing (RNA-Seq) data for isoform discovery and abundance estimation. *Proc. Natl Acad. Sci. USA*, **108**, 19867–19872.

Li, L., Zhang, X. and Zhao, H. (2012b). eQTL. In *Quantitative Trait Loci (QTL)*. Springer, pp. 265–279.

Li, Y., Álvarez, O.A., Gutteling, E.W., *et al.* (2006). Mapping determinants of gene expression plasticity by genetical genomics in *C. elegans*. *PLoS Genet.*, **2**, e222.

Li, Y., Willer, C.J., Ding, J., Scheet, P. and Abecasis, G.R. (2010b). MaCH: using sequence and genotype data to estimate haplotypes and unobserved genotypes. *Genet. Epidemiol.*, **34**, 816–834.

Listgarten, J., Kadie, C., Schadt, E.E. and Heckerman, D. (2010). Correction for hidden confounders in the genetic analysis of gene expression. *Proc. Natl Acad. Sci. USA*, **107**, 16465–16470.

Mackay, T.F., Stone, E.A. and Ayroles, J.F. (2009). The genetics of quantitative traits: challenges and prospects. *Nature Rev. Genet.*, **10**, 565–577.

Manolio, T.A. (2013). Bringing genome-wide association findings into clinical use. *Nature Rev. Genet.*, **14**, 549–558.

McCarthy, M.I., Abecasis, G.R., Cardon, L.R., *et al.* (2008). Genome-wide association studies for complex traits: consensus, uncertainty and challenges. *Nature Rev. Genet.*, **9**, 356–369.

Michaelson, J.J., Loguercio, S. and Beyer, A. (2009). Detection and interpretation of expression quantitative trait loci (eQTL). *Methods*, **48**, 265–276.

Mortazavi, A., Williams, B.A., McCue, K., Schaeffer, L. and Wold, B. (2008). Mapping and quantifying mammalian transcriptomes by RNA-Seq. *Nature Meth.*, **5**, 621–628.

Nica, A.C. and Dermitzakis, E.T. (2008). Using gene expression to investigate the genetic basis of complex disorders. *Hum. Molec. Genet.*, **17**, R129–R134.

Obozinski, G., Wainwright, M.J. and Jordan, M.I. (2011). Support union recovery in high-dimensional multivariate regression. *Ann. Statist.*, **39**, 1–47.

Pastinen, T., Ge, B. and Hudson, T.J. (2006). Influence of human genome polymorphism on gene expression. *Hum. Molec. Genet.*, **15**, R9–R16.

Price, A.L., Patterson, N.J., Plenge, R.M., *et al.* (2006). Principal components analysis corrects for stratification in genome-wide association studies. *Nature Genet.*, **38**, 904–909.

Purcell, S., Neale, B., Todd-Brown, K., *et al.* (2007). Plink: a tool set for whole-genome association and population-based linkage analyses. *Am. J. Hum. Genet.*, **81**, 559–575.

Richard, H., Schulz, M.H., Sultan, M., *et al.* (2010). Prediction of alternative isoforms from exon expression levels in RNA-Seq experiments. *Nucleic Acids Res.*, **38**, e112–e112.

Robertson, G., Schein, J., Chiu, R., *et al.* (2010). *De novo* assembly and analysis of RNA-Seq data. *Nature Meth.*, **7**, 909–912.

Rockman, M.V. and Kruglyak, L. (2006). Genetics of global gene expression. *Nature Rev. Genet.*, **7**, 862–872.

Salzman, J., Jiang, H. and Wong, W.H. (2011). Statistical modeling of RNA-Seq data. *Statist. Sci.*, **26**, 62–83.

Schadt, E.E., Lamb, J., Yang, X., *et al.* (2005). An integrative genomics approach to infer causal associations between gene expression and disease. *Nature Genet.*, **37**, 710–717.

Shabalin, A.A. (2012). Matrix eQTL: ultra fast eQTL analysis via large matrix operations. *Bioinformatics*, **28**, 1353–1358.

Simon, N., Friedman, J., Hastie, T. and Tibshirani, R. (2013). A sparse-group lasso. *J. Comput. Graph. Statist.*, **22**, 231–245.

Smyth, G.K. (2005). Limma: linear models for microarray data. In R. Gentleman, V. Carey, S. Dudoit, R. Irizarry and W. Huber (Eds.), *Bioinformatics and Computational Biology Solutions Using R and Bioconductor*. Springer, New York, NY, pp. 397–420.

Stojmirovi, A. and Yu, Y.-K. (2009). ITM probe: analyzing information flow in protein networks. *Bioinformatics*, **25**, 2447–2449.

Stojmirovi, A. and Yu, Y.-K. (2012). Information flow in interaction networks II: channels, path lengths, and potentials. *J. Comput. Biol.*, **19**, 379–403.

Sun, T. and Zhang, C.-H. (2012). Scaled sparse linear regression. *Biometrika*, **99**, 879–898.

Sun, W. and Hu, Y. (2013). eQTL mapping using RNA-seq data. *Statist. Biosci.*, **5**, 198–219.

Suthram, S., Beyer, A., Karp, R.M., Eldar, Y. and Ideker, T. (2008). eQED: an efficient method for interpreting eQTL associations using protein networks. *Molec. Syst. Biol.*, **4**, 162.

Tibshirani, R. (1996). Regression shrinkage and selection via the lasso. *J. R. Statist. Soc. Ser. B (Method.)*, **58**, 267–288.

Trapnell, C.,Williams, B.A., Pertea, G., *et al.* (2010). Transcript assembly and quantification by RNA-seq reveals unannotated transcripts and isoform switching during cell differentiation. *Nature Biotechnol.*, **28**, 511–515.

Tu, Z., Wang, L., Arbeitman, M.N., Chen, T. and Sun, F. (2006). An integrative approach for causal gene identification and gene regulatory pathway inference. *Bioinformatics*, **22**, e489–e496.

Verbeke, L.P., Cloots, L., Demeester, P., Fostier, J. and Marchal, K. (2013). Epsilon: an eQTL prioritization framework using similarity measures derived from local networks. *Bioinformatics*, **29**, 1308–1316.

Voevodski, K., Teng, S.-H. and Xia, Y. (2009). Spectral affinity in protein networks. *BMC Syst. Biol.*, **3**, 112.

Wang, X., Qin, L., Zhang, H., *et al.* (2015). A regularized multivariate regression approach for eQTL analysis. *Statist. Biosci.*, **7**, 129–146.

Witte, J.S. (2010). Genome-wide association studies and beyond. *Annu. Rev. Publ. Health*, **31**, 9–20.

Xia, Z., Wen, J., Chang, C.-C. and Zhou, X. (2011). NSMAP: A method for spliced isoforms identification and quantification from RNA-Seq. *BMC Bioinform.*, **12**, 162.

Yang, C., Wang, L., Zhang, S. and Zhao, H. (2013). Accounting for non-genetic factors by low-rank representation and sparse regression for eQTL mapping. *Bioinformatics*, **29**, 1026–1034.

Yin, J. and Li, H. (2011). A sparse conditional gaussian graphical model for analysis of genetical genomics data. *Ann. Appl. Statist.*, **5**, 2630.

Zerbino, D.R. and Birney, E. (2008). Velvet: algorithms for *de novo* short read assembly using de Bruijn graphs. *Genome Res.*, **18**, 821–829.

Zou, H. and Hastie, T. (2005). Regularization and variable selection via the elastic net. *J. R. Statist. Soc. Ser. B (Statist. Method.)*, **67**, 301–320.

Zou, W., Aylor, D.L. and Zeng, Z.-B. (2007). eQTL viewer: visualizing how sequence variation affects genome-wide transcription. *BMC Bioinform.*, **8**, 7.

Part IV

Next-generation sequencing technology and pharmaco-genomics

15 Next-generation sequencing for rare diseases

Elena Bosch and Ferran Casals

15.1. Introduction

Most rare diseases have a genetic base, and are inherited in a Mendelian fashion. They are usually monogenic disorders segregating in families, or are sporadic, being autosomal or sex-linked and dominant or recessive. Before the advent of the new next-generation sequencing (NGS) technologies, positional cloning was the most commonly used technique for the analysis of the genetic basis of Mendelian diseases. Usually the first step of such studies was linkage analysis in pedigrees with multiples cases. In 1986, the genes underlying chronic granulomatous disease, Duchenne muscular dystrophy, and retinoblastoma were mapped using the linkage approach assisted by the identification of patients with structural changes or cytogenetically detectable deletions (Collins, 1995). Three years later, cystic fibrosis represented the first case where the gene of a Mendelian disease was mapped based only on linkage analysis and positional cloning (Kerem *et al.*, 1989; Riordan *et al.*, 1989). Another classical example is Huntington's disease, although in this case the gene was identified 10 years after the disease was mapped in 1983 (Gusella *et al.*, 1983; Huntington *et al.*, 1993). Cystic fibrosis and Huntington's disease are examples that fulfilled the most important requisites for successful linkage studies: a set of families with the disease segregating and minimal ambiguity of the cohort member status. On the other hand, misdiagnosis, incomplete penetrances, or allelic heterogeneity have often hindered linkage studies (Botstein and Risch, 2003). Very rare diseases, with the availability of only a few affected individuals from different pedigrees, have also remained unapproachable following this methodology. Another extensively used tool for recessive diseases is homozygosity mapping (Lander and Botstein, 1987), which is based on the location of the disease gene in tracks of homozygosity in the affected consanguineous children. These regions are expected to be homozygous across all the patients, and not in the unaffected relatives. Linkage and homozygosity mapping have benefitted from the increasing densities of genetic and physical mapping, improving their power with the

Genome-Wide Association Studies: From Polymorphism to Personalized Medicine, ed. K. Appasani. Published by Cambridge University Press.

availability of more complete human genetic variation maps. Paradoxically, the discovery rate of new genes related to Mendelian diseases did not increase after the release of the human genome sequence, and genetic research efforts shifted preferentially to the genetics of complex disorders (Antonarakis and Beckmann, 2006).

15.2. Exome sequencing for rare disease study

Exome sequencing has revolutionized the field of genomics of rare diseases, with the description of about 180 causal genes using this methodology since 2009 (Boycott *et al.*, 2013). It has been estimated that about 85% of the disease-causing genetic variants fall into the coding fraction of the genome (the exome; Choi *et al.*, 2009), although this percentage is probably overestimated because of publication bias. It has to be noted, however, that even when sequencing the whole genome to detect all the genetic variation in an individual, some of the causal variants outside the coding regions would remain unidentified because of the difficulties of functional variant annotation of non-coding variants. Although not free of some caveats, such as the existence of multiple transcripts per gene, the assignation of a functional category to each coding variant (synonymous, missense, nonsense, in frame or frameshift indels) is quite straightforward, while functional annotation in non-coding regions presents many more limitations (Alexander *et al.*, 2010). This, together with the costs and complexity reduction compared to whole-genome sequencing, makes exome sequencing a cost-efficient and promising tool for the study of the genetic etiology of rare diseases.

Exome sequencing was first presented by Ng *et al.* as a valuable tool for the study of rare disorders (Ng *et al.*, 2009, 2010), and represented a great advance in the research on Mendelian diseases. These two works present what has become the standard approach to detect genetic variants at the origin of rare diseases from exome sequencing (Figure 15.1), and has been successfully applied in tens of works since then (Bamshad *et al.*, 2011; Gilissen *et al.*, 2011; Ku *et al.*, 2011). Essentially, the approach consists of reporting all coding functional variants by sequencing the exome in a set of patients with a given disease, and identifying genes shared across all patients harboring functional variants that are absent or segregating at very low frequencies in the population. Synonymous variants are discarded and only functional variants including stop codons (nonsense mutations), splice-site mutations, missense mutations, and frame-shift indels (1000 Geneomes Project Consortium, 2010) are considered for subsequent analyses. For an autosomal dominant disorder, the number of candidate genes will be higher, because any gene with one or more functional variants has to be considered. For recessive disorders, only genes with a homozygous variant or two or more variants will be considered as candidates. In the rare syndromes, the patients are usually found to be compound heterozygotes (i.e., they have received one different causal variant in the same gene from each parent), but in cases of consanguinity they are expected to be homozygous for the causal variant (Figure 15.1). Finally, these features and the genetic methodologies used to unravel their genetic origin can also be applied to the monogenic forms of complex disease.

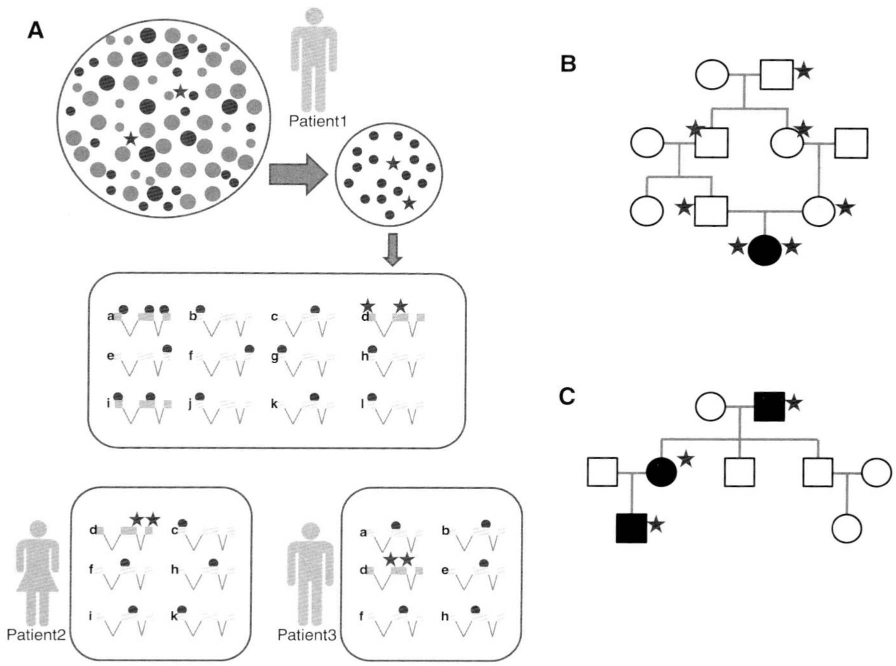

Figure 15.1. Three approaches to the study of rare disorders by whole-exome sequencing. (A) Every individual's genome harbors thousands of functional (red) and non-functional variants (blue). Only rare functional variants (small circles) are considered as candidate causal variants, while common (big circles) and non-functional variants are discarded. Only two of the variants are the causal mutations for the disease (stars). The list of candidate genes (orange) includes all *loci* with two or more functional variants for which the patient is a compound heterozygote. The comparison of candidate genes list across different patients allows detection of the causal gene. (B) In the case of inbreeding, most probably the causal variant will have been inherited twice from the same relative and found in homozygosis in the patient. (C) In a linkage analysis, the causal variant is expected to be present in all the patients from the same family and not present in the unaffected relatives of the pedigree. A black and white version of this figure will appear in some formats. For the color version, please refer to the plate section.

The critical step in these studies is how to reduce the number of functional variants in an individual's exome to narrow down the list of candidate genes. Sequencing a human exome typically yields about 10,000 functional variants, resulting in several hundred candidate genes. Considering that a causal variant for a rare disease cannot be found at high frequencies in the population, the first filter commonly applied consists of the elimination of all the variants above a given frequency. Initially, the filter was based on the presence or absence of the discovered variants in public databases such as dbSNP (Sherry *et al.*, 2001), HapMap (International HapMap Consortium, 2003), the 1000 Genomes Project (1000 Genomes Project Consortium, 2010), and in-house databases of genomic variation. However, as the amount of sequencing data in humans is increasing, this procedure cannot exclude the risk of eliminating some causal variants detected in heterozygous state in healthy carriers in the general population. Now, with the availability of deep catalogs of human variation in several human populations (Abecasis *et al.*, 2012), a low-frequency threshold according to the prevalence of the disease in the population can be used. The 1000 Genomes Project presents resequencing data in 1092 individuals from 14 different populations. By combining low-coverage whole-genome with exome sequencing, it consists of a validated haplotype map of more than 38 million SNPs, 1.4 million short indels, and 14,000 large deletions (Abecasis *et al.*, 2012). However, because of their recent exponential growth (Keinan and Clark, 2012), human populations show a reduced level of allele sharing of rare variants (Gravel *et al.*, 2011): populations in the same continent share 70–80% of rare variants, and populations from different continents only 10–30% (Nelson *et al.*, 2012; Tennessen *et al.*, 2012), which makes the catalog of rare genetic variation from one population of limited use for other populations. These findings have emphasized the need for expanding deep catalogs of genetic variation to other worldwide populations (Bustamante *et al.*, 2011).

After filtering functional variants according to their frequency in the population, an individual still harbors an excessive number of functional variants to allow candidate gene identification, except in cases of very rare diseases with several patients available (Figure 15.1). The effect of functional mutations can vary from drastic, as expected for a frameshift indel or a stop gain codon, to neutral, as those producing conservative amino acid changes. Distinguishing for these effects would help to focus the analysis in the variants with a greater functional effect, and discard the less detrimental ones. A number of computational tools have been designed to predict the effect of a given mutation on the protein function, based on evolutionary constraints and biochemical or structural properties. The fact that they present some discrepancies among them – that is, the same variant can be predicted as damaging or benign using different methods (Cooper and Shendure, 2011) – emphasizes the need of not using these predictions as stringent filters to discard genetic variants to avoid excluding causal variants. As an example, in the seminal paper on the use of exome sequencing for the study of rare disease by Ng *et al.* (2009), one of the six causal variants was predicted to be benign with Polyphen (Adzhubei *et al.*, 2010). Therefore, it seems recommendable to use these predictions first as an exclusive filter that can be relaxed in case of no evident candidate genes, or as a tool to prioritize variants rather than excluding them from the analysis. However, the development of accurate tools to predict the phenotypic effect of genetic variants is an active research field and computational predictions are constantly improving. Some approaches, such as Condel (González-Pérez and López-Bigas, 2011), improve their accuracy by integrating weighted predictions from different methods to produce a unified classification with a consensus deleteriousness score for each missense variant.

One important limitation of functional effect prediction arises from the fact that it is performed at the protein level, which might differ in effect at the organism level. To illustrate that, we refer to the analysis of polymorphic loss of function (LoF) variants in humans: on average, every human carries about 100 LoF mutations including stop gain and splice-site variants, frameshift indels and large deletions, with about 20 of them in homozygosis (MacArthur *et al.*, 2012). LoF variants are those with an more drastic expected effect, because they disrupt the protein sequence. However, it seems logical to conclude from the above results that most of these LoF variants in humans do not have such important consequences for the organism, highlighting the difference between the predicted effect at the molecular level and the true effect on organism viability and fitness. These results also suggest an important level of redundancy in molecular function and the robustness of biological pathways that need to be integrated in the predicted effect. Some proteins are less essential than others, and their loss of function can be more easily compensated by other genes or biological pathways. Thus, phenotypic effect prediction also needs to integrate information on aspects such as the essentiality, connectivity, pathway, and function and gene redundancy (Khurana *et al.*, 2013). However, a precise prediction of the variant (or gene) effect at the organism level will ultimately need to include functional validation, which is a major challenge for genetics in the coming years and still difficult to conceive at a high-throughput level.

The final list of candidate genes can still be reduced by incorporating sequence information from parents and siblings or linkage analysis. Gene expression patterns can be also considered given the functional knowledge on the disease, which may help together with phenotypical and clinical information to prioritize as candidate genes those participating in certain biological pathways or known to interact with already described causal genes.

15.3. Use of genetic information from relatives

In an analysis of an autosomal recessive Mendelian disease, information on the paternal or maternal origin of the variants allows us to infer their phase; for example, if the two functional variants are located in the same chromosome and therefore coming from the same parent. In one half of the cases a compound heterozygote for a given gene will have received one functional variant from each parent (each parent being a healthy carrier for only one mutation), while in the other half both are from the same parent. In the latter case this gene can be excluded as a candidate, because the father or the mother were also carriers for the two mutations and did not develop the disease. Sequence information from parents does not necessarily need to come from whole-exome sequencing, and it can be originated at the validation step by Sanger sequencing of genetic variants in a set of candidate genes, which will reduce costs considerably. In fact, only information from one of the two parents is needed to determine their origin, although sequencing in both parents also has the advantage of allowing the detection of *de novo* mutations.

With an estimated rate of 1.8×10^{-8} mutations per site and generation (Conrad *et al.*, 2011), every individual is expected to harbor about 80 *de novo* mutations, and approximately one located in the coding fraction of the genome (Veltman and Brunner, 2012). Functional *de novo* mutations have not been screened by natural selection, except for mutations producing early development defects, and are therefore good candidates to explain some cases of sporadic diseases (Awadalla *et al.*, 2010). Two works in 2010 firstly identified *de novo* variants at the origin of two rare diseases, the Schinzel–Giedion and the Kabuki syndromes (Hoischen *et al.*, 2010; Ng *et al.*, 2010). This *de novo* approach has also been successfully applied to complex disease, in particular to neurological disorders with an early onset, such as autism (O'Roak *et al.*, 2011) and mental retardation (Vissers *et al.*, 2010) or even schizophrenia (Girard *et al.*, 2011). Typically, studies aiming to detect *de novo* mutations perform whole-exome sequencing both in patients and parents, to identify mutations in the proband not present in the parents, tripling the cost of the experiment. Alternatively, a set of candidate mutations prioritized by their frequency, functional predicted impact, or gene function can be selected to be validated in the patient and the parents, which reduces cost at the expense of limiting the number of tested variants. Among *de novo* events, copy number variants (CNVs) have been shown to be at the origin of neurological disorders such as autism, schizophrenia, and bipolar disorder (Itsara *et al.*, 2010; Malhotra *et al.*, 2011; Sanders *et al.*, 2011). Although their mutation rate is much lower than for SNPs or indels, these mutations are expected to have on average a stronger phenotypic effect as shown by the higher effect of purifying selection on them, especially for the larger CNVs (Itsara *et al.*, 2010).

Linkage analysis incorporating pedigree information is used in combination with exome sequencing, because causal variants need to cosegregate with the phenotype, being shared between patients and absent in healthy relatives (Wang *et al.*, 2010) (Figure 15.1). This pedigree-based approach, in combination with frequency and functional filters, is especially useful in cases of clinical and genetic heterogeneity where the causal variant and gene are probably different across patients. Instead, patients in the same pedigree are expected to harbor the same causal variant. The simplest approach consists of including in the study two or more siblings, and is mostly used for Mendelian recessive diseases where siblings represent sporadic cases in the pedigree. Any variant that is not shared by the siblings can be automatically discarded. However, because siblings still share 50% of their genome, this strategy is often combined with the inclusion of other patients (Ng *et al.*, 2009). Including several and distant patients in a pedigree confers more power to discard functional variants not present in all the patients, because the genome fraction shared by all of them will be substantially reduced compared to the use of siblings. This strategy was successfully applied to study an autosomal dominant spinocerebellar ataxia in a four-generation pedigree (Wang *et al.*, 2010). Once an 8.4 Mb region was identified by linkage mapping in three generations of the pedigree, the exome of four affected patients was sequenced. Filtering of the functional variants in this region by their presence in the public databases and predicted functional impact with Polyphen detected the causal variant at the *TGM6* gene. Subsequently, this missense variant was shown to completely cosegregate with the disease in the pedigree.

In cases of consanguinity, where Mendelian recessive inheritance with the presence of a variant in homozygosity is the most plausible model, homozygosity mapping is also successfully used to narrow down the list of genetic variants.

Typically, homozygosity mapping in the patients is combined with exome sequencing, and the search for causal variants is reduced to the homozygosity tracks. This combination of exome sequencing and homozygosity mapping has allowed to unravel the molecular origin of Mendelian disease by exome sequencing even with a single patient (Bolze *et al.*, 2010; Walsh *et al.*, 2010). Bolze *et al.* (2010) used this approach to identify the causal mutation of a newly described clinical syndrome with symptoms of autoimmune lymphoproliferative syndrome (ALPS), recurrent severe viral and bacterial infections, hepatopathy, encephalopathy, and cardiac malformations. The large consanguineous pedigree included four patients and several members in the previous generation who died in early childhood with the same clinical symptoms. Homozygosity mapping was performed by genotyping a SNP array with 500,000 SNPs in three of the patients and their parents and one healthy sibling, identifying two large regions that were homozygous in the patients and heterozygous in the healthy relatives. Exome sequencing in one of the patients identified only one previously unreported missense variant mapping to one of the homozygosity regions. The variant was checked to cosegregate with the disease along the pedigree, and not found in 282 individuals from the same population.

15.4. Limitations of exome-sequencing approaches for rare disease

It has been proposed that the causal genes are identified in approximately 50% of the exome sequencing studies of rare well-defined Mendelian diseases (Gilissen *et al.*, 2011). Different reasons can explain the unsuccessful exome-sequencing efforts where no candidate variants or genes are identified. In some cases, the fault might lie with the experimental design and the limitations of exome sequencing, whereas in others the reason might be analytical. Obviously, the most evident limitation of exome sequencing is the restriction to the coding fraction of the genome. The probability of non-coding variants being associated with the disease increases with the prevalence of the disease, but some causal variants for rare diseases may be located in non-coding regions not included in the standard exome-capture arrays. However, it has to be noted that arrays are designed to capture important functional positions as splicing sites and are recurrently modified to also include regulatory regions in the neighboring regions. While the approximately 180,000 exons in the human genome include 35 Mb (1–2% of the genome), because of the addition of some non-coding regions, commercially available exome-capture arrays include more than 50 Mb of targeted regions. However, promoters and enhancers regulating gene expression can be located at long distances from the transcription unit (Kleinjan and van Heyningen, 2005). Finally, the Encyclopedia of DNA Elements (ENCODE) consortium, aiming to identify all the functional elements in the human genome, has assigned biochemical functions to 80% of the genome (Dunham *et al.*, 2012), challenging the concept of most of the human genome being junk DNA. The ENCODE's definition of functionality and the exact fraction of the genome including functionally important sites in the human genome and its implications are hotly debated

(Doolittle, 2013), but regardless of the actual number of functional sites, their location in regions distant from exons may explain some of the negative results in exome-sequencing studies.

The success of exome sequencing in mapping variants and genes to disease greatly depends on its genetic etiology. The possibilities of success increase with rarity, severity and early age of onset, because these features are more probably related to genetic variants with high penetrance and limited to a single or a small number of genes. The higher the prevalence of the disease, the higher the probability of having a non-Mendelian transmission, with several variants in different genes conferring susceptibility to the disease. Filtering by frequency relies on the idea that purifying selection will keep at low frequencies the deleterious genetic variants that affect to the fitness of the individuals. Genetic variants causing a Mendelian disease but with incomplete penetrance or late-onset are more prone to be found at higher frequencies than those with full penetrance and/or early-onset. Exome-sequencing power will also be diminished in the case of a polygenic disease, where the disease is the result of the combined action of several genes, a scenario that would require much higher sample sizes. The mode of inheritance also affects the success rate of these studies. In the schema described in Figure 15.1, the number of candidate genes (two or more functional mutations required) for a recessive disease can be quite low, especially if several patients are available. In contrast, for a dominant syndrome the number of genes with at least one functional variant can be 50-fold higher (Bamshad *et al.*, 2011), hindering candidate gene identification. Finally, mosaic mutations will remain undetected if DNA from only one tissue is sequenced (Boycott *et al.*, 2013), as is commonly performed. In these cases, bioinformatic pipelines need to be optimized to detect variants usually present in a low proportion of reads, which could be removed from the analysis as sequencing errors (Lindhurst *et al.*, 2011).

The study design can be inappropriate because of misdiagnosis that may lead to the inclusion of patients with different diseases, although with phenotypic similarities. In this case no candidate genes in common are expected to be found across patients. Similarly, under allelic heterogeneity, where a similar phenotype is produced by mutations in different genes, pooling different patients to identify common genes with two or more rare functional mutations may be unfruitful. In these cases, pooling by pathway or function instead of single genes may be of utility, although it will increase the number of candidate genes.

Several technical reasons can also explain study failure, either because of false-negative or false-positive results. A false-negative result – that is, a real causal gene exists but it is not detected in the study – can arise from miscalling of genetic variants. First, genetic variants might remain undetected because of poor coverage. To avoid this, a minimum 100× average coverage is recommended to compensate the bias in the coverage produced by the capture and sequencing steps. Second, another reason for miscalling, which can also originate false variant callings, is the presence of homopolymers (stretches of identical bases in a sequence) which result in higher sequencing error rates, especially for indels (Nielsen *et al.*, 2011). Exome sequencing also presents a limited resolution to detect CNVs, although some methods are being developed (Krumm *et al.*, 2012). A critical point for CNV detection from exome data will be the availability of several samples captured and sequenced with exactly the same technologies, to reduce unequal coverage bias across chromosomal regions and samples. Last, the causal variants can also be lost if too stringent frequency and functional predicted effects filters are applied. On the other hand, false positives will mainly arise from sequencing or mapping errors, which represent an important challenge for genes related to paralagous sequences including pseudogenes, repetitive regions and segmental duplications (Treangen and Salzberg, 2012). However, these errors should be detected in the sequence validation step for putative causal variants.

15.5. NGS for molecular diagnosis

One straightforward application of exome sequencing is the screening for known or new functional variants in previously discovered candidate genes, or genes functionally related to them. This genetic screening has usually been performed using Sanger technology for a few genes. Now, any patient included in an exome-sequencing study can be automatically scanned for functional known or new genetic variants in candidate genes (Yang *et al.*, 2013). Above a certain number of genes, depending on their number of exons and length, NGS represents a more cost-efficient approach. In addition, exome sequencing will provide information on other functionally related genes, playing both roles as a discovery and a diagnostic tool (Ku *et al.*, 2012). A diagnostic application of whole-exome sequencing was first presented by Choi *et al.* (2009), where a misdiagnosed patient was diagnosed with congenital chloride-losing diarrhea thanks to the identification of a homozygous functional mutation in a gene related to this disease. In several cases, exome sequencing has resolved or improved misdiagnoses thanks to the description of functional variants in genes related to Mendelian disease (Ku *et al.*, 2012). Interestingly, in the study by Ng *et al.*, in which the cause of the Miller syndrome was unraveled, exome sequencing revealed that two affected siblings also had primary ciliary dyskinesia because of functional mutations in the *DNAH5* gene (Ng *et al.*, 2009).

Alternatively, NGS sequencing can be performed to limit the genetic screening for several known candidate genes thanks to different target capture platforms, as has been done for the diagnostic testing of 24 genes known to cause congenital disorders of glycosylation (Jones *et al.*, 2011). However, these approaches seem to be more reasonable when several individuals can be screened simultaneously using the bar-coding systems to avoid unnecessarily excessive coverage (Ku *et al.*, 2012). Therefore, the decision must consider the cost-efficiency of the different platforms, taking into account the number of samples (which for diagnosis might be quite reduced), the predicted average coverage, or the need for a rapid diagnosis. The utility of exome sequencing for diagnosis and treatment is well exemplified by the successful case of inflammatory bowel disease in a child, which facilitated subsequent decisions on treatment (Worthey *et al.*, 2011). Finally, whole-genome and exome sequencing can also be extended to applications in non-invasive prenatal or postnatal disease screening. Prenatal screening can be performed using NGS with a non-invasive methodology from cell-free fetal DNA present in maternal plasma (Lo *et al.*, 1997). This technique has already been used to scan for certain single genetic disorders such as Huntington's disease or beta-thalasemias, chromosomal aneuploidies, and sex determination (Stumm *et al.*, 2012). However, whole-genome or exome prenatal screening seems costly and unnecessary because most of the genetic information would be useless for genetic counseling. A targeting approach, where the analysis is restricted to a subset of genes or variants with a well-known relationship with disease, is a more realistic and affordable possibility (Snyder *et al.*, 2013).

Acknowledgments

We thank Francesc Calafell for critical reading of the manuscript. Elena Bosch and Ferran Casals' work is funded by grants SAF2011-29239 and SAF2012-35025, respectively, from the Spanish Ministerio de Economía y Competitividad as well as by grant 2009SGR-1101 from Direcció General de Recerca, Generalitat de Catalunya. Ferran Casals holds a Beatriu de Pinós (2010-BP-B-00128) fellowship from AGAUR (Generalitat de Catalunya).

REFERENCES

1000 Genomes Project Consortium. (2010). A map of human genome variation from population-scale sequencing. *Nature*, **467**(7319), 1061–1073.

Abecasis, G.R., Auton, A., Brooks, L.D., *et al.* (2012). An integrated map of genetic variation from 1,092 human genomes. *Nature*, **491**(7422), 56–65.

Adzhubei, I.A., Schmidt, S., Peshkin, L., *et al.* (2010). A method and server for predicting damaging missense mutations. *Nature Meth.*, **7**(4), 248–249.

Alexander, R.P., Fang, G., Rozowsky, J., *et al.* (2010). Annotating non-coding regions of the genome. *Nature Rev. Genet.*, **11**(8), 559–571.

Antonarakis, S.E. and Beckmann, J.S. (2006). Mendelian disorders deserve more attention. *Nature Rev. Genet.*, **7**(4), 277–282.

Awadalla, P., Gauthier, J., Myers, R.A., *et al.* (2010). Direct measure of the *de novo* mutation rate in autism and schizophrenia cohorts. *Am. J. Hum. Genet.*, **87**(3), 316–324.

Bamshad, M.J., Ng, S.B., Bigham, A.W., *et al.* (2011). Exome sequencing as a tool for Mendelian disease gene discovery. *Nature Rev. Genet.*, **12**(11), 745–755.

Bolze, A., Byun, M., McDonald, D., *et al.* (2010). Whole-exome-sequencing-based discovery of human FADD deficiency. *Am. J. Hum. Genet.*, **87**(6), 873–881.

Botstein, D. and Risch, N. (2003). Discovering genotypes underlying human phenotypes: past successes for mendelian disease, future approaches for complex disease. *Nature Genet.*, **33**(Suppl.), 228–237.

Boycott, K.M., Vanstone, M.R., Bulman, D.E., *et al.* (2013). Rare-disease genetics in the era of next-generation sequencing: discovery to translation. *Nature Rev. Genet.*, **14**(10), 681–691.

Bustamante, C.D., Burchard, E.G. and De la Vega, F.M. (2011). Genomics for the world. *Nature*, **475**(7355), 163–165.

Choi, M., Scholl, U.I., Ji, W., *et al.* (2009). Genetic diagnosis by whole exome capture and massively parallel DNA sequencing. *Proc. Natl Acad. Sci. USA*, **106**(45), 19096–19101.

Collins, F.S. (1995). Positional cloning moves from perditional to traditional. *Nature Genet.*, **9**(4), 347–350.

Conrad, D.F., Keebler, J.E., Depristo, M.A., *et al.* (2011). Variation in genome-wide mutation rates within and between human families. *Nature Genet.*, **43**(7), 712–714.

Cooper, G.M. and Shendure, J. (2011). Needles in stacks of needles: finding disease-causal variants in a wealth of genomic data. *Nature Rev. Genet.*, **12**(9),628–640.

Doolittle, W.F. (2013). Is junk DNA bunk? A critique of ENCODE. *Proc. Natl Acad. Sci. USA*, **110**(14), 5294–5300.

Dunham, I., Kundaje, A., Aldred, S.F., *et al.* (2012). An integrated encyclopedia of DNA elements in the human genome. *Nature*, **489**(7414), 57–74.

Gilissen, C., Hoischen, A., Brunner, H.G., *et al.* (2011). Unlocking Mendelian disease using exome sequencing. *Genome Biol.*, **12**(9), 228.

Girard, S.L., Gauthier, J., Noreau, A., *et al.* (2011.) Increased exonic de novo mutation rate in individuals with schizophrenia. *Nature Genet.*, **43**(9), 860–863.

González-Pérez, A. and López-Bigas, N. (2011). Improving the assessment of the outcome of nonsynonymous SNVs with a consensus deleteriousness score, Condel. *Am. J. Hum. Genet.*, **88**(4), 440–449.

Gravel, S., Henn, B.M., Gutenkunst, R.N., *et al.* (2011). Demographic history and rare allele sharing among human populations. *Proc. Natl Acad. Sci. USA*, **108**(29), 11983–11988.

Gusella, J., Wexler, N., Conneally, P., *et al.* (1983) A polymorphic DNA marker genetically linked to Huntington's disease. *Nature*, **306**, 234–238.

Hoischen, A., van Bon, B.W.M, Gilissen, C., *et al.* (2010). *De novo* mutations of *SETBP1* cause Schinzel–Giedion syndrome. *Nature Genet.*, **42**(6), 483–485.

Huntington, T., Macdonald, M.E., Ambrose, C.M., *et al.* (1993). A novel gene containing a trinucleotide that is expanded and unstable on Huntington's disease chromosomes. *Cell*, **72**, 971–983.

International HapMap Consortium. (2003). The International HapMap Project. *Nature*, **426** (6968), 789–796.

Itsara, A., Wu, H., Smith, J.D., *et al.* (2010). *De novo* rates and selection of large copy number variation. *Genome Res.*, **20**(11), 1469–1481.

Jones, M., Bhide, S., Chin, E., *et al.* (2011). Targeted PCR-based enrichment and next generation sequencing for diagnostic testing of congenital disorders of glycosylation (CDG). *Genet. Med.*, **13**(11), 921–932.

Keinan, A. and Clark, A.G. (2012). Recent explosive human population growth has resulted in an excess of rare genetic variants. *Science*, **336**(6082), 740–743.

Kerem, A.B., Rommens, J.M., Buchanan, J.A., *et al.* (1989). Identification of the cystic fibrosis gene: genetic analysis. *Science*, **245**(4922), 1073–1080.

Khurana, E., Fu, Y., Chen, J., *et al.* (2013). Interpretation of genomic variants using a unified biological network approach. *PLoS Comput. Biol.*, **9**(3), e1002886.

Kleinjan, D.A. and van Heyningen, V. (2005). Long-range control of gene expression: emerging mechanisms and disruption in disease. *Am. J. Hum. Genet.*, **76**(1), 8–32.

Krumm, N., Sudmant, P.H., Ko, A., *et al.* (2012). Copy number variation detection and genotyping from exome sequence data. *Genome Res.*, **22**(8), 1525–1532.

Ku, C.S., Naidoo, N. and Pawitan, Y. (2011). Revisiting Mendelian disorders through exome sequencing. *Hum. Genet.*, **129**(4), 351–370.

Ku, C.-.S, Cooper, D.N., Polychronakos, C., *et al.* (2012). Exome sequencing: dual role as a discovery and diagnostic tool. *Ann. Neurol.*, **71**(1), 5–14.

Lander, E. and Botstein, D. (1987). Homozygosity mapping: a way to map human recessive traits with the DNA of inbred children. *Science*, **236**, 1567–1570.

Lindhurst, M.J., Sapp, J.C., Teer, J.K., *et al.* (2011) A mosaic activating mutation in *AKT1* associated with the Proteus syndrome. *New Engl. J. Med.*, **365**(7), 611–619.

Lo, Y.M.D., Corbetta, N., Chamberlain, P.F., *et al.* (1997). Presence of fetal DNA in maternal plasma and serum. *Lancet*, 350, 485–487.

MacArthur, D.G., Balasubramanian, S., Frankish, A., *et al.* (2012). A systematic survey of loss-of-function variants in human protein-coding genes. *Science (New York, N.Y.)*, **335** (6070), 823–828.

Malhotra, D., McCarthy, S., Michaelson, J.J., *et al.* (2011). High frequencies of *de novo* CNVs in bipolar disorder and schizophrenia. *Neuron*, **72**(6), 951–963.

Nelson, M.R., Wegmann, D., Ehm, M.G., *et al.* (2012). An abundance of rare functional variants in 202 drug target genes sequenced in 14,002 people. *Science*, **337**(6090), 100–104.

Ng, S.B., Buckingham, K.J., Lee, C., *et al.* (2009). Exome sequencing identifies the cause of a mendelian disorder. *Nature Genet.*, **42**(1), 30–35.

Ng, S.B., Bigham, A.W., Buckingham, K.J., *et al.* (2010). Exome sequencing identifies *MLL2* mutations as a cause of Kabuki syndrome. *Nature Genet.*, **42**(9), 790–793.

Nielsen, R., Paul, J.S., Albrechtsen, A., *et al.* (2011). Genotype and SNP calling from next-generation sequencing data. *Nature Rev. Genet.*, **12**(6), 443–451.

O'Roak, B.J., Deriziotis, P., Lee, C., *et al.* (2011). Exome sequencing in sporadic autism spectrum disorders identifies severe *de novo* mutations. *Nature Genet.*, **43**(6), 585–589.

Riordan, J.R., Rommens, J.M., Kerem, B., *et al.* (1989). Identification of the cystic fibrosis gene: cloning and characterization of complementary DNA. *Science (New York, N.Y.)*, **245**(4922), 1066–1073.

Sanders, S.J., Ercan-Sencicek, A.G., Hus, V., *et al.* (2011). Multiple recurrent *de novo* CNVs, including duplications of the 7q11.23 Williams syndrome region, are strongly associated with autism. *Neuron*, **70**(5), 863–885.

Sherry, S.T., Ward, M.H., Kholodov, M., *et al.* (2001) dbSNP: the NCBI database of genetic variation. *Nucleic Acids Res.*, **29**(1), 308–311.

Snyder, M.W., Simmons, L.E., Kitzman, J.O., *et al.* (2013). Noninvasive fetal genome sequencing: a primer. *Prenat. Diagn.*, **33**(6), 547–554.

Stumm, M., Entezami, M., Trunk, N., *et al.* (2012). Noninvasive prenatal detection of chromosomal aneuploidies using different next generation sequencing strategies and algorithms. *Prenat. Diagn.*, **32**(6), 569–577.

Tennessen, J.A., Bigham, A.W., O'Connor, T.D., *et al.* (2012). Evolution and functional impact of rare coding variation from deep sequencing of human exomes. *Science*, **337**(6090), 64–69.

Treangen, T.J. and Salzberg, S.L. (2012). Repetitive DNA and next-generation sequencing: computational challenges and solutions. *Nature Rev. Genet.*, **13**(1), 36–46.

Veltman, J.A. and Brunner, H.G. (2012). *De novo* mutations in human genetic disease. *Nature Rev. Genet.*, **13**(8), 565–575.

Vissers, L.E., de Ligt, J., Gilissen, C., *et al.* (2010). A *de novo* paradigm for mental retardation. *Nature Genet.*, **42**(12), 1109–1112.

Walsh, T., Shahin, H., Elkan-Miller, T., *et al.* (2010). Whole exome sequencing and homozygosity mapping identify mutation in the cell polarity protein GPSM2 as the cause of nonsyndromic hearing loss *DFNB82*. *Am. J. Hum. Genet.*, **87**(1), 90–94.

Wang, J.L., Yang, X., Xia, K., *et al.* (2010). *TGM6* identified as a novel causative gene of spinocerebellar ataxias using exome sequencing. *Brain*, **133**(Pt 12), 3510–3518.

Worthey, E.A., Mayer, A.N., Syverson, G.D., *et al.* (2011). Making a definitive diagnosis: successful clinical application of whole exome sequencing in a child with intractable inflammatory bowel disease. *Genet. Med.*, **13**(3), 255–262.

Yang, Y., Muzny, D.M., Reid, J.G., *et al.* (2013). Clinical whole-exome sequencing for the diagnosis of mendelian disorders. *New Engl. J. Med.*, **369**(16), 1502–1511.

16 Next-generation sequencing for complex disorders

Ferran Casals and Elena Bosch

16.1. Complex diseases and GWAS

Complex or multifactorial diseases are, by definition, determined by a number of genetic and environmental factors. Unlike the mutations that underlie Mendelian diseases, which involve single mutations with strong phenotypic effects, complex diseases are believed to be caused by multiple genetic variants along the genome, implying that each of them has a weak effect on its own. The most common strategy for identifying such risk alleles has been through association studies, in which allele frequencies from patients and controls were initially compared in candidate genes, although now this is usually done on a genome-wide scale, in the so-called genome-wide association studies (GWASs). In the last years, GWASs have successfully identified hundreds of common risk variants associated with the most common human diseases such as diabetes, Alzheimer's, Parkinson's, Crohn's disease, schizophrenia, and various types of cancer. This notable achievement was not reached until it became possible to compare, in large cohorts of patients and controls (more recently in the range of tens of thousands individuals), the allele frequencies at several hundred thousand single nucleotide polymorphisms (SNPs) which are assayed on arrays that are able to capture the most common variations in the human genome (usually those with minor allele frequencies, or MAF, > 5%).

Although GWASs have remarkably facilitated insights into understanding complex disease, it is clearly recognized that most of the disease-associated SNPs identified in GWASs usually imply a very small increase in risk and explain only a modest fraction of the genetic component (commonly referred as heritability) of most complex traits (Manolio *et al.*, 2009). Many factors have been suggested to explain such "missing heritability": the existence of much larger numbers of still-undetected small-effect variants; the possible involvement of structural variants, inadequately picked up by SNP arrays; potential gene–gene interactions; and heritability overestimates due to gene–environment correlations. GWASs may

Genome-Wide Association Studies: From Polymorphism to Personalized Medicine, ed. K. Appasani. Published by Cambridge University Press.

have also partially failed because the common variant–common disease model was uncritically assumed, and rarer variants (those with MAF < 5%, and therefore not included in SNP arrays used for GWASs) with moderate to high penetrance could actually play an important role in common, complex diseases. Thanks to the development of next-generation sequencing technologies (NGS), most of the focus to find the missing heritability in complex disease is now going to the low-frequency (defined here as those in the 0.5–5% range) and rare (MAF < 0.5%) genetic variants. Indeed, recent large-scale studies of human variation have provided further interest for investigating the role of rare variants in disease susceptibility, mainly for two reasons. First, they all consistently report that most of the genetic variants occur at very low frequency in human populations; a pattern in the site frequency spectrum (SFS) that probably results from the explosive population growth in recent human history (Coventry *et al.*, 2010; Keinan and Clark, 2012; Tennessen *et al.*, 2012). Therefore, although the frequency of any single rare variant is low, collectively, rare variants represent a substantial proportion of human variation, which has not been properly investigated in GWASs. Second, rare coding variation has been shown to be particularly enriched for deleterious alleles, as a result of both the cumulative action of purifying selection on past existing functional variation and selection not having had sufficient time to purge recently arisen variants (Fu *et al.*, 2012). Thus, rare variants could be influencing complex disease outcome.

At the same time, there are only a few examples in which the actual causative variant of a GWAS signal has been identified, namely, the true causal genetic variant contributing to a complex phenotype or susceptibility to complex disease. Of note, a SNP associated to particular GWAS signal informs us of allele frequencies differences at a particular polymorphic genetic site between control and disease cohorts or between the extremes of the distribution of a complex phenotype, but does not imply that variation at this particular site is functionally relevant for causing such disease risk. Indeed, any functional susceptibility variant in linkage disequilibrium with the genotyped SNP could be responsible for such association signal. Although recently, GWAS have often been coupled with expression and methylation array data to understand the mechanistic nature of risk alleles, only by sequencing and subsequently analyzing all possible linked variation to each GWAS signal in extended experimental follow-ups will determination of the true causative susceptibility variant behind each case be possible.

16.2. A new approach: NGS in complex diseases

Sequencing technologies allow the direct uncovering of the full spectrum of DNA variation including common, rare, and indel variation either along particular genomic regions through targeting sequencing strategies, or at genome-wide level. Moreover, additional forms of structural variation can be effectively interrogated using NGS technologies when using adequate experimental methods and performing much more elaborate subsequent computational analysis of sequencing data through read-pairs, split-reads, read depth, and sequence assembly (Alkan *et al.*, 2011). Although providing further complete knowledge than SNP chip arrays, sequencing costs have not been yet affordable to apply NGS technologies at genome-wide level in the huge sample sizes that GWASs have interrogated. However, the increasing throughput of the new NGS instruments together with improved strategies for targeting regions of interest and for tagging multiple samples is starting to change the almost exclusive predominance of high-throughput SNP genotyping array-based technologies for NGS technologies in complex disease association studies. Such new technology inclusion (but probably not replacement, as once new genetic variants have been identified, we still require them to be genotyped and/or replicated in large samples sizes) has been mainly prompted by the benefits of not only allowing the detection of true common causal variants behind GWAS signals, but also the identification of rare variations at larger scales than previously assayed. Furthermore, the availability of sequencing data around GWAS signals will also allow investigation of whether rare variants are found co-occurring in particular common backgrounds and, therefore, whether previously observed associations with common variants could be attributed to the aggregate, synthetic effect of rare variants (Dickson *et al.*, 2010).

In the last years, a growing number of studies has demonstrated that resequencing can pinpoint disease-causing genes as well as a significant burden of rare variants in genomic regions initially identified by GWAS. For example, rare variants that lower type 1 diabetes (T1D) risk were discovered in *IFIH* (Nejentsev *et al.*, 2009), a gene located in a region previously associated with T1D by GWAS. In the case of hypertriglyceridemia, the resequencing of candidate genes in loci found to be associated by GWAS revealed that they also harbor a significant burden of rare variants (Johansen *et al.*, 2010). High-throughput targeted sequencing (although in DNA pools) was also applied to search for low-frequency variants influencing susceptibility to Crohn's disease in 63 GWAS-identified positional candidate genes (Momozawa *et al.*, 2011). These findings led to the suggestion that a proportion of the heritability attributable to GWAS-identified loci will be substantially greater than the estimates from the common variants in the initial GWASs (Gloyn and McCarthy, 2010). Additionally, rare and low-frequency variations as well as structural variants in other susceptibility genomic regions yet to be identified and possibly additional common variants with smaller effect sizes than the ones already detected in GWAS will probably add to the remaining missing heritability.

A variety of different study designs have been suggested for identifying rare variant associations from NGS data by using either traditional control and disease cohorts but also targeting individuals with extreme phenotype traits, extreme disease phenotypes regarding early onset of the disease, rapid progression and poor survival, or by using affected probands in families (i.e., family-based sequencing of affected individuals). As an example of an extreme phenotype trait-based study, Ahituv *et al.* (2007) found an enrichment of rare non-synonymous (and more deleterious) variants in obese individuals when comparing sequence data in genes associated with monogenic obesity between the extremes of human body mass (i.e., using an obese and a lean group of more than 300 individuals each from the total distribution). Similarly, in one of the earliest known rare-variant disease associations, Cohen *et al.* (2004) sequenced three candidate genes that cause Mendelian forms of low high-density lipoprotein cholesterol (HDL-C) levels in the upper and lower 5% extremes of the HDL-C level distribution of the Dallas Heart Study population, and demonstrated that multiple rare variants contributed to low plasma levels of HDL-C.

Most of the initial studies investigating the role of rare variants through NGS technologies were not genome-wide analyses, but focussed on specific candidate regions either from a previous GWAS or other gene prioritizations, such as those resulting from expression data, functional annotation, or biomedical literature, among others. In these designs, regions of interest need to be specifically captured from the whole genome and then subsequent sequencing of targeted regions is performed in pools of DNA from multiple individuals. Sometimes, to further reduce reagent costs, even the enrichment step is done in DNA pools. However, any pooling of DNA samples requires accurate estimation of DNA concentrations as well as additional validation procedures including replicates and readjustments of particular tagged pools before variants can be finally called and allele frequencies determined. This strategy allows complete use to be made of the high-throughput capabilities of NGS instruments, especially when applying a custom targeting strategy, but importantly, depending on whether samples are tagged before their pooling, the possible genetic variations to be found may be directly traced back to each particular individual sample. Initially, the number of tags or indices was rather limited and different strategies were developed for estimating disease associations with rare variants from pooled sequencing data (Wang *et al.*, 2010). However, the recent development of new pre-enrichment sample pooling options and improved tagging systems allows tracing sequences from an increasing number of different individuals in parallel. In addition to the different custom-enrichment kits commercially available, which can be easily designed through the corresponding commercial software, a number of commercial disease-specific targeting kits have also been developed.

As NGS costs continue to decrease and given the initial success of candidate gene studies, future studies will probably move from these custom targeting strategies which focus on predefined candidate regions (usually based on the imperfect understanding of the pathology of the disease) to more hypothesis-free approaches; that is, genome-wide strategies interrogating either the whole human genome (WGS, whole-genome sequencing), or those parts of the genome we know are functional (ES, exome sequencing). However, besides the sequencing costs, the analysis of genome-wide sequencing data still has important analytical and statistical challenges. Recent improvements on exome enrichment kits allow targeting not only all human coding exons but also their regulatory regions, including UTR regions, miRNAs and miRNA targets, and other conserved non-coding regions, among others. Also, it is worth noting that the different commercial exome capture platform technologies available differ considerably in their corresponding targeted regions as well in the efficiency to provide enough coverage for variant detection (Kiezun *et al.*, 2012). Therefore, when compiling ES data from independent studies involving different platforms, or even different versions of the same commercial platform, such factors should be carefully considered as they may provide a significant loss of information in these regions being differentially targeted. On the contrary, comparison of different WGS data sets in principle should be more straightforward. Thus, special care should be taken into account when planning possible meta-analyses of existing data sets or the shared use of particular control samples for different disease association studies when those have been generated through different sequencing and enrichment platforms.

16.3. New statistical approaches

Notwithstanding the promising potential contribution of rare variants into complex disease, the statistical analysis of rare variants presents notable challenges. Unfortunately, the usual association tests for common variants are clearly underpowered for detecting variants of lower frequency with small to moderate effect sizes (Ladouceur *et al.*, 2012) due to the small number of observations for any given variant (even in very large samples) and a need for more stringent multiple-test correction than for common variants (Price *et al.*, 2010). This has motivated the emergence of several alternative association statistical strategies (Asimit and Zeggini, 2010; Bansal *et al.*, 2010) including (i) collapsing approaches that test the combined effect of multiple rare variants within and across genomic regions; (ii) approaches based on the similarity or diversity of DNA sequences; and (iii) regression models that consider collapsed sets of variants and other factors as predictors of a phenotype (Wu *et al.*, 2011), among others. The performance of such rare variant association methods has been demonstrated to depend upon the underlying assumption of the relationship between rare variants and disease status. Given that in most cases the particular biological scenario of each complex disease (numbers and frequencies of variants involved, presence of synergistic effects within and across genomic regions, etc.) is unknown, assessing the association between rare variants and complex disease will be a challenging task (Ladouceur *et al.*, 2012). Indeed, complementary and independent analyses for low-frequent and rare-variant association have been suggested as a desirable approach to surveying large WGS and ES data sets (Neale *et al.*, 2011).

The general reasoning behind most burden or collapsing methods is that if a genomic region is critical to disease pathogenesis, rare mutations may modulate disease susceptibility. Then, many affected individuals may have rare mutations more frequently in that region (usually a gene) in comparison to controls, although the mutations may be different from and unrelated to one another (Figure 16.1). Rather than testing each variant individually, multiple variations within genes, pathways, or conserved or *a priori* selected regions which may influence disease risk (such as previous GWAS signals) are thus analyzed together by comparing their global numbers between disease and control cohorts in such genomic units. While the most simple burden tests such as the Cohort Allelic

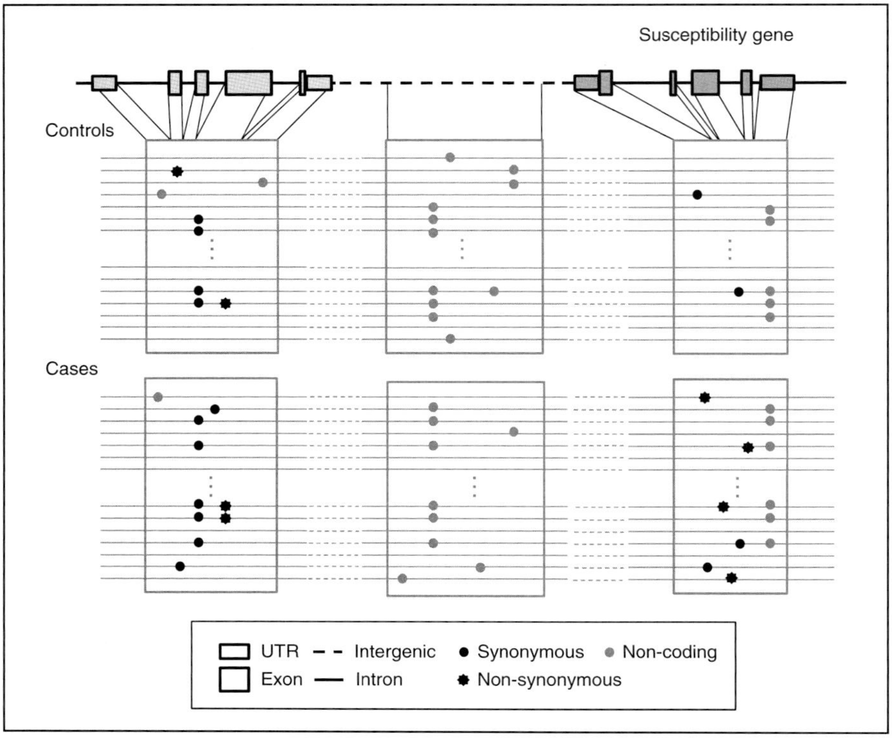

Figure 16.1. An excess of rare variants in cases can signal the involvement of a genomic region in a complex disease. If rare variants contribute to a complex disease, it is expected that in the genome regions associated with the disease, patients would be more likely than controls to present different sets of rare (functional) variants, each of them increasing individual liability to the disease in different ways.

Sums Test (CAST; Morgenthaler and Thilly, 2007) compare the number of individuals with one or more mutations in a given region between affected and unaffected individuals, more sophisticated strategies allow weighting variants according to their frequencies in controls (Madsen and Browning, 2009) or to their predicted *in silico* potential damaging properties (Sul *et al.*, 2011). Some applications only consider rare variation or just rare variants with clear potential functional relevance, whereas others take into account both rare and common variation. For example, the Combined Multivariate and Collapsing Method (CMC; Li and Leal, 2008) only collapses rare variants and these are treated as a single common variant, which is analyzed together with the common variants of the region using multivariate analysis. However, in the Weighted-Sum method (Madsen and Browning, 2009) or in the Data Adaptive Sum Test (Han and Pan, 2010), mutations of all frequencies are considered but weighted according to their frequency in the unaffected individuals. Further elaborated testing procedures, such as the replication-based strategy (Ionita-Laza *et al.*, 2011) or the C-alpha statistic (Neale *et al.*, 2011), account for the possibility that the sequenced region contains both protective and risk-conferring mutations. Notably, while rare variant burden methods assume that an overall enrichment of rare variants could be influencing the disease or complex phenotype, regression procedures (Hoggart *et al.*, 2008) rely on an alternative scenario in which some causal SNPs should have effect sizes large enough to noticeably impact the phenotype (King *et al.*, 2010).

Although filtering rare variants according to their potential functional relevance (for example, by selecting just non-synonymous variants, variants at splice acceptor or donor sites, or in predicted regulatory sequences) may increase the power of rare variant aggregating tests (Sul *et al.*, 2011), it should also be noted that extended weighting or prioritization according to predicted damaging properties in protein function or structure is not always straightforward, as congruency among functional prediction methods such as SIFT, Polyphen and others often fails. Additionally, weighting according to conservation has also been taken into account. For instance, multiple rare functional variants in three renal salt-handling genes were successfully identified as contributing to blood pressure variation when applying phylogenetic conservation and rare allele frequency criteria (Ji *et al.*, 2008). Notably, none of the aforementioned classical aggregating tests exploits the knowledge about the frequency spectrum of genetic variants that has been gathered after decades of studies on molecular evolution. However, the population genetics theory has also been successfully applied to identify candidate genes enriched for rare functional variants from resequencing data and with sample sizes far smaller than those required by GWAS (Myers *et al.*, 2011). For example, the estimation of the population selection parameter γ (Bustamante *et al.*, 2002) in brain-expressed genes detected enrichment for missense variants in two neurological disease cohorts of autism and schizophrenia when compared to controls (Myers *et al.*, 2011). Finally, the Evolutionary Mixed Model for Pooled Associated Testing (EMMPAT) associated variation in serum triglycerides and the *ANGPTL4* gene (King *et al.*, 2010) by specifically considering disease impact, selective pressure, and SNP frequencies.

16.4. Power matters

Custom targeting sequencing, ES, and WGS explore substantially different scales of sequence variation. Besides the differential run sequencing capacities and costs that these different approaches require, they will also imply substantially differential power for testing rare variant association to complex disease. In fact, it has been noted that no published candidate gene study reported *P* values would be significant in the context of the complete exome and that, indeed, detecting the effects of rare variants using ED data will indeed require very large sample sizes (Kiezun *et al.*, 2012; Tennessen *et al.*, 2012). Similarly, because the prior probability of having an association in the ~2940 Mb non-coding genome is reasonably lower than in the ~60 Mb exome, a WGS approach would probably lower the statistical power to detect disease risk and increase the false-positive rate. Other reasons may additionally favor ED over WGS: alleles associated with complex disease have been found to be underrepresented in intergenic regions and overrepresented in non-synonymous sites and 5 kb promoter regions (Hindorff *et al.*, 2009), and we have much greater understanding of the impact of variation in coding than in the non-functional remaining parts of our genome. Nonetheless, WGS could also provide important insight on structural polymorphism contributing to disease, which is more difficult to assess when applying capturing strategies.

Besides gene-level approaches that seek to identify multiple rare risk alleles within particular genes of interest, other strategies focus on testing whether a substantial proportion of cases for a complex phenotype can be explained by individual, moderately rare variants with strong effects. In that case, a two-stage discovery approach using ES in affected individuals and controls (in one or two hundreds of individuals) and a subsequent large-scale follow-up genotyping of identified candidate variants in a larger cohort of affected subjects and controls (comprising around 1000 or 2000 samples in each group) failed to identifying single rare variants of large effect in idiopathic generalized epilepsy (Heinzen *et al.*, 2012), but indeed suggested a limited role for moderately rare risk factors of strong effect in schizophrenia (Need *et al.*, 2012). Similarly, targeting NGS data in pools of DNA from 400 probands from late-onset Alzheimer's disease (AD) families coupled with additional genotyping in sporadic AD cases and elderly non-demented controls and comparison with public population sequencing data revealed that rare coding variants in *APP, PSEN1*, and *PSEN2* increased risk for late-onset AD (Cruchaga *et al.*, 2012). Low-frequency coding variants conferring risk for AD were also identified in the *PLD3* gene (Cruchaga *et al.*, 2014) when following the multistage strategy of including an initial ES phase in a limited number of selected members from families with multiple affected individuals, subsequent validation of a single coding variant in several large case-control data sets plus targeted resequencing of *PLD3* and gene-burden analysis in two case-control cohorts of moderate size of European (around 2000 samples in each group) and African descent (with 130 cases and 172 controls). However, the contribution of low-frequency and rare coding variants to overall risk in other complex disease traits such as psoriasis has been recognized as rather limited (Tang *et al.*, 2014). In that case, a multistage screen of risk coding variants predisposing to psoriasis including ES in hundreds of both cases and controls and subsequent follow-up validation by targeted sequencing in thousands of individuals revealed only two moderate-effect low-frequency variants but no low-frequency or rare coding variants with strong genetic effect on disease risk.

WGS data obtained from hundreds of Icelandic individuals has also been used to complement SNP genotyping data in thousands of samples from the same population through imputation. Such a strategy has allowed the identification of low-frequency variants associated with gout and serum uric acid levels (Sulem *et al.*, 2011) and also of a rare variant associated with high risk of sick sinus syndrome in a previously unidentified susceptibility gene (Holm *et al.*, 2011). Likewise, Auer *et al.* (2012) used imputation of ES variants initially sequenced in 761 individuals into a larger population-based sample set of 13,000 individuals and successfully identified several genome-wide significant associations with blood cell traits in African Americans for variants that are clearly not covered in traditional genotyping assays. Thus, ES followed by imputation in larger sample sets is demonstrated to be a powerful approach for identifying low-frequency variants associated with quantitative traits and complex disease.

16.5. NGS challenges

Data from NGS platforms and the analysis of rare variants for association with complex disease present further additional challenges to be considered in the planning phase of any study design. For example, the higher error rates than traditional Sanger sequencing and unequal coverage across sequenced regions may result in substantial uncertainty in allele calls or large amounts of missing data in the analysis. Additionally, while imputation works for common and low-frequency variants, the low MAF and the low LD of rare variants with other variants make their imputation more problematic. As in a high-throughput genotyping chip assay, when applying any of the NGS applications in an association study of complex disease, the importance of preventing technical biases or other confounding factors in order to avoid distorting association results it should be especially noted. For example, the design strategy of sequencing an excess of cases for variant discovery and subsequently genotyping these variants in controls can inflate false-positive rates (Li and Leal, 2009). In addition, it should be preferable that sequencing data generated from case and control cohorts are generated in parallel and under the same experimental conditions because data may be imbalanced by several technical confounders, such as source of sample material, storage, DNA preparation, capturing technology, commercial kits and reagent batches, sequencing instrument, read length, depth of coverage, SNP calling algorithm, and subsequent quality control filters (Kiezun *et al.*, 2012). Furthermore, genotyping or new independent sequencing data around identified SNPs should be used to validate true rare variants and to help optimize SNP filtering and calling procedures.

While it is not straightforward to assign causality to the common susceptibility variants detected in GWASs, the greater functional impact and stronger effect sizes found for the low-frequency and rare variants (Gorlov *et al.*, 2010) should in theory facilitate their association as well as the understanding of their role in complex disease etiology. As in GWASs, particular gene associations arising from WGS or ES should be subsequently replicated in new samples in order to confirm association, although clearly such a secondary replication phase could be limited to a few candidate genes resulting from a first genome-wide discovery phase. Although the finding of an excess of rare variants in a particular genomic region may help to associate the influence of that genomic region to a complex phenotype, it does not directly translate that all the rare variants found in that genomic region are functionally relevant for the phenotype. Then, beyond replicating any new association signal to complex disease, *in silico* prediction of possible phenotypical relevance and proper functional validation should be performed before assigning causality or susceptibility to a particular set of rare or low-frequency variants. Besides most of the human variability being rare, such rare variation has been shown to exhibit little sharing among diverging populations, which translates as particular associations with rare variants being difficult to replicate across populations (Gravel *et al.*, 2011). However, the aggregation of rare variants in a particular gene should be replicated if that particular gene is important for the etiology of the disease, and independent population-specific rare susceptibility variants influence the disease outcome.

16.6. Final remarks

We are indeed living in exciting times as NGS technologies allow characterization of the full genetic spectrum of our genome at scales and sample sizes greater than ever. As discussed here, these new advances no doubt increase our chances of understanding the etiology of rare disorders and identifying possible common functional variants behind GWAS signals, as well as of unravelling the role that less common (even structural) variation not appropriately captured by existing strategies can play in complex disease. Importantly, together with NGS advances, new *in silico* approaches for variant functional prioritization and cataloguing as well as novel statistical approaches for appropriately capturing the relevance of rare variants in both complex traits and disease phenotypes are emerging in the scientific community.

The sequencing of the human genome in 2000 started a long run to understand the instruction manual for human biology (Collins and McKusick, 2001). Indeed, as more sequencing data are gathered from multiple individuals and diverse phenotypes, further insights into the understanding of healthy and both Mendelian and complex disease biology will be possible in the near future. However, as occurred more than 10 years ago with the advent of the human genome, new medical advances will not be a straightforward direct enterprise because new analytical tools need to be developed and optimized for understanding and analyzing the unprecedented scale and depth of new re-sequencing data. Finally, only with the integration of additional complementary next-generation technologies interrogating other biological dimensions such as the transcriptome, the proteome, and the epigenome will we obtain further comprehension of human disease etiology.

Acknowledgments

We thank Francesc Calafell for critical reading of the manuscript. Elena Bosch and Ferran Casals work is funded by grants SAF2011-29239 and SAF2012-35025, respectively, from the Spanish Ministerio de Economía y Competitividad as well as by grant 2009SGR-1101 from Direcció General de Recerca, Generalitat de Catalunya. Ferran Casals holds a Beatriu de Pinós (2010-BP- B-00128) fellowship from AGAUR (Generalitat de Catalunya).

REFERENCES

Ahituv, N., Kavaslar, N., Schackwitz, W., *et al.* (2007). Medical sequencing at the extremes of human body mass. *Am. J. Hum. Genet.*, **80**, 779–791.

Alkan, C., Coe, B.P. and Eichler, E.E. (2011). Genome structural variation discovery and genotyping. *Nature Rev. Genet.*, **12**, 363–376.

Asimit, J. and Zeggini, E. (2010). Rare variant association analysis methods for complex traits. *Annu. Rev. Genet.*, **44**, 293–308.

Auer, P.L., Johnsen, J.M., Johnson, A.D., *et al.* (2012). Imputation of exome sequence variants into population-based samples and blood-cell-trait-associated loci in African Americans: NHLBI GO Exome Sequencing Project. *Am. J. Hum. Genet.*, **91**, 794–808.

Bansal, V., Libiger, O., Torkamani, A. and Schork, N.J. (2010). Statistical analysis strategies for association studies involving rare variants. *Nature Rev. Genet.*, **11**, 773–785.

Bustamante, C.D., Nielsen, R., Sawyer, S.A., *et al.* (2002). The cost of inbreeding in *Arabidopsis*. *Nature*, **416**, 531–534.

Cohen, J.C., Kiss, R.S., Pertsemlidis, A., *et al.* (2004). Multiple rare alleles contribute to low plasma levels of HDL cholesterol. *Science*, **305**, 869–872.

Collins, F.S. and McKusick, V.A. (2001). Implications of the Human Genome Project for medical science. *J. Am. Med. Ass.*, **285**, 540–544.

Coventry, A., Bull-Otterson, L.M., Liu, X., *et al.* (2010). Deep resequencing reveals excess rare recent variants consistent with explosive population growth. *Nature Commun.*, **1**, 131.

Cruchaga, C., Haller, G., Chakraverty, S., *et al.* (2012). Rare variants in *APP, PSEN1* and *PSEN2* increase risk for AD in late-onset Alzheimer's disease families. *PLoS ONE*, **7**, e31039.

Cruchaga, C., Karch, C.M., Jin, S.C., *et al.* (2014). Rare coding variants in the phospholipase D3 gene confer risk for Alzheimer's disease. *Nature*, **505**, 550–554.

Dickson, S.P., Wang, K., Krantz, I., Hakonarson, H., Goldstein, D.B. (2010). Rare variants create synthetic genome-wide associations. *PLoS Biol.*, **8**, e1000294.

Fu, W., O'Connor, T.D., Jun, G., *et al.* (2012). Analysis of 6,515 exomes reveals the recent origin of most human protein-coding variants. *Nature*, **493**, 216–220.

Gloyn, A.L., McCarthy, M.I. (2010). Variation across the allele frequency spectrum. *Nature Genet.*, **42**, 648–650.

Gorlov, I.P., Gorlova, O.Y., Frazier, M.L., Spitz, M.R., Amos, C.I. (2010). Evolutionary evidence of the effect of rare variants on disease etiology. *Clin. Genet.*, **79**, 199–206.

Gravel, S., Henn, B.M., Gutenkunst, R.N., *et al.* (2011). Demographic history and rare allele sharing among human populations. *Proc. Natl Acad. Sci. USA*, **108**, 11983–11988.

Han, F. and Pan, W. (2010). A data-adaptive sum test for disease association with multiple common or rare variants. *Hum. Hered.*, **70**, 42–54.

Heinzen, E.L., Depondt, C., Cavalleri, G.L., *et al.* (2012). Exome sequencing followed by large-scale genotyping fails to identify single rare variants of large effect in idiopathic generalized epilepsy. *Am. J. Hum. Genet.*, **91**, 293–302.

Hindorff, L.A., Sethupathy, P., Junkins, H.A., *et al.* (2009). Potential etiologic and functional implications of genome-wide association loci for human diseases and traits. *Proc. Natl Acad. Sci. USA*, **106**, 9362–9367.

Hoggart, C.J., Whittaker, J.C., De Iorio, M. and Balding, D.J. (2008). Simultaneous analysis of all SNPs in genome-wide and re-sequencing association studies. *PLoS Genet.*, **4**, e1000130.

Holm, H., Gudbjartsson, D.F., Sulem, P., *et al.* (2011). A rare variant in *MYH6* is associated with high risk of sick sinus syndrome. *Nature Genet.*, **43**, 316–320.

Ionita-Laza, I., Buxbaum, J.D., Laird, N.M. and Lange, C. (2011). A new testing strategy to identify rare variants with either risk or protective effect on disease. *PLoS Genet.*, **7**, e1001289.

Ji, W., Foo, J.N. and O'Roak, B.J. (2008). Rare independent mutations in renal salt handling genes contribute to blood pressure variation. *Nature Genet.*, **40**, 592–599.

Johansen, C.T., Wang, J., Lanktree, M.B., *et al.* (2010). Excess of rare variants in genes identified by genome-wide association study of hypertriglyceridemia. *Nature Genet.*, **42**, 684–687.

Keinan, A. and Clark, A.G. (2012). Recent explosive human population growth has resulted in an excess of rare genetic variants. *Science*, **336**, 740–743.

Kiezun, A., Garimella, K., Do, R., *et al.* (2012). Exome sequencing and the genetic basis of complex traits. *Nature Genet.*, **44**, 623–630.

King, C.R., Rathouz, P.J. and Nicolae, D.L. (2010). An evolutionary framework for association testing in resequencing studies. *PLoS Genet.*, **6**, e1001202.

Ladouceur, M., Dastani, Z., Aulchenko, Y.S., Greenwood, C.M. and Richards, J.B. (2012). The empirical power of rare variant association methods: results from Sanger sequencing in 1,998 individuals. *PLoS Genet.*, **8**, e1002496.

Li, B. and Leal, S.M. (2008). Methods for detecting associations with rare variants for common diseases: application to analysis of sequence data. *Am. J. Hum. Genet.*, **83**, 311–321.

Li, B. and Leal, S.M. (2009). Discovery of rare variants via sequencing: implications for the design of complex trait association studies. *PLoS Genet.*, **5**, e1000481.

Madsen, B.E. and Browning, S.R. (2009). A groupwise association test for rare mutations using a weighted sum statistic. *PLoS Genet.*, **5**, e1000384.

Manolio, T.A., Collins, F.S., Cox, N.J., *et al.* (2009). Finding the missing heritability of complex diseases. *Nature*, **461**, 747–753.

Momozawa, Y., Mni, M., Nakamura, K., *et al.* (2011). Resequencing of positional candidates identifies low frequency IL23 R coding variants protecting against inflammatory bowel disease. *Nature Genet.*, **43**, 43–47.

Morgenthaler, S. and Thilly, W.G. (2007). A strategy to discover genes that carry multi-allelic or mono-allelic risk for common diseases: a cohort allelic sums test (CAST). *Mutat. Res.*, **615**, 28–56.

Myers, R.A., Casals, F., Gauthier, J., *et al.* (2011). A population genetic approach to mapping neurological disorder genes using deep resequencing. *PLoS Genet.*, **7**, e1001318.

Neale, B.M., Rivas, M.A., Voight, B.F., *et al.* (2011). Testing for an unusual distribution of rare variants. *PLoS Genet.*, **7**, e1001322.

Need, A.C., McEvoy, J.P., Gennarelli, M., *et al.* (2012). Exome sequencing followed by large-scale genotyping suggests a limited role for moderately rare risk factors of strong effect in schizophrenia. *Am. J. Hum. Genet.*, **91**, 303–312.

Nejentsev, S., Walker, N., Riches, D., Egholm, M. and Todd, J.A. (2009). Rare variants of *IFIH1*, a gene implicated in antiviral responses, protect against type 1 diabetes. *Science*, **324**, 387–389.

Price, A.L., Kryukov, G. V, de Bakker, P.I., *et al.* (2010). Pooled association tests for rare variants in exon-resequencing studies. *Am. J. Hum. Genet.*, **86**, 832–838.

Sul, J.H., Han, B., He, D. and Eskin, E. (2011). An optimal weighted aggregated association test for identification of rare variants involved in common diseases. *Genetics*, **188**, 181–188.

Sulem, P., Gudbjartsson, D.F., Walters, G.B., *et al.* (2011). Identification of low-frequency variants associated with gout and serum uric acid levels. *Nature Genet.*, **43**, 1127–1130.

Tang, H., Jin, X., Li, Y., *et al.* (2014). A large-scale screen for coding variants predisposing to psoriasis. *Nature Genet.*, **46**, 45–50.

Tennessen, J.A., Bigham, A.W., O'Connor, T.D., *et al.* (2012). Evolution and functional impact of rare coding variation from deep sequencing of human exomes. *Science*, **337**, 64–69.

Wang, T., Lin, C.Y., Rohan, T.E. and Ye, K. (2010). Resequencing of pooled DNA for detecting disease associations with rare variants. *Genet. Epidemiol.*, **34**, 492–501.

Wu, M.C., Lee, S., Cai, T., *et al.* (2011). Rare-variant association testing for sequencing data with the sequence kernel association test. *Am. J. Hum. Genet.*, **89**, 82–93.

17 Chromosomal breakpoints in breast cancer co-localize with differentially methylated regions

Man-Hung Eric Tang, Vinay Varadan, Sitharthan Kamalakaran, Michael Q. Zhang, James Hicks and Nevenka Dimitrova

17.1. Introduction

Breast cancer is the most common type of malignancy among women in many countries around the world. It is well established that multiple genetic and epigenetic factors play an important role in breast cancer. In the last decade, seminal work by Perou *et al.* (2000) and Sørlie *et al.* (2003) showed that breast cancer tumors consist of five gene expression-based molecular subtypes with different clinical outcomes. In addition, as cancers evolve, their genomes undergo massive alterations at the architectural level including rearrangements, deletions, and amplifications. Genome-wide high-resolution copy number profiling allowed characterizaiton of breast cancer tumors to be studied with unprecedented detail (Lucito *et al.*, 2003). This type of genome instability in breast cancer has also been extensively characterized, first using array comparative genomic hybridization (CGH)-based methods such as in Hicks *et al.* (2006) Kamalakaran *et al.* (2009), Bergamaschi *et al.* (2006), Chin *et al.* (2006), and André *et al.* (2009), and more recently using high-resolution 500 k SNP arrays in, for example Haverty *et al.* (2008). These studies showed that cancer genomes are highly unstable, with recurrent, subtype-specific rearrangements, defining groups that are consistent with existing molecular subtypes (Weigman *et al.*, 2012). Genome rearrangements occur frequently with copy number gains in 1q, 8q11, 11q, 17q, and 20q, and losses in 5q, 6q, and 8p. Very importantly, these regions harbor cancer-related genes such as *TP53, CDKN2A, ERBB2, KRAS*, and *PTEN*, and are therefore extensively cataloged. In Hicks *et al.* (2006), three patterns were defined to qualitatively classify genome rearrangement profiles of breast tumors and measure correlations with patient survival. One of these patterns is characterized by multiple closely spaced amplicons, called firestorms, affecting single chromosome arms which are correlated with poor survival. A formalization of the model

Genome-Wide Association Studies: From Polymorphism to Personalized Medicine, ed. K. Appasani. Published by Cambridge University Press.

was proposed in Russnes *et al.* (2010), using scores to quantify the complexity of genome-wide architectural distortion.

Epigenetic characterization of cancer using DNA methylation profiling of tumors and their corresponding normal profiles has shown that the methylation landscapes are quite disrupted in cancer. For example, the *BRCA1* gene promoter is often hypermethylated in hereditary breast cancers (Tapia *et al.*, 2008). Epigenetic profiling (Kamalakaran *et al.*, 2010) showed that Luminal and non-Luminal breast cancer tumors have different methylation patterns and that differentially methylated genes are associated with relapse risk and overall survival. Breast cancer subtype-specific changes in DNA methylation are associated with expression-based subtypes Luminal A, B, *HER2* positive and basal-like tumors (Bediaga *et al.*, 2010). Furthermore, the analysis of cancer-related genes showed that Luminal A, B and basal-like tumors had distinct methylation patterns (Holm *et al.*, 2010), with a lower methylation frequency in basal and higher methylation frequency in the Luminal B subtype.

In the recent large-scale integrated analysis of 2000 breast tumors (Curtis *et al.*, 2012), novel molecular subgroups were defined based on the inter-relationship between inherited genomic variants, somatic copy number alterations, and their impact on the transcriptome. These methodologies are important in defining predictive or prognostic signatures, or common aberrations, in each type of cancer. However, it is natural to ask whether there is mechanistic cross-talk between epigenetic modifications, genome instability and transcriptional programs within breast cancers. Existing methodologies provide little insight into the mechanisms that drive these epigenetic and genetic changes on a genome-wide scale. In order to investigate these mechanisms we have analyzed published data in search of relationships between epigenetic gene regulation and the genomic alterations associated with cancer (Tang *et al.*, 2012). In this chapter, we have asked several questions. (1) What is the relationship between the genomic rearrangements and DNA methylation? (2) If correlation exists, what is the overlap between these regions with genomic regions prominent in cancer? (3) What is the relationship between breakpoint dense and differentially methylated regions and repetitive elements across the genome?

In this chapter we present our methodology and results that tackle this problem. We used data from tumor and adjacent normal tissue samples from a Norwegian breast cancer cohort (Wiedswang *et al.*, 2003) using in-house analysis platforms: methylation oligonucleotide microarray analysis (MOMA; Kamalakaran *et al.*, 2009) and representational oligonucleotide microarray analysis (ROMA; Lucito *et al.*, 2003). We performed systematic identification of differentially methylated regions (DMRs) and Alu-enriched loci in relation to major genome rearrangements and breakpoint-enriched regions (BERs).

17.2. Material and methods

17.2.1. Tumor sample set

We used the 119 Norwegian breast cancer data set (the Oslo Metastases Study) described in Wiedswang *et al.* (2003). Each patient of the study is further classified into one of the following subgroups: Luminal A tumor subtype (40 patients), Luminal B (15), *ERBB2*-positive (19), basal-like (12), normal-like (14), and 8 undefined. The normal tissue data set consisted of 11 adjacent breast tissue samples. For each sample, DNA methylation and copy number variation (CNV) analysis was performed. We used the DNA methylation MOMA analysis data previously published in Kamalakaran *et al.* (2010) and the copy number ROMA analysis data previously published in Hicks *et al.* (2006). The MOMA and ROMA experimental platforms are described below.

17.2.2. CNV analysis across tumor samples

To measure CNV across the genome, we used the ROMA platform described in Lucito *et al.* (2003). The genome is covered by regularly spaced 82055 probes printed on an array, providing a coverage of the genome at 40,000 nucleotides resolution. We used the circular binary segmentation (CBS) algorithm (Venkatraman and Olshen, 2007) to build segmented copy number profiles.

We partitioned the genome into variable windows such that each sample is observed in a single segmented copy number state (amplified, deleted, normal copy number). The three levels of amplifications were defined by empirically chosen thresholds on copy number ratios (greater than 1.1 for amplifications, less then 0.9 for deletions), taking into account noise around 1. Windows are determined by all the breakpoints obtained by segmentation of the copy number values in each sample using the CBS algorithm. Longer intervals describe regions that have very little copy number change across all the patients, while short intervals correspond to regions with high copy number changes, that is, many breaks across different samples. The above-defined intervals are used to determine differentially methylated regions.

17.2.3. Breakpoint-enriched region detection

We used the segments' start and end defined by the CBS algorithm for the CNV profile of each sample to define our breakpoints. We then calculated the density function using the R function with a bandwidth of 1 Mb and defined the center of the breakpoint dense region as the local maxima of the density.

17.2.4. Differentially methylated region detection

We surveyed the methylome of each tumor sample using the MOMA platform (Kamalakaran *et al.*, 2009). Each CpG island is covered by one or several MOMA fragments that undergo MspI cleavage and McrBC or mock digestion. In total, the 27,000 CpG islands annotated by the UCSC genome browser (hg17 build) are covered by 159,436 MOMA fragments. The data are normalized by converting the hybridization log-ratios into the probabilistic space using an expectation-maximization (EM) method (Kamalakaran *et al.*, 2010) and each MOMA fragment is assigned one of the following states: high methylation (+1), low methylation (–1), and 0 state for cases falling in neither of the two categories.

To identify local variations of DNA methylation in the 108 breast cancer samples, we compared the distribution of methylations calls within each of the intervals defined by all the copy number breakpoints with the one observed across the genome. Each MOMA fragment within these intervals is surveyed and we can associate to each fragment a triplet of observations accounting for the number of "+1"s, "0"s, and "−1"s seen across all samples. For example, a window can be seen 30 times as "+1," 3 times as "0," and 7 times as "−1."

To identify local changes in DNA methylation across the genome, we use Hotelling's T^2-test, a generalization of Student's T-test for multivariate hypothesis testing. The null hypothesis H_0 is defined as the observed distribution of "+1"s, "0"s, and "−1"s observed at each fragment across the MOMA platform. It is calculated based on 159,436 observations. It has an expectation $\mu_0 = (\mu_{01}, \mu_{02}, \mu_{03})$ and covariance B. If a window contains n MOMA fragments, let $X_1, X_2, \ldots, X_n$ be n independent three-dimensional vectors, $n - 1 \geq 3$. $X_1, X_2, \ldots, X_n$ follows the normal law $N(\mu,B)$. Then, the T^2 statistic can be expressed as:

$$T^2 = n(\mu_X - \mu_0)^T S^{-1}(\mu_X - \mu_0) \tag{17.1}$$

where

$$\mu_X = \frac{1}{n}\sum_{i=1}^{n} X_i \tag{17.2}$$

and

$$S^{-1} = \frac{1}{n-1}\sum_{i=1}^{n}(X_i - \mu_X)(X_i - \mu_X)^T \tag{17.3}$$

are the sample maximum likelihood estimators of μ and B.

Then T^2 has the Hotelling's T-square distribution and the statistic

$$F = \frac{n-p}{p(n-1)}T^2 \tag{17.4}$$

has a Fisher's F distribution with p and $n - p$ degrees of freedom, $p = 3$ and parameter $(\mu - \mu_0)^T B^{-1} (\mu - \mu_0)$.

To test whether the null hypothesis $H_0{:}\mu = \mu_0$ is rejected, we compute the F statistics using the observations $X_1, X_2, \ldots, X_n$ of the three-dimensional normal law $N(\mu,B)$ and derive the associated p-value. We then perform a Benjamini and Hochberg false discovery rate (FDR) correction (Benjamini and Hochberg, 1995) on the obtained statistics. A window is considered to have significant deviation in its methylation pattern if its BH p-value is smaller than 10^{-2}.

17.2.5. Breakpoint-enriched differentially methylated region detection

To detect breakpoint-enriched differentially methylated regions (BEDMRs), we detect association between breakpoint-enriched regions (BERs) and differentially methylated regions (DMRs). For this, we evaluate the cumulative number of DMRs as a function of the distance to the nearest BER and compare the distance distributions of the observed occurrences in tumor with randomized locations using a null model. We evaluated three different null models (uniform, normal, and gamma) with differing degrees of similarity to the observed distribution of distances between locations of methylation deviation, and zeroed in on the gamma model because it is the most stringent model. Then we compute the mean cumulative distributions of the randomized locations based on the individual null models (after 1000 randomizations). In addition, we applied an FDR-corrected Wilcoxon test to identify locations that have maximum difference between the observed distribution with the one generated by the gamma null model.

17.2.6. Repeat-enriched loci detection

To identify local changes of Alu repeat frequencies, we used the Repeatmasker database (hg17) as reference and compared the repeat-enrichment frequencies in our regions of interest with the one observed across the genome. An FDR-corrected Wilcoxon test was performed for each repeat type (AluJ, AluS, AluY), in every non-overlapping sliding window of 100 kb. An empirical threshold of $p < 0.001$ was used to decide whether the tested region was significantly enriched or not.

17.3. Results

In Figure 17.1, the conceptual integration model presents how we integrate copy number and DNA methylation patterns in order to determine if there is a mechanistic association between the location of major chromosomal breakpoints and

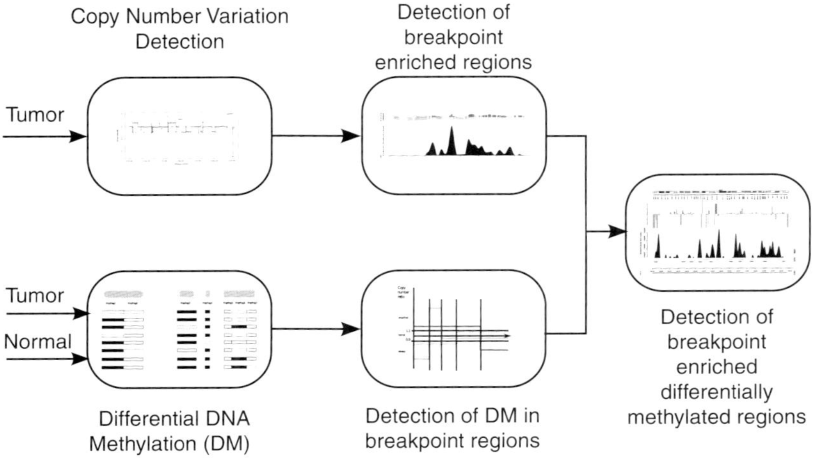

Figure 17.1. Analysis method in order to find DMRs associated with BERs. ROMA genome-wide copy number profiles from breast tumors were combined to partition the genome into variable intervals of stable copy number state in which we estimate DNA methylation levels using MOMA measurements from tumor and normal samples. A Hotelling's T^2 test is performed to identify significant DMRs. On the other track, the locations of BERs are obtained from the ROMA profiles and the list of both significant DMRs and BERs are further evaluated for statistical association (Figure 17.3). A black and white version of this figure will appear in some formats. For the color version, please refer to the plate section.

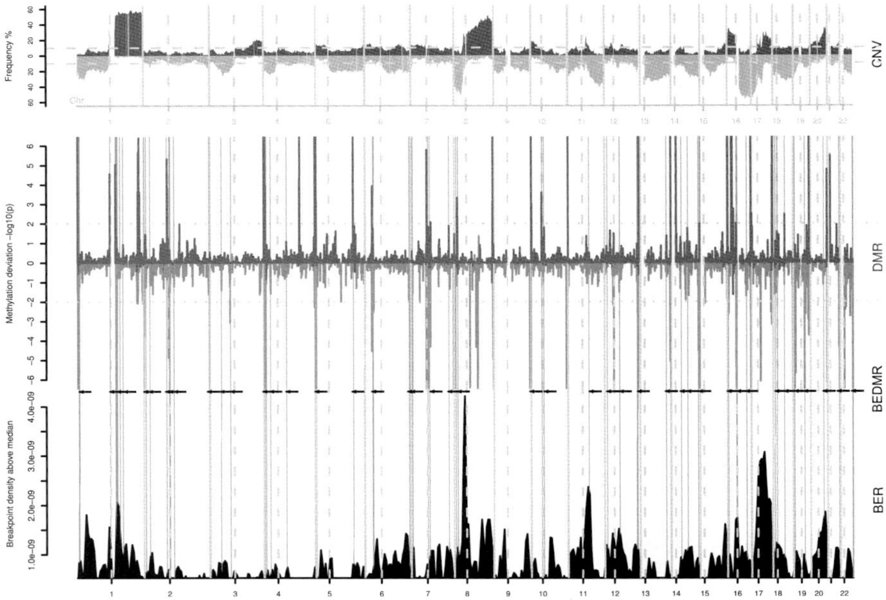

Figure 17.2. Significant DMRs tend to co-localize with breakpoint-enriched regions. The copy number profile of all 108 breast tumors is shown on the top track (CNV). The middle track (methylation) shows the amplitude of the DNA methylation level change compared to normal across genome. Hypomethylated regions are assigned a negative score, defined as $\log_{10}(p)$, while hypermethylated regions take a score equal to $-\log(p)$. Significant DMRs are marked by peaks with a score greater than ±2. The bottom track (breakpoint density) shows the locations of BERs. Breakpoint-enriched DMRs, i.e., DMRs occurring in the vicinity of a BER, are marked by vertical yellow lines and black arrows. Figure reproduced from Tang *et al.* (2012). A black and white version of this figure will appear in some formats. For the color version, please refer to the plate section.

local DNA methylation changes. We first determine the genomic regions where the breakpoint versus differential methylation associations can be tested. Using the copy number profiles of the 108 breast tumors, we partition the genome into variable intervals, and calculate the density of breakpoint observations, the so-called breakpoint-enriched regions (BERs). Next, we identify frequently differentially methylated regions (DMRs) in tumor samples compared to normal samples (using MOMA and Hotelling's T^2 test, BH $p < 0.01$, see Methods). Locations of BERs and DMRs are then statistically assessed.

In Figure 17.2, top track (CNV) recapitulates the copy number gains and losses as frequencies among all the tumor samples, while the next track below, shows scores and locations of significant DMRs. In this track, a positive score means that the surveyed window is hypermethylated compared to the baseline for normal samples, while a negative score indicates a local hypomethylation. We identified 217 DMRs in all tumor samples (BH $p < 0.01$). The first observation is that the detected DMRs are spread across the genome and 145 of them are hypomethylated, while 72 are hypermethylated loci compared to the normal. Finally, we combine these DMRs with the BERs, shown in the bottom track, in order to test their associations. We highlight BEDMRs with vertical yellow lines and arrows to visually link DMRs that appear to co-localize with BERs, within a distance of 1 Mb. In the next section, we provide a more objective measure of this association.

17.3.1. Significant DMRs in tumors co-localize with breakpoints

We measured the cumulative number of DMRs as a function of the distance to the nearest BER and compared the distance distributions of the observed occurrences in tumor with randomized locations derived using a null model (see Methods). Figure 17.3A presents the mean cumulative distributions of the randomized locations based on the gamma null model alongside the observed data for DMRs determined with all tumor samples, and each individual subtype. We also observed that the co-localization of DMR and BERs was insensitive to particular subsets of samples because DMRs determined from Luminal A, Luminal B, *ERBB2*+ Basal samples also exhibit co-localization between DMRs and BERs within 1 Mb.

The cumulative frequency curve obtained with the data set of all tumor samples shows that DMRs occur more frequently than random events generated by the gamma model in the 0–10 Mb distance range away from a BER, suggesting the existence of a positional bias (Wilcoxon test, FDR corrected $p < 0.05$; Figure 17.3B). The best p-value score ($p = 0.039$), that is, the lowest value on the blue curve, was reached at a distance of 1 Mb, where 91 (42.8%) of the observed DMRs were found, while 73% occur within 2 Mb. In contrast,

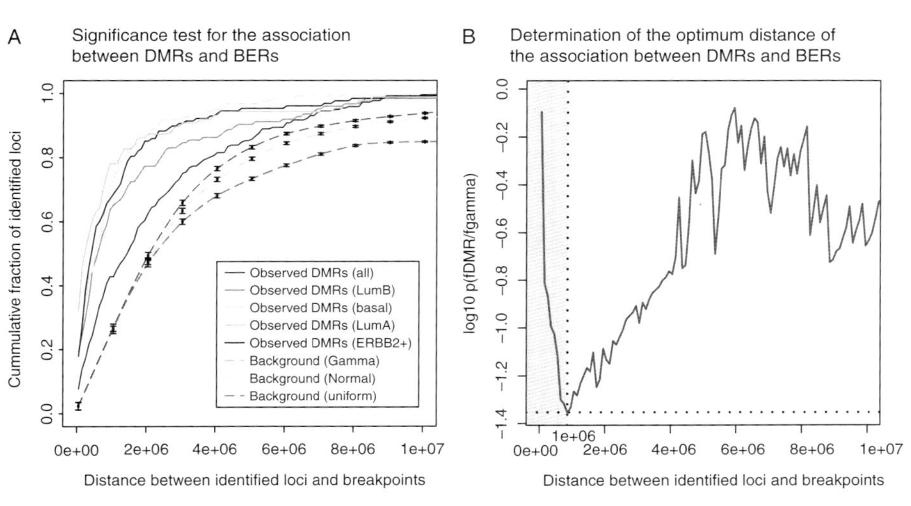

Figure 17.3. Differentially methylated regions co-localize with breakpoint-enriched regions. (A) DMRs tend to be more proximal to BER than expected. (B) The most significant distance of the association between DMRs and BERs occurs at a distance of 1 Mb (shown in yellow). A black and white version of this figure will appear in some formats. For the color version, please refer to the plate section.

only 23.7% of the randomly generated regions with the gamma null model are within 1 Mb of BERs.

This result provides the evidence of a likely association between differentially methylated and breakpoint-enriched regions within a distance of 1 Mb (shaded yellow in Figure 17.3B). In the following, we will use the shorthand BEDMR (breakpoint-enriched DMR) to designate these regions.

We repeated our analysis on 533 breast tumors and 98 normal samples from the The Cancer Genome Atlas (TCGA; The Cancer Genome Atlas Network, 2012). CNV was inferred from Illumina SNP6 Copy number platform and differential methylation was inferred from the Infinium human methylation 450 k platform. We applied the same analytical methods and statistical assessment and refined our parameters. The higher resolution of these newer platforms required the adjustment of some parameters in our model (bandwidth set to 500 kb to account the finer segmentation of CNV); preliminary results on all samples confirmed our findings that significant DMRs co-localize with major chromosomal BERs, 56% of the identified DMRs occurred within 1 Mb of a breakpoint, and 77% within 2 Mb. The optimal distance of the association remained stable, evaluated at 1.3 Mb in this new data set (data not shown).

This result shows additional statistical evidence that the significant association between DMRs and BERs is insensitive to the change of platforms and data sets, suggesting that this may play an important role in genome instability.

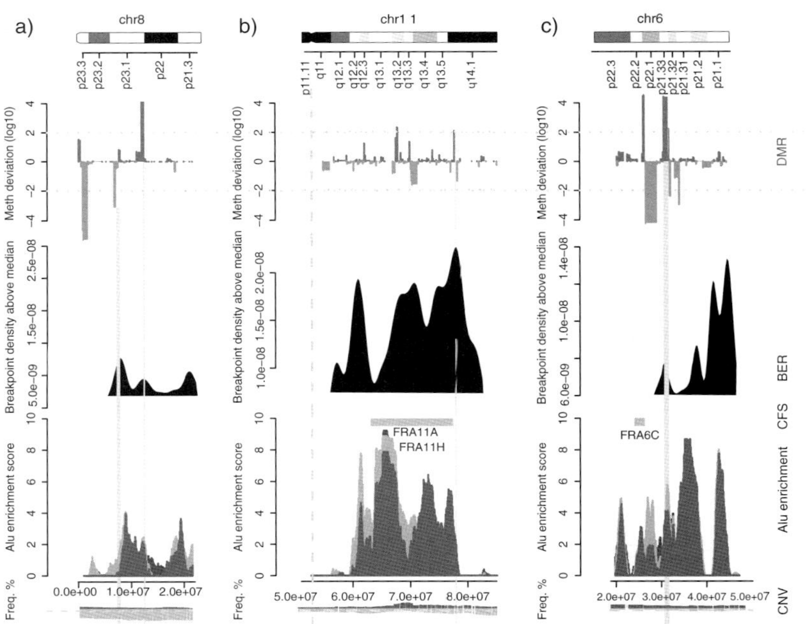

Figure 17.4. Localization of BEDMRs in the genome. BEDMRs tend to occur in genomic contexts. For example A: 8p23.1 (*DUB3*); B: 11q14.1 (*GAB2*); C: 6p21.33 (*DDR1*). These regions contain important cancer-related genes and can be both deleted and hypermethylated or amplified and demethylated. A black and white version of this figure will appear in some formats. For the color version, please refer to the plate section.

17.3.2. Functional insights into BEDMRs

Figure 17.4 illustrates the genomic context in which BEDMRs can occur: telomeric (Figure 17.4A), pericentromeric (Figure 17.4B), and intra-chromosomal regions (Figure 17.4C). We found that BEDMRs were ubiquitous and not biased towards repetitive regions such as telomeres or centromeres.

We observed that 69 of 91 BEDMR regions were genic and that many of the discovered regions have been previously linked with breast cancer and contained known oncogenes. For example, Figure 17.4A shows two BEDMRs located in the p-arm of chromosome 8. The BEDMR at 8p23.1 contains the *DUB3* gene, which regulates DNA damage response by controlling deubiquination (Delgado-Diaz *et al.*, 2014); this region is deleted in 40% of the breast cancer samples. Figure 17.4B shows the q-arm of chromosome 11, which undergoes intense copy number changes, in particular the deletion of the *GAB2* locus (11q14.1) and amplification of *CCND1* (11q13.3). The *GAB2* gene, located within a BEDMR, was reported to inhibit E-cadherin expression and to enhance the expression of *ZEB1*, a transcription factor involved in epithelial-to-mesenchymal transition, cell migration, and invasion through the activation of the PI3 K pathway (Wang *et al.*, 2012). Finally, Figure 17.4C shows a gene dense locus on chr6 with many interesting functional features: the BEDMR marked here with the yellow line contains *DDR1*, involved in the regulation of cell growth and metabolism, and *PPP1R10*, which is involved in the cell cycle. Interestingly, upstream of this locus, we observed a hypomethylated histone gene cluster on 6p22.2. Hypomethylation of histone genes seems to be an important mechanism because we found hypomethylated DMRs containing histone genes in 1q42.13 and 6p22.2. Deregulation of histone genes could contribute to genome instability in cancer by affecting chromatin structure (Tang *et al.*, 2012).

In addition to these examples, we investigated the connection with cancer-related genes such as the genes reported in the Cancer Gene Census (Futreal *et al.*, 2004), which reported a list of 473 genes (at the time of the study) with mutations that have been causally associated with cancer. We found that 6 of our 91 regions contained such types of genes, including *AKT1, ARNT, PMS2,* and the oncogenic ubiquitin hydrolase, *DUB3*, for which we previously reported abnormal demethylation in our integrated study of ovarian cancer (Wrzeszczynski *et al.*, 2011). The described results provided lines of evidence that many BEDMRs were proximal to important cancer genes, although there was no strong statistical association.

17.3.3. Alu repeat enrichment in the vicinity of significant methylation changes and breakpoints

Recent studies showed that short interspersed elements (SINE) and long interspersed elements (LINE) could have a large impact on genome instability, increasing local recombination rates (Witherspoon *et al.*, 2009). Alu repeats are the most numerous transposable elements (one insertion every 3 kb), and Alu-mediated non-allelic homologous recombination (NAHR) are more frequent than other transposable element-mediated NAHR (Konkel and Batzer, 2010). Naturally, the question is whether the BEDMR loci that we identified can be linked to the presence of repeat elements such as Alu repeats.

We observed a significant association between Alu-enriched regions and breakpoint-dense regions where the repeat enrichment is localized and ocurrs within 3 Mb of a BER (Tang *et al.*, 2012). Furthermore, when compared to the background, 32 of 91 BEDMRs have significant Alu-repeat enrichment (in 100 kb sliding windows, Wilcoxon test, FDR-corrected, $p < 0.001$). A large majority of them (22/31) are hypomethylated compared to normal (Tang *et al.*, 2012).

17.4. Discussion

Our analysis on two independent data sets and different platforms provides evidence that there is a statistically significant association between the locations of differentially methylated regions (DMRs) and breakpoints-enriched regions (BERs). In particular, 91 DMR regions occur within a distance as short as 1 Mb from BERs, which we call breakpoint-enriched DMRs (BEDMRs). It is interesting that DMRs are found to be pervasive and often hypomethylated: 145 hypomethylated and 72 hypermethylated loci compared to the normal. This result is in line with what has been observed previously: global DNA hypomethylation associated with punctual hypermethylations in cancer genomes (tumor suppressor genes). Stratifying our analysis based on molecular subtypes (Luminal A and B, *ERBB2+*, basal-like), we found that the statistical association between DMRs and BERs was more significant than expected in each individual subtype (with somewhat different statistical strength: stronger in basal-like, weaker in *ERBB2+* and all combined). More interestingly, we observed potential subtype-specific BEDMR position patterns. For example, 58 BEDMRs occurred only in basal-like samples, in particular in chromosomes 6 and 18, which undergo frequent copy number alterations. We repeated the analysis on publicly available data sets from the TCGA. Preliminary results on a data set of 533 tumors analyzed with Infinium human methylation 450 k and SNP6 platforms provided concordant results with our initial findings, demonstrating that the statistical association between DMRs and BERs was insensitive to changes of data sets and platforms (data not shown). Results on the different subtypes may be further refined in another study with a larger number of samples.

We focused on the regions where breakpoints and methylation pattern deviations co-localize. The analysis of a few important loci (Figure 17.4) allowed us to understand better the functional aspect of the BEDMR events. We detected changes in the entire chromosome, indicating that there was no positional preferences on the chromosome and possibly also no bias towards centromeres or telomeres. Subtelomeric regions are potentially unreliable for array-based methods due to their highly repetitive DNA composition and high C+G content, and associated high methylation levels (Lee *et al.*, 2009). Furthermore, BEDMR regions seem to occur in genic regions, in particular near genes with interesting cancer-related functions. Figure 17.4 shows, for example in chromosome 11, a BEDMR was found near the *GAB2* gene that inhibits E-cadherin and promotes cell migration and invasion, and in chromosome 8 a BEDMR was found near *GAB3*, which is involved in DNA damage repair. More interestingly, we found hypomethylated BEDMRs and DMRs affecting histone gene clusters in chromosomes 1q21.2 and 6p22.1. The deregulation of members of histone gene family and chromatin remodeling factors such as the histone H3 demethylase *JMJD1B* may affect the chromatin structure and disrupt the coordinate signaling and repair, contributing to genome instability in cancer.

We investigated the relationship between the density of retro-transposable SINE elements (Alu) and genome instability. We suspect that there might be a mechanistic relationship between hypomethylation, the presence of these repeat elements, and genome instability. Observing the enrichment levels of the 91 BEDMRs, we found that in a significant fraction of cases (32/91), Alu-repeat enrichment occurs in the vicinity of frequent recombination area. The role of Alu-repeat elements in non-allelic homologous recombination events has been well described in the literature, but many aspects are still unclear. Furthermore, the presence of SINE and LINE elements affects DNA methylation. It has been reported that the promoter regions of methylation-resistant genes are twice as frequently enriched with SINEs and LINEs than those of methylation-prone genes (Estécio *et al.*, 2010). Furthermore, in a recent article (Li *et al.*, 2012), it was shown that segments repeated in low-copy number regions (LCRs) were associated with genome instability and hypomethylation in the germline, and interestingly it was found that homebox, cadherin and histone families were highly enriched in regions with low methylation levels.

One important question is whether BEDMRs are associated with fragile sites and associated genes. Fragile sites are frequently deleted or rearranged in many cancer cells and other studies have demonstrated their importance in genome instability in cancer (Debacker *et al.*, 2007). Of the 91 BEDMRs detected in breast tumor samples, 38 overlap with fragile sites, while only 18 had an overlap with both Alu-enriched regions and fragile sites (Tang *et al.*, 2012). In these BEDMRs that overlap with Alu-enriched regions and fragile sites, we found 35 genes with significant DNA methylation and copy number state in at least 20% of the patients. A remarkable locus is 1q21.3, overlapping with the fragile site *FRA1A*, in which *SETDB1* and *ARNT* are amplified and unmethylated in 51% of the samples. *SETDB1* is a histone methyltransferase and was previously shown to have oncogenic functions in melanoma, accelerating its formation (Ceol *et al.*, 2011). Furthermore, *ARNT* is a regulator involved in TF-miRNA feed-forward loop in cancer (Yan *et al.*, 2012). Interestingly, the 1q21.3 locus has also been reported to be a melanoma-susceptibility locus (MacGregor *et al.*, 2011), suggesting that BEDMRs might target regions that are frequently fragilized or susceptible to deregulation in cancer. We should note that the fraction of BEDMRs overlapping with fragile sites is indeed not statistically significant, suggesting that other elements may contribute to the mechanism of association between BERs and DMRs. Furthermore, BEDMRs provide a much higher resolution insight into the relationship between breakpoints and differential methylation.

Combining the different clues obtained throughout our work, we can sketch a tentative model that describes the relationship between the epigenetic and genetic changes in the genome associated with cancer and try to address the several questions that we asked in introduction. First, we showed that breakpoint occurrences seem to co-occur with local hypomethylation and breakpoint-enriched regions within 1 Mb. These regions, hereby called BEDMR, were found in the presence of retro-transposable SINE Alu elements in 35% of these cases within a distance of 3 Mb. A second interrogation focused on the functional aspects of these structural and epigenetic changes and whether they had an impact on genomic regions which are prominent in cancer.

It is important to summarize that 6 of 91 BEDMRs were co-located with regions containing genes causally linked with cancer in the Cancer Gene Census definition, and another half-dozen of them contained cancer-related genes (e.g., *PCDH* family, *SETDB1, ARNT, PRDX6, ADAMTS4, EGR1, CTNN1A*) and genes involved in the chromatin structure such as histone gene families and remodeling factors. Our result was in agreement with other studies suggesting that transposable elements' insertions, combined with abnormal hypomethylation and increased genome instability, provide a selective advantage in tumorigenesis. Our contribution is to strengthen this association; although no causal relationships can be inferred, we can postulate that each feature contributes partially to the preferential choice of certain loci for genome rearrangement.

We recognize that further study is required to tease out how much these elements really contribute to the genome instability and whether the presence of oncogenes, change of methylation state, or local sequence repeat enrichment prevail in the mechanism.

REFERENCES

André, F., Job, B., Dessen, P., *et al.* (2009). Molecular characterization of breast cancer with high-resolution oligonucleotide comparative genomic hybridization array. *Clin. Cancer Res.*, **15**(2), 441–451.

Bediaga, N.G., Acha-Sagredo, A., Guerra, I., *et al.* (2010). DNA methylation epigenotypes in breast cancer molecular subtypes. *Breast Cancer. Res.*, **12**(5), R77.

Benjamini, Y. and Hochberg, Y. (1995). Controlling the false discovery rate: a practical and powerful approach to multiple testing. *J. R. Statist. Soc. B*, **57**, 289–300.

Bergamaschi, A., Kim, Y.H., Wang, P., *et al.* (2006). Distinct patterns of DNA copy number alteration are associated with different clinicopathological features and gene-expression subtypes of breast cancer. *Genes Chromosomes Cancer*, **45**(11), 1033–1040.

Ceol, C.J., Houvras, Y., Jane-Valbuena, J., *et al.* (2011). The histone methyltransferase *SETDB1* is recurrently amplified in melanoma and accelerates its onset. *Nature*, **471** (7339), 513–517.

Chin, K., DeVries, S., Fridlyand, J., *et al.* (2006). Genomic and transcriptional aberrations linked to breast cancer pathophysiologies. *Cancer Cell*, **10**(6), 529–541.

Curtis, C., Shah, S.P., Chin, S.F., *et al.* (2012). The genomic and transcriptomic architecture of 2,000 breast tumours reveals novel subgroups. *Nature*, **486**(7403), 346–352.

Debacker, K. and Kooy, R.F. (2007). Fragile sites and human disease. *Hum. Molec. Genet.*, **16** (Spec No. 2), R150–158.

Delgado-Díaz, M.R., Martín, Y., Berg, A., Freire, R. and Smits V.A. (2014). Dub3 controls DNA damage signalling by direct deubiquitination of *H2AX*. *Molec. Oncol.*, **8**(5), 884–893.

Estécio, M.R., Gallegos, J., Vallot, C., *et al.* (2010). Genome architecture marked by retrotransposons modulates predisposition to DNA methylation in cancer. *Genome Res.*, **20** (10), 1369–1382.

Futreal, P.A., Coin, L., Marshall, M., *et al.* (2004). A census of human cancer genes. *Nature Rev. Cancer*, **4**(3), 177–183.

Haverty, P.M., Fridlyand, J., Li, L., *et al.* (2008). High-resolution genomic and expression analyses of copy number alterations in breast tumors. *Genes Chromosomes Cancer*, **47**(6), 530–542.

Hicks, J., Krasnitz, A., Lakshmi, B., *et al.* (2006). Novel patterns of genome rearrangement and their association with survival in breast cancer. *Genome Res.*, **16**(12), 1465–1479.

Holm, K., Hegardt, C., Staaf, J., *et al.* (2010). Molecular subtypes of breast cancer are associated with characteristic DNA methylation patterns. *Breast Cancer Res.*, **12**(3), R36.

Kamalakaran, S., Kendall, J., Zhao, X., *et al.* (2009). Methylation detection oligonucleotide microarray analysis: a high-resolution method for detection of CpG island methylation. *Nucleic Acids Res.*, **37**(12), e89.

Kamalakaran, S., Varadan, V., Giercksky Russnes, H.E., *et al.* (2010). DNA methylation patterns in luminal breast cancers differ from non-luminal subtypes and can identify relapse risk independent of other clinical variables. *Molec. Oncol.*, **5**(1), 77–92.

Konkel, M.K. and Batzer, M.A. (2010). A mobile threat to genome stability: the impact of non-LTR retrotransposons upon the human genome. *Semin. Cancer Biol.*, **20**(4), 211–221.

Lee, M.E., Rha, S.Y., Jeung, H.C., Chung, H.C. and Oh, B.K. (2009). Subtelomeric DNA methylation and telomere length in human cancer cells. *Cancer Lett.*, **281**(1), 82–91.

Li, J., Harris, R.A., Cheung, S.W., *et al.* (2012). Genomic hypomethylation in the human germline associates with selective structural mutability in the human genome. *PLoS Genet.*, **8**(5), e1002692.

Lucito, R., Healy, J., Alexander, J., *et al.* (2003). Representational oligonucleotide microarray analysis: a high-resolution method to detect genome copy number variation. *Genome Res.*, **13**(10), 2291–2305.

MacGregor, S., Montgomery, G.W., Liu, J.Z., *et al.* (2011). Genome-wide association study identifies a new melanoma susceptibility locus at 1q21.3. *Nature Genet.*, **43**(11), 1114–1118.

Perou, C.M., Sørlie, T., Eisen, M.B., *et al.* (2000). Molecular portraits of human breast tumours. *Nature*, **406**(6797), 747–752.

Russnes, H.G., Vollan, H.K., Lingjaerde, O.C., *et al.* (2010). Genomic architecture characterizes tumor progression paths and fate in breast cancer patients. *Sci. Transl. Med.*, **2**(38), 38–47.

Sørlie, T., Tibshirani, R., Parker, J., *et al.* (2003). Repeated observation of breast tumor subtypes in independent gene expression data sets. *Proc. Natl Acad. Sci. USA*, **100**(14), 8418–8423.

Tang, M.H., Varadan, V., Kamalakaran, S., *et al.* (2012). Major chromosomal breakpoint intervals in breast cancer co-localize with differentially methylated regions. *Front. Oncol.*, **2**, 197.

Tapia, T., Smalley, S.V., Kohen, P., *et al.* (2008). Promoter hypermethylation of *BRCA1* correlates with absence of expression in hereditary breast cancer tumors. *Epigenetics*, **3**(3), 157–163.

The Cancer Genome Atlas Network. (2012). Comprehensive molecular portraits of human breast tumours. *Nature*, **490**, 61–70.

Venkatraman, E.S. and Olshen, A.B. (2007). A faster circular binary segmentation algorithm for the analysis of array CGH data. *Bioinformatics*, **23**(6), 657–663.

Wang, Y., Sheng, Q., Spillman, M.A., Behbakht, K. and Gu, H. (2012). Gab2 regulates the migratory s and E-cadherin expression via activation of the PI3 K pathway in ovarian cancer cells. *Oncogene*, **31**(20), 2512–2520.

Weigman, V.J., Chao, H.H., Shabalin, A.A., *et al.* (2012). Basal-like breast cancer DNA copy number losses identify genes involved in genomic instability, response to therapy, and patient survival. *Breast Cancer Res. Treat.*, **133**(3), 865–880.

Wiedswang, G., Borgen, E., Karesen, R. and Naume, B. (2003). Detection of isolated tumor cells in BM from breast-cancer patients: significance of anterior and posterior iliac crest aspirations and the number of mononuclear cells analyzed. *Cytotherapy*, **5**(1), 40–45.

Witherspoon, D.J., Watkins, W.S., Zhang, Y., *et al.* (2009). Alu repeats increase local recombination rates. *BMC Genom.*, **10**, 530.

Wrzeszczynski, K.O., Varadan, V., Byrnes, J., *et al.* (2011). Identification of tumor suppressors and oncogenes from genomic and epigenetic features in ovarian cancer. *PLoS ONE*, **6** (12), e28503.

Yan, Z., Shah, P.K., Amin, S.B., *et al.* (2012). Integrative analysis of gene and miRNA expression profiles with transcription factor-miRNA feed-forward loops identifies regulators in human cancers. *Nucleic Acids Res.*, **40**(17), e135.

18 Signaling network analysis of genomic alterations predicts breast cancer drug targets

Naif Zaman and Edwin Wang

18.1. Introduction

Thousands upon thousands of tumors have been sequenced so far, representing over 20 different cancer types. These efforts allowed scientists to take a closer look at the genomic alterations (i.e., mutations and copy number variations) within a tumor's genome, in order to be able to potentially explain the underlying mechanism that drives cancer cell survival and proliferation (Banerji *et al.*, 2012; Cancer Genome Atlas Network, 2012; Stephens *et al.*, 2012). However, extracting useful information from a vast source of various data types to establish a link between genomic alterations and the driving force behind cancer cells remains a challenge (Chin *et al.*, 2011).

Over the past decades of cancer research, scientists have learned that during the evolution of normal cells to cancer cells, different genomic alterations are compiled. These alterations can impact gene expression and protein function to modulate certain fundamental characteristics (i.e., cancer hallmarks) of a cancer cell. Cell survival, proliferation, and apoptosis are among the most primitive cancer hallmarks (Hanahan and Weinberg, 2011). The accumulation of genomic alterations allows cancer cells to reach a neoplastic state that enables them to proliferate indefinitely and become nearly immortal. However, these changes appear to be random, with no patterns that can be used for classifications of patients or identifying drug targets for treatment.

Recent studies (Schlabach *et al.*, 2008; Silva *et al.*, 2008) have gone on to identify genes that are required for cancer cell survival and proliferation (i.e., essential genes). They accomplished this by performing a genome-wide RNAi knockdown screening for different cancer cell lines from three different cancer types, whereupon a gene was considered to be an essential gene if the knockdown of that gene reduced the cell's survival and proliferation based on p-values. A key observation to note in these RNAi knockdown

Genome-Wide Association Studies: From Polymorphism to Personalized Medicine, ed. K. Appasani. Published by Cambridge University Press.

studies is that different cancer cell lines had different sets of essential genes, implying that the cancer hallmark traits, such as survival and proliferation, can be affected by different sets of genes. This was true for cell lines that belong to the same cancer type. Therefore, no two lung cancer cell lines, for example, had identical sets of essential genes, although there was some overlap. In addition, there was no one gene that appeared to be essential across all the different lung cancer cell lines. This may explain why the same drug target does not work on different patients that have the same type of cancer.

These essential gene findings also concur with a previous study (Gerlinger *et al.*, 2012) that has shown tumors to be heterogeneous using sequencing information. They sequenced different regions of a tumor and identified different sets of genomic alterations. This is quite fundamental in the sense that it suggests a tumor to be heterogeneous in nature and that there could be different existing subpopulations of cancer cells with unique genomic alterations. Therefore, it is imperative that we integrate genomic alteration and functional genetic (i.e., RNAi knockdown) data to pursue the identification of the underlying mechanisms that drive cancer cell survival and proliferation. Towards this end, in our study (Zaman *et al.*, 2013), we performed genome-wide exome sequencing for a panel of breast cancer cell lines and matched their corresponding genome-wide RNAi screening data (Marcotte *et al.*, 2012) to carry out an integrated network analysis, to gain insights on the role of genomic alterations on the underlying mechanism of cancer cell survival and proliferation.

18.2. Breast cancer as a model

Breast cancers are classified into three molecular subtypes – luminal, *HER2*, and basal (basal A and basal B; van 't Veer *et al.*, 2002) – using a 50-gene expression signature (*PAM50*) (Parker *et al.*, 2009). The *HER2* subtype is often characterized by the expression of mutated or amplified *ERBB2* receptor. This subtype has had some degree of clinical success because *ERBB2* can be targeted by means of effective therapeutics (Slamon *et al.*, 1987). The luminal subtype is often characterized by the expression of the estrogen receptor (ER+), which is not expressed in the basal subtype. The luminal subtype also has some degree of varying drug responses because ER can be targeted using drugs. However, the basal subtype, considered to be the most aggressive among the subtypes, lacks the expression of *HER2*, ER and progesterone receptors. It is also referred to as triple-negative breast cancer, and has had very limited chemotherapy or other molecularly targeted drug treatment options available. Our goal was to use breast cancer as a model so that we can further characterize the luminal and basal subtypes by gaining insights into their fundamental network wiring differences and identify key subtype-specific genes that could lead to better treatment options in the near future.

18.3. Exome sequencing

For the panel of breast cancer cell lines used in our study, a genome-wide cell survival and proliferation RNAi screening has previously been conducted (Marcotte *et al.*, 2012). In addition, because 5 cell lines have been sequenced previously (Sjoblom *et al.*, 2006), we exome-sequenced the remaining 11 cell lines. From the sequencing data, we removed naturally occurring genetic polymorphisms using the dbSNP database and 1000 Genomes Project and identified 3817 somatic point mutations.

Using the Annovar program (http://www.openbioinformatics.org/annovar/), which predicts potential functional mutations (i.e., mutations that affect protein function), we obtained 1630 potential driver-mutated genes (i.e., cancer-causing genes) for all 11 cell lines. From the sequencing data, all of the cell lines in the panel had mutations among the MAP kinase family, and as expected, *TP53* was mutated across 80% of the basal subtype cell lines. These results are in agreement with the results of genome sequencing of nearly 1000 breast cancer samples (Banerji *et al.*, 2012; Cancer Genome Atlas Network, 2012; Stephens *et al.*, 2012). We also compared the driver-mutated genes in our study to those derived from COSMIC database and found 45 novel driver-mutating genes in at least one cell line. Three genes among them (*ZBTB18, TENM4*, and *TMEM178A*) were found in two cell lines.

18.4. Cancer hallmark network framework

Early cancer research focused on identifying individual genes that could be essential to cancer cell survival. This approach, however, was confined to a small pool of predefined, well studied pathways or protein-to-protein interactions. In addition, secondary redundant pathways often allow for the build-up of resistance to anticancer treatment by the tumor, rendering the treatment ineffective. In the recent past, different signaling pathways have been identified that have been shown to be used by various cancer types for its wide range of biological processes. These pathways may encompass different groups of proteins (e.g. MAPK pathways) and biological processes, such as apoptosis. Genomic alterations can affect these signaling pathways in ways that may result in cancer phenotypes. However, the cell is a complex dynamic system with tens of thousands of protein interactions and crosstalk between hundreds of dynamic pathways. New protein and gene interactions are also being discovered every year. Therefore, a more holistic approach is required that can account for all the different interactions that we know of, not simply those that have been catalogued to predefined existing pathways (Wang, 2010).

In our study, we used an integrative network approach that is not dependent on pathways. In terms of biology, it transforms proteins/genes and their interactions into nodes and edges, respectively. The edges can be classified as activating, inhibiting, or physical interactions. The network is the skeleton upon which we can integrate different types of genomic information, such as gene expression and mutation, and use computer modeling to analyze the cell from a systems perspective. This frees us from being confined to predefined pathways. In fact, using a network, we can discover new pathways, or add to existing ones.

Another advantage of using a network approach is that we can identify key regions (i.e., subnetworks) that are densely connected, called modules, using different unsupervised clustering algorithms. These modules are important – because they are not there by random, they must have a functional significance. A group of genes/proteins that have a high degree of interaction with each other often share similar functions and have intertwined pathways. The network approach and modules can also help us understand which cancer hallmarks are more enriched in our cancer network and playing a major role in the cancer under study. Finally, we have developed a cancer hallmark network framework (Wang *et al.*, 2013a, 2013b, 2015) for studying cancer problems at the systems level.

18.5. Subtype-specific survival signaling networks highlight the evolutionary convergence of selective genomic alterations

With the aid of genome-wide RNAi screening experiments, we are able to uncover genes that are essential for cancer cell survival and proliferation. This screening information also allowed us to identify other genes (i.e., proliferation-influencing genes) that may influence survival and proliferation. Knocking down a proliferation-influencing gene will not necessarily lead to cell death, but will greatly reduce cell growth. In our study, genes that are mutated or amplified and are involved in the regulation of cell growth are defined as "driving-regulators." Previously, we identified high-quality cancer biomarkers by modeling with and perturbing the signaling networks (Cui *et al.*, 2007; Barabasi *et al.*, 2011; Rozenblatt-Rosen *et al.*, 2012) and cancer hallmark networks. To gain greater insight into the underlying mechanism of cancer cell proliferation, and the effect of genomic alteration on it, we used the cancer hallmark network framework to integrate all the information to study the problem. To do so, for each cell line in our panel, we mapped essential, driving-regulator and proliferation-influencing genes onto a manually curated human signaling network (containing ~6000 genes and ~50,000 interactions; Awan *et al.*, 2007; Cui *et al.*, 2007; Li *et al.*, 2012), to build cell-line specific networks (i.e., survival networks; Figure 18.1) (Zaman *et al.*, 2013). These networks highlight the signaling mechanism for cell growth. Detailed information on the different parameters used to define the genes for the cell-line specific networks can be found in the supplemental information of our previous publication (Zaman *et al.*, 2013).

In network graph theory, genes that have a high interaction with many other genes, called hubs, have the crucial role of acting as a global integrator or regulator for multiple signaling pathways. Therefore, we wanted to see how the hubs in our cell-line specific survival networks are behaving. We defined the top 10% of highly connected genes as hubs. If we isolate the driving-regulator (p-value = 0.12) or

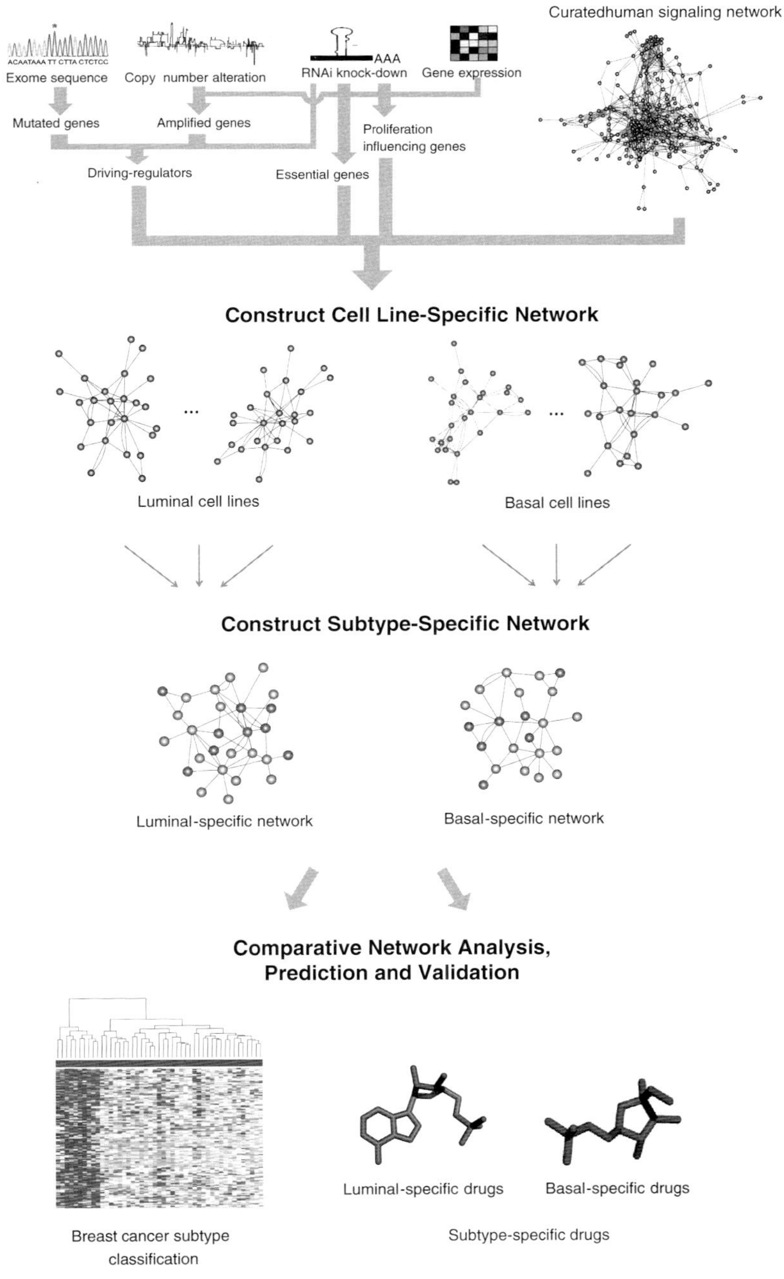

Figure 18.1. Integrated networks for breast cancer cell survival and proliferation framework. The data of genome sequencing, genome-wide RNAi screening, copy number variations, and gene expression profiles of individual lines were used for constructing an integrated network for each individual cell line. Cell-line specific networks across each of the breast cancer subtypes were used for constructing subtype-specific networks for cancer cell survival and proliferation. Comparative and differential analysis of the subtype-specific networks allowed us to predict subtype-specific treatments and significantly classify breast tumor samples. A black and white version of this figure will appear in some formats. For the color version, please refer to the plate section.

essential (p-value = 1.0) hubs, they are not able to classify the individual cell lines into their respective luminal or basal subtypes. However, if we combine both the driving-regulator and essential hubs, the cell lines can then be significantly better classified into luminal and basal subtypes (p-value = 0.03). The p-values were calculated using fuzzy k-mean clustering and Fisher's test. We also tested the hubs using the top 15% as a cut-off in all the analyses and found that both cut-offs generated similar results. Permutation tests were also done to show that our classification of luminal and basal subtypes by the network hubs cannot be obtained at random ($p = 9.0 \times 10^{-4}$). These results suggested that although essential and driving-regulator genes may be quite different between cancer cells, they are complementary and converge to similar signaling mechanisms within their respective subtypes.

To explore our observation in greater detail, we constructed subtype-specific networks containing ~200 genes that appear across ≥ 50% of a subtype's cell lines. In conjunction with our observation of convergence from the cell-line specific network analysis, we noticed that nearly all the genes (> 95%) in a subtype-specific network act as an essential gene in one cell line, but act as a driving-regulator in another line (Figure 18.2A–C). Randomization tests were performed to show that the recurrent usage of the genes in the luminal and basal subtype-specific networks is not random (p-value $< 1.0 \times 10^{-4}$). Our analyses showed that the subtype-specific network genes are used recurrently, suggesting that cancer cells are "addicted" to their respective subtype-specific network for survival and proliferation. These genes represent the core survival signaling mechanisms and could explain the convergent evolutionary events that provide functional constraints for selecting genomic alterations that can offer competitive growth advantage for cancer cells. This selective pressure led to the emergence of distinct network hubs in the luminal and basal subtypes (Figure 18.3A–C). For example, as expected, *AKT1, PIK3*, and *ESR1* are dominantly selected in luminal subtypes, whereas *TP53* and *SRC* are genetically dominant in the basal subtypes.

To investigate the different functional modules and hallmark-related modules that may exist in the subtype-specific networks, we used the gene ontology-guided Markov cluster (MCL) algorithm. In the basal subtype-specific network, three functional modules were identified: cell cycle, proliferation, and apoptosis modules, centered around *CDK1, TP53*, and *EGFR/MAPK* pathway components, respectively (Figure 18.3A, B; Zaman *et al.*, 2013). In the luminal subtype-specific network, two functional modules were identified: cell cycle and proliferation modules, centered around *CDK1/MYC* and *AKT/PIK3CA/MAPK* pathway components, respectively (Figure 18.3C; Zaman *et al.*, 2013). In addition, to interpret these findings from a different angle, we conducted pathway enrichment analysis for each subtype-specific network using the essential, proliferation-influencing, and driving-regulator genes. Signaling pathways of cell-cycle, apoptosis, MAPK/growth factors (i.e., MET), and transcription processes were found in both luminal and basal lines. This highlights the fact that these cancer subtypes share a set of core survival pathways commonly used by breast cancers. Furthermore, the basal subtypes (basal A and

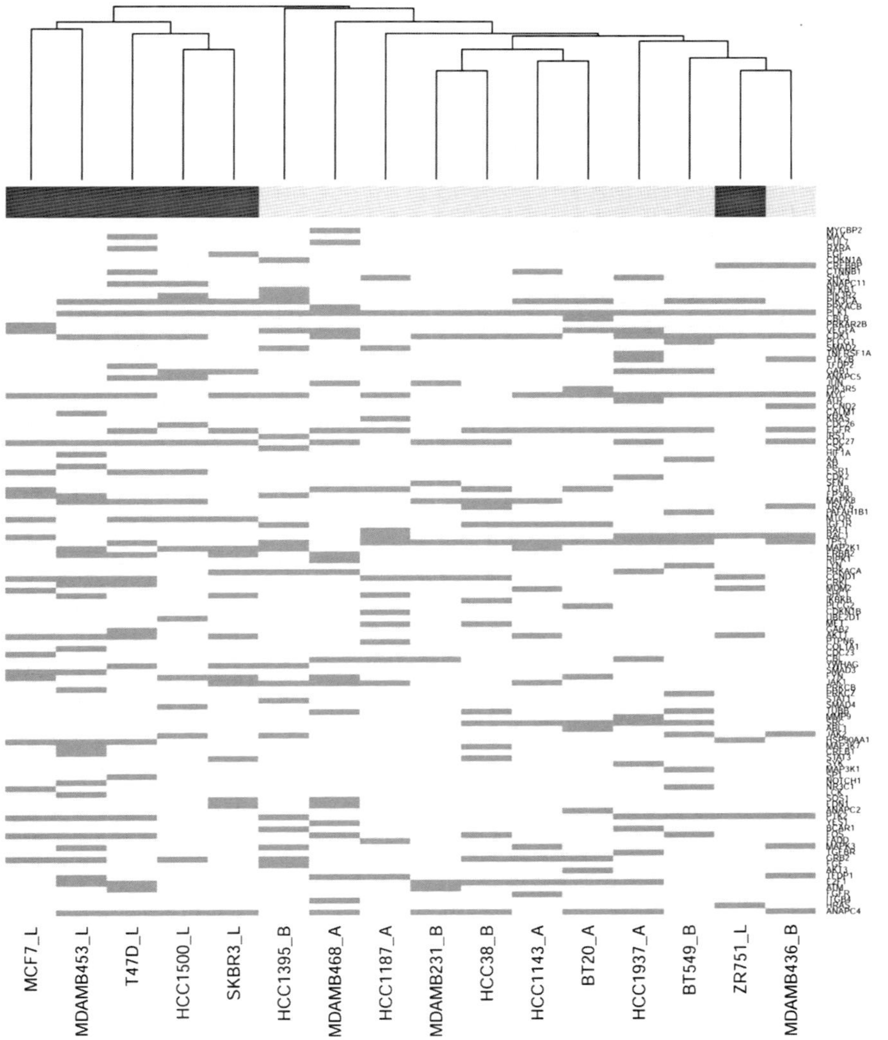

Figure 18.2. Subtype-specific survival signaling networks. Subtype-specific survival signaling networks for basal A (A), basal B (B), and luminal (C) subtypes. Nodes represent genes while links represent regulation (directed links) or interaction (neutral links) between genes. A node is represented by a pie chart that shows each gene's distribution as an essential gene (red), a driving-regulator (blue), or a proliferation-influencing gene (cream) in its subtype. The background color behind the clusters represents a cluster's function in relation to one of the cancer hallmarks: apoptosis (pink), cell proliferation (green), and cell cycle (blue). Cytoscape (Saito *et al.*, 2012) was used to present and visualize the networks. A black and white version of this figure will appear in some formats. For the color version, please refer to the plate section.

basal B) share the signaling pathways for genome instability, such as P53, DNA repair, and telomere extension and maintenance, which were not commonly used by the luminal subtype. A majority of the essential genes affecting genome instability pathways are relatively unique for the basal subtypes, emphasizing that the basal subtype requires unique drug targets for these aggressive triple-negative groups.

A

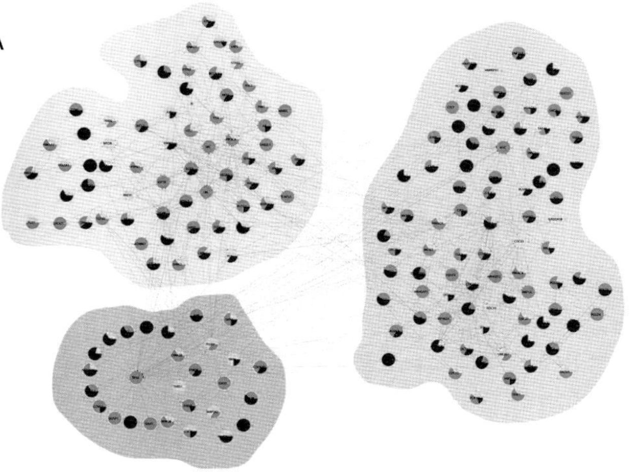

B

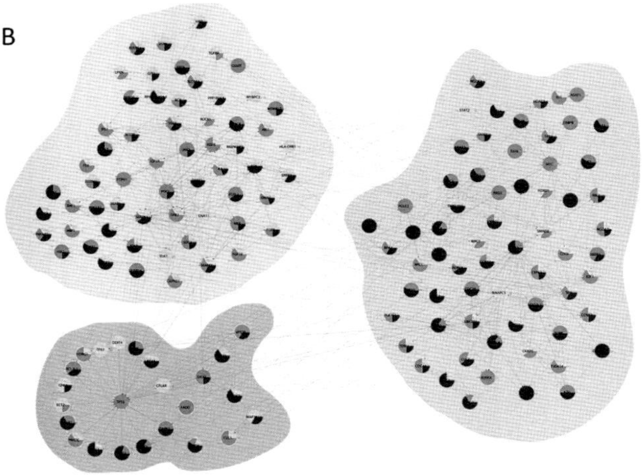

C

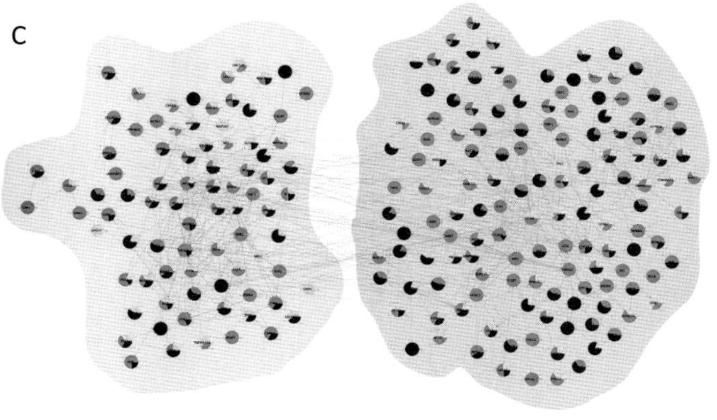

Figure 18.3. Clustering of 402 breast tumor samples using the hubs from subtype-specific networks. Hierarchical clustering of the 402 breast tumor samples using the differential hubs from the subtype-specific networks of luminal and basal subtypes. In the heatmap, red represents mutated genes or amplified genes that are among the top 50% of the expressed genes for tumor samples; otherwise, it appears in beige. On the side bar, gray and yellow represent luminal and basal cell lines, respectively. A black and white version of this figure will appear in some formats. For the color version, please refer to the plate section.

18.6. Subtype-specific survival signaling networks provide predictive power

Cancer essential and driving-regulator genes converge into their respective subtype-specific survival networks. This indicates that for breast cancer subtypes, there lies a "deterministic" path for cell proliferation driven by genomic alterations and could therefore have "predictable" power to accurately identify breast cancer tumor subtypes. To test this, we identified differential hubs between the luminal and basal subtype-specific networks, and classified our panel of cell lines (p-value = 1.2×10^{-4}). These results suggested that amplification or mutation of a few top hub genes could activate the entire network for cancer cell survival and proliferation. Next, we extended this analysis to 402 breast tumor samples, to demonstrate that these differential hub genes' genomic alteration profiles (amplification and mutation status) are able to significantly classify the samples into the basal and luminal subtypes (Figure 18.4; p-value = 2.2×10^{-16}; Zaman *et al.*, 2013). These results highlight the convergent and deterministic nature of the selective genomic alterations that exploit distinct core survival signaling networks (i.e.,

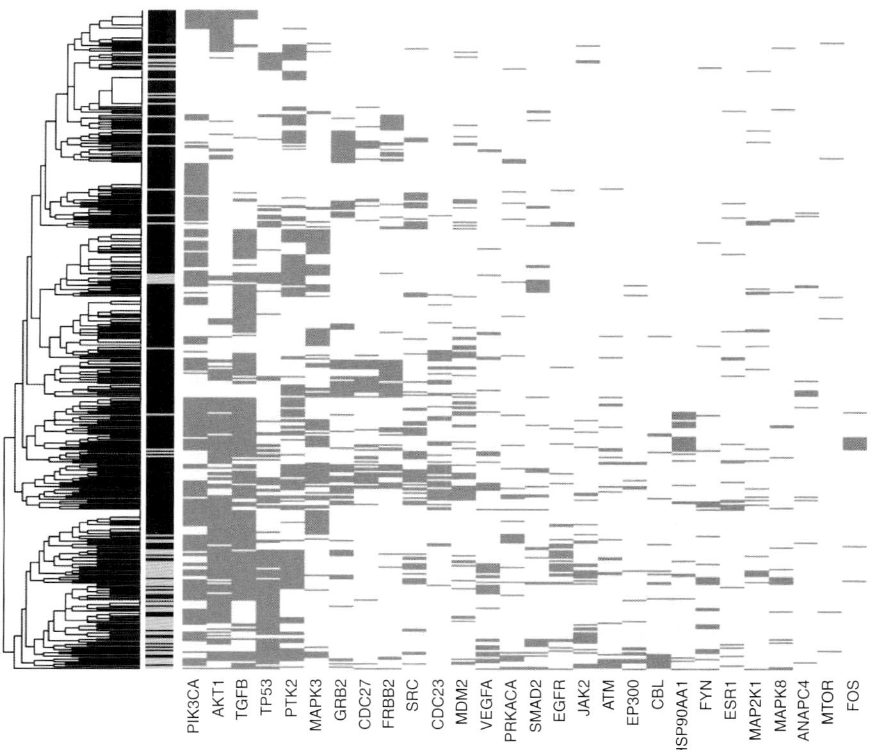

Figure 18.4. Clustering of 402 breast tumor samples using the hubs from subtype-specific networks. Hierarchical clustering using the differential hubs from the subtype-specific networks of luminal and basal subtypes. In the heatmap, for a given sample, red represents mutated genes or amplified genes that are among the top 50% of the expressed genes for tumor samples; otherwise, it appears in beige. On the side bar, gray and yellow represent luminal and basal cell lines, respectively. For the color version, please refer to the plate section.

Table 18.1. Validation of the predicted subtype-specific drug targets.

Compound	Predicted subtype-specific drug target	Basal vs. luminal (*p*- value)	Subtype specificity
Sigma AKT12 inhibitor	*AKT1, AKT2* (luminal)	5.04E-04	Luminal
Tamoxifen	*ESR1* (luminal)	3.92E-02	Luminal
Nutlin3a	*MDM2* (luminal)	3.13E-02	Luminal
Rapamycin	*mTOR* (luminal)	1.78E-03	Luminal
17-AAG	*HSP90* (luminal)	3.98E-02	Luminal
Bosutinib	*SRC* (basal)	1.08E-02	Basal
Docetaxel	*TUBB1* (basal)	1.27E-02	Basal
BMS.536924	*IGF1R* (basal)	4.95E-02	Basal
VX-680	*JAK2* (basal)	4.95E-02	Basal
Erlotinib	*EGFR* (basal)	2.33E-02	Basal
RDEA119	*MAP2K1/MEK12* (luminal)	2.04E-02	Basal
TCS 2312 dihydrochloride	*CHEK1* (luminal)	1.46E-01	Not significant

subtype-specific networks) for cancer cell proliferation. These genomic alterations could occur gradually or suddenly accumulate through genome duplication, which can then be selected during cancer evolution. Early detection of the genomic alterations of these differential hub genes could help in early diagnosis of breast tumors. Recent studies have shown that plasma genome sequencing approach can detect copy number variations and mutations of plasma DNA, which is comparable between cancer patients and healthy individuals (Leary *et al.*, 2012; Chan *et al.*, 2013). With sequencing costs declining, these genes could be used to develop new non-invasive tests (Murtaza *et al.*, 2013) to screen for early stages of breast cancer in patients or even distinguish between subtypes.

Next, to further demonstrate the predictive power of our differential genes from the subtype-specific networks, we sought to use them to predict subtype-specific therapeutic interventions (Table 18.1) (Zaman et al., 2013). If a hub gene is specific to either luminal or basal subtype-specific networks, we expected them to be a specific drug target for its subtype. We isolated *AKT1, mTOR, MET, MDM2, HSP90AA1*, RAF1, *SFN, FYN, CHEK1,* and *ESR1* to potentially be luminal-specific drug targets, while *TGF-β, IGF1* R, *MAPK3, GRB2, SRC, TUBB, JAK2,* and *EGFR* were predicted to be potentially basal-specific drug targets (hubs that are undruggable, like P53 were not considered). We obtained drug screening data for cancer cell lines, which included over 40 breast cancer cell lines (Garnett *et al.*, 2012; Heiser *et al.*, 2012), and statistically evaluated each drug's sensitivity towards luminal and basal subtype. All our predicted targets were in agreement with the drug screening data set (except for *MAP2K1* and *CHEK1*).

18.7. Conclusion

In summary, using a network analysis approach that integrated genomic alteration and functional genetic data for breast cancer, we have shown that there exist core signaling pathways for cell cycle, apoptosis, and growth. Some of these pathways are shared between the subtypes, while the signaling pathways for P53 and genome instability and telomere maintenance are specifically exploited by genomic alterations in the basal subtype. Moreover, the convergence of the essential and driving-regulator genes to a set of core signaling pathways leads to the emergence of the subtype-specific survival networks, where these genes recurrently switch their roles between essential and driving-regulator genes in cancer cells. These networks elucidate the underlying signaling mechanisms that govern cancer cell survival and proliferation and imply selective pressures for evolutionary convergence of cancer genomic alterations. In addition, it is also clear that the signaling mechanisms of the two subtypes are distinct. This is evident by the two subtypes' differential genes, whose genomic alteration profiles can significantly distinguish our panel of breast cancer cell lines and tumor samples into luminal and basal subtypes. Finally, we showed the subtype-specific network's predictive power with experimental validation (i.e., drug screening) with high accuracy (~80%). In conclusion, our findings highlight the importance of amplification of genes and the profound clinical implications that it may have in personalized treatment of cancer patients and the screening of early stage breast cancer patients.

REFERENCES

Awan, A., Bari, H., Yan, F., *et al.* (2007). Regulatory network motifs and hotspots of cancer genes in a mammalian cellular signalling network. *IET Syst. Biol.*, **1**, 292–297.

Banerji, S., Cibulskis, K., Rangel-Escareno, C., *et al.* (2012). Sequence analysis of mutations and translocations across breast cancer subtypes. *Nature*, **486**, 405–409.

Barabasi, A.L., Gulbahce, N. and Loscalzo, J. (2011). Network medicine: a network-based approach to human disease. *Nature Rev. Genet.*, **12**, 56–68.

Cancer Genome Atlas Network. (2012). Comprehensive molecular portraits of human breast tumours. *Nature*, **490**, 61–70.

Chan, K.C., Jiang, P., Zheng, Y.W., *et al.* (2013). Cancer genome scanning in plasma: detection of tumor-associated copy number aberrations, single-nucleotide variants, and tumoral heterogeneity by massively parallel sequencing. *Clin. Chem.*, **59**, 211–224.

Chin, L., Hahn, W.C., Getz, G. and Meyerson, M. (2011). Making sense of cancer genomic data. *Genes Develop.*, **25**, 534–555.

Cui, Q., Ma, Y., Jaramillo, M., *et al.* (2007). A map of human cancer signaling. *Molec. Syst. Biol.*, **3**, 152.

Garnett, M.J., Edelman, E.J., Heidorn, S.J., *et al.* (2012). Systematic identification of genomic markers of drug sensitivity in cancer cells. *Nature*, **483**, 570–575.

Gerlinger, M., Rowan, A.J., Horswell, S., *et al.* (2012). Intratumor heterogeneity and branched evolution revealed by multiregion sequencing. *New. Engl. J. Med.*, *366*, 883–892.

Hanahan, D. and Weinberg, R.A. (2011). Hallmarks of cancer: the next generation. *Cell*, **144**, 646–674.

Heiser, L.M., Sadanandam, A., Kuo, W.L., *et al.* (2012). Subtype and pathway specific responses to anticancer compounds in breast cancer. *Proc. Natl Acad. Sci. USA*, **109**, 2724–2729.

Leary, R.J., Sausen, M., Kinde, I., *et al.* (2012). Detection of chromosomal alterations in the circulation of cancer patients with whole-genome sequencing. *Sci. Transl. Med.*, **4**, 162ra154.

Li, L., Tibiche, C., Fu, C., *et al.* (2012). The human phosphotyrosine signaling network: evolution and hotspots of hijacking in cancer. *Genome Res.*, **22**, 1222–1230.

Marcotte, R., Brown, K.R., Suarez, F., *et al.* (2012). Essential gene profiles in breast, pancreatic, and ovarian cancer cells. *Cancer Discov.*, **2**, 172–189.

Murtaza, M., Dawson, S.J., Tsui, D.W., *et al.* (2013). Non-invasive analysis of acquired resistance to cancer therapy by sequencing of plasma DNA. *Nature*, **497**, 108–112.

Parker, J.S., Mullins, M., Cheang, M.C., *et al.* (2009). Supervised risk predictor of breast cancer based on intrinsic subtypes. *J. Clin. Oncol.*, **27**, 1160–1167.

Rozenblatt-Rosen, O., Deo, R.C., Padi, M., *et al.* (2012). Interpreting cancer genomes using systematic host network perturbations by tumour virus proteins. *Nature*, **487**, 491–495.

Saito, R., Smoot, M.E., Ono, K., *et al.* (2012). A travel guide to Cytoscape plugins. *Nature Meth.*, **9**, 1069–1076.

Schlabach, M.R., Luo, J., Solimini, N.L., *et al.* (2008). Cancer proliferation gene discovery through functional genomics. *Science*, **319**, 620–624.

Silva, J.M., Marran, K., Parker, J.S., *et al.* (2008). Profiling essential genes in human mammary cells by multiplex RNAi screening. *Science*, **319**, 617–620.

Sjoblom, T., Jones, S., Wood, L.D., *et al.* (2006). The consensus coding sequences of human breast and colorectal cancers. *Science*, **314**, 268–274.

Slamon, D.J., Clark, G.M., Wong, S.G., *et al.* (1987). Human breast cancer: correlation of relapse and survival with amplification of the *HER-2/neu* oncogene. *Science*, **235**, 177–182.

Stephens, P.J., Tarpey, P.S., Davies, H., *et al.* (2012). The landscape of cancer genes and mutational processes in breast cancer. *Nature*, **486**, 400–404.

van 't Veer, L.J., Dai, H., van de Vijver, M.J., *et al.* (2002). Gene expression profiling predicts clinical outcome of breast cancer. *Nature*, **415**, 530–536.

Wang, E. (2010). *Cancer Systems Biology*. CRC Press, Boca Raton, FL.

Wang, E., Zou, J., Zaman, N., *et al.* (2013a). Cancer systems biology in the genome sequencing era: part 1, dissecting and modeling of tumor clones and their networks. *Semin. Cancer Biol.*, **23**, 279–285.

Wang, E., Zou, J., Zaman, N., *et al.* (2013b). Cancer systems biology in the genome sequencing era: part 2, evolutionary dynamics of tumor clonal networks and drug resistance. *Semin. Cancer Biol.*, **23**, 286–292.

Wang, E., Zaman, N., McGee, S., *et al.* (2015). Predictive genomics: a cancer hallmark network framework for predicting tumor clinical phenotypes using genome sequencing data. *Semin. Cancer Biol.*, **30**, 4–12.

Zaman, N., Li, L., Jaramillo, M.L., *et al.* (2013). Signaling network assessment of mutations and copy number variations predict breast cancer subtype-specific drug targets. *Cell Rep.*, **5**, 216–223.

19 Pharmacogenetic studies in pediatric acute myeloid leukemia

Neha S. Bhise, Lata Chauhan and Jatinder Kaur Lamba

19.1. Introduction

Acute myeloid leukemia (AML) is a clonal disorder characterized by appearance of immature, abnormal myeloid cells in bone marrow and other organs. AML accounts for around 15–20% of childhood leukemias. AML is a very heterogeneous disease with various subtypes classified based on the morphology, immunophenotype, and cytogenetics. In spite of advances in recent years, the 5-year survival rates for AML are ~68% for children younger than 15 years and ~57% for children between 15 and 19 years. The utilization of multiple clinical, cytogenetic, and other molecular features that are associated with response has helped in the identification of a patient being more or less likely to respond. Additionally, minimal residual disease (MRD after induction 1) has been identified as a powerful predictor of poor outcome. The nucleoside analog cytarabine (ara-C) has been the mainstay of AML chemotherapy for more than 40 years. However, extensive inter-patient variation in treatment response, development of resistance, and inadequate response to first-line therapy remain the major hurdles to effective chemotherapy. Patients within standard and high-risk categories often experience induction failure and have early relapse, warranting the need for better diagnostic and therapeutic strategies. One of the critical components contributing to the efficacy of the chemotherapeutic agents is variability in the expression and/or activity of genes involved in drug pharmacokinetics and pharmacodynamics. This chapter summarizes the recent advances in pediatric AML pharmacogenomics.

19.2. Pediatric acute myeloid leukemia

Leukemia is the most common cancer among children, with acute lymphocytic leukemia being the most common and AML being the second most common

Genome-Wide Association Studies: From Polymorphism to Personalized Medicine, ed. K. Appasani. Published by Cambridge University Press.

leukemia in children. Approximately 800 new cases of childhood AML are diagnosed in the US annually (Meshinchi and Arceci, 2007; Pui *et al.*, 2011). AML is a clonal disorder originating from a hematopoietic stem cell or lineage-specific progenitor cells (Jordan, 2007; Lane *et al.*, 2009). The malignant transformation of the stem or progenitor cells results in the accumulation of immature myeloid cells in the bone marrow and other organs. Prognostic factors help in strategizing the treatment regimens for patients with less or more likelihood of response. As indicated before, AML is a very heterogeneous disease with several subtypes that differ from one another in morphology, immunophenotye, and cytogenetics. The French–American–British (FAB) classification system was developed by FAB Cooperative group and it categorized AML into major subtypes based on morphology and immunohistochemical detection of lineage markers – M0, acute myeloblastic leukemia without differentiation; M1, acute myeloblastic leukemia with minimal differentiation but with the expression of MPO; M2, acute myeloblastic leukemia with differentiation; M3, acute promyelocytic leukemia; M4, acute myelomonocytic leukemia; M4Eo, M5, acute monocytic leukemia, M6, acute erythroid leukemia; and M7, acute megakaryocytic leukemia (Bennett *et al.*, 1976, 1980, 1985a, 1985b, 1991; Kaleem and White, 2001; Cheson *et al.*, 2003). In 2001, the FAB system was replaced by the World Health Organization (WHO) system of classification, which utilized cytogenetic information for more reliable prediction of outcome. In this classification, patients with t(8:21), inv (16), t(15:17), or MLL translocation were classified as AML with recurrent cytogenetic abnormalities. Later additional cytogenetic markers and *CEBPA* and *NPM* mutations were also included in the classification system (Vardiman *et al.*, 2009). Rubnitz and Inaba (2012) have very comprehensively reviewed the genetics and biology of AML along with risk stratification. Overall, although all AML subtypes share abnormalities in pathways leading to differentiation arrest and enhanced self-renewal, there are distinctive features that are important predictors of clinical outcome (Rubnitz and Inaba, 2012). Table 19.1 summarizes the prognostic risk groups with genetic abnormalities of clinical relevance. A cytogenetically normal karyotype comprises the major part of the intermediate-risk group; however, within this group, the occurrence of various mutations in genes such as *NPM1* (Rau and Brown, 2009; Falini and Martelli, 2011), *CEBPA* (Bacher *et al.*, 2010), *IDH1/2* (Abdel-Wahab *et al.*, 2011), and *WT1* are associated with prognosis (Rockova *et al.*, 2011; Hou *et al.*, 2014). An internal tandem duplication (ITD) in the juxtamembrane domain of fms-like tyrosinekinase receptor-3 (FLT3-ITD) results in constitutive kinase activity influencing *RAS/MAPK*, *PI3* K/*Akt*/*mTOR*, and *JAK/STAT* signaling and is associated with poor prognosis (Sallmyr *et al.*, 2008; Gu *et al.*, 2011; Tasian *et al.*, 2014a). The use of tyrosine kinase inhibitors such as sorafenib, crenolanib, etc., are currently being explored in FLT3-ITD AML (Tasian *et al.*, 2014a).

19.3. Treatment strategy for *de novo* patients with AML

The most common treatment strategy includes induction therapy, which consists of a 3+7 regimen with daunorubicin (45 mg/m^2 per day for 3 days) and

Table 19.1. Cytogenetic risk groups in pediatric AML.

Prognostic risk group	Cytogenetic abnormality and affected gene	Incidence (%)
Favorable	t(15;17)(q24;q21) –PMA-RARA	8–11
	t(8;21)(q22;q22) –RUNX1-ETO	11–13
	inv(16)(p13.1q22)/ t(16;16)(p13.1;q22) – MYH11-CBFB	3–6
	t(1;11)(q21;q23)– MLL-MLLT3	3
Intermediate	t(9;11)(p21;q23) – MLL-MLLT4	10
	t(1;22)(p13;q13) – RBM15-MKL1	2–3
	normal karyotype	20–25
	< 3 abnormalities	6–7
Adverse	inv(3)(q21q26)/t(3;3)(q21;q26)	1–2
	–5/ del(5q)/ add(5q)	5–6
	–7/ del(7q)/ add(7q)	1–2
	t(11q23) [excluding t(9;11)]	18–22
	t(6;9)(p23;q34) – DEK-NUP214	1
	t(9;22)(q34;q11.2)	1
	–17/ abn(17p)	2
	Complex (≥ 3 unrelated abnormalities)	10
	Normal with FLT3-ITD	12

cytarabine 100 mg/m^2 for 7 days). This induction regimen has been shown to induce remission in 60–70% of patients (Yates *et al.*, 1982; Rubnitz and Inaba, 2012). Cytarabine in combination with varying doses of anthracycline and the addition of a third agent (cladribine, gemtuzumab, ozogamicin, etc.) has been the standard of practice for a few years. Although the intensive cytotoxic chemotherapy consisting of anti-leukemic agents listed above has been shown to induce remission in 60–70% of pediatric leukemia patients, more than half of these patients relapse. The development of both intrinsic and acquired cellular resistance to cytarabine-containing chemotherapy has been a major hurdle to the successful treatment of AML. Improving the treatment outcome for patients with AML remains a major clinical challenge. Although the prognostic factors listed above, including AML heterogeneity, contribute to the observed variation in clinical response, one of the potentially significant contributors is inter-patient variation in genes involved in drug pharmacokinetics (PK) and pharmacodynamics (PD). The field of pharmacogenomics focuses on identifying genetic variation in PK/PD genes and how these influence treatment outcomes. Understanding the variability in response to drugs is of very high stakes, especially when the therapeutic index is very narrow and the toxicities are severe. In addition to the efforts in genetics, significant efforts are also being focused on developing molecular targeted approaches to improve the outcome of refractory or relapsed AML patients (Tasian *et al.*, 2014b). In the following sections we will summarize the efforts in pharmaco-genomics of the major anti-leukemic agents used in AML.

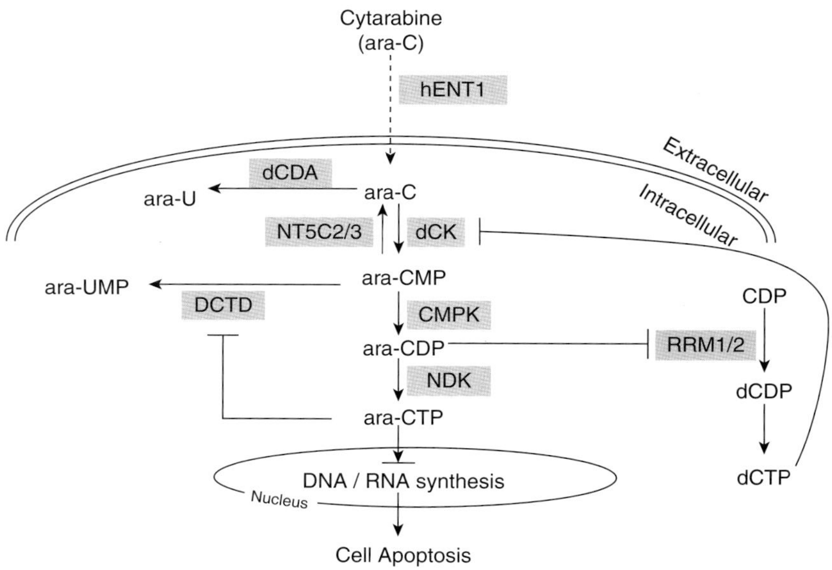

Figure 19.1. Metabolic pathway of cytarabine (Ara-C).

19.4. Pharmacogenomics of cytarabine

Cytarabine (1-β-D-arabinofuranosylcytosine, ara-C) has been the mainstay of AML chemotherapy for more than four decades. One of the most common and severe toxicities associated with cytarabine is myelosuppression. Cytarabine is a prodrug requiring conversion to its active metabolite, ara-C triphosphate (ara-CTP). The metabolic pathway of ara-C is shown in Figure 19.1. An equilibrative nucleoside transporter hENT1 encoded by the gene *SLC29A1* is involved in cellular uptake of cytarabine. Once inside the cell, deoxycytidine kinase (DCK) catalyzes the first rate-limiting step of conversion of ara-C to ara-C monophosphate (ara-CMP). Ara-CMP is then phosphorylated by deoxycytidylate kinase (dCMPK) to ara-CMP followed by phosphorylation by nucleoside diphosphate kinase (NDK) to form the active 5′-triphosphate derivative, ara-CTP. The triphosphate form of ara-C inhibits DNA polymerase-α and -β and competes with deoxycytidine triphosphate (dCTP) for incorporation into DNA and RNA, thereby inhibiting DNA synthesis and triggering leukemic cell death (Kufe *et al.*, 1980; Major *et al.*, 1981; Raza *et al.*, 1992). Among the genes involved in cytarabine inactivation, the most significant are cytidine deaminase (*CDA*) and deoxycytidylate deaminase (*DCTD*), which convert ara-C to ara-U and ara-CMP to ara-UMP, respectively. 5′-Nucleotidases (*NT5C2* and *NT5C3*) can dephosphorylate ara-CMP back to ara-C. Insufficient intracellular levels of the ara-CTP due to reduced uptake/activation or increased inactivation can have a significant impact on the therapeutic efficacy of cytarabine. Additionally, intracellular dCTP levels can influence ara-C sensitivity, by competing with ara-CTP for incorporation into DNA, and also inhibiting DCK activity through feedback mechanism. Also, because the enzyme ribonucleotide reductase (consisting of *RRM1* and *RRM2* subunits) regulates intracellular levels of dCTPs, the expression levels of *RRM1* and *RRM2* could also have an impact on ara-C sensitivity.

Single nucleotide polymorphisms (SNPs) in genes involved in transport, activation, and inactivation of cytarabine can influence the intracellular ara-CTP levels by influencing the expression and activity of these genes, and this in turn can also have an effect on the clinical outcome of the patients treated with ara-C. Table 19.2 summarizes the potential genes and SNPs within them as it relates to the pharmacogenomics of cytarabine in AML. A brief description on SNPs of importance to cytarabine pharmacogenomics is also described in subsequent sections.

19.4.1. Human equilibrative nucleoside transporter *hENT1* (*SLC29A1*)

hENT1 mRNA expression levels show wide variation in AML patients; AML patients with lower *hENT1* expression had a shorter disease-free survival and higher risk of early relapse than *hENT1*-positive patients (7.7 months vs. 13 months; Galmarini *et al.*, 2002b). Several *in vitro* studies have shown significant correlation between *hENT1* expression levels and ara-C sensitivity (Gati *et al.*, 1997, 1998). With respect to SNPs, the coding region of *SLC29A1* (*hENT1*) has been sequenced in several ethnic groups (Caucasians, African Americans, Asians, and others), and exceptionally low mutability was observed as compared with the other transporters. Four synonymous and two non-synonymous changes (Ile216Thr, Glu391Lys) were identified; however, no functional effects of these were observed in *in vitro* uptake assays (Osato *et al.*, 2003). Although there was lack of functional SNPs, the expression of *SLC29A1* has been reported to affect sensitivity to ara-C, indicating the observed variability in expression to be due to differences in transcriptional regulation of *SLC29A1*. Screening of the *SLC29A1* proximal promoter identified SNPs at −1345C>G, −1050 G>A and −706 G>C). Some of these SNPs were associated with loss of binding sites for transcription factors FTF/LHR-1 (−1345C>G), MyT1 zinc finger transcription factor, AhR nuclear translocator, and X-box binding protein (−706 G>C) (Myers *et al.*, 2006).

19.4.2. Deoxycytidine kinase (DCK)

Lower levels of DCK activity have been observed in cell lines resistant to ara-C, confirming its importance in cytarabine activation (Verhoef *et al.*, 1981; Hapke *et al.*, 1996; Dumontet *et al.*, 1999). Within AML patients, DCK mRNA expression has been shown to positively correlate with treatment outcome in AML patients; patients expressing higher DCK expression had longer event-free survival than patients with low DCK expression (Galmarini *et al.*, 2002a). DCK also demonstrates approximately 50-fold variation in its mRNA expression in patient leukemic cells (Kakihara *et al.*, 1998), and around 35-fold variation in primary AML cells (Lotfi *et al.*, 2006).

The sequencing of DCK proximal promoter region and coding exons in Chinese individuals identified two regulatory SNPs (−360 C>G and −201 C>T) in linkage

Table 19.2. Major SNPs in genes of significant relevance to cytarabine chemotherapy.

Gene	Gene symbol	SNPs	Relevance	Reference
Deoxycytidine kinase	*DCK*	rs2306744	SNPs −360 C>G and −201 C>T (rs2306744) in DCK were associated with higher mRNA expression and higher transcriptional activity. In AML patients, it was found that −360 G/−201 T haplotype was associated with favorable response	Shi *et al.* (2004)
Deoxycytidine kinase	*DCK*	rs4694362	In Korean AML patients, the CC genotype was a significant poor prognostic factor for poor overall survival	Kim *et al.* (2013)
Deoxycytidine kinase	*DCK*	rs111454937	AA genotype was more frequent in Chinese AML patients with higher platelet count and A allele frequency was significantly higher in the group <40 years, lower WBC count patients group and group with platelet count > 60×10^9/L	Xu *et al.* (2012)
Deoxycytidine kinase	*DCK*	rs72552079	In Chinese AML patients, the TC genotype showed significantly improved ara-C based chemotherapy response	Xu *et al.* (2012)
Deoxycytidine kinase	*DCK*	rs11554389	In Chinese AML patients, the TC genotype showed significantly improved ara-C based chemotherapy response	Xu *et al.* (2012)
Cytidine deaminase	*CDA*	rs2072671 rs532545	In FLT3-ITD positive AML patients, the CDA 79 C/C (rs2072671) and CDA −451 T/T (rs532545) genotypes were associated with shorter overall survival compared to other genotypes	Falk *et al.* (2013)
Cytidine deaminase	*CDA*	rs2072671	Plasma CDA activity was significantly decreased in samples homozygous for the A79 C SNP (CC) compared with samples homozygous for the ancestral allele (AA)	Mahfouz *et al.* (2013)
Cytidine deaminase	*CDA*	rs60369023	Significantly improved ara-C based chemotherapy response in Chinese AML patients.	Xu *et al.* (2012)
Cytidine deaminase	*CDA*	rs3215400	In AML patients, there was a significantly higher CDA RNA expression in patients homozygous for −33/−31 delC mutant as compared with those with heterozygous or wild-type genotypes	Abraham *et al.* (2012)
Cytidine deaminase	*CDA*	rs75720390	AML patients with intron 2 (TCAT) 5/5 and (TCAT) 5/4 genotypes showed significantly higher CDA expression as compared to those with wild type (TCAT) 4/4 repeats	Abraham *et al.* (2012)

Cytidine deaminase	*CDA*	rs71864371 rs1048977	The 3′-UTR 816 in/delC variants in LD with synonymous coding SNP 435 C>T in exon 4 was found to be significantly associated with lower CDA RNA expression compared to the wild-type in AM patients	Abraham *et al.* (2012)
Cytidine deaminase	*CDA*	rs532545	AML patients with the CDA C-451 T/promoter SNP TT-genotype had significantly higher lactate dehydrogenase levels when compared with the CC or CT genotypes ($P = 0.01$)	Mahlknecht *et al.* (2009)
Cytidine deaminase	*CDA*	rs2072671 rs602950	AML patients with AA genotype for CDA A79 C SNP and AA genotype for CDA A-92 G SNP had lower incidences for grade III/IV liver toxicity when compared with the other respective genotypes	Mahlknecht *et al.* (2009)
Nucleoside diphosphate kinase 1	*NME1*	rs2302254	Caucasian AML patients with the T_T genotype for promoter SNP −835 C/T had a significantly lower platelet count and better ECOG performance status compared to the patients with C_C genotype. This study also identified an increased risk of neurotoxicity for T_T genotype	Braunagel *et al.* (2012)
Ribonucleotide reductase M1	*RRM1*	rs1042919 rs1561876	In AML patients receiving cytarabine and cladribine, these SNPs are associated with intracellular CTP level, response after remission induction therapy, risk of relapse, and overall survival	Cao *et al.* (2013)
Solute carrier family 29 (nucleoside transporters), member 1	*SLC29A1*	rs3734703	In AML patients, the AA and AC genotype for SLC29A1 SNP rs3734703 in combination with TYMS rs2612100 (AA genotype) was found to be significantly associated with shorter relapse-free survival	Kim *et al.* (2013)
ATP-binding cassette, subfamily C, member 3	*ABCC3*	rs4148405	The SNP rs4148405 was found to be associated with significantly shorter disease-free survival in AML patients, with the minor allele (G) being associated with shorter time to relapse	Yee *et al.* (2013)
Glutathione *S*-transferase	*GSTM1–GSTM5*	rs3754446	This SNP in the non-European ancestry AML patients was found to be associated with shorter disease-free survival	Yee *et al.* (2013)

disequilibrium, which were associated with higher mRNA expression of DCK. *In vitro* reporter assay also confirmed that the –360 G/–201 T genotype had higher transcriptional activity as compared to the –360 C/–201 C genotype. In addition, in AML patients, the –360 G/–210 T haplotype was associated with a favorable response as compared to the AML patients homozygous for the –360 C/–210 C haplotype (Shi *et al.*, 2004). In another study performed in healthy Caucasian volunteers, six novel SNPs in *DCK* (–243 G>T, –135 G>C, 261 G>A, 364 C>T (Pro121Ser), 727A>C (Lys242Gln), and intron 6 T>A) were reported (Joerger *et al.*, 2006). A total of 64 genetic polymorphisms, including 3 non-synonymous coding SNPs (Ile24Val, Ala119Gly, and Pro122Ser) were identified when the DCK proximal promoter along with the 7 coding exons of DCK were sequenced in the International HapMap Project panels with European (Centre d' Etude du Polymorphisme Humain [CEPH], Paris, France, panel, $n = 90$) and African (Yoruba people in Ibadan, Nigeria [YRI], $n = 90$) ancestry. The variant forms of these non-synonymous coding SNPs (24Val, 119Gly, and 122Ser) demonstrated reduced DCK activity (85%, 66%, and 43%, respectively) as compared to the wild-type (WT) isoform (Lamba *et al.*, 2007). A 3′UTR SNP rs4643786 (that occurs in LD with 13 intronic and one other 3′UTR SNP) was found to be associated with DCK mRNA expression in both European and African ancestry panels, as well as with lower leukemic blast cell intracellular levels of ara-CTP (Lamba *et al.*, 2007). A follow-up resequencing and functional study of *DCK* in 240 DNA samples from subjects with 4 different ethnicities (African American, Caucasian American, Han Chinese American, and Mexican American) identified 28 polymorphisms in *DCK* including the three non-synonymous changes indicated earlier (Kocabas *et al.*, 2008). In accordance with the previous study, functional analysis of DCK isoforms with these SNPs showed lower activities (Val24: 70%; Gly119: 42%; and Ser122: 66%, ;T 100%). Detailed substrate kinetics studies in different cell types and with multiple substrates as well as study in patients with AML is needed in order to determine the functional and clinical relevance of the coding SNPs in *DCK*.

19.4.3. Cytosolic 5′ nucleotidase

The 5′ nucleotidases (*NT5C2* and *NT5C3*) are a group of enzymes involved in dephosphorylation of common ribo- and deoxyribonucleoside phosphates, and thus maintain balanced nucleotide pools in cells (Bianchi *et al.*, 1986). Higher expression of *NT5C2* is associated with a lower disease-free and overall survival in adult patients with AML undergoing treatment with ara-C and with the development of resistance to clinically important nucleoside analogs, such as ara-C, cladribine, and gemcitabine in both *in vivo* and *in vitro* studies (Galmarini *et al.*, 2002b, 2005; Hunsucker *et al.*, 2005). Intracellular ara-CTP production as well as ara-C sensitivity has been shown to be significantly associated with the *DCK: NT5C2* ratio in the HL-60 leukemic cell line and its ara-C-resistant variants (Yamauchi *et al.*, 2009).

We recently reported 41 genetic variants including 3 non-synonymous SNPs (Y3A, K47 R, and Q136 R) in *NT5C2* (Mitra *et al.*, 2011). We observed very strong LD patterns in both Caucasian and African cohorts for *NT5C2* SNPs. The most interesting SNPs in *NT5C2* that were associated with its mRNA expression and *in vitro* ara-C sensitivity in both control lymphoblast cell lines from HapMap samples as well as from primary leukemic blasts from patients with AML included: rs4917996, rs11191558, rs11598702, rs4917384, and rs1163075. All these SNPs occur in LD with multiple other SNPs. *NT5C2* SNP rs4917384 was associated with minimal residual disease at day 22 (MRD22; Mitra *et al.*, 2011).

19.4.4. Cytidine deaminase

Cytidine deaminase (CDA) is the major player in inactivation of cytarabine and other nucleoside analogs. CDA converts ara-C to ara-U by deamination (Figure 19.1). Higher levels of CDA have been linked with development of ara-C resistance as well as disease recurrence, whereas low CDA levels have been linked to longer remission period (Jahns-Streubel *et al.*, 1997; Grant, 1998; Schroder *et al.*, 1998a, 1998b; Maring *et al.*, 2005).

Most commonly studied SNPs in CDA, namely A79 C/Lys27Gln-rs2072671, G208A/Ala70Thr-rs60369023, T435 C), were originally discovered in the Japanese population (Yue *et al.*, 2003), and later in Caucasians and African Americans (Gilbert *et al.*, 2006). Although altered deamination ability was observed for Lys27Gln in *in vitro* studies (Kirch *et al.*, 1998), recent biochemical characterization of Lys27Gln and Ala70Thr SNPs reported higher catalytic efficiency towards natural substrate as well as ara-C for the 27Gln as compared to 27 Lys isoform; whereas lower catalytic efficiency was observed for the Thr70 variant as compared to Ala70 (Micozzi *et al.*, 2014). A higher risk of treatment-related mortality has been seen in pediatric AML patients' homozygous for CDA Gln27 polymorphism (which is considered the low-activity allele; Bhatla *et al.*, 2009).

With respect to regulatory SNPs in CDA, three SNPs (–92 A>G-rs602950, –451 C>T-rs532545, and –897 C>A) have been associated with altered promoter and protein function (Gilbert *et al.*, 2006). In patients with AML, the –451 C>T SNP has been identified as an independent prognostic parameter for survival (Mahlknecht *et al.*, 2009).

Simultaneous evaluation of multiple genes in the ara-C metabolic pathway is currently underway. In *de novo* patients with AML with normal karyotype, CDA 79A>C and –451 C>T SNPs were associated with shorter overall survival in FLT3-ITD positive patients (Falk *et al.*, 2013). CDA SNP rs6036023 G>A SNP along with DCK SNP rs72552079 has been shown to be predictive of a better response in Chinese patients with AML, whereas DCK SNP rs11554389 showed poor response in patients treated with ara-C based chemotherapy (Xu *et al.*, 2012).

19.4.5. Other genes of relevance to cytarabine pharmacogenomics

Other less-studied enzymes/genes influencing ara-C metabolism include *DCTD, CMPK, CTPS* and ribonucleotide reductases. *DCTD* catalyzed deamination of ara-CMP to form ara-UMP. Although *DCTD* levels have been associated with ara-C response/chemosensitivity, the data supporting its role are not very well defined (Liliemark and Plunkett, 1986; Fridland and Verhoef, 1987; Capizzi *et al.*, 1991). Asn58Asp (rs35932500) is the only significant SNP identified in *DCTD* that results in loss of enzyme activity in *in vitro* assays (Gilbert *et al.*, 2006).

Cytidine-5′-triphosphate synthetase (*CTPS*) is critical for the maintenance of cellular CTP/dCTP pools. Chemical modulation of *CTPS* levels has been shown to influence ara-C sensitivity in multiple cell lines (Verschuur *et al.*, 2000, 2001, 2002; Bierau *et al.*, 2003, 2006). SNPs within *CTPS* have not been evaluated for functional/clinical relevance in AML or any other disease.

Ribonucleotide reductases (RR) regulate the production of deoxyribonucleotides and hence are of potential interest in cancer chemotherapy. *RRM1* and *RRM2* subunits are part of the ribonucleotide reductases holoenzyme. RR regulates intracellular pools of dCTP, which have an impact on ara-C resistance (Chabner *et al.*, 1979; Liliemark and Plunkett, 1986; Chiba *et al.*, 1990; Shao *et al.*, 2006) by feedback inhibition of *DCK*; allosteric activation of the inactivating enzyme *CDA* (Chiba *et al.*, 1990), and competing with the active metabolite ara-CTP for incorporation into DNA (Meuth, 1989). Our recent study on resequencing identified 44 genetic variants (8 coding SNPs) in *RRM1* and 15 SNPs including one coding SNP in *RRM2*. Higher *RRM1* (but not *RRM2*) expression levels were seen in patients with AML with high-risk features such as -7, FLT3-ITD, (t6:9), megakryoblastic leukemia, treatment-related AML, or AML arising from myelodysplastic syndrome (Cao *et al.*, 2013). *RRM1* promoter SNPs rs11030918, rs12806698, and rs1042927 and *RRM2* SNP rs1138729 were associated with the respective gene's expression levels in patients with AML. We also observed the association of *RRM1* SNP rs2898950 and *RRM2B* SNP rs1265138 with inferior response after induction therapy. *RRM2* SNPs rs1130609 and rs5030743 demonstrated association with poor event-free survival (Cao *et al.*, 2013). However, this study was exploratory in nature with limited sample size, and follow-up confirmatory studies in a bigger cohort are warranted to validate these results.

Although the above-indicated genes are involved in cytarabine pharmacokinetics, few GWAS studies have been performed to identify potential candidate genes of pharmacodynamic relevance. Discussion on those is beyond the scope of this chapter, as most of those studies have been performed using lymphoblast cell lines (LCLs) from normal subjects as models for pharmacogenomics discovery. Basically, LCLs have been treated with varying concentrations of cytarabine followed by genome-wide association analysis to identify the genetic markers (*FKBP5, GCAT, MYBBP1A, TLE4, ZNF278, GIT1, RAD51*, and SLC25A37) predictive of cytarabine *in vitro* sensitivity (Hartford *et al.*, 2009; Li *et al.*, 2009). Most of these results have not yet been validated in the AML patient population except for a recent report in which SNPs (rs9883101, rs6550826, and rs1533140) flanking the *NR1D2* gene (nuclear receptor subfamily 1, group D member 2) were associated with treatment-related mortality (Gamazon *et al.*, 2013).

19.5. Pharmacogenomics of other anti-leukemic agents utilized in AML chemotherapy

Anthracyclins such as doxorubicin and daunorubicin are used in combination with cytarabine. These agents undergo detoxification by glutathione-*S*-transfereases (GSTs). GSTs are multi-gene isoenzymes that are involved in the cellular detoxification of both xenobiotic and endobiotic compounds. The subclasses of GSTs include alpha, mu, pi, and theta. Deletion of *GSTM1* and *GSTT1* has been previously shown to influence outcome in patients with AML (Naoe *et al.*, 2002; Voso *et al.*, 2002). The null genotype of *GSTT1* has been shown to be associated with worse prognosis and reduced survival in pediatric patients with AML (Davies *et al.*, 2001). A recent meta-analysis of 1837 patients enrolled in 11 studies revealed that the *GSTT1* null genotype and the *GSTT1*/*GSTM1* double-null genotype was predictive of worse outcome in patients with AML (Xiao *et al.*, 2014).

19.6. Overall conclusions

The field of pharmacogenomics in the area of AML/pediatric AML is still in its infancy. AML is a very heterogeneous disease with multiple subgroups that impact therapeutic outcome. Pharmacogenomics has the potential for improving clinical outcomes in AML; although there are limited studies, results so far indicate potential for SNPs in activating and inactivating enzymes as well as drug transporters to be of clinical relevance. The integration of pharmacogenetic markers with prognostic markers in larger clinical cohorts would advance our ability to design personalized therapy in patients to achieve the greatest therapeutic benefit.

REFERENCES

Abdel-Wahab, O., Patel, J. and Levine, R.L. (2011). Clinical implications of novel mutations in epigenetic modifiers in AML. *Hematol. Oncol. Clin. North Am.*, **25**, 1119–1133.

Abraham, A., Varatharajan, S., Abbas, S., *et al.* (2012). Cytidine deaminase genetic variants influence RNA expression and cytarabine cytotoxicity in acute myeloid leukemia. *Pharmacogenomics*, **13**, 269–282.

Bacher, U., Schnittger, S. and Haferlach, T. (2010). Molecular genetics in acute myeloid leukemia. *Curr. Opin. Oncol.*, **22**, 646–655.

Bennett, J.M., Catovsky, D., Daniel, M.T., *et al.* (1976). Proposals for the classification of the acute leukaemias. French–American–British (FAB) co-operative group. *Br. J. Haematol.*, **33**, 451–458.

Bennett, J.M., Catovsky, D., Daniel, M.T., *et al.* (1980). A variant form of hypergranular promyelocytic leukaemia (M3). *Br. J. Haematol.*, **44**, 169–170.

Bennett, J.M., Catovsky, D., Daniel, M.T., *et al.* (1985a). Criteria for the diagnosis of acute leukemia of megakaryocyte lineage (M7). A report of the French–American–British Cooperative Group. *Ann. Int. Med.*, **103**, 460–462.

Bennett, J.M., Catovsky, D., Daniel, M.T., *et al.* (1985b). Proposed revised criteria for the classification of acute myeloid leukemia. A report of the French–American–British Cooperative Group. *Ann. Int. Med.*, **103**, 620–625.

Bennett, J.M., Catovsky, D., Daniel, M.T., *et al.* (1991). Proposal for the recognition of minimally differentiated acute myeloid leukaemia (AML-MO). *Br. J. Haematol.*, **78**, 325–329.

Bhatla, D., Gerbing, R.B., Alonzo, T.A., *et al.* (2009). Cytidine deaminase genotype and toxicity of cytosine arabinoside therapy in children with acute myeloid leukemia. *Br. J. Haematol.*, **144**, 388–394.

Bianchi, V., Pontis, E. and Reichard, P. (1986). Interrelations between substrate cycles and *de novo* synthesis of pyrimidine deoxyribonucleoside triphosphates in 3T6 cells. *Proc. Natl Acad. Sci. USA*, **83**, 986–990.

Bierau, J., Van Gennip, A.H., Leen, R., *et al.* (2003). Cyclopentenyl cytosine primes SK-N-BE(2)c neuroblastoma cells for cytarabine toxicity. *Int. J. Cancer*, **103**, 387–392.

Bierau, J., van Gennip, A.H., Leen, R., *et al.* (2006). Cyclopentenyl cytosine-induced activation of deoxycytidine kinase increases gemcitabine anabolism and cytotoxicity in neuroblastoma. *Cancer Chemother. Pharmacol.*, **57**, 105–113.

Braunagel, D., Schaich, M., Kramer, M., *et al.* (2012). The T_T genotype within the *NME1* promoter single nucleotide polymorphism -835 C/T is associated with an increased risk of cytarabine induced neurotoxicity in patients with acute myeloid leukemia. *Leuk. Lymph.*, **53**, 952–957.

Cao, X., Mitra, A.K., Pounds, S., *et al.* (2013). *RRM1* and *RRM2* pharmacogenetics: association with phenotypes in HapMap cell lines and acute myeloid leukemia patients. *Pharmacogenomics*, **14**, 1449–1466.

Capizzi, R.L., White, J.C., Powell, B.L. and Perrino, F. (1991). Effect of dose on the pharmacokinetic and pharmacodynamic effects of cytarabine. *Semin. Hematol.*, **28**, 54–69.

Chabner, B.A., Hande, K.R. and Drake, J.C. (1979). Ara-C metabolism: implications for drug resistance and drug interactions. *Bull. Cancer*, **66**, 89–92.

Cheson, B.D., Bennett, J.M., Kopecky, K.J., *et al.* (2003). Revised recommendations of the International Working Group for Diagnosis, Standardization of Response Criteria, Treatment Outcomes, and Reporting Standards for Therapeutic Trials in Acute Myeloid Leukemia. *J. Clin. Oncol.*, **21**, 4642–4649.

Chiba, P., Tihan, T., Szekeres, T., *et al.* (1990). Concordant changes of pyrimidine metabolism in blasts of two cases of acute myeloid leukemia after repeated treatment with ara-C in vivo. *Leukemia*, **4**, 761–765.

Davies, S.M., Robison, L.L., Buckley, J.D., *et al.* (2001). Glutathione *S*-transferase polymorphisms and outcome of chemotherapy in childhood acute myeloid leukemia. *J. Clin. Oncol.*, **19**, 1279–1287.

Dumontet, C., Fabianowska-Majewska, K., Mantincic, D., *et al.* (1999). Common resistance mechanisms to deoxynucleoside analogues in variants of the human erythroleukaemic line K562. *Br. J. Haematol.*, **106**, 78–85.

Falini, B. and Martelli, M.P. (2011). *NPM1*-mutated AML: targeting by disassembling. *Blood*, **118**, 2936–2938.

Falk, I.J., Fyrberg, A., Paul, E., *et al.* (2013). Decreased survival in normal karyotype AML with single-nucleotide polymorphisms in genes encoding the AraC metabolizing enzymes cytidine deaminase and 5′-nucleotidase. *Am. J. Hematol.*, **88**, 1001–1006.

Fridland, A. and Verhoef, V. (1987). Mechanism for ara-CTP catabolism in human leukemic cells and effect of deaminase inhibitors on this process. *Semin. Oncol.*, **14**, 262–268.

Galmarini, C.M., Cros, E., Thomas, X., Jordheim, L. and Dumontet, C. (2005). The prognostic value of cN-II and cN-III enzymes in adult acute myeloid leukemia. *Haematologica*, **90**, 1699–1701.

Galmarini, C.M., Thomas, X., Calvo, F., *et al.* (2002a). Potential mechanisms of resistance to cytarabine in AML patients. *Leuk. Res.*, **26**, 621–629.

Galmarini, C.M., Thomas, X., Calvo, F., *et al.* (2002b). *In vivo* mechanisms of resistance to cytarabine in acute myeloid leukaemia. *Br. J. Haematol.*, **117**, 860–868.

Gamazon, E.R., Lamba, J.K., Pounds, S., *et al.* (2013). Comprehensive genetic analysis of cytarabine sensitivity in a cell-based model identifies polymorphisms associated with outcome in AML patients. *Blood*, **121**, 4366–4376.

Gati, W.P., Paterson, A.R., Larratt, L.M., Turner, A.R. and Belch, A.R. (1997). Sensitivity of acute leukemia cells to cytarabine is a correlate of cellular es nucleoside transporter site content measured by flow cytometry with SAENTA-fluorescein. *Blood*, **90**, 346–353.

Gati, W.P., Paterson, A.R., Belch, A.R., *et al.* (1998). Es nucleoside transporter content of acute leukemia cells: role in cell sensitivity to cytarabine (araC). *Leuk. Lymph.*, **32**, 45–54.

Gilbert, J.A., Salavaggione, O.E., Ji, Y., *et al.* (2006). Gemcitabine pharmacogenomics: cytidine deaminase and deoxycytidylate deaminase gene resequencing and functional genomics. *Clin. Cancer Res.*, **12**, 1794–1803.

Grant, S. (1998). Ara-C: cellular and molecular pharmacology. *Adv. Cancer Res.*, **72**, 197–233.

Gu, T.L., Nardone, J., Wang, Y., *et al.* (2011). Survey of activated FLT3 signaling in leukemia. *PloS ONE*, **6**, e19169.

Hapke, D.M., Stegmann, A.P. and Mitchell, B.S. (1996). Retroviral transfer of deoxycytidine kinase into tumor cell lines enhances nucleoside toxicity. *Cancer Res.*, **56**, 2343–2347.

Hartford, C.M., Duan, S., Delaney, S.M., *et al.* (2009). Population-specific genetic variants important in susceptibility to cytarabine arabinoside cytotoxicity. *Blood*, **113**, 2145–2153.

Hou, H.A., Lin, C.C., Chou, W.C., *et al.* (2014). Integration of cytogenetic and molecular alterations in risk stratification of 318 patients with *de novo* non-M3 acute myeloid leukemia. *Leukemia*, **28**, 50–58.

Hunsucker, S.A., Mitchell, B.S. and Spychala, J. (2005). The 5′-nucleotidases as regulators of nucleotide and drug metabolism. *Pharmacol. Therapeut.*, **107**, 1–30.

Jahns-Streubel, G., Reuter, C., Auf der Landwehr, U., *et al.* (1997). Activity of thymidine kinase and of polymerase alpha as well as activity and gene expression of deoxycytidine deaminase in leukemic blasts are correlated with clinical response in the setting of granulocyte-macrophage colony-stimulating factor-based priming before and during TAD-9 induction therapy in acute myeloid leukemia. *Blood*, **90**, 1968–1976.

Joerger, M., Bosch, T.M., Doodeman, V.D., *et al.* (2006). Novel deoxycytidine kinase gene polymorphisms: a population screening study in Caucasian healthy volunteers. *Eur. J. Clin. Pharmacol.*, **62**, 681–684.

Jordan, C.T. (2007). The leukemic stem cell. *Best Pract. Res. Clin. Haematol.*, **20**, 13–18.

Kakihara, T., Fukuda, T., Tanaka, A., *et al.* (1998). Expression of deoxycytidine kinase (dCK) gene in leukemic cells in childhood: decreased expression of dCK gene in relapsed leukemia. *Leuk. Lymph.*, **31**, 405–409.

Kaleem, Z. and White, G. (2001). Diagnostic criteria for minimally differentiated acute myeloid leukemia (AML-M0). Evaluation and a proposal. *Am. J. Clin. Pathol.*, **115**, 876–884.

Kim, K.I., Huh, I.S., Kim, I.W., *et al.* (2013). Combined interaction of multi-locus genetic polymorphisms in cytarabine arabinoside metabolic pathway on clinical outcomes in adult acute myeloid leukaemia (AML) patients. *Eur. J. Cancer*, **49**, 403–410.

Kirch, H.C., Schroder, J., Hoppe, H., *et al.* (1998). Recombinant gene products of two natural variants of the human cytidine deaminase gene confer different deamination rates of cytarabine *in vitro*. *Exp. Hematol.*, **26**, 421–425.

Kocabas, N.A., Aksoy, P., Pelleymounter, L.L., *et al.* (2008). Gemcitabine pharmacogenomics: deoxycytidine kinase and cytidylate kinase gene resequencing and functional genomics. *Drug Metab. Dispos.*, **36**, 1951–1959.

Kufe, D.W., Major, P.P., Egan, E.M. and Beardsley, G.P. (1980). Correlation of cytotoxicity with incorporation of ara-C into DNA. *J. Biol. Chem.*, **255**, 8997–9000.

Lamba, J.K., Crews, K., Pounds, S., *t al.* (2007). Pharmacogenetics of deoxycytidine kinase: identification and characterization of novel genetic variants. *J. Pharmacol. Exp. Therapeut.*, **323**, 935–945.

Lane, S.W., Scadden, D.T. and Gilliland, D.G. (2009). The leukemic stem cell niche: current concepts and therapeutic opportunities. *Blood*, **114**, 1150–1157.

Li, L., Fridley, B.L., Kalari, K., *et al.* (2009). Gemcitabine and arabinosylcytosin pharmacogenomics: genome-wide association and drug response biomarkers. *PloS ONE*, **4**, e7765.

Liliemark, J.O. and Plunkett, W. (1986). Regulation of 1-beta-D-arabinofuranosylcytosine 5′-triphosphate accumulation in human leukemia cells by deoxycytidine 5′-triphosphate. *Cancer Res.*, **46**, 1079–1083.

Lotfi, K., Karlsson, K., Fyrberg, A., *et al.* (2006). The pattern of deoxycytidine- and deoxyguanosine kinase activity in relation to messenger RNA expression in blood cells from untreated patients with B-cell chronic lymphocytic leukemia. *Biochem. Pharmacol.*, **71**, 882–890.

Mahfouz, R.Z., Jankowska, A., Ebrahem, Q., *et al.* (2013). Increased CDA expression/activity in males contributes to decreased cytidine analog half-life and likely contributes to worse outcomes with 5-azacytidine or decitabine therapy. *Clin. Cancer Res.*, **19**, 938–948.

Mahlknecht, U., Dransfeld, C.L., Bulut, N., *et al.* (2009). SNP analyses in cytarabine metabolizing enzymes in AML patients and their impact on treatment response and patient survival: identification of CDA SNP C-451 T as an independent prognostic parameter for survival. *Leukemia*, **23**, 1929–1932.

Major, P.P., Egan, E.M., Beardsley, G.P., Minden, M.D. and Kufe, D.W. (1981). Lethality of human myeloblasts correlates with the incorporation of arabinofuranosylcytosine into DNA. *Proc. Natl Acad. Sci. USA*, **78**, 3235–3239.

Maring, J.G., Groen, H.J., Wachters, F.M., Uges, D.R. and de Vries, E.G. (2005). Genetic factors influencing pyrimidine-antagonist chemotherapy. *Pharmacogenom. J.*, **5**, 226–243.

Meshinchi, S. and Arceci, R.J. (2007). Prognostic factors and risk-based therapy in pediatric acute myeloid leukemia. *Oncologist*, **12**, 341–355.

Meuth, M (1989). The molecular basis of mutations induced by deoxyribonucleoside triphosphate pool imbalances in mammalian cells. *Exp. Cell Res.*, **181**, 305–316.

Micozzi, D., Carpi, F.M., Pucciarelli, S., *et al.* (2014). Human cytidine deaminase: a biochemical characterization of its naturally occurring variants. *Int. J. Biol. Macromolec.*, **63**, 64–74.

Mitra, A.K., Crews, K.R., Pounds, S., *et al.* (2011). Genetic variants in cytosolic 5′-nucleotidase II are associated with its expression and cytarabine sensitivity in HapMap cell lines and in patients with acute myeloid leukemia. *J. Pharmacol. Exp. Therapeut.*, **339**, 9–23.

Myers, S.N., Goyal, R.K., Roy, J.D., *et al.* (2006). Functional single nucleotide polymorphism haplotypes in the human equilibrative nucleoside transporter 1. *Pharmacogenet. Genom.*, **16**, 315–320.

Naoe, T., Tagawa, Y., Kiyoi, H., *et al.* (2002). Prognostic significance of the null genotype of glutathione *S*-transferase-T1 in patients with acute myeloid leukemia: increased early death after chemotherapy. *Leukemia*, **16**, 203–208.

Osato, D.H., Huang, C.C., Kawamoto, M., *et al.* (2003). Functional characterization in yeast of genetic variants in the human equilibrative nucleoside transporter, *ENT1*. *Pharmacogenetics*, **13**, 297–301.

Pui, C.H., Carroll, W.L., Meshinchi, S. and Arceci, R.J. (2011). Biology, risk stratification, and therapy of pediatric acute leukemias: an update. *J. Clin. Oncol.*, **29**, 551–565.

Rau, R. and Brown, P. (2009). Nucleophosmin (*NPM1*) mutations in adult and childhood acute myeloid leukaemia: towards definition of a new leukaemia entity. *Hematol. Oncol.*, **27**, 171–181.

Raza, A., Gezer, S., Anderson, J., *et al.* (1992). Relationship of [3 H]Ara-C incorporation and response to therapy with high-dose Ara-C in AML patients: a Leukemia Intergroup study. *Exp. Hematol.*, **20**, 1194–1200.

Rockova, V., Abbas, S., Wouters, B.J., *et al.* (2011). Risk stratification of intermediate-risk acute myeloid leukemia: integrative analysis of a multitude of gene mutation and gene expression markers. *Blood*, **118**, 1069–1076.

Rubnitz, J.E. and Inaba, H. (2012). Childhood acute myeloid leukaemia. *Br. J. Haematol.*, **159**, 259–276.

Sallmyr, A., Fan, J., Datta, K., *et al.* (2008). Internal tandem duplication of *FLT3* (FLT3/ITD) induces increased ROS production, DNA damage, and misrepair: implications for poor prognosis in AML. *Blood*, **111**, 3173–3182.

Schroder, J.K., Kirch, C., Seeber, S. and Schutte, J. (1998a). Structural and functional analysis of the cytidine deaminase gene in patients with acute myeloid leukaemia. *Br. J. Haematol.*, **103**, 1096–1103.

Schroder, J.K., Seidelmann, M., Kirch, H.C., Seeber, S. and Schutte, J. (1998b). Assessment of resistance induction to cytosine arabinoside following transfer and overexpression of the deoxycytidylate deaminase gene *in vitro*. *Leukemia Res.*, **22**, 619–624.

Shao, J., Zhou, B., Chu, B. and Yen, Y. (2006). Ribonucleotide reductase inhibitors and future drug design. *Curr. Cancer Drug Targets*, **6**, 409–431.

Shi, J.Y., Shi, Z.Z., Zhang, S.J., *et al.* (2004). Association between single nucleotide polymorphisms in deoxycytidine kinase and treatment response among acute myeloid leukaemia patients. *Pharmacogenetics*, **14**, 759–768.

Tasian, S.K., Pollard, J.A. and Aplenc, R. (2014a). Molecular therapeutic approaches for pediatric acute myeloid leukemia. *Front. Oncol.*, **4**, 55.

Tasian, S.K., Teachey, D.T. and Rheingold, S.R. (2014b). Targeting the PI3 K/mTOR pathway in pediatric hematologic malignancies. *Front. Oncol.*, **4**, 108.

Vardiman, J.W., Thiele, J., Arber, D.A., *et al.* (2009). The 2008 revision of the World Health Organization (WHO) classification of myeloid neoplasms and acute leukemia: rationale and important changes. *Blood*, **114**, 937–951.

Verhoef, V., Sarup, J. and Fridland, A. (1981). Identification of the mechanism of activation of 9-beta-D-arabinofuranosyladenine in human lymphoid cells using mutants deficient in nucleoside kinases. *Cancer Res.*, **41**, 4478–4483.

Verschuur, A.C., van Gennip, A.H., Leen, R., Voute, P.A. and van Kuilenburg, A.B. (2000). Cyclopentenyl cytosine increases the phosphorylation and incorporation into dna of arabinofuranosyl cytosine in a myeloid leukemic cell-line. *Adv. Exp. Med. Biol.*, **486**, 311–317.

Verschuur, A.C., Brinkman, J., Van Gennip, A.H., *et al.* (2001). Cyclopentenyl cytosine induces apoptosis and increases cytarabine-induced apoptosis in a T-lymphoblastic leukemic cell-line. *Leukemia Res.*, **25**, 891–900.

Verschuur, A.C., Van Gennip, A.H., Leen, R., *et al.* (2002). Cyclopentenyl cytosine increases the phosphorylation and incorporation into DNA of 1-beta-D-arabinofuranosyl cytosine in a human T-lymphoblastic cell line. *Int. J. Cancer*, **98**, 616–623.

Voso, M.T., D'Alo, F., Putzulu, R., *et al.* (2002). Negative prognostic value of glutathione *S*-transferase (*GSTM1* and *GSTT1*) deletions in adult acute myeloid leukemia. *Blood*, **100**, 2703–2707.

Xiao, Q., Deng, D., Li, H., *et al.* (2014). *GSTT1* and *GSTM1* polymorphisms predict treatment outcome for acute myeloid leukemia: a systematic review and meta-analysis. *Ann. Hematol.*, **93**, 1381–1390.

Xu, P.P., Chen, B.A., Feng, J.F., *et al.* (2012). Association of polymorphisms of cytosine arabinoside-metabolizing enzyme gene with therapeutic efficacy for acute myeloid leukemia. *Chin. Med. J.*, **125**, 2137–2143.

Yamauchi, T., Negoro, E., Kishi, S., *et al.* (2009). Intracellular cytarabine triphosphate production correlates to deoxycytidine kinase/cytosolic 5′-nucleotidase II expression ratio in primary acute myeloid leukemia cells. *Biochem. Pharmacol.*, **77**, 1780–1786.

Yates, J., Glidewell, O., Wiernik, P., *et al.* (1982). Cytosine arabinoside with daunorubicin or adriamycin for therapy of acute myelocytic leukemia: a CALGB study. *Blood*, **60**, 454–462.

Yee, S.W., Mefford, J.A., Singh, N., *et al.* (2013). Impact of polymorphisms in drug pathway genes on disease-free survival in adults with acute myeloid leukemia. *J. Hum. Genet.*, **58**, 353–361.

Yue, L., Saikawa, Y., Ota, K., *et al.* (2003). A functional single-nucleotide polymorphism in the human cytidine deaminase gene contributing to ara-C sensitivity. *Pharmacogenetics*, **13**, 29–38.

20 Pharmaco-genomics of antiretroviral drugs

Chonlaphat Sukasem, Apichaya Puangpetch and Sadeep Medhasi

20.1. Introduction

Currently several million people take chronic combination antiretroviral (ARV) therapy to suppress human immunodeficiency virus (HIV) infection. There are around 34 million people living with the HIV infection, most of whom reside in developing or under-developed nations. The combined antiretroviral therapy (cART) has prolonged life and prevented the progression of the disease. The cART has lessened the burden of the disease worldwide (Verma *et al.*, 2014). Because of the long-term and multi-drug therapy, several complexities may arise which might affect adherence and tolerability in patients. Serious efforts must be made in order to make the pharmacotherapy of HIV infection efficacious with minimal adverse events. The use of ARV drugs has been associated with different types of toxicities including hypersensitivity reactions, metabolic disturbances, peripheral neuropathy, hyperbilirubinemia, neuropsychiatry disorders, and myelosupression (Subbaraman *et al.*, 2007). Among several variables, such as age, sex, disease state, and drug–drug interactions influencing inter-patient variability in the spectrum of effectiveness and toxic effects of ARV drugs, host genetic factors may have a significant role in predisposing the patient to the variation in treatment outcome (Vidal *et al.*, 2010). Polymorphisms in genes encoding drug-metabolizing enzymes, drug transporters, and other targets make the patient susceptible to variation in drug response. It is imperative to predict the genetic markers to optimize the therapy and select the best combination of drugs to avoid toxicity. The application of pharmacogenomics in clinical practice might help us to develop the notion of personalized medicine. Pharmacogenetic tests can provide the data on genetically determined inter-individual differences in pharmacokinetics and genotype–phenotype associations (Shah and Shah, 2012).

This chapter reviews the pharmaco-genetic aspects of variation in ARV drug response in HIV-infected patients. The most overlapping examples of the

Genome-Wide Association Studies: From Polymorphism to Personalized Medicine, ed. K. Appasani. Published by Cambridge University Press.

heterogeneity in the efficacy and toxic effects of ARV drugs as found between populations will be discussed here.

20.2. Antiretroviral drugs

ARV drugs are commonly used to prevent transmission or acquisition of HIV-1 infection. Current practice uses a combination of at least three drugs referred to as "highly active antiretroviral therapy" (HAART) in the standard-of-care therapy for HIV infection. Zidovudine was the first drug to be approved by Food and Drug Administration (FDA) to treat HIV infection in 1987. There has been remarkable improvement in the prognosis of HIV infection since 1995 with the use of HAART (Pirmohamed and Back, 2001).

Based on their mechanism of action, there are six groups of ARV drugs classified into the following categories: (1) nucleoside (or nucleotide) reverse transcriptase inhibitors (NRTIs), (2) non-nucleoside reverse transcriptase inhibitors (NNRTIs), (3) protease inhibitors (PIs), (4) fusion inhibitors, (5) CCR5 antagonist, and (6) integrase inhibitors (InIs) (Table 20.1; Chaponda and Pirmohamed, 2011). If left untreated, HIV infection results in severe CD4+ T-cell count depletion. Lymphocytes and macrophages are the primary targets of HIV, and CD4+ T-cells act as a receptor and means of infection. ARV drugs target the different phases of the HIV replication cycle. NRTIs inhibit the activity of reverse transcriptase enzymes and cause the termination of transcription, thereby blocking viral replication (Michaud *et al.*, 2012). NNRTIs act as non-competitive inhibitors of the reverse transcriptase of HIV-1. NNRTIs do not integrate into the growing DNA chain, unlike the NRTIs (Gotte and Wainberg, 2000). PIs inhibit the activity of HIV-1 protease, which is required for the replication and maturation of new HIV-1 particles (Tozzi, 2010). Enfuvirtide is the only currently available fusion inhibitor for the treatment of HIV infection. Fusion inhibitors block the fusion of the virus with the outer cell membrane of the host CD4+ T cell. Chemokine co-receptor 5 (CCR5) antagonist binds to the co-receptor CCR5 and inhibits the binding of the HIV glycoprotein gp 120 to CD4 cell membrane (Clercq, 2009). Fusion inhibitors and CCR5 antagonists are also collectively

Table 20.1. Drugs used for the treatment of HIV-1 infection.

NRTIs	NNRTIs	PIs	Fusion inhibitors	CCR5 antagonist	Integrase inhibitors
Abacavir	Delavirdine	Atazanavir	Enfuvirtide	Maraviroc	Raltegravir
Didanosine	Efavirenz	Darunavir			
Emtricitabine	Etravirine	Fosamprinavir			
Lamivudine	Nevirapine	Indinavir			
Stavudine	Rilpivirine	Nelfinavir			
Tenofovir		Ritonavir			
Zidovudine		Saquinavir			
		Tipranavir			

called entry inhibitors, as they block virus attachment to receptors of the cell surface (Bazzoli et al., 2010). Integrase inhibitor, raltegravir, inhibits HIV-1 integrase and blocks the incorporation of the proviral HIV-1 DNA into the host cell DNA (Tozzi, 2010).

20.3. HIV mutations and drug resistance

ARV drug resistance is one of the problems evolving as an inevitable consequence in the treatment among HIV-1 patients. The emergence of ARV drug resistance is the result of the capacity of the HIV-1 strain to replicate rapidly, incomplete viral suppression, and the error-prone nature of HIV-1 reverse transcriptase (HIV-1 RT) in DNA synthesis (Adamson and Freed, 2008). ARV drugs currently in widespread use include NRTIs, NNRTIs, and PIs. Resistance can be evaluated through genotyping and phenotypic testing, with genotyping analysis more sensitive in detecting resistant virus, faster, and less expensive than phenotypic testing (Gallant, 2005). Genotyping assays determine the mutations of the nucleotide sequence of regions that confer decreased drug susceptibility, whereas phenotypic assays determine the drug susceptibility of HIV-1 isolates by estimating the inhibitory concentration of an ARV agent that reduces viral replication by 50% (IC_{50}) in tissue culture (Hanna and D'Aquila, 2001). A detailed understanding and determination of ARV drug resistance is essential for the early detection of treatment failure and to assist in the therapeutic decision-making process.

Resistance to NRTIs occurs due to interference with the incorporation of the NRTIs or its removal from the elongating viral DNA chain. M184V mutation of the RT polymerase active site is a classic example with high-level resistance to lamivudine and emtricitabine, and low resistance to didanosine and abacavir. Other mutations that affect susceptibility to NRTIs include Q151M complex, K65 R, and L74 V. The Q151M complex mutation is associated with variable decrease in susceptibility to zidovudine, didanosine, stavudine, and abacavir. Steady-state kinetic studies have shown the resistance to didanosine and abacavir with *L74* V mutation. K65 R mutation is a common mutation of tenofovir; it can also confer resistance to didanosine and abacavir. Resistance among the NNRTIs users is due to the alterations in the size, shape, and polarity of NNRTI binding pocket (NNIBP), or affecting access to the pocket, and is associated with a high level of cross-resistance. Common NNRTI-associated resistance mutations include *K103* N, L100I, V106A, Y181 C, Y188 C/L, and G190A (Singh *et al.*, 2010). A retrospective study among Thai HIV-infected patients found mutations including M184 V/I and K65 R associated with virologic failure. High prevalence of etravirine-resistance-associated mutations (ETV-RAMs) was found which included Y181 C, G190A, and K101E. Nevirapine was associated with four times the risk of intermediate or reduced response to etravirine as compared to using efavirenz (Kiertiburanakul *et al.*, 2010). The study of resistance to HIV drugs among Thai HIV-infected patients after the National Access to Antiretroviral Programs for People who have AIDS (NAPHA) during 2002–2005 found the increased prevalence of resistance mutations in reverse transcriptase genes like M184 V/I and K103S/H (Sukasem *et al.*, 2007). Resistance to PIs involves the mutations in the protease coding region. Drug resistance is brought about by structural changes in the substrate cleft resulting in a decreased drug-binding affinity to the mutant target molecule (Shafer *et al.*, 2000). Major mutations include G48 V and *L90* M for saquinavir, *I84* V, I50 L, and N88S for atazanavir, V82A/F/T/S and I84 V for ritonavir, D30 N and L90 M for nelfinavir, and V32I and I47 V/A for lopinavir (Saskova *et al.*, 2009). The minor mutations in the protease (PR) sequences did not reduce PI susceptibility after initiating PI-based HAART among Thai HIV-infected patients (Sukasem *et al.*, 2008). An *in silico* experiment showed D29 V mutation increasing the probability of resistance to PIs by forming unstable PR complexes at the HIV-1 PR active site (Mata-Munguia *et al.*, 2014).

20.4. Pharmacokinetic variability

ARV drugs after oral administration undergo pharmacokinetic processes of absorption, distribution, metabolism, and elimination (ADME). The processes regulating drug ADME are major factors contributing to the plasma and cellular concentration of drugs. It is apparent that ADME genes have functional variants with different phenotypic consequences. Marked inter-patient variability in plasma level from the same regimen of ARV drug is one of the critical factors contributing to the difference in ART in HIV-infected patients. There are ethnic and genetic differences in activity of the transporter proteins that have complex effects on drug disposition and response (Cropp *et al.*, 2008). Below, we focus on the significant associations of genetic polymorphisms in genes encoding metabolizing enzymes and transporters.

20.5. Metabolizing enzymes

NRTIs have to be phosphorylated to active triphosphate anabolite, so the plasma level of the parent nucleoside does not cast the concentration of the intracellular active drug (Pirmohamed and Back, 2001), and also they are not metabolized as extensively by cytochrome P450 enzymes as are PIs and NNRTIs (Telenti and Zanger, 2008). As a result, much research has been focused on PIs and NNRTIs. Polymorphic variants of cytochrome P450 isoenzymes play a significant role in displaying distinct phenotypes affecting pharmacokinetics, efficacy, and toxicity of ARV drugs.

The human *CYP2B6* gene is highly polymorphic and several studies have shown the high relevance of this enzyme in the metabolism of ARV drugs, such as efavirenz and nevirapine (Ingelman-Sundberg *et al.*, 2007; Zhou *et al.*, 2009). The *CYP2B6*6* haplotype (516 G>T) leads to diminished metabolic activity and a significant decrease in protein expression (Lang i, 2001). The *CYP2B6*6* mutant allele is prevalent among various ethnic populations with differing frequencies: 15–40% in Asians, 25% in white populations, more than 50% in Afro-Americans and Africans (Guan *et al.*, 2006; Mehlotra *et al.*, 2006). Rotger and colleagues

(2005a) evaluated the *CYP2B6* 516 G>T (Gln172His) relation with efavirenz and nevirapine toxicity in HIV-infected patients. The group found the association of *CYP2B6* 516 T/T with greater plasma and intracellular exposure to efavirenz, and greater plasma exposure to nevirapine. The Adult AIDS Clinical Trials Group examined the prediction of plasma exposure of efavirenz and nevirapine among the different genetic variants. The median values of the area under the 24-h efavirenz concentration–time curve were almost double in TT homozygotes than in GG homozygotes of *CYP2B6* 516 G>T polymorphism (Haas *et al.*, 2005). Another study also showed the *CYP2B6* poor metabolizer genotypes with significant elevated efavirenz plasma levels (Rotger *et al.*, 2007). *CYP2B6* 983 T>C improved the predictive ability of *CYP2B6* 516 G>T for efavirenz pharmacokinetics (Ribaudo *et al.*, 2010), and *CYP2B6* 983 T>C SNP have significant association with higher efavirenz and nevirapine plasma concentrations (Wyen *et al.*, 2008). A retrospective genome-wide association study (GWAS) to identify novel variants associated with efavirenz pharmacokinetics in AIDS Clinical Trials Group protocols found significant associations between *CYP2B6* 516 G>T, 983 T>C, and rs4803419 C>T with increased median efavirenz estimated C_{min} values (Holzinger *et al.*, 2012). The plasma level of efavirenz in mutant *CYP2B6* 516 T/T genotype was found to be significantly greater than G/T and G/G genotypes in HIV/TB co-infected Thai adults (Uttayamakul *et al.*, 2010). Several other studies found a significant association between 516 G>T polymorphism and higher plasma efavirenz concentration (Gounden *et al.*, 2010; Sukasem *et al.*, 2012, 2014a; Manosuthi *et al.*, 2013; Lee *et al.*, 2014) with the recommendation of genetic screening for patients at risk of developing efavirenz-induced neurotoxicity in clinical practice (Sukasem and Sungkanuparph, 2013). A study among Thai HIV-infected adults found *CYP2B6* 18492 T>C polymorphism significantly associated with lower plasma efavirenz concentration (Sukasem *et al.*, 2014b, 2014c). A recent study had similar findings with low efavirenz concentration significantly associated with *CYP2B6* 1848 T>C among Thai HIV patients co-infected with tuberculosis carrying the *CYP2B6* haplotype *1/*1 (Manosuthi *et al.*, 2014b). HIV-infected patients carrying the *CYP2B6* haplotype *6/*6 are prone to develop hepatotoxicity with efavirenz-based ARV regimen (Manosuthi *et al.*, 2014a).

CYP2C19 is another enzyme of interest for HIV drugs due to its polymorphic characteristics influencing drug response. The *CYP2C19**2 allele (681 G>A) produces a decreased CYP2C19 activity and the poor metabolizer *CYP2C19**2 allele is more frequent in Chinese (30%) as compared to Caucasians (~15%) and African Americans (~17%) (Desta *et al.*, 2002). Among subjects receiving nelfinavir, *CYP2C19* 681 G>A was significantly associated with plasma exposure to nelfinavir in a study conducted by the Adult AIDS Clinical Trials Group (Haas *et al.*, 2005). Evaluation of HIV-1-infected children receiving nelfinavir as a component of HAART showed the significant difference of nelfinavir oral clearance (CL/F) among the *CYP2C19* 681 G>A genotypes ($P < 0.001$); children with the *CYP2C19* 681A/A had lower median nelfinavir CL/F as compared with the G/G genotype. The study also observed the significant virologic suppression in subjects with A/A and G/A genotypes as compared to G/G genotype (Saitoh *et al.*, 2010).

20.6. Transporters and their genetic polymorphisms

There is increasing evidence to suggest that genetic heterogeneity in drug transporters affects disposition of ARV drugs. Multi-drug resistance protein 1 gene (*MDR1*), also known as ATP-binding cassette subfamily B member 1 gene (*ABCB1*), has been extensively studied in identification and characterization of single-nucleotide polymorphisms (SNPs). A study of the impact of *ABCB1* c.4036A>G (rs3842) on single-dose efavirenz pharmacokinetics in healthy Ugandan subjects found mutant homozygous (G/G) and heterozygous (A/G) subjects with 26% higher relative bioavailability of efavirenz compared with the wild genotype (A/A) (Mukonzo *et al.*, 2009). Two hundred eighty-two South African HIV/AIDS patients were evaluated for *ABCB1* polymorphisms on plasma efavirenz levels and CD4 cell count (Swart *et al.*, 2012). Patients with *ABCB1* 4036A/G and G/G genotypes were significantly associated with reduced efavirenz levels ($P = 0.0236$), while 1236C/T and T/T genotypes ($P = 0.0282$) were associated with increased plasma efavirenz levels as compared to the wild genotype of the *ABCB1* 1236 C>T polymorphism. In a recent parallel-group prospective cohort study evaluating the importance of *ABCB1* c.4036A>G in determining efavirenz pharmacokinetics in two African populations from Ethiopia and Tanzania, the study group demonstrated that *ABCB1* c.4036A>G genotype was significantly associated with higher plasma efavirenz concentration (Ngaimisi *et al.*, 2013). Interestingly, there has been significant improvement in CD4+ T-cell count in the discovery subset with variation in *ABCB1* SNPs (rs1045642, 3435 T>C and rs2032582, 2677 T>G), and the results were subsequently confirmed in a validation subset of the cohort (Zhu *et al.*, 2013). Additionally, an exploratory analysis showed that *ABCB1* variations may be mediating response to treatment with nevirapine or lamivudine. Therefore, close monitoring of plasma efavirenz level in genotyping practice in HIV clinics is suggested for variations in *ABCB1*. A prospective study among Thai HIV-infected patients under a tenofovir-containing ARV regimen designed to study two SNPs ABCC2*1 C c.–24 C>T and ABCB1*6 c.3435C>T to correlate with plasma tenofovir concentrations and glomerular filtration rate found patients carrying *ABCC2*1 C* genotype C/C had higher plasma tenofovir concentrations along with decreased estimated glomerular filtration rate (eGFR) (Manosuthi *et al.*, 2014c). However, they found no association of ABCB1*6 c.3435C>T with higher tenofovir concentration and decreased eGFR.

Another important class of drug uptake transporter, organic anion transporting polypeptide (OATP), coded by the *SLCO* genes, has also been associated with variability in drug disposition, including the PIs. Patients with the *SLCOB1* 521 T>C polymorphism had a significant trend for increasing trough concentration of lopinavir (Kohlrausch *et al.*, 2010). In a prospective cohort study in African American HIV-infected children on ART with lopinavir/ritonavir, a significant association between *SLCO1B1* 521 T>C and lopinavir AUC ($P = 0.042$) was observed (Rakhmanina *et al.*, 2011). *SLCO1B1* 521 T>C was also significantly associated with higher lopinavir plasma concentration, which adds to our understanding of the factors contributing to the variability in the plasma concentrations of lopinavir (Hartkoorn *et al.*, 2010).

20.7. Pharmacogenetics of antiretroviral drug toxicity

20.7.1. Hypersensitivity reactions

Most ARV drugs have been related with delayed hypersensitivity reactions manifested by the onset of multiple symptoms, including erythematous, maculopapular, pruritic and confluent rash with or without fever (Chaponda and Pirmohamed, 2011). The pathogenesis of hypersensitivity is unknown. The common ARV drugs significantly associated with hypersensitivity reactions include abacavir and nevirapine. The *HLA-B*57:01* allele is the most extensively studied pharmaco-genetic marker, and there has been earlier association between abacavir hypersensitivity and the *HLA-B*57:01* allele (Hetherington *et al.*, 2002). A study from an Australian HIV cohort also found a significant association of abacavir hypersensitivity with *HLA-B*57:01* (Mallal *et al.*, 2002). With a 100% sensitivity of *HLA-B*57:01* as a predictor for immunologically confirmed abacavir hypersensitivity reactions in both US white and black participants (Saag *et al.*, 2008) and the significantly lower incidence of hypersensitivity reaction due to *HLA-B*57:01* in the prospective-screening group (Mallal *et al.*, 2008), a pharmacogenetic test in abacavir treatment is suggested to prevent the toxic effect of the drug. Two *HLA-B* 57:01*-tagging SNPs, HCP5 rs2395029 and TNF rs3093726, can be used as alternative tests for screening abacavir-induced hypersensitivity reaction in the unavailability of sequence-based *HLA* typing (Sukasem *et al.*, 2014b).

Nevirapine hypersensitivity is associated with different HLA class I and class II alleles. A stepwise case-control study in HIV-infected Thai patients found a strong association between *HLA-B*35:05* and nevirapine induced skin rash (Chantarangsu *et al.*, 2009). Association between nevirapine-associated rash and *HLA-Cw*04* allele was reported in a case-control study in Thai HIV-infected patients (Likanonsakul *et al.*, 2009). In an HIV French Caucasian cohort study, cases with isolated rash were significantly associated with the *HLA-DRB1*01* allele in HIV patients treated with nevirapine and efavirenz (Vitezicaa *et al.*, 2008). Two SNPs, rs1265112 and rs746647, in the *CCHCR1* locus were found to have significant associations ($P = 1.2 \times 10^{-8}$) with nevirapine-induced rash in a GWAS and replication studies of candidate SNPs among Thai HIV-infected patients (Chantarangsu *et al.*, 2011).

20.7.2. Hyperbilirubinemia

Indinavir and atazanavir have been found to be associated with increased unconjugated bilirubin levels (Cressey and Lallemant, 2007) due to inhibition of the enzyme uridine diphosphate-gluronosyltransferase 1A1 (UGT1A1) (Zucker *et al.*, 2001; Pavlos and Phillips, 2012), an enzyme responsible for the metabolism of bilirubin. Hyperbilirubinemia affects significant proportions of patients in atazanavir (20–50%) and indinavir (5–25%) therapy (Cholewinska, 2007). Genetic polymorphisms of the *UGT1A1* gene are associated with variations in the enzyme activity. *UGT1A1*6* and *UGT1A1*28* are the most relevant alleles associated with the incidence of hyperbilirubinemia in patients exposed to indinavir and atazanavir. *UGT1A1*28* is less frequent in Asian populations (Japanese 11% and Chinese 16%) as compared to whites (36–39%) and African Americans (43%) (Park *et al.*, 2010; Michaud *et al.*, 2012). *UGT1A1*6* (211 G>A) is more prevalent in Asians (13–23%) as compared to whites (less than 1%) (Takeuchi *et al.*, 2004; Kaniwa *et al.*, 2005).

The risk for hyperbilirubinemia due to the administration of ARV drugs, indinavir and atazanavir, was demonstrated by the presence of homozygous *UGT1A1*28* in 67% of patients who had increased levels of bilirubin as compared with patients having the common allele and not receiving either of those protease inhibitors (Rotger *et al.*, 2005b). The genetic susceptibility of *UGT1A1*28* as a risk factor in developing hyperbilirubinemia has been studied by other groups as well (Huang *et al.*, 2002; Huang *et al.*, 2005; Rodriguez-Novoa *et al.*, 2007). Notably, a study in Korean HIV-infected patients also found the *UGT1A1*28* allele to be a potential risk factor for severe atazanavir-associated hyperbilirubinemia, although the Koreans have lower frequency of the **28* allele (Park *et al.*, 2010). In a study involving Thai HIV-infected patients treated with indinavir, hyperbilirubinemia was associated with the presence of the *UGT1A1*6* (G71 R) allele (Boyd *et al.*, 2006).

20.7.3. Lipodystrophy and metabolic disturbances

Lipodystrophy is an umbrella term characterized by a set of morphological changes in the body-fat distribution and classified into three categories: lipohypertrophy, lipoatrophy, and mixed syndrome. Lipohypertrophy is the accumulation of fat in one or more anatomical sites (e.g., abdomen, dorsocervical spine, and breast). Lipoatrophy or peripheral atrophy is the loss of subcutaneous fat mainly in the face, buttocks, and extremities. Mixed syndrome refers to the combination of lipoatrophy and lipohypertrophy (Justina *et al.*, 2014; Lana *et al.*, 2014). Patients with lipodystrophy also tend to suffer from metabolic complications, including dyslipidemia, insulin resistance, and hyperglycemia (Fiorenza *et al.*, 2011). The ratio of small-dense low-density lipoprotein cholesterol (sdLDL-C) and large-buoyant low-density lipoprotein cholesterol (lbLDL-C) was identified as a significant marker of lipodystrophy ($P < 0.001$) in HIV-infected patients receiving stavudine containing ARV regimen (Srisawasdi *et al.*, 2013). The long-term use of ARV drugs, the use of regimens containing NRTIs (stavudine and, to a lesser extent, zidovudine), have been described as risk factors for lipodystrophy due to mitochondrial toxicity (Villarroya *et al.*, 2005). Genetic factors may be associated in developing lipodystrophy, because these complications do not occur in all treated patients, with a large inter-individual variability in the emergence and severity of the symptoms (Tozzi, 2010).

In a genetic case-control association study focusing on a promoter polymorphism in the tumour necrosis factor-alpha gene (*TNF-α*), TNF-α −238 G>A SNP was significantly associated ($P = 0.01$) with lipodystrophy in HIV-infected patients

(Maher *et al.*, 2002). Similar findings with the presence of TNF-α −238 G>A polymorphism in 13.1% of the cohort and associated with rapid progression of lipodystrophy was reported (Nolan *et al.*, 2003). In a case-control, cross-sectional study of HIV-infected patients from Thailand receiving stavudine-containing ARV regimen, *HLA-B*40:01* was significantly associated with stavudine-associated lipodystrophy. *HLA-B*40:01* can be used as a genetic marker to predict stavudine-associated lipodystrophy which needs to be studied further extensively (Wangsomboonsiri *et al.*, 2010).

Dyslipidemia is frequent in patients using HAART and is characterized by hypercholesterolemia, elevation of low-density lipoprotein (LDL-C), lower high-density lipoprotein (HDL-C) and hypertriglyceridemia. In a Multileft AIDS Cohort study, biogeographical ancestry (BGA) was significantly associated with lipid levels (Nicholaou *et al.*, 2013). Although the exact mechanism for dyslipidemia is unknown and is multifactorial, recent candidate gene studies have identified certain SNPs associated with a significant variation in lipid levels suggesting an important role for genetic factors in the pathogenesis of ART-induced dyslipidemia. Exposure to PIs is thought to play an important role in developing dyslipidemia (Mallon, 2006). In the first pharmaco-genetic study analyzing the effect of race/ethnicity and PI on lipid profile, PI exposure was a significant predictor of triglyceride and LDL-C. Participants receiving ritonavir-containing PI regimen had an estimated 1.46-fold greater triglyceride levels and 1.25-fold greater LDL-C compared to participants receiving non-PI based regimen (Foulkes *et al.*, 2006). In a cross-sectional study with HIV-1 infected patients, SNPs of apolipoprotein E gene (*APOE*) 334 T>C (E4; rs429358) and 472C>T (E2; rs7412) were associated with increased plasma triglyceride and LDL-C level and apolipoprotein A-V gene (*APOA5*) −1131 T>C (rs662799) associated with increased triglyceride level. In addition, *APOA5* −1131 T>C (rs662799) and sterol regulatory element-binding factor cleavage-activating protein gene (*SCAP*) 2386A>G (rs12487736) SNPs were significantly associated with lower plasma HDL-C levels (Lazzaretti *et al.*, 2013). Polymorphism in *APOA5* and carriers of *APOA5* c.553 G>T and −1131 T>C among Taiwanese HIV-1-infected patients receiving PI-based HAART were significantly associated with the risk of developing hypertriglyceridemia (Chang *et al.*, 2009). Carriers of the variant *APOC3* rs10892151 A allele have been associated with higher serum triglyceride concentrations and lower HDL-C levels during PI treatment (Aragones *et al.*, 2011). However, a recent study did not find any significant association of *ApoE* genotype and HAART-induced lipodystrophy in a case-control study conducted among Thai HIV-infected patients receiving HAART (Suwalak *et al.*, 2013).

20.7.4. Peripheral neuropathy

"D-drug" use, that is, didanosine (ddI), dalcitabine (ddC), and stavudine (d4T), classified as first-generation NRTIs, has been associated with the risk for sensory neuropathy that has debilitating effects on quality of life (Dalakas, 2001). In addition, several other ARV drugs including PIs such as indinavir sulphate, saquinavir, and ritonavir can induce neuropathy (Chen *et al.*, 2013). The neuronal and axonal mitochondrial injury resulting from mitochondrial DNA damage could possibly be the pathophysiological mechanism of sensory neuropathy in patients undergoing ART (Gabbai *et al.*, 2013). In a case-control study of the Adult AIDS Clinical Trials Group (ACTG) study, 384 white ACTG study participants belonging to the specific mitochondrial haplogroup T developed peripheral neuropathy, particularly among the whites receiving ddI plus d4T (Hulgan *et al.*, 2005). Mutations in the hemochromatosis gene (*HFE*), C282Y, are significantly associated with a decreased risk of neuropathy than non-carriers during treatment with ddI/d4T ($P = 0.04$) (Kallianpura *et al.*, 2006). In a randomized clinical trial of HIV-infected patients exposed to NRTIs, two non-synonymous mitochondrial DNA polymorphisms, MTND1*LHON4216C (4216C) and MTND2*LHON4917G (4917G), were analyzed for their association with peripheral neuropathy, and it was found that 4917G allele was related with NRTI-induced peripheral neuropathy (Canter *et al.*, 2008).

20.8. Conclusions and future perspectives

Genetic variations responsible for the inter-individual responses to ARV will offer understanding into the disease pathology and drug action. Over the past decade, a vast amount of literature has been published with the significant pharmacogenetic associations with ARV drugs use and our knowledge has grown steadily. The field of personalized medicine clutches the tremendous commitment in improving patient care by customizing the treatment of each patient to his or her individual biological make-up. Several pharmaco-genetic tests have managed to achieve acceptance in clinical practice, with more drugs on the process of evaluation. *HLA-B*5701* screening before abacavir prescription has already entered into the routine clinical practice for preventing hypersensitivity reactions. Other notable genetic associations include the *CYP2B6* gene and plasma level of efavirenz and nevirapine, *UGT1A1* alleles and risk for hyperbilirubinemia among indinavir- and atazanavir-exposed patients, and *HLA-B*4001* with stavudine-associated lipodystrophy.

The clinical adoption of pharmaco-genomics testing is lagging behind due to the lack of educating the health professionals regarding both the theory and specifics of pharmaco-genomics. The vision of pharmacogenomics to make drug discovery, development, and delivery to patients more rational and efficient has to be given more effort and it requires the different sections of the healthcare system to work together in an attempt to better understand and coordinate policy. The impact of pharmacogenomics screening in the clinical set up must be considered both on the patients and health practitioners. Cost-effective analysis must be considered in tailoring the therapy for the patient. The responsibilities of the healthcare professionals might need to be revised with the presence of a large number of pharmacogenomics drugs. Genotype–phenotype associations will require the use of upgraded advanced technologies for genome-wide screening and statistical analysis in order to draw a definitive comprehensive association. A remarkable combination of genome technology, computational and informatics support in large multi-ethnic pharmacogenomics study among ARV drug users is warranted to achieve the translation of research findings into clinical practice.

REFERENCES

Adamson, C.S. and Freed, E.O. (2008). Recent progress in antiretrovirals-lessons from resistance. *Drug Discov. Today*, **13**(9–10), 424–432.

Aragones, G., Alonso-Villaverde, C., Pardo-Reche, P., *et al.* (2011). Antiretroviral treatment-induced dyslipidemia in HIV-infected patients is influenced by the APOC3-related rs10892151 polymorphism. *BMC Med. Genet.*, **12**, 120.

Bazzoli, C., Jullien, V., Tiec, C.L., *et al.* (2010). Intracellular pharmacokinetics of antiretroviral drugs in HIV-infected patients, and their correlation with drug action. *Clin. Pharmacokinet.*, **49**(1), 17–45.

Boyd, M.A., Srasuebkul, P., Ruxrungtham, K., *et al.* (2006). Relationship between hyperbilirubinaemia and UDPglucuronosyltransferase 1A1 (UGT1A1) polymorphism in adult HIV-infected Thai patients treated with indinavir. *Pharmacogenet. Genom.*, **16**, 321–329.

Canter, J.A., Haas, D.W., Kallianpur, A.R., *et al.* (2008). The mitochondrial pharmacogenomics of haplogroup T: MTND2*LHON4917G and antiretroviral therapy-associated peripheral neuropathy. *Pharmacogenom. J.*, **8**(1), 71–77.

Chang, S.-Y., Ko, W.-S., Kao, J.-T., *et al.* (2009). Association of single-nucleotide polymorphism 3 and c.553G1T of APOA5 with hypertriglyceridemia after treatment with highly active antiretroviral therapy containing protease Inhibitors in HIV-infected individuals in Taiwan. *Clin. Infect. Dis.*, **48**, 832–835.

Chantarangsu, S., Mushiroda, T., Mahasirimongkol, S., *et al.* (2009). HLA-B*3505 allele is a strong predictor for nevirapine-induced skin adverse drug reactions in HIV-infected Thai patients. *Pharmacogenet. Genom.*, **19**(2), 139–146.

Chantarangsu, S., Mushiroda, T., Mahasirimongkol, S., *et al.* (2011). Genome-wide association study identifies variations in 6p21.3 associated with nevirapine-induced rash. *Clin. Infect. Dis.*, **53**(4), 341–348.

Chaponda, M. and Pirmohamed, M. (2011). Hypersensitivity reactions to HIV therapy. *Br. J. Clin. Pharmacol.*, **71**(5), 659–671.

Chen, H., Clifford, D.B., Deng, L., *et al.* (2013). Peripheral neuropathy in ART-experienced patients: prevalence and risk factors. *J. NeuroVirol.*, **19**(6), 557–564.

Cholewinska, G. (2007). Pharmacogenetics in HIV clinical practice. *HIV AIDS Rev.*, **7**(2), 9–14.

Clercq, E.D. (2009). Anti-HIV drugs: 25 compounds approved within 25 years after the discovery of HIV. *Int. J. Antimicrob. Agents*, **33**, 307–320.

Cressey, T.R. and Lallemant, M. (2007). Pharmacogenetics of antiretroviral drugs for the treatment of HIV-infected patients: an update. *Infect. Genet. Evol.*, **7**(2), 333–342.

Cropp, C., Yee, S. and Giacomini, K. (2008). Genetic variation in drug transporters in ethnic populations. *Clin. Pharmacol. Therapeut.*, **84**(3), 412–416.

Dalakas, M.C. (2001). Peripheral neuropathy and antiretroviral drugs. *J. Periph. Nerv. Syst.*, **6**(1), 14–20.

Desta, Z., Zhao, X., Shin, J.G., *et al.* (2002). Clinical significance of the cytochrome P450 2C19 genetic polymorphism. *Clin. Pharmacokinet.*, **41**(12), 913–958.

Fiorenza, C.G., Chou, S.H. and Mantzoros, C.S. (2011). Lipodystrophy: pathophysiology and advances in treatment. *Nature Rev. Endocrinol.*, **7**, 137–150.

Foulkes, A.S., Wohl, D.A., Frank, I., *et al.* (2006). Associations among race/ethnicity, ApoC-III genotypes, and lipids in HIV-1-infected individuals on antiretroviral therapy. *PLoS Med.*, **3**(3), 337–347.

Gabbai, A.A., Castelo, A. and Oliveira, A.S. (2013). HIV peripheral neuropathy. *Handbk Clin. Neurol.*, **115**, 515–529.

Gallant, J.E. (2005). Antiretroviral drug resistance and resistance testing. *Top. HIV Med.*, **13** (5), 138–142.

Gotte, M. and Wainberg, M.A. (2000). Biochemical mechanisms involved in overcoming HIV resistance to nucleoside inhibitors of reverse transcriptase. *Drug Resist. Updates*, **3**, 30–38.

Gounden, V., Niekerk, C. v., Snyman, T., *et al.* (2010). Presence of the CYP2B6 516 G >T polymorphism, increased plasma Efavirenz concentrations and early neuropsychiatric side effects in South African HIV-infected patients. *AIDS Res. Ther.*, **7**, 32.

Guan, S., Huang, M., Li, X., *et al.* (2006). Intra- and inter-ethnic differences in the allele frequencies of cytochrome P450 2B6 gene in Chinese. *Pharmaceut. Res.*, **23**(9), 1983–1990.

Haas, D.W., Smeaton, L.M., Shafer, R.W., *et al.* (2005). Pharmacogenetics of long-term responses to antiretroviral regimens containing efavirenz and/or nelfinavir. An Adult AIDS Clinical Trials Group Study. *J. Infect. Dis.*, **192**, 1931–1942.

Hanna, G.J. and D'Aquila, R.T. (2001). Clinical use of genotypic and phenotypic drug resistance testing to monitor antiretroviral chemotherapy. *Clin. Infect. Dis.*, **32**, 774–782.

Hartkoorn, R.C., Kwan, W.S., Shallcross, V., *et al.* (2010). HIV protease inhibitors are substrates for OATP1A2, OATP1B1 and OATP1B3 and lopinavir plasma concentrations are influenced by SLCO1B1 polymorphisms. *Pharmacogenet. Genom.*, **20**(2), 112–120.

Hetherington, S., Hughes, A.R., Mosteller, M., *et al.* (2002). Genetic variations in HLA-B region and hypersensitivity reactions to abacavir. *Lancet*, **359**(9312), 1121–1122.

Holzinger, E.R., Grady, B., Ritchie, M.D., *et al.* (2012). Genome-wide association study of plasma efavirenz pharmacokinetics in AIDS Clinical Trials Group protocols implicates several CYP2B6 variants. *Pharmacogenet. Genom.*, **22**(12), 858–867.

Huang, C.S., Huang, M.J., Lin, M.S., *et al.* (2005). Genetic factors related to unconjugated hyperbilirubinemia amongst adults. *Pharmacogenet. Genom.*, **15**(1), 43–50.

Huang, M.J., Yang, Y.C., Yang, S.S., *et al.* (2002). Coinheritance of variant UDP-glucuronosyl transferase 1A1 gene and glucose-6-phosphate dehydrogenase deficiency in adults with hyperbilirubinemia. *Pharmacogenetics*, **12**(8), 663–666.

Hulgan, T., Haas, D.W., Haines, J.L., *et al.* (2005). Mitochondrial haplogroups and peripheral neuropathy during antiretroviral therapy: an adult AIDS clinical trials group study. *AIDS*, **19**(13), 1341–1349.

Ingelman-Sundberg, M., Sim, S.C., Gomez, A., *et al.* (2007). Influence of cytochrome P450 polymorphisms on drug therapies: pharmacogenetic, pharmacoepigenetic and clinical aspects. *Pharmacol. Therapeut.*, **116**(3), 496–526.

Justina, L.B., Luiz, M.C., Maurici, R., *et al.* (2014). Prevalence and factors associated with lipodystrophy in AIDS patients. *Rev. Soc. Bras. Med. Trop.*, **47**(1), 30–37.

Kallianpura, A.R., Hulgan, T., Canter, J.A., *et al.* (2006). Hemochromatosis (HFE) gene mutations and peripheral neuropathy during antiretroviral therapy. *AIDS*, **20**, 1503–1513.

Kaniwa, N., Kurose, K., Jinno, H., *et al.* (2005). Racial variability in haplotype frequencies of UGT1A1 and glucuronidation activity of a novel single nucleotide polymorphism 686 C >T (P229L) found in an African-American. *Drug Metab. Dispos.*, **33**(3), 458–465.

Kiertiburanakul, S., Wiboonchutikul, S., Sukasem, C., *et al.* (2010). Using of nevirapine is associated with intermediate and reduced response to etravirine among HIV-infected patients who experienced virologic failure in a resource-limited setting. *J. Clin. Virol.*, **47**, 330–334.

Kohlrausch, F. B., Estrela, R. d. C., Barroso, P.F., *et al.* (2010). The impact of SLCO1B1 polymorphisms on the plasma concentration of lopinavir and ritonavir in HIV-infected men. *Br. J. Clin. Pharmacol.*, **69**(1), 95–98.

Lana, L.G.C., Junqueira, D.R.G., Perini, E., *et al.* (2014). Lipodystrophy among patients with HIV infection on antiretroviral therapy: a systematic review protocol. *BMJ Open*, **4**, e004088.

Lang, T., Klein, K., Fischer, J., *et al.* (2001). Extensive genetic polymorphism in the human *CYP2B6* gene with impact on expression and function in human liver. *Pharmacogenetics*, **11**(5), 399–415.

Lazzaretti, R.K., Gasparotto, A.S., Sassi, M.G., *et al.* (2013). Genetic markers associated to dyslipidemia in HIV-infected individuals on HAART. *Sci. World J.*, 2013, 608415.

Lee, K.-Y., Lin, S.-W., Sun, H.-Y., *et al.* (2014). Therapeutic drug monitoring and pharmacogenetic study of HIV-infected ethnic chinese receiving efavirenz-containing antiretroviral therapy with or without rifampicin-based anti-uberculous therapy. *PLoS ONE*, **9**(2), e88497.

Likanonsakul, S., Rattanatham, T., Feangvad, S., *et al.* (2009). HLA-Cw*04 allele associated with nevirapine-induced rash in HIV-infected Thai patients. *AIDS Res. Ther.*, **6**, 22.

Maher, B., Alfirevic, A., Vilar, F.J., *et al.* (2002). TNF-α promoter region gene polymorphisms in HIV-positive patients with lipodystrophy. *AIDS*, **16**, 2013–2018.

Mallal, S., Nolan, D., Witt, C., *et al.* (2002). Association between presence of HLA-B*5701, HLA-DR7, and HLA-DQ3 and hypersensitivity to HIV-1 reverse-transcriptase inhibitor abacavir. *Lancet*, **359**, 727–732.

Mallal, S., Phillips, E., Carosi, G., *et al.* (2008). HLA-B*5701 screening for hypersensitivity to abacavir. *New Engl. J. Med.*, **358**(6), 568–579.

Mallon, P.W.G. (2006). Antiretroviral therapy and dyslipidaemia: unlocking the code. *PLoS Med.*, **3**(3), e85.

Manosuthi, W., Sukasem, C., Lueangniyomkul, A., *et al.* (2013). Impact of pharmacogenetic markers of *CYP2B6*, clinical factors, and drug–drug interaction on efavirenz concentrations in HIV/tuberculosis-coinfected patients. *Antimicrob. Agents Chemother.*, **57**(2), 1019–1024.

Manosuthi, W., Sukasem, C., Lueangniyomkul, A., *et al.* (2014a). *CYP2B6* haplotype and biological factors responsible for hepatotoxicity in HIV-infected patients receiving efavirenz-based antiretroviral therapy. *Int. J. Antimicrob. Agents*, **43**(3), 292–296.

Manosuthi, W., Sukasem, C., Thongyen, S., *et al.* (2014b). *CYP2B6* 18492 T→C polymorphism compromises efavirenz concentration in coinfected HIV and tuberculosis patients carrying CYP2B6 haplotype *1/*1. *Antimicrob. Agents Chemother.*, **58**(4), 2268–2273.

Manosuthi, W., Sukasem, C., Thongyen, S., *et al.* (2014c). ABCC2*1C and plasma tenofovir concentration are correlated to decreased glomerular filtration rate in patients receiving a tenofovir-containing antiretroviral regimen. *J. Antimicrob. Chemother.*, 69(8), 2195–2201.

Mata-Munguia, C., Escoto-Delgadillo, M., Torres-Mendoza, B., *et al.* (2014). Natural polymorphisms and unusual mutations in HIV-1 protease with potential antiretroviral resistance: a bioinformatic analysis. *BMC Bioinform.*, **15**, 72.

Mehlotra, R.K., Ziats, M.N., Bockarie, M.J., *et al.* (2006). Prevalence of *CYP2B6* alleles in malaria-endemic populations of West Africa and Papua New Guinea. *Eur. J. Clin. Pharmacol.*, **62**(4), 267–275.

Michaud, V., Bar-Magen, T., Turgeon, J., *et al.* (2012). The dual role of pharmacogenetics in HIV treatment: mutations and polymorphisms regulating antiretroviral drug resistance and disposition. *Pharmacol. Rev.*, **64**(3), 803–833.

Mukonzo, J.K., Röshammar, D., Waako, P., *et al.* (2009). A novel polymorphism in *ABCB1* gene, *CYP2B6**6 and sex predict single-dose efavirenz population pharmacokinetics in Ugandans. *Br. J. Clin. Pharmacol.*, **68**(5), 690–699.

Ngaimisi, E., Habtewold, A., Minzi, O., *et al.* (2013). Importance of ethnicity, *CYP2B6* and *ABCB1* genotype for efavirenz pharmacokinetics and treatment outcomes: a parallel-group prospective cohort study in two sub-Saharan Africa populations. *PLoS ONE*, **8**(7), e67946.

Nicholaou, M.J., Martinson, J.J., Abraham, A.G., *et al.* (2013). HAART-associated dyslipidemia varies by biogeographical ancestry in the multileft AIDS cohort study. *AIDS Res. Hum. Retrovir.*, **29**(6), 871–879.

Nolan, D., Moore, C., Castley, A., *et al.* (2003). Tumour necrosis factor-alpha gene -238 G/A promoter polymorphism associated with a more rapid onset of lipodystrophy. *AIDS*, **17**(1), 121–123.

Park, W.B., Choe, P G., Song, K.-H., *et al.* (2010). Genetic factors influencing severe atazanavir-associated hyperbilirubinemia in a population with low UDP-glucuronosyltransferase 1A1*28 allele frequency. *Clin. Infect. Dis.*, **51**(1), 101–106.

Pavlos, R. and Phillips, E.J. (2012). Individualization of antiretroviral therapy. *J. Pharmacogenom. Person. Med.*, **5**, 1–17.

Pirmohamed, M. and Back, D.J. (2001). The pharmacogenomics of HIV therapy. *Pharmacogenom. J.*, **1**, 243–253.

Rakhmanina, N.Y., Neely, M.N., Van Schaik, R.H., *et al.* (2011). *CYP3A5, ABCB1,* and *SLCO1B1* polymorphisms and pharmacokinetics and virologic outcome of lopinavir/ritonavir in HIV-infected children. *Therapeut. Drug Monit.*, **33**(4), 417–424.

Ribaudo, H.J., Liu, H., Schwab, M., *et al.* (2010). Impact of *CYP2B6, ABCB1* and *CYP3A5* polymorphisms on efavirenz pharmacokinetics and treatment response: an AIDS Clinical Trials Group study. *J. Infect. Dis.*, **202**(5), 717–722.

Rodriguez-Novoa, S., Martin-Carbonero, L., Barreiro, P., *et al.* (2007). Genetic factors influencing atazanavir plasma concentrations and the risk of severe hyperbilirubinemia. *AIDS*, **21**, 41–46.

Rotger, M., Colombo, S., Furrer, H., *et al.* (2005a). Influence of *CYP2B6* polymorphism on plasma and intracellular concentrations and toxicity of efavirenz and nevirapine in HIV-infected patients. *Pharmacogenet. Genom.*, **15**(1), 1–5.

Rotger, M., Taffe, P., Bleiber, G., *et al.* (2005b). Gilbert syndrome and the development of antiretroviral therapy-associated hyperbilirubinemia. *J. Infect. Dis.*, **192**(8), 1381–1386.

Rotger, M., Tegude, H., Colombo, S., *et al.* (2007). Predictive value of known and novel alleles of *CYP2B6* for efavirenz plasma concentrations in HIV-infected individuals. *Clin. Pharmacol. Therapeut.*, **81**(4), 557–566.

Saag, M., Balu, R., Phillips, E., *et al.* (2008). High sensitivity of human leukocyte antigen–B*5701 as a marker for immunologically confirmed abacavir hypersensitivity in white and black patients. *Clin. Infect. Dis.*, **46**, 1111–1118.

Saitoh, A., Capparelli, E., Aweeka, F., *et al.* (2010). *CYP2C19* genetic variants affect nelfinavir pharmacokinetics and virologic response in HIV-1-infected children receiving highly active antiretroviral therapy. *J. AIDS*, **54**(3), 285–289.

Saskova, K.G., Kozisek, M., Rezacova, P., *et al.* (2009). Molecular characterization of clinical isolates of human immunodeficiency virus resistant to the protease inhibitor darunavir. *J. Virol.*, **83**(17), 8810–8818.

Shafer, R.W., Kantor, R. and Gonzales, M.J. (2000). The genetic basis of HIV-1 resistance to reverse transcriptase and protease inhibitors. *AIDS Rev.*, **2**(4), 211–228.

Shah, R.R. and Shah, D.R. (2012). Personalized medicine: is it a pharmacogenetic mirage? *Br. J. Clin. Pharmacol.*, **74**(4), 698–721.

Singh, K., Marchand, B., Kirby, K.A., *et al.* (2010). Structural aspects of drug resistance and inhibition of HIV-1 reverse transcriptase. *Viruses*, **2**(2), 606–638.

Srisawasdi, P., Suwalak, T., Sukasem, C., *et al.* (2013). Small-dense LDL cholesterol/large-buoyant LDL cholesterol ratio as an excellent marker for indicating lipodystrophy in HIV-infected patients. *Am. J. Clin. Pathol.*, **140**(4), 506–515.

Subbaraman, R., Chaguturu, S.K., Mayer, K.H., *et al.* (2007). Adverse effects of highly active antiretroviral therapy in developing countries. *Clin. Infect. Dis.*, **45**, 1093–1101.

Sukasem, C. and Sungkanuparph, S. (2013). Would a *CYP2B6* test help HIV patients being treated with efavirenz? *Pharmacogenomics*, **14**(9), 999–1001.

Sukasem, C., Churdboonchart, V., Chasombat, S., *et al.* (2007). Surveillance of genotypic resistance mutations in chronic HIV-1 treated individuals after completion of the National Access to Antiretroviral Program in Thailand. *Infection*, **35**(2), 81–88.

Sukasem, C., Churdboonchart, V., Sukeepaisarncharoen, W., *et al.* (2008). Genotypic resistance profiles in antiretroviral-naive HIV-1 infections before and after initiation of first-line HAART: impact of polymorphism on resistance to therapy. *International Journal of Antimicrobial Agents*, **31**(3), 277–281.

Sukasem, C., Cressey, T.R., Prapaithong, P., *et al.* (2012). Pharmacogenetic markers of *CYP2B6* associated with efavirenz plasma concentrations in HIV-1 infected Thai adults. *Br. J. Clin. Pharmacol.*, **74**(6), 1005–1012.

Sukasem, C., Chamnanphon, M., Koomdee, N., *et al.* (2013). High plasma efavirenz concentration and *CYP2B6* polymorphisms in Thai HIV-1 infections. *Drug Metab. Pharmacokinet.*, **28**(5), 391–397.

Sukasem, C., Chamnanphon, M., Koomdee, N., *et al.* (2014a). Pharmacogenetics and clinical biomarkers for subtherapeutic plasma efavirenz concentration in HIV-1 infected Thai adults. *Drug Metab. Pharmacokinet.*, 29(4), 289–295.

Sukasem, C., Gatrungsei, M., Promso, S., *et al.* (2014b). Evaluation of a pharmacogenetic test in Thailand for abacavir hypersensitivity screening in human immunodeficiency virus infection. *Curr. Pharmacogenom. Person. Med.*, **11**(3), 231–236.

Sukasem, C., Manosuthi, W., Koomdee, N., *et al.* (2014c). Low level of efavirenz in HIV-1-infected Thai adults is associated with the CYP2B6 polymorphism. *Infection*, 42(3), 469–474.

Suwalak, T., Srisawasdi, P., Puangpetch, A., *et al.* (2013). The distribution of apolipoprotein E gene polymorphism and their influences in dyslipidemias in HIV-1 infections. 15th International Workshop on Co-morbidities and Adverse Drug Reactions in HIV. Brussels, Belgium.

Swart, M., Ren, Y., Smith, P., *et al.* (2012). *ABCB1* 4036A >G and 1236C >T polymorphisms affect plasma efavirenz levels in South African HIV/AIDS patients. *Front. Genet.*, **3**, 236.

Takeuchi, K., Kobayashi, Y., Tamaki, S., *et al.* (2004). Genetic polymorphisms of bilirubin uridine diphosphate-glucuronosyltransferase gene in Japanese patients with Crigler–Najjar syndrome or Gilbert's syndrome as well as in healthy Japanese subjects. *J. Gastroenterol. Hepatol.*, **19**, 1023–1028.

Telenti, A. and Zanger, U.M. (2008). Pharmacogenetics of anti-HIV drugs. *Annu. Rev. Pharmacol. Toxicol.*, **48**, 227–256.

Tozzi, V. (2010). Pharmacogenetics of antiretrovirals. *Antiviral Res.*, **85**, 190–200.

Uttayamakul, S., Likanonsakul, S., Manosuthi, W., *et al.* (2010). Effects of *CYP2B6* G516T polymorphisms on plasma efavirenz and nevirapine levels when co-administered with rifampicin in HIV/TB co-infected Thai adults. *AIDS Res. Ther.*, **7**, 8.

Verma, A.S., Singh, I.G., Bansal, R., *et al.* (2014). HIV and antiretroviral drugs. Animal Biotechnology. New York, NY: Elsevier Inc.

Vidal, F., Gutierrez, F., Gutierrez, M., *et al.* (2010). Pharmacogenetics of adverse effects due to antiretroviral drugs. *AIDS Rev.*, **12**(1), 15–30.

Villarroya, F., Domingo, P. and Giralt, M. (2005). Lipodystrophy associated with highly active anti-retroviral therapy for HIV infection: the adipocyte as a target of anti-retroviral-induced mitochondrial toxicity. *Trends Pharmacol. Sci.*, **26**(2), 88–93.

Vitezicaa, Z.G., Milpied, B., Lonjoua, C., *et al.* (2008). HLA-DRB1*01 associated with cutaneous hypersensitivity induced by nevirapine and efavirenz. *AIDS*, **22**(4), 540–541.

Wangsomboonsiri, W., Mahasirimongkol, S., Chantarangsu, S., *et al.* (2010). Association between HLA-B*4001 and lipodystrophy among HIV-infected patients from Thailand who received a stavudine-containing antiretroviral regimen. *Clin. Infect. Dis.*, **50**, 597–604.

Wyen, C., Hendra, H., Vogel, M., *et al.* (2008). Impact of *CYP2B6* 983 T >C polymorphism on non-nucleoside reverse transcriptase inhibitor plasma concentrations in HIV-infected patients. *J. Antimicrob. Chemother.*, **61**(4), 914–918.

Zhou, S.F., Liu , J.P. and Chowbay, B. (2009). Polymorphism of human cytochrome P450 enzymes and its clinical impact. *Drug Metab. Rev.*, **41**(2), 89–95.

Zhu, P., Zhu, Q., Zhang, Y., *et al.* (2013). *ABCB1* variation and treatment response in AIDS patients: initial results of the Henan cohort. *PLoS ONE*, **8**(1), e55197.

Zucker, S.D., Qin, X., Rouster, S.D., *et al.* (2001). Mechanism of indinavir-induced hyperbilirubinemia. *Proc. Natl Acad. Sci. USA*, **98**(22), 12671–12676.

Part V

Population genetics and personalized medicine

21 Population stratification and its implications: lessons from genome-wide studies

Sheikh Nizamuddin, Rakesh Tamang and Kumarasamy Thangaraj

21.1. Introduction

It is a well-established fact that the anatomically modern human originated in Africa about 200 thousand years ago (KYA) and migrated out-of-Africa about 60–70 KYA. In this process, several populations arose and each of them has their own evolutionary history. Genetic drift, endogamy practices, admixture, and natural selection are examples of a few evolutionary phenomena leading to genetic diversity among populations around the world including susceptibility and resistance to genetic diseases, infectious diseases, therapeutic response to drugs, etc. Understanding these phenomena is much more relevant in a country like India, which has the richest ethnic, cultural, linguistic, and social diversity in the world and harbors 4635 anthropologically well-defined populations, among which 532 are tribes, including 72 primitive tribes (36 hunters and gatherers). In simple terms, there are populations in India who are phenotypically similar to Africans (Andamanese), some are similar to Europeans, some are similar to southeast Asians, and some are not related to any populations outside India. Most interestingly, Indian populations differ from each other with respect to their social structure, language, marriage practices, dress and food habits, physical appearance, etc. They are inhabitated in a variety of the geographical realms, ranging from scorching desert to evergreen forest, fertile plains to the southern dry plateaus, lowlands to the breath-taking high Himalayas. Broadly, Indians speak four major families of languages: Indo-European, Dravidian, Austroasiatic, and Tibeto-Burman. In addition, India has two distinguished isolated languages, namely Andamanese and Nihalis spoken in Andaman and Nicobar Islands, and pockets of Madhya Pradesh and Maharashtra, respectively. In addition to the indigenous populations, India experienced migration of several populations in the recent past. With everything mentioned above about India, it is tempting to say that India is an "ethnic museum." This chapter

Genome-Wide Association Studies: From Polymorphism to Personalized Medicine, ed. K. Appasani. Published by Cambridge University Press.

is intended to highlight the Indian population structure and its important implications.

21.2. Population stratification: India as an example

India is known for its social stratifications that include castes, tribes, and religious groups. The *Rig Veda*, which was written between 1700 and 1100 BC (Rao *et al.*, 2009), is considered to be the oldest written evidence about the existence and organization of the caste system. The precise date of origin of the caste system in India is still unclear. Initially, the newborn individual was devoid of any predefined caste. Later, one's caste was assigned based on the profession followed in the course of their life. As time passed, descendants started proclaiming the caste of their parents as their own, and this is how the strong demarcation of the caste system came into existence and started governing all social, religious, and economic activities of the people. The broad castes in Hindu society are Brahmin, Kshatriya, Vaishya, and Sudra. Brahmins are a group of people performing rituals and are in charge of teaching the society; Kshatriyas are rulers and warriors, and are involved in ruling and defending the territories; Vaishyas are cultivators and businessmen; Sudras rank last in society and are laborers by profession. Each caste is further subdivided into smaller units generally known as subcastes, which in turn are further divided into multiple purely exogamous clans. The strong social barrier emerging from the caste system, the long-term endogamy practice along with geographic and cultural differences gave rise to many smaller endogamous groups (Reich *et al.*, 2009), which were proved by several genetic studies in different eras and discussed further.

21.2.1. Understanding population stratification during pre genome-wide study era

Majumdar and Rao (1960) efficiently described that the Indian populations are structured based on geography and ethnicity. Their work was based on a systematic survey of anthropometric traits and is considered to be the first extensive and well-organized study reflecting Indian scenario. This has been well supported by later investigations using morphological and protein data sets (Roychoudhury *et al.*, 1985; Das et al. 1986). Later, protein and enzyme polymorphisms were used in building human genetic relationships (Nei, 1982). With the help of the classical markers, Papiha (1996) suggested a distinct demarcation between Indian caste and tribal populations. This is followed by studies using uniparental markers, namely Y chromosome and mitochondrial DNA (mtDNA) markers, which eventually became efficient and handy tools in understanding human phylogeny (Kivisild *et al.*, 1999, 2003; Metspalu *et al.*, 2004; Thangaraj *et al.*, 2005, 2006a; Sahoo and Kashyap, 2006; Sahoo *et al.*, 2006; Sengupta *et al.*, 2006; Kumar *et al.*, 2007b; Chaubey *et al.*, 2008; Palanichamy *et al.*, 2010; Underhill *et al.*, 2010; Rai *et al.*, 2012; Sharma *et al.*, 2012). These markers are still widely used worldwide because they are passed uniparently and more or less intact from one generation to the next. Previously, variations in HVSI and HVSII regions of mtDNA were used extensively for determining maternal ancestry; subsequently, the sequencing of complete mtDNA came into practice. Similarly for dissecting paternal ancestry, studies with a few SNPs and short tandem repeats (STRs) were succeeded by studies coupled with higher numbers of STRs and SNPs. The uniparental marker approach along with hundreds of thousands of biparental (autosomal) markers to deduce population structure gained popularity (Reich *et al.*, 2009; Behar *et al.*, 2010; Chaubey *et al.*, 2011; Metspalu *et al.*, 2011; Shah *et al.*, 2011; Moorjani *et al.*, 2013b). Thus, using all the available tools, the most accepted and proven views about Indian populations based on uniparental and biparental markers are: (1) the peopling of India is very ancient, and (2) there is evidence of recent geneflow from west and east Eurasia (Kivisild *et al.*, 1999; Bamshad *et al.*, 2001; Misra, 2001; Thangaraj *et al.*, 2005, 2006a; Sahoo and Kashyap 2006; Sahoo *et al.*, 2006; Sengupta *et al.*, 2006; Chaubey *et al.*, 2008; Underhill *et al.*, 2010).

21.2.1.1. Genetic footprints of early modern humans in the Andaman and Nicobar Islands

The Andaman and Nicobar Islands are located in the Bay of Bengal. It is a group of more than 500 small islands which are inhabited by the enigmatic tribes, who have remained isolated from the rest of the world for a considerable length of time. The people of this island can be divided into two groups. One group shares physical features with African pygmies and other Asian Negrito people, and the second group shares their physical features with southeast Asians.

To understand the peopling of the Andaman and Nicobar Islands in detail, a combined study using Y chromosome and mtDNA markers was undertaken (Thangaraj *et al.*, 2005). The complete mtDNA genomes of Onges, Great Andamanese, and Nicobarese revealed novel variations in their mtDNA that had helped in defining two new deep-rooted mtDNA haplogroups, M31 and M32, which were not observed in the worldwide samples until then (Thangaraj *et al.*, 2005). The coalescent age of this group was estimated to be about 65,000 YBP (years before present). Therefore, the Onges and Great Andamanese were predicted to be the descendants of the early modern humans, who ventured out-of-Africa, for the first time, via the southern coastal route. There was a massive search for these new haplogroups (M31 and M32) in mainland India and no one was found to harbor it. Interestingly, later, two individuals from Rajbhansi (West Bengal) were found to harbor the M31 haplogroup (Palanichamy *et al.*, 2006), but this can be attributed to their recent migration from the Andaman Islands. An elaborate study with high-density markers suggested their unique genetic component of the Onges that was absent outside India (Reich *et al.*, 2009). Recently, a strong southeast Asian affinity of the Andaman Islanders was established (Chaubey and Endicott, 2013).

21.2.2. Understanding population stratification in the era of Genome-Wide Association Study (GWAS)

Combinations of higher diversity, sociocultural barriers, and small population sizes in Indians have presented a challenging task for scholars of different fields.

The complex structure of the Indian population adds further difficulties in understanding diseases and preventive measures. Despite technological advances from low-resolution markers to high-throughput whole-genome sequencing, genome-wide studies in India are scarce. Very few groups in India have made rigorous efforts to understand the genetic structure of Indian populations using high-density genetic markers to understand the complexities of the Indian genetic structure and its implications for better understanding of human health and diseases. The clear-cut existence of multilayered genetic structure of Indian populations came into focus lately when scientists across the globe started using a comprehensive and high-density marker (genome-wide study) approach.

With the aim of understanding the complex issues related to origin, migrations, diversity, and predisposition to diseases, a combined attempt was initiated with a collaboration between six laboratories of the Council of Scientific and Industrial Research (CSIR), which includes (Indian Genome Variation Consortium, 2008): the Institute of Genomics and Integrative Biology (IGIB), Delhi; the Centre for Cellular and Molecular Biology (CCMB), Hyderabad; the Indian Institute of Chemical Biology (IICB), Kolkata; the Central Drug Research Institute (CDRI), Lucknow; the Industrial Toxicological Research Centre (ITRC), Lucknow; and the Institute of Microbial Technology (IMTECH), Chandigarh. This project was given the name "Indian Genome Variation (IGV) consortium," funded by CSIR, the Government of India, and was projected to use repeats and SNPs to meet the objectives among extensive samples collected across the country. Studying 405 SNPs from 75 disease or drug-response related genes and a 5.2 Mb chromosome, 22 genomic region among 1871 individuals from 55 endogamous Indian populations, they observed high levels of genetic divergence between groups of studied populations, clustering on a basis of ethnicity and languages, but the study lacked sufficient statistical power due to fewer SNPs (Indian Genome Variation Consortium, 2008). Pioneering the extensive and comprehensive genome-wide study coupled with powerful statistical tools, Reich *et al.* undertook a large-scale population survey in India (Reich *et al.*, 2009). Using an Affymetrix (SNP 6.0) array on a total of 132 individuals from 25 diverse groups representing all 4 major language groups (Dravidian, Indo-European, Austroasiatic, and Tibeto-Burman), they predicted that the present-day Indian populations probably originated from a relatively small group of isolated ancestors. This study also revealed that the contemporary Indian populations received varied genomic contribution from two hypothetical ancestral groups of pre-historic India. These hypothetical ancestral populations were termed "ancestral North Indian (ANI)," which shared genetic affinity with the populations of the Middle East, Central Asia and Europe (30–70%), and an "ancestral South Indian (ASI)," which has no relation with any population outside India (Figure 21.1). Further, the present-day Indian populations are the admixture of both ANI and ASI groups. The aboriginal Andaman Islanders were found to harbor only ASI-related genetic component without ANI ancestry (Reich *et al.*, 2009).

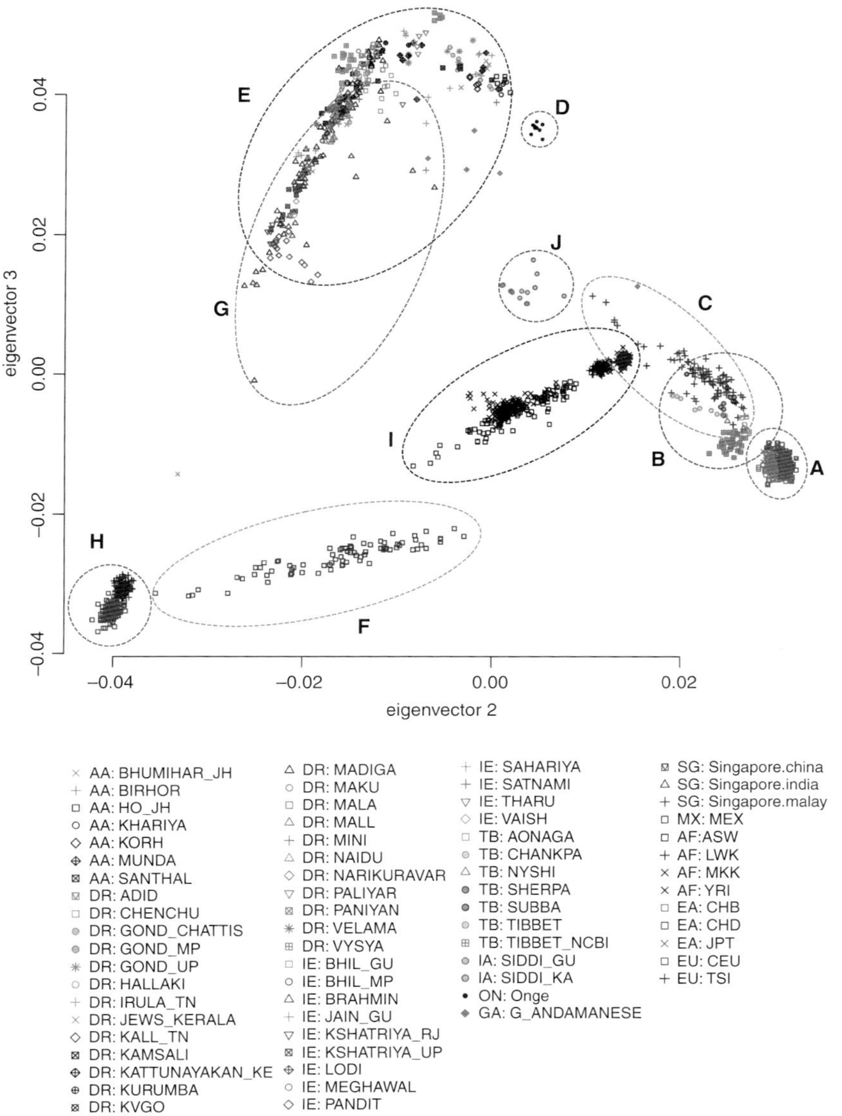

Figure 21.1. Principal component analysis (PCA) of the Indian population with HapMap samples revealed that the genetic architecture of the Indian population is different and the gradient of genetic relatedness with Europeans and the Onge population. Individuals from the same population making a tight cluster suggests endogamy and points toward founder events in the majority of the population. Cluster A: HapMap East Asians (JPT, CHB, CHS) and Singaporean Chinese; cluster B: Tibeto-Burmans of India; cluster C: Singaporean Malay; cluster D: Onge; cluster E: Indo-Europeans, Dravidians and Austroasiatic; cluster F: HapMap Mexican population (MXL); cluster G: Singaporean Indians; cluster H: HapMap European population (CEU and TSI); cluster I: HapMap African population (MKK, YRI and ASW); cluster J: Indo-Africans. (AA, Austroasiatic; DR, Dravidians; IE, Indo-Europeans; TB, Tibeto-Burmans; IN, Gujarati Indians of Houston; EA, East Asians; AF, Africans; EU: Europeans.) A black and white version of this figure will appear in some formats. For the color version, please refer to the plate section.

21.2.2.1. Ancient genetic admixture in India

Mainland India lacked a population with only ASI ancestry like the aboriginal inhabitants of Andaman Islands. When the Indian data set was examined in depth, the allele frequency differences between groups in India were found to be larger than those in Europe. This suggests the occurrence of strong founder events whose signatures have been maintained for thousands of years owing to long-term practised strict endogamy. This study also predicted that there will be an excess of recessive diseases in India. In the follow-up of the above study, Moorjani *et al.* (2013b). have analyzed linkage disequilibrium (LD) to estimate ANI–ASI mixture dates. The exact date of admixture was estimated ranging from about 1900 to 4200 years ago. Interestingly, this result showed the occurrence of a major demographic change in India several thousand years ago; that is, India was transformed from a region of major population admixture to the one in which admixture even between closely related groups became rare because of a shift to strict endogamy practices within the same caste (Moorjani *et al.*, 2013b).

21.3. Recent migrations into and out of India: genetic evidence

21.3.1. Recent migration out-of-India: the story of Roma

The Roma people, living throughout Europe and West Asia, are a diverse population linked by the Romani language and culture. According to Rai *et al.* (2012), the name by which Roma designate themselves is *Rroma* (singular *Rrom*), whereby the double rr in Romani orthography represents a uvular "r" [R] as opposed to an apical "r" [r]. The autonym *Rroma* is held to be cognate with *Ḍoma*, a collective term for the ancient aboriginal populations of the Indian subcontinent (Rai *et al.*, 2012). Many such *Ḍoma* remained as outcastes or tribals, and some were amalgamated into the lower strata of the caste system (Traill, 1851; Driem, 2001). The Indian origin of the Roma populations living in Europe has been established by linguists and geneticists (Turner, 1927; Ali *et al.*, 2009; Mendizabal *et al.*, 2011). However, a conclusive remark on the exact timing of their movement out-of-India and the original parental population was lacking. With scanty historical documentation of the Roma, comparative linguistic studies were the first to recognize their Indian origin (Pott, 1844; Morgenstierne, 1973). Molecular studies based on disease-causing mutations and haploid DNA markers (i.e., mtDNA and Y chromosome) well supported the linguistic view (Gresham *et al.*, 2001; Ali *et al.*, 2009; Klaric *et al.*, 2009). The idea that the proto-Romani founder population of the Roma must have been in northwestern India has been debated (Fraser, 1992; Hancock and Centre de recherches tsiganes (Universit©* Ren©* Descartes), 2002), even though their own origin myth suggests an origin in the Gangetic plain. The preliminary studies with fewer marker and population numbers were able to link the Roma with the Rajput and the Punjabi populations from Rajasthan and Punjab, respectively (Mastana and Papiha, 1992; Mendizabal *et al.*, 2011). In an elaborate study using Y chromosomal haplogroup H1a1a-M82 data across the globe among more than 10,000 samples, Rai *et al.* suggested an early origin of this haplogroup in the Indian subcontinent and its further expansion to other regions (Rai *et al.*, 2012). The STRs of H1a1a-M82 lineages revealed the closest connection of Romani haplotypes with the traditional scheduled caste and scheduled tribe populations of northwestern India (Rai *et al.*, 2012). The average age estimate of Roma founders, considering their distance from Northwest Indian founders, was estimated to be 1405 ± 688 YBP (Rai *et al.*, 2012). Using hundreds of thousands of single nucleotide polymorphisms (SNPs) from 6 Roma groups, it was estimated that the Roma harbored about 80% West Eurasian ancestry (a combination of European and South Asians), and the date of admixture of South Asian and European ancestry was about 850 years before present (Moorjani *et al.*, 2013a). By computing allele sharing as a measure of LD, it was shown that the migration of the Roma out of the Indian subcontinent was accompanied by a severe founder event, followed by a major demographic expansion in Europe (Moorjani *et al.*, 2013a).

The period of the Roma's settlement overlaps with the major epidemic events in the European continent, most notably the plague that took many lives. Populations experience selective pressure and, intriguingly, convergent evolution occurred in both Roma and European populations under the same selective pressure despite their ethnic and genetic differences. It is well known that *TLR2* cluster genes are involved in the recognition of *Yersinia pestis* and trigger the immune system through the release of cytokines (Akira *et al.*, 2006). Laayouni *et al.* found innate immune pathways and *TLR1/TLR6/TLR10* under natural selection, which itself explains natural selection against plague in both populations (Laayouni *et al.*, 2014).

21.3.2. Recent migration into the mainland India: the story of the Siddis

The Siddis are tribal people living mainly in three Indian states, namely Gujarat, Karnataka, and Telangana (formerly Andhra Pradesh). They have typical African features such as dark skin, curly hair, broad nose, etc. They were brought to India by Portuguese traders between the seventeenth and nineteenth centuries and sold to the Nawabs and the Sultans of India to serve as soldiers and slaves. The African-specific marker (*Alu* insertion) was observed among 40% of Siddis (Thangaraj *et al.*, 1999). The African affinity of the Siddis was also confirmed by mtDNA-HVS region analysis (Ramana *et al.*, 2001; Gauniyal *et al.*, 2008, 2011). Despite these findings, a high-resolution approach was lacking to firmly establish the precise genetic origin of the Siddis and their affinity with global populations. To resolve these issues together, a comprehensive study was undertaken on the Siddi group living in the Junagarh district of Gujarat and the Uttara Kannad district of Karnataka using uniparental and high-density biparental markers. To capture the scenario of gene flow, six populations living in close geographical proximity with Siddis were included. Y chromosome and mtDNA data showed the presence of African-specific haplogroups (Shah *et al.*, 2011). To exactly pinpoint the ancestral population, extensive investigation on Y-STRs was carried out, where it was revealed that the Siddis are the direct descendants of the Bantu-speakers of sub-Saharan Africa. The Siddi population was found to have constituted approximately 70% African, and 30% Indian and European genome (Shah et al. 2011). Another study on the same population predicted 58.7% African genomic ancestry, but it was based on only 18,534 SNPs (Narang *et al.*, 2011). Nevertheless, it is well established that the Siddi is an admixed population. To qualitatively demonstrate a history of admixture and estimate a precise date for the admixture event, the ROLLOFF method (Moorjani *et al.*, 2011) was used. The genome of an admixed population contains chromosomal segments from ancestral populations, whose length is inversely proportional to the date of admixture. This piece of information is explored in the ROLLOFF method that uses the information of admixture LD and estimates the time since admixture. The ROLLOFF estimated that the Siddis have admixed with the neighboring Indian populations for about 200 years ago (8 generations). This age of admixture more or less coincides with the historical record of the arrival of Siddi people in India.

An admixture event in the Siddi caused two different genomes that are optimized in different conditions to blend together. It challenged their genome to adapt in new environmental conditions. Thus, during the course of time, the genome accumulated various changes to adapt to the new habitat. The A-variant of the *G6PD* gene protects against malaria and is in high frequency (~60%) among Africans, but absent in our 26 ethnic Indian populations. It is beneficial in the Indian subcontinent also, as malaria is a common disease here, but due to an admixture event it is diluted to ~10% in the Siddi (Shah *et al.*, 2011). Additionally, the Siddis have significant enrichment of variants in genes related to ion channel- and cadherin-related processes close to Indo-Europeans, and kinase- and immune-related processes close to Africans (Narang *et al.*, 2011). It will be intriguing to study the enrichment effect of specific processes in this population.

21.4. Population stratification: implication for health and diseases

21.4.1. Heterogeneity in etiology of disease: prediction comes true

The genetic heterogeneity arising from strict sociocultural, religious, geographical and linguistic demarcations, and endogamy practices along with evolutionary forces has presented an immense challenge in studying human health and diseases in India. The long-term isolation that existed for thousands of years has resulted in higher differences in allele frequency between the groups in India (Reich *et al.*, 2009), and thus, an excess of recessive diseases in India has been predicted (Reich *et al.*, 2009). This is why we have witnessed several regional as well as population-specific diseases in India. To name a few: Handigodu disease (Agarwal *et al.*, 1994; Badadani *et al.*, 2008, 2010), Madras motor neuron disease (Nalini *et al.*, 2006, 2008; Govindaraj *et al.*, 2013), and pseudocholinesterase deficiency among Vyshyas (Rao and Gopalam, 1979; Pandit *et al.*, 2011), etc. Further, the genetic makeup of the Indian populations greatly varies from the rest of the world. Due to this, the designated causal or associated genetic locus (or loci) in western or southeastern Eurasian populations may not be associated with the disease phenotype among Indians. Several genetic studies have proved this

(Table 21.1). One of the best examples is the 25bp deletion in *MYBPC3* (Dhandapany *et al.*, 2009). In phase I, Dhandapany *et al.* selected 354 cardiomyopathy patients and 238 ethnically matched controls, and found the 25bp deletion associated with the phenotype (p-value = 2×10^{-6} and OR = 5.3). Even in the replication study (phase II) with 446 case and 466 controls, the 25bp deletion was in association (p-value = 2×10^{-8} and OR = 8.59) with the phenotype. Moreover, in a recent study, the Indian project team of the International Cancer Genome Consortium (ICGC) reported new genes with recurrent amplifications (*DROSHA, YAP1*) or homozygous deletions (*DDX3X*) in Indian gingivo-buccal oral squamous cell carcinomas which were not reported in other world populations with the same phenotype (India Project Team of the International Cancer Genome Consortium, 2013). Moreover, out of 49 SNPs associated (revealed by meta-analysis) with ulcerative colitis (UC) in populations of European origin, only 5 (rs2395185 of HLA-DRA, rs3024505 of IL10, rs6426833 of RNF186, rs3763313 of BTNL2, and rs2066843 of NOD2) were replicated in a North Indian UC cohort (648 cases and 850 controls; Juyal *et al.*, 2011). It is noteworthy to observe that Caucasian-based meta-analyses have limited replication among Indians. Besides these, few successful replication studies are also reported in Indian population, which makes for a complex scenario (Table 21.2). Therefore, an efficient multidisciplinary section encompassing pharmaco-genomics, evolutionary genomics, and computer science is the immediate need of Indian science to overcome the huge task of understanding Indian genome diversity and its health implications.

21.4.2. Signatures of selection in Indian genome

India is surrounded by Nepal, China, and Bhutan in the north; Burma and Bangladesh in the east; and Pakistan in the west. It has various geographical landscapes. Interestingly, all these habitats have been well explored by different tribes, ancestral tribes and some caste groups. Thus, Indian populations must have acquired changes in their genome, which in turn might have helped them to sustain life in their immediate environment. The detection of such recent natural selection in the human population has profound implications in understanding human history and health conditions.

Using 405 genome-wide SNPs, the Indian Genome Variation Consortium was first to predict the importance of selection in Indian genomes. They observed a number of SNPs with extreme mean heterozygosity and F_{ST} distributions. Subsequently, a comprehensive study on Ayurvedic Prakritis was carried out. Ayurvedic Prakriti is determined on the basis of phenotypic characters in normal individuals. The three Prakriti are Vata, Pitta, and Kapha, and vary with respect to phenotypes as well as their response to the external environment. A study using 141 tag SNPs encompassing 30 genes selected from the 251 differentially expressed genes among 3 Prakriti types observed differential expression of the *EGLN1* gene (Aggarwal *et al.*, 2010). They observed a significant link between

Table 21.1. Genetic heterogeneity in disease causing gene/mutation among Indian and world populations.

Phenotype	Gene	Genetic variation	Ethnicity	Assoc.	*P*-value	Number of cases	Number of controls	References
Glaucoma	*ZP4*	rs547984	Japanese	✓	0.00006	418	300	Nakano *et al.*, 2009
Glaucoma	*ZP4*	rs540782	Japanese	✓	0.00006			
Glaucoma	*ZP4*	rs693421	Japanese	✓	0.00004			
Glaucoma	*ZP4*	rs2499601	Japanese	✓	0.00009			
Glaucoma	*DKFZp762A217*	rs7081455	Japanese	✓	0.00001			
Glaucoma	*PlxDC2*	rs7961953	Japanese	✓	0.00007			
Glaucoma	*ZP4*	rs547984	Indians	✗	0.271	140 + 111	219	Rao *et al.*, 2009
Glaucoma	*ZP4*	rs540782	Indians	✗	0.255			
Glaucoma	*ZP4*	rs693421	Indians	✗	0.19			
Glaucoma	*ZP4*	rs2499601	Indians	✗	0.168			
Glaucoma	*DKFZp762A217*	rs7081455	Indians	✗	0.676			
Glaucoma	*PlxDC2*	rs7961953	Indians	✗	0.124			
Parkinson's disease	*LRRK2*	–	Italian	✓	40 (1.59%)	2523	–	Cilia *et al.*, 2014
Parkinson's disease	*LRRK2*	6 tag SNPs	Indians	✗	0.29–1.0	249	231	Sadhukhan *et al.*, 2012
Parkinson's disease	*LRRK2*	5 SNPs at 12q12	Japanese	✓	2.72×10^{-8}	1078	2628	Satake *et al.*, 2009
T2D	*SGCG*	rs952911	Indians	✓	1.82×10^{-8}	12,535	–	Saxena *et al.*, 2013
T2D	*TMEM163*	rs998451	Indians	✓	6.3×10^{-12}			
Infertility	*AR*	CAG repeats	Spanish	✓	0.033	102	96	Mengual *et al.*, 2003
Infertility	*AR*	CAG repeats	Indian	✗	0.967	280	201	Thangaraj *et al.*, 2002
Infertility	*DAZL*	A396G	Italian	✓	Absent	92	63	Bartoloni *et al.*, 2004
Infertility	*DAZL*	A396G	Indian	✗	Absent	1010	–	Thangaraj *et al.*, 2006b
Infertility	*DAZL*	A396G	Taiwanese	✗	0.003	160	–	Teng *et al.*, 2002
Infertility	*APOB*	Deletion	Solvene	✓	0.03	195	250	Peterlin *et al.*, 2006
Infertility	*APOB*	Deletion	Indian	✗	0.45	102	251	Khattri *et al.*, 2009a, 2009b
Infertility	*Erß*	rs1256049	Caucasians	✓	0.01	106	186	Aschim *et al.*, 2005

Infertility	*Erß*	rs1256050	Indians	✘	0.769	443	252	Khattri, Pandey et al. 2009a, 2009b
Obesity	*INSIG2*	rs756605	Europeans: FHS	✓	0.0026	3996	–	Herbert *et al.*, 2006
Obesity	*INSIG2*	rs756605	Indians	✘	0.04 R/ 0.67 D	1577	–	Kumar *et al.*, 2007a
Obesity	*NPY*	L7P-non-obese	Swedish	✓	0.005	1246	–	Ding *et al.*, 2005
Obesity	*NPY*	L7P obese	Indians	✘	TC-0.515, CC-0.718	93	170	Bhaskar *et al.*, 2010
Low birth weight and T2D	*CCNL1*	rs900400	Europeans	✓	3×10^{-26}	10623+27591	–	Freathy *et al.*, 2010
Low birth weight and T2D	*ADCY5*	rs9883204	Europeans	✓	3×10^{-9}			
Low birth weight and T2D	*CCNL1*	rs900400	Indians	✘	0.87	2151	–	Vasan *et al.*, 2011
Low birth weight and T2D	*ADCY5*	rs9883204	Indians	✘	0.54			
Noonan syndrome	*PTPN11*	Various mutations	Europeans	✓	–	–	–	Lee *et al.*, 2010
Noonan syndrome	*PTPN11*	Absent of Mutation	Indians	✘	Absent	–	–	Rani *et al.*, 2010
Visceral Leishmanias	*SLC11A1*	GTn	Sudanians	✓	0.0145	59 Family (177 cases)	–	Mohamed *et al.*, 2004
Visceral Leishmanias	*SLC11A1*	274C/T	Sudanians	✓	0.0037			
Visceral Leishmanias	*SLC11A1*	469+14C/G	Sudanians	✓	0.0049			
Visceral Leishmanias	*SLC11A1*	GTn	Indians	✓	0.052	Phase I: 176 Family (313 cases)	–	Mehrotra *et al.*, 2011

Table 21.1. (cont.)

Phenotype	Gene	Genetic variation	Ethnicity	Assoc.	*P*-value	Number of cases	Number of controls	References
Visceral Leishmanias	*SLC11A1*	274C/T	Indians	✘	0.282			
Visceral Leishmanias	*SLC11A1*	469+14C/G	Indians	✘	0.148			
Visceral Leishmanias	*SLC11A1*	GTn	Indians	✘	0.585	Phase II: 941 cases	992	
Visceral Leishmanias	*SLC11A1*	274C/T	Indians	✘	0.529			
Visceral Leishmanias	*SLC11A1*	469+14C/G	Indians	✘	0.987			
Prostate cancer	*SRD5A2*	TAn	Australian	✘	Absent	736	827	Hayes *et al.*, 2007
Prostate cancer	*SRD5A2*	A49T	Australian	✓	0.02			
Prostate cancer	*SRD5A2*	V89L	Australian	✘	0.5			
Prostate cancer	*SRD5A2*	TAn	French	✘	Absent	803	802	Cussenot *et al.*, 2007
Prostate cancer	*SRD5A2*	A49T	French	✘	Absent			
Prostate cancer	*SRD5A2*	V89L	French	✓	0.0017			
Prostate cancer	*SRD5A2*	TAn	Indian	✓	–	127	96	Rajender *et al.*, 2009
Prostate cancer	*SRD5A2*	A49T	Indian	✘	Absent			
Prostate cancer	*SRD5A2*	V89L	Indian	✘	0.86/1			
Gall bladder cancer	*DCC*	rs4078299	Japanese	✓		30	898	Cha *et al.*, 2012
Gall bladder cancer	*DCC*	rs7504990	Japanese	✓	7.46×10^{-8}			
Gall bladder cancer	*DCC*	rs2229080	Japanese	–	–			
Gall bladder cancer	*DCC*	rs714	Japanese	–	–			
Gall bladder cancer	*DCC*	rs4078299	Indians	✘	0.101	406	260	Rai *et al.*, 2013
Gall bladder cancer	*DCC*	rs7504990	Indians	✘	0.644			
Gall bladder cancer	*DCC*	rs2229080	Indians	✓	0.002			
Gall bladder cancer	*DCC*	rs714	Indians	✓	0.007			

Migrain	*LRP1*	rs11172113	Europeans	✓	4.3×10^{-9}	5122	18,108	Chasman *et al.*, 2011
Migrain	*PRDM16*	rs2651899	Europeans	✓	3.8×10^{-9}			
Migrain	*LRP1*	rs11172113	Indians	✓	0.019	340	200	Ghosh *et al.*, 2013
Migrain	*PRDM16*	rs2651899	Indians	✓	0.017			
Level of Vit.B12	*FUT2*	rs602662	Mostly Europeans	✓	5.36×10^{-17}	1658+1059	–	Hazra *et al.*, 2008
Level of Vit.B12	*FUT2*	rs602662	Indians	✓	<0.0001	1146	–	Tanwar *et al.*, 2013
Lipid parametes	*APOA5*	rs662799	Europeans	✓	2.4×10^{-15}	8816	–	Willer *et al.*, 2008
Lipid parametes	*LPL*	rs10503669	Europeans	✓	4.1×10^{-19}			
Lipid parametes	*GCKR*	rs780094	Europeans	✓	6.1×10^{-32}			
Lipid parametes	*APOA5*	rs662799	Indians	✓	1.97×10^{-7}	3178 sib-pairs	–	Rafiq *et al.*, 2012
Lipid parametes	*LPL*	rs10503669	Indians	✓	1×10^{-4}			
Lipid parametes	*GCKR*	rs780094	Indians	✓	0.001			
T2D	*KCNQ*	rs231362	Japanese	✓	2.8×10^{-13}	8130	38987	Voight *et al.*, 2010
T2D	*KCNQ*	rs231362	Indians	✓	0.002	879	–	Been *et al.*, 2011

FHS, Framingham heart study; T2D, type 2 diabetes; Vit.B12, vitamin B12; ✓, associated; ✘, not associated.

Table 21.2. Frequency distribution of some variants with therapeutic importance in Indian and world populations.

Chromosome	rsIDs	Effect on therepeutic response	Population	Allele-I	Allele-I	Allele-I frequency	Allele-I count	Total Allele	Chromosome $^{rh}{}_{c}$	rsIDs	Effect on therepeutic response	Population	Allele-I	Allele-I	Allele-I frequency	Allele-I count	Total allele
22	rs9332377	T allele is associated with	Indians	T	C	0.1107	60	542	3	rs4686799	T allele is associated with	Indians	T	C	0.1305	71	544
22	rs9332377	increased risk of ototoxicity	Siddi	T	C	0.2857	8	28	3	rs4686799	increased aldosterone	Siddi	T	C	0.3214	9	28
22	rs9332377	when treated with cisplatin in children as compared to allele C	Tibeto-Burman + Tibetans	T	C	0.01724	2	116	3	rs4686799	response when treated with candesartan and hydrochlorothiazide	Tibeto-Burman + Tibetans	T	C	0.4211	48	114
22	rs9332377		ASW	T	C	0.3673	36	98	3	rs4686799		ASW	T	C	0.1875	18	96
22	rs9332377		CEU	T	C	0.1741	39	224	3	rs4686799		CEU	T	C	0.2027	45	222
22	rs9332377		CHB	T	C	0.00595	1	168	3	rs4686799		CHB	T	C	0.3735	62	166
22	rs9332377		CHD	T	C	0	0	170	3	rs4686799		CHD	T	C	0.4471	76	170
22	rs9332377		GIH	T	C	0.1761	31	176	3	rs4686799		GIH	T	C	0.1534	27	176
22	rs9332377		JPT	T	C	0	0	172	3	rs4686799		JPT	T	C	0.407	70	172
22	rs9332377		LWK	T	C	0.3444	62	180	3	rs4686799		LWK	T	C	0.2557	45	176
22	rs9332377		MEX	T	C	0.08	8	100	3	rs4686799		MEX	T	C	0.24	24	100
22	rs9332377		MKK	T	C	0.3732	106	284	3	rs4686799		MKK	T	C	0.2413	69	286
22	rs9332377		TSI	T	C	0.1609	28	174	3	rs4686799		TSI	T	C	0.2216	39	176
22	rs9332377		YRI	T	C	0.3584	81	226	3	rs4686799		YRI	T	C	0.1549	35	226
22	rs9332377		Singapore. china	T	C	0	0	192	3	rs4686799		Singapore. china	T	C	0.3073	59	192
22	rs9332377		Singapore. Indians	T	C	0.2048	34	166	3	rs4686799		Singapore. Indians	T	C	0.1807	30	166
22	rs9332377		Singapore. malay	T	C	0.00562	1	178	3	rs4686799		Singapore. malay	T	C	0.2416	43	178
2	rs887829	T allele is associated with	Indians	T	C	0.4267	233	546	1	rs1801131	G allele is associated with	Indians	G	T	0.4007	222	554
2	rs887829	increased bilirubin	Siddi	T	C	0.5	14	28	1	rs1801131	side effect of	Siddi	G	T	0.2143	6	28

2	rs887829	concentration on treatment with atazanavir in patients with HIV infections	Tibeto-Burman + Tibetans	T	C	0.1207	14	116	1	rs1801131	methotrexate and increased response of fluorouracil	Tibeto-Burman + Tibetans	G	T	0.1638	19	116
2	rs887829		ASW	T	C	0.4694	46	98	1	rs1801131		ASW	G	T	0.1837	18	98
2	rs887829		CEU	T	C	0.317	71	224	1	rs1801131		CEU	G	T	0.3393	76	224
2	rs887829		CHB	T	C	0.1131	19	168	1	rs1801131		CHB	G	T	0.2202	37	168
2	rs887829		CHD	T	C	0.1412	24	170	1	rs1801131		CHD	G	T	0.2118	36	170
2	rs887829		GIH	T	C	0.4253	74	174	1	rs1801131		GIH	G	T	0.392	69	176
2	rs887829		JPT	T	C	0.1802	31	172	1	rs1801131		JPT	G	T	0.186	32	172
2	rs887829		LWK	T	C	0.5278	95	180	1	rs1801131		LWK	G	T	0.1833	33	180
2	rs887829		MEX	T	C	0.31	31	100	1	rs1801131		MEX	G	T	0.21	21	100
2	rs887829		MKK	T	C	0.4825	138	286	1	rs1801131		MKK	G	T	0.2657	76	286
2	rs887829		TSI	T	C	0.2614	46	176	1	rs1801131		TSI	G	T	0.3182	56	176
2	rs887829		YRI	T	C	0.531	120	226	1	rs1801131		YRI	G	T	0.115	26	226
2	rs887829		Singapore. china	T	C	0.08333	16	192	1	rs1801131		Singapore. china	G	T	0.2552	49	192
2	rs887829		Singapore. Indians	T	C	0.4398	73	166	1	rs1801131		Singapore. Indians	G	T	0.4458	74	166
2	rs887829		Singapore. malay	T	C	0.1067	19	178	1	rs1801131		Singapore. malay	G	T	0.3933	70	178

high-altitude adaptation and common variations rs479200 (C/T) and rs480902 (T/C) of the *EGLN1* gene. The TT genotype of rs479200 was found to be more frequent in Kapha types and was associated with patients suffering from high-altitude pulmonary edema. Its lower frequency was found in Pitta types and it was nearly absent in high-altitude natives. The ancestral allele (T) of rs480902 was found in a higher frequency in Pitta and positively correlated among people living in high altitudes globally ($P < 0.001$), including in India. Thus, polymorphisms in the *EGLN1* gene are associated with high-altitude adaptation and a genotype rare in highlanders but frequent in normal lowlanders may confer increased risk for high-altitude pulmonary edema (Aggarwal *et al.*, 2010).

Malaria is one of the major infectious diseases, and is caused by *Plasmodium* sp. It is the major reason for mortality in tropical/subtropical countries such as India. With 430 cases and ethnically matched controls, Jha *et al* (2012) found that –590C>T, –34C>T (P-value < 0.001) and intron-3 VNTR R2 (P-value $= 0.003$) of the *IL-4* gene are associated with reduced malaria risk (P-value $= 0.009$). Intriguingly, a social cluster in the Indian population has a significantly different frequency spectrum. The ancestral tribal population has 40% TTR2 haplotype, while both nomadic and castes have the lowest, 21.6% (Jha *et al.*, 2012).

Using thousands of autosomal markers, Narang *et al.* (2011) showed selective enrichment of biological process among the admixed population. They found significant enrichment of cadherins, potassium channels, membrane proteins, solute carriers, protein kinases, and immune-related genes in the studied admixed group (Narang *et al.*, 2011). Metspalu *et al.* (2011) showed a higher proportion of the Indian genome with regional-specific signals of high haplotype homozygosity as compared to other populations. Positive selections among Indians were observed in *MSTN* and *DOK5* genes. These genes have potential implications in lipid metabolism and the etiology of type 2 diabetes (Metspalu *et al.*, 2011). Basu Mallick *et al.* (2013) found the positive selection of *SLC24A5* gene among populations of Middle East, Central Asia, Pakistan, and North India, but not in South India.

21.5. Population stratification: implications for pharmaco-genomics

Inter-population variation in therapeutic response to drugs is the major hurdle in the treatment of diseases, as a lack of response or hypersensitive reactions have severe medical consequences. It is well documented that significant differences in the frequency of pharmaco-genomic variants among different ethnic populations are one of the reasons for modulation of drug response (Yasuda *et al.*, 2008; Loh *et al.*, 2013). For example, individuals with European ancestry need higher doses of warfarin in comparison to people with African ancestry and Asians, which is due to variation in the frequency of *CYP2C9*2* (0.08–0.18 in Europeans, rare in Africans and Asians), *CYP2C9*3* (0.05–0.13 in Europeans, 0.01–0.02 in Africans, and 0.02–0.05 in Asians), and *VKORC1* variants (0.35–0.45 in Europeans, 0.08–0.1 in Africans, and 0.9–0.95 in Asians). Similarly,

*UGT1A1*6*, which has reduced catalytic activity against irinotecan, is absent in Europeans and Africans, while in Japanese its frequency is 0.04; and the frequency of *SCLO1B1*15*, which has reduced activity against pravastatin and rosuvastatin, is 0.027 in Europeans, 0.14 in Chinese, and 0.03–0.103 in Japanese (Yasuda *et al.*, 2008). Several reports are available for representing the frequency of pharmacologically important variants in most of the populations worldwide. Unfortunately, Indian populations have not been explored in this manner.

In order to understand the distribution of pharmacologically important SNPs, we performed PCA with 54 Indian populations from diverse ethnic, linguistic (4 Tibeto-Burmans, 1 semi-nomadic Tibetans, 1 Tibetan-refugees, 13 Indo-Europeans, 21 Dravidians, 10 Austroasiatic, 2 Indo-Africans, 1 Great-Andamanese, and 1 Onge), and geographical locations. To compare the diversity and natural selection, we selected 10 populations from HapMap samples (Europeans: CEU, TSI; East Asians: CHB, CHD, JPT; Americans: MEX; Africans: MKK, YRI, ASW; and admixed Indian population: GIH), 3 from the Singapore genome diversity project (Malaysians, Chinese, and Indians), and 1 Tibetan population from the NCBI-GEO database. We selected a total of 811 genes from 102 pharmacokinetic and pharmacodynamic important pathways from PharmGKB database. Besides these, we have also selected 1 core and 165 extended ADME genes from the PharmaADME database, which are not in the pathway but were previously reported to be associated with drug metabolism. PCA generated using 17,358 SNPs revealed that Indian populations have different genetic compositions, pharmaco-genomics in particular, compared to populations from the rest of the world (Figure 21.2). the Within Indian subcontinent, Tibetan-refugees, Tibeto-Burman and Indo-Africans population are different compared to mainland Indo-European, Dravidian and Austroasiatic populations. Besides this, Singapore-Indians were observed not clustering with the Indian population, suggesting that this population should not be used as representative of the Indian population. In order to reflect the frequency distribution of pharmacologically important variants among Indians and world population, we have presented some examples in Table 21.2.

The above prediction came true in a recent study on 24 ethnically diverse populations with 2680 individuals for rs1799853 (*CYP2C9*3*), rs1057910 (*CYP2C9*3*), rs2108622 (*CYP4F2*3*), and rs7294, rs9923231, and rs9934438 of *VKORC1* (related to warfarin dose). The frequency of rs2108622 was much higher in Indo-Europeans (0.3–0.44) in comparison to the world population, that is, Caucasian (0.34), African-American (0.12), and Hispanic (0.23). rs9923231 was much more prevalent in Tibeto-Burmans (>0.7) in comparison to Indo-Europeans (as low as 0.065), whereas all three variations of *CYP2C9* showed a considerable amount of variation within Indo-Europeans. Moreover, the frequency of rs1057910 (0.9–0.12) was higher in comparison to other Asian populations (0.4). Based on this, it can be hypothesized that the major Indian subpopulation needs a higher dose of warfarin (as known earlier), but care should be taken in the case of Tibeto-Burmans (Giri *et al.*, 2014).

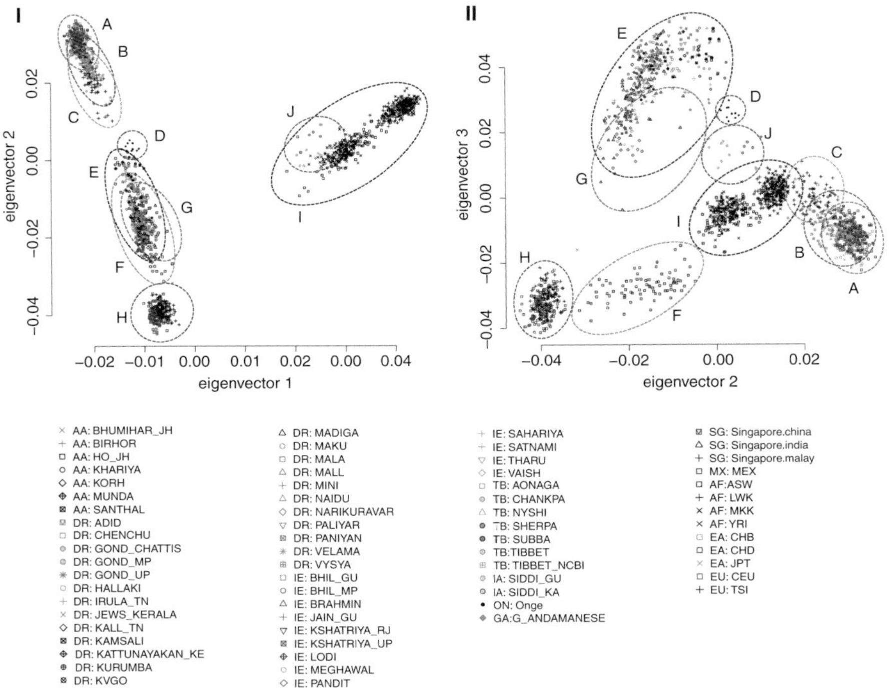

Figure 21.2. Principle component analysis (PCA) of different ethnic populations, using pharmacogenomically important SNPs. (I) On eigenvector 1 the Indian population splits out into two clusters; one consists of IA population while the rest of the population is in the second cluster. On eigenvector 2, Tibeto-Burmans split from the second cluster. (II) Mexican and Singaporean Indians which are clustering on both eigenvectors 1 and 2 with the Indian population split on eigenvector 3, which reveals that Singaporean Indians cannot be used as representative of Indians. Notations are the same as in Figure 21.1. A black and white version of this figure will appear in some formats. For the color version, please refer to the plate section.

21.5.1. Implication of ancestry studies in pharmaco-genomics

Systematic allele frequency differences between cases and controls due to ancestry are one of the major hurdles in the association studies as they could confound genetic mapping results. The F_{st} value among the Indian population is very high (>0.0033; average F_{st} value among 23 European groups), even after correction for inbreeding and consideration of the same linguistic population living at a nearby location – for example, Mala has 0.014 F_{st} value with Vysya and 0.004 with Adi-Dravidar populations, while Vysya and Mala live nearby the region in the same Andhra Pradesh state and Adi-Dravidars are living in Tamil Nadu. After an admixture event between ANI and ASI, founder events resulted in the rise of the several small genetic clusters of populations which has remained isolated to date through strict endogamy practices (Reich *et al.*, 2009).

The F_{st} value directly points to the higher allele frequency among the populations due to ancestry. Besides Mala and Vysya, we also considerd two Indo-European populations, Brahmin and Kshatriya, living nearby and calculated the p-value

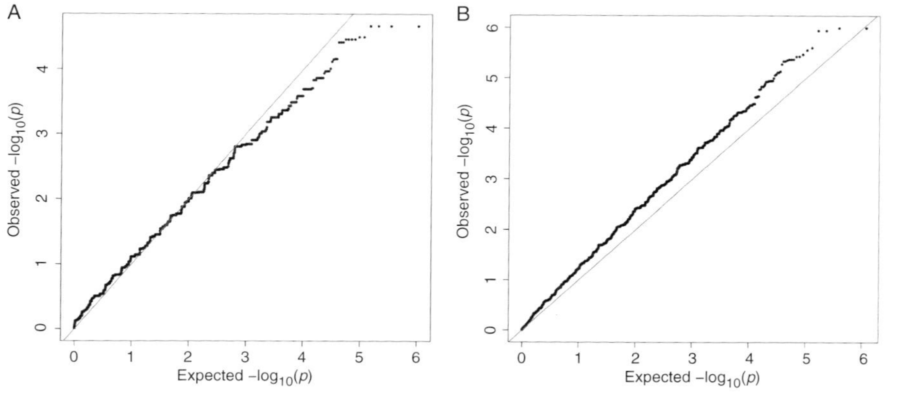

Figure 21.3. Stratification in population living at nearby location and with same ancestry, revealed by QQ plot. Deviation of observed $-\log_{10}$ *p*-value of allele frequency differences between (A) Brahmin and Kshatriya populations (Indo-Europeans, living in Uttar Pradesh), (B) Mala and Vysya (Dravidians, living in Andhra Pradesh) revealed stratification.

(chi-square statistics) of allele frequency differences. A quantile–quantile (QQ) plot revealed that several SNPs analyzed had a $-\log_{10}$ *p*-value significantly deviated from their expected value. Hence, stratification is more dominant in the Indian population (Figure 21.3). It is strongly advisable to correct it before performing association analysis, and this could be achieved with a SMARTPCA tool (Patterson *et al.*, 2006).

21.5.2. Imputation in pharmaco-genomics: obstacle in the Indian population

A major obstacle of genome-wide association analysis in pharmaco-genomics is sample size, as non-responder and hypersensitive reactions for a drug are very rare in nature. Although some studies have managed to collect a sufficient number of individuals, there is a need to increase the statistical power. SNP arrays are designed to represent the whole genome, but due to experimental error we may lose some of the informative markers, and the power of association studies further decreases. Moreover, it will be more detrimental if the lost marker is in LD with the functional SNP. These markers can be statistically imputed, for which many tools such as PHASE, fastPHASE, BEAGLE, IMPUTE, and MACH are available. PHASE and fastPHASE are based on the expectation-maximization algorithm (computationally extensive), while MACH, IMPUTE, and BEAGLE use a hidden Markov Model (computationally intensive; needs reference population; Browning and Browning, 2007). With dense markers (low density = 1 SNP per 10 kb while high density = 1 SNP per 3 kb) and large number of samples (sample size: small = 100; medium = 1000; large = 5000), PHASE and fastPHASE are not practically feasible, as they require days to finish the imputation (150.6 seconds per SNP for highly dense and large number of SNPs; imputation time is calculated for fastPHASE only; Browning and Browning, 2007). In this case, BEAGLE, IMPUTE, and MACH are the best choice, not only because of their speed but also for accuracy. However,

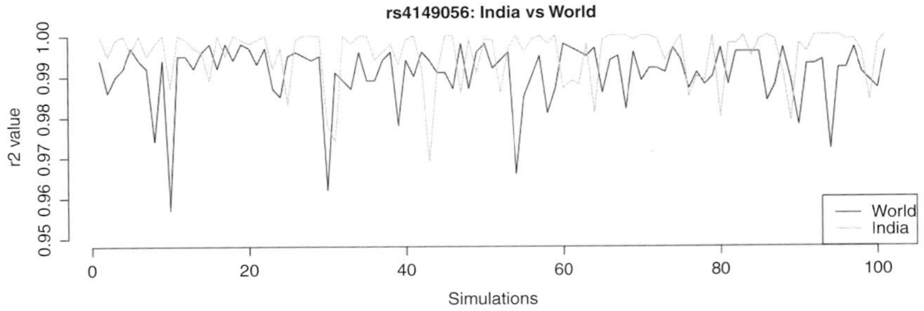

Figure 21.4. Performance of imputation in Indian genotypic data is always better with Indian genotype reference compared to populations from the rest of the world.

these tools consider the genotypic reference collectively from world population (Europeans, Asian, Americans, and Africans) for imputation. Because Indians clustered separately (which pinpoint different genetic architecture) in PCA, the correct imputation with reference of world population is skeptical. As an example; with 2%, 5% masked data of chromosome 22 in Indian samples, we performed 110 simulations and found that the Indian genotypic reference is more appropriate for evaluation compared to the populations from the rest of the world. It is explained with rs4149056, a variant present in *SLCO1B1* gene which is associated with statin-induced myopathy (P-value = 2.00e^{-9}; OR = 4.5). In 110 simulations, for imputing the genotype of Indian samples, the reference of India outperformed in comparison to populations from the rest of the world (Figure 21.4). Incorrect imputed genotype could lead to either a spurious association or no association. Hence, we strongly recommend the use of the Indian data set as a reference to analyse Indian genotypic data.

21.6. Why personalized therapy: a statistical overview

Traditionally, patients are treated based on clinical symptoms, medical history, and laboratory evaluation, even though this approach fails sometime. Different therapeutic responses are the major hurdle in disease treatment, because a lack of response and hypersensitive reactions have major economic and medical consequences. The main reason behind this is not taking into consideration the genetic difference while treating patients. The genetic bases of diseases are totally different from drug metabolism, their absorption, and excretion, and hence there is the need for consideration of these facts by physicians.

The importance of personalized therapy can be explained statistically. Suppose a phenotype x is influenced by alleles $a_1, a_2, a_3, \ldots, a_n$ and the probability of occurrence of these markers in population are $o_1, o_2, o_3, \ldots, o_i$, respectively. Then, the probability P of finding the same combination of alleles (if any alleles are not in LD) is

$$P\alpha_{i=1}\Pi^{i=n} o_i$$

Because o_i always has a value < 1:

$$P\alpha 1/i$$

Therefore, the probability of finding the same combination is inversely proportional to the number of alleles affecting the phenotype. We should also note that here, we are talking about population-based association studies, and this statistic is true only for population- and not for family-based association studies. If the alleles have penetrance m_1, m_2, m_3,..., m_n, then outcomes Xs_1, Xs_2, Xs_3, ..., Xs_j will be different for set s_1, s_2, s_3, ..., s_j, which depends upon composition and can be given as if we are not considering epistasis:

$$Xs_j = \text{Constant} +_{i=1}\Pi^{i=n} m_i$$

We have seen earlier that individuals have different combinations so their clinical outcome will be different, and this concept has much importance if the drug has a narrow therapeutic index, such as warfarin.

21.7. Concluding remarks

The era of modern human migration, settlement, social structure, habitat, etc., has many hidden answers for modern-day issues including resistance or susceptibility to disease, response to drugs, and different physiological and physical features. The first wave of migration into the Indian subcontinent, the presence of different genetic ancestors, their multiple admixture and subsequent endogamy practices in Vedic times makes the Indian subcontinent an ethnic museum. Long-term isolation of Indian populations with unique genetic histories provide the basis for genetic heterogeneity. The different genetic architecture of Indian population poses problems in the correct imputation of genotype. Moreover, founder events and endogamy practices have strong influences on population stratification, which is a serious concern for statistical association studies. It is evident that ethnic differences in the frequency of pharmacologically important variation are related to different therapeutic responses. Indian populations were explored and striking variation was found compared to the rest of the world. Researchers should focus on Indian-specific variations for better understanding health, disease, drug response, adaptation, etc.

Acknowledgments

The authors thank all our national and international collaborators for their constant cooperation and efficient teamwork. Students of various universities across the country are greatly acknowledged for their valuable contributions in sampling. SN was supported by ICMR-SRF, Government of India. RT was supported by the CSIR-SRF, Government of India and European Social Funds Doctoral Studies and Internationalisation Programme DoRa. KT was supported by CSIR Network project (GENESIS – BSC0121), Government of India.

REFERENCES

Agarwal, S.S., Phadke, S.R., Phadke, R.V., *et al.* (1994). Handigodu disease: a radiological study. A new variety of spondyloepi(meta)physeal dysplasia of the autosomal dominant type. *Skel. Radiol.*, **23**(8), 611–619.

Aggarwal, S., Negi, S., Jha, P., *et al.* (2010). EGLN1 involvement in high-altitude adaptation revealed through genetic analysis of extreme constitution types defined in Ayurveda. *Proc. Natl Acad. Sci. USA*, **107**(44), 18961–18966.

Akira, S., Uematsu, S. and Takeuchi, O. (2006). Pathogen recognition and innate immunity. *Cell*, **124**(4), 783–801.

Ali, M., McKibbin, M., Booth, A., *et al.* (2009). Null mutations in *LTBP2* cause primary congenital glaucoma. *Am. J. Hum. Genet.*, **84**(5), 664–671.

Aschim, E.L., Giwercman, A., Stahl, O., *et al.* (2005). The RsaI polymorphism in the estrogen receptor-beta gene is associated with male infertility. *J. Clin. Endocrinol. Metab.*, **90**(9), 5343–5348.

Badadani, M., Shetty, K.T. and Agarwal, S. (2010). Hypocalcitonemia in Handigodu Disease: a spondylo epi (meta) physeal dysplasia. *Int. J. Clin. Exp. Med.*, **3**(2), 115–121.

Badadani, M., Shetty, K.T., Babu, S.V. and Agarwal, S. (2008). Metabolic status of magnesium and ceruloplasmin in Handigodu Joint Disease: a variety of spondylo epi (meta) physeal dysplasia. *Clin. Chim. Acta*, **395**(1–2), 170–171.

Bamshad, M., Kivisild, T., Watkins, W.S., *et al.* (2001). Genetic evidence on the origins of Indian caste populations. *Genome Res.*, **11**(6), 994–1004.

Bartoloni, L., Cazzadore, C., Ferlin, A., Garolla, A. and Foresta, C. (2004). Lack of the T54A polymorphism of the *DAZL* gene in infertile Italian patients. *Molec. Hum. Reprod.*, **10** (8), 613–615.

Basu Mallick, C., Iliescu, F.M., Mols, M., *et al.* (2013). The light skin allele of SLC24A5 in South Asians and Europeans shares identity by descent. *PLoS Genet.*, **9**(11), e1003912.

Been, L.F., Ralhan, S., Wander, G.S., *et al.* (2011). Variants in *KCNQ1* increase type II diabetes susceptibility in South Asians: a study of 3,310 subjects from India and the US. *BMC Med. Genet.*, **12**, 18.

Behar, D.M., Yunusbayev, B., Metspalu, M., *et al.* (2010). The genome-wide structure of the Jewish people. *Nature*, **466**(7303), 238–242.

Bhaskar, L.V., Thangaraj, K., Pardhasaradhi, G., *et al.* (2010). Neuropeptide Y gene polymorphisms are not associated with obesity in a South Indian population. *Eur. J. Clin. Nutr.*, **64**(8), 868–872.

Browning, S.R. and Browning, B.L. (2007). Rapid and accurate haplotype phasing and missing-data inference for whole-genome association studies by use of localized haplotype clustering. *Am. J. Hum. Genet.*, **81**(5), 1084–1097.

Cha, P.C., Zembutsu, H., Takahashi, A., *et al.* (2012). A genome-wide association study identifies SNP in DCC is associated with gallbladder cancer in the Japanese population. *J. Hum. Genet.*, **57**(4), 235–237.

Chasman, D.I., Schurks, M., Anttila, V., *et al.* (2011). Genome-wide association study reveals three susceptibility loci for common migraine in the general population. *Nature Genet.*, **43**(7), 695–698.

Chaubey, G. and Endicott, P. (2013). The Andaman Islanders in a regional genetic context: reexamining the evidence for an early peopling of the archipelago from South Asia. *Hum. Biol.*, **85**(1–3), 153–172.

Chaubey, G., Karmin, M., Metspalu, E., *et al.* (2008). Phylogeography of mtDNA haplogroup R7 in the Indian peninsula. *BMC Evol. Biol.*, **8**, 227.

Chaubey, G., Metspalu, M., Choi, Y., *et al.* (2011). Population genetic structure in Indian Austroasiatic speakers: the role of landscape barriers and sex-specific admixture. *Molec. Biol. Evol.*, **28**(2), 1013–1024.

Cilia, R., Siri, C., Rusconi, D., *et al.* (2014). *LRRK2* mutations in Parkinson's disease: confirmation of a gender effect in the Italian population. *Parkins. Rel. Disord.*, **20**(8), 911–914.

Cussenot, O., Azzouzi, A.R., Nicolaiew, N., *et al.* (2007). Low-activity V89L variant in SRD5A2 is associated with aggressive prostate cancer risk: an explanation for the adverse effects observed in chemoprevention trials using 5-alpha-reductase inhibitors. *Eur. Urol.*, **52**(4), 1082–1087.

Das, B.M., Das, P.B., Das, R., Walter, H. and Danker-Hopfe, H. (1986). Anthropological studies in Assam, India. *Anthropol. Anz.*, **44**(3), 239–248.

Dhandapany, P.S., Sadayappan, S., Xue, Y., *et al.* (2009). A common *MYBPC3* (cardiac myosin binding protein C) variant associated with cardiomyopathies in South Asia. *Nature Genet.*, **41**(2), 187–191.

Ding, B., Kull, B., Liu, Z., *et al.* (2005). Human neuropeptide Y signal peptide gain-of-function polymorphism is associated with increased body mass index: possible mode of function. *Regul. Pept.*, **127**(1–3), 45–53.

Driem, G. v. (2001). *Languages of the Himalayas: An Ethnolinguistic Handbook of the Greater Himalayan Region, Containing an Introduction to the Symbiotic Theory of Language*. Brill, Leiden.

Fraser, A.M. (1992). *The Gypsies*. Blackwell, Oxford.

Freathy, R.M., Mook-Kanamori, D.O., Sovio, U., *et al.* (2010). Variants in *ADCY5* and near *CCNL1* are associated with fetal growth and birth weight. *Nature Genet.*, **42**(5), 430–435.

Gauniyal, M., Chahal, S.M. and Kshatriya, G.K. (2008). Genetic affinities of the Siddis of South India: an emigrant population of East Africa. *Hum. Biol.*, **80**(3), 251–270.

Gauniyal, M., Aggarwal, A. and Kshatriya, G.K. (2011). Genomic structure of the immigrant Siddis of East Africa to southern India: a study of 20 autosomal DNA markers. *Biochem. Genet.*, **49**(7–8), 427–442.

Ghosh, J., Pradhan, S. and Kshatriya, G.K. (2013). Genome-wide-associated variants in migraine susceptibility: a replication study from North India. *Headache*, **53**(10), 1583–1594.

Giri, A.K., Khan, N.M., Grover, S., *et al.* (2014). Genetic epidemiology of pharmacogenetic variations in *CYP2C9, CYP4F2* and *VKORC1* genes associated with warfarin dosage in the Indian population. *Pharmacogenomics*, **15**(10), 1337–1354.

Govindaraj, P., Nalini, A., Krishna, N., *et al.* (2013). Mitochondrial DNA variations in Madras motor neuron disease. *Mitochondrion*, **13**(6), 721–728.

Gresham, D., Morar, B., Underhill, P.A., *et al.* (2001). Origins and divergence of the Roma (gypsies). *Am. J. Hum. Genet.*, **69**(6), 1314–1331.

Hancock, I.F. and Centre de recherches tsiganes (Universit©* Ren©* Descartes). (2002). *We are the Romani People = Ame sam e Rromane d¿*ene*. Centre de Recherches Tsiganes; University of Hertfordshire Press, Hatfield.

Hayes, V.M., Severi, G., Padilla, E.J., *et al.* (2007). 5alpha-Reductase type 2 gene variant associations with prostate cancer risk, circulating hormone levels and androgenetic alopecia. *Int. J. Cancer*, **120**(4), 776–780.

Hazra, A., Kraft, P., Selhub, J., *et al.* (2008). Common variants of *FUT2* are associated with plasma vitamin B12 levels. *Nature Genet.*, **40**(10), 1160–1162.

Herbert, A., Gerry, N.P., McQueen, M.B., et al. (2006). A common genetic variant is associated with adult and childhood obesity. *Science*, **312**(5771), 279–283.

India Project Team of the International Cancer Genome Consortium. (2013). Mutational landscape of gingivo-buccal oral squamous cell carcinoma reveals new recurrently-mutated genes and molecular subgroups. *Nature Commun.*, **4**, 2873.

Indian Genome Variation Consortium. (2008). Genetic landscape of the people of India: a canvas for disease gene exploration. *J. Genet.*, **87**(1), 3–20.

Jha, A.N., Singh, V.K., Kumari, N., *et al.* (2012). IL-4 haplotype -590T, -34T and intron-3 VNTR R2 is associated with reduced malaria risk among ancestral Indian tribal populations. *PLoS ONE*, **7**(10), e48136.

Juyal, G., Prasad, P., Senapati, S., *et al.* (2011). An investigation of genome-wide studies reported susceptibility loci for ulcerative colitis shows limited replication in north Indians. *PLoS ONE*, **6**(1), e16565.

Khattri, A., Pandey, R.K., Gupta, N.J., *et al.* (2009). APOB gene signal peptide deletion polymorphism is not associated with infertility in Indian men. *J. Androl.*, **30**(6), 734–738.

Khattri, A., Pandey, R.K., Gupta, N.J., *et al.* (2009). CA repeat and RsaI polymorphisms in ERbeta gene are not associated with infertility in Indian men. *Int. J. Androl.*, **32**(1), 81–87.

Kivisild, T., Bamshad, M.J., Kaldma, K., *et al.* (1999). Deep common ancestry of Indian and western-Eurasian mitochondrial DNA lineages. *Curr. Biol.*, **9**(22), 1331–1334.

Kivisild, T., Rootsi, S., Metspalu, M., et al. (2003). The genetic heritage of the earliest settlers persists both in Indian tribal and caste populations. *Am. J. Hum. Genet.*, **72**(2), 313–332.

Klaric, I.M., Salihovic, M.P., Lauc, L.B., *et al.* (2009). Dissecting the molecular architecture and origin of Bayash Romani patrilineages: genetic influences from South-Asia and the Balkans. *Am. J. Phys. Anthropol.*, **138**(3), 333–342.

Kumar, J., Sunkishala, R.R., Karthikeyan, G. and Sengupta, S. (2007a). The common genetic variant upstream of *INSIG2* gene is not associated with obesity in Indian population. *Clin. Genet.*, **71**(5), 415–418.

Kumar, V., Reddy, A.N., Babu, J.P., *et al.* (2007b). Y-chromosome evidence suggests a common paternal heritage of Austro-Asiatic populations. *BMC Evol. Biol.*, **7**, 47.

Laayouni, H., Oosting, M., Luisi, P., *et al.* (2014). Convergent evolution in European and Rroma populations reveals pressure exerted by plague on Toll-like receptors. *Proc. Natl Acad. Sci. USA*, **111**(7), 2668–2673.

Lee, I., Pecinova, A., Pecina, P., *et al.* (2010). A suggested role for mitochondria in Noonan syndrome. *Biochim. Biophys. Acta*, **1802**(2), 275–283.

Loh, M., Chua, D., Yao, Y., *et al.* (2013). Can population differences in chemotherapy outcomes be inferred from differences in pharmacogenetic frequencies? *Pharmacogenom. J.*, **13**(5), 423–429.

Majumdar, D.N. and Rao, C.R. (1960). *Race Elements in Bengal; A Quantitative Study*. Asia Publishing House, London.

Mastana, S.S. and Papiha, S.S. (1992). Origin of the Romany gypsies – genetic evidence. *Zeitschr. Morphol. Anthropol.*, **79**(1), 43–51.

Mehrotra, S., Oommen, J., Mishra, A., *et al.* (2011). No evidence for association between *SLC11A1* and visceral leishmaniasis in India. *BMC Med. Genet.*, **12**, 71.

Mendizabal, I., Valente, C., Gusmão, A., *et al.* (2011). Reconstructing the Indian origin and dispersal of the European Roma: a maternal genetic perspective. *PLoS ONE*, **6**(1), e15988.

Mengual, L., Oriola, J., Ascaso, C., Ballescà, J.L. and Oliva, R. (2003). An increased CAG repeat length in the androgen receptor gene in azoospermic ICSI candidates. *J. Androl.*, **24**(2), 279–284.

Metspalu, M., Kivisild, T., Metspalu, E., *et al.* (2004). Most of the extant mtDNA boundaries in south and southwest Asia were likely shaped during the initial settlement of Eurasia by anatomically modern humans. *BMC Genet.*, **5**, 26.

Metspalu, M., Romero, I.G., Yunusbayev, B., *et al.* (2011). Shared and unique components of human population structure and genome-wide signals of positive selection in South Asia. *Am. J. Hum. Genet.*, **89**(6), 731–744.

Misra, V.N. (2001). Prehistoric human colonization of India. *J. Biosci.*, **26**(4 Suppl), 491–531.

Mohamed, H.S., Ibrahim, M.E., Miller, E.N., *et al.* (2004). *SLC11A1* (formerly *NRAMP1*) and susceptibility to visceral leishmaniasis in The Sudan. *Eur. J. Hum. Genet.* **12**(1): 66–74.

Moorjani, P., Patterson, N., Hirschhorn, J.N., *et al.* (2011). The history of African gene flow into Southern Europeans, Levantines, and Jews. *PLoS Genet.*, **7**(4), e1001373.

Moorjani, P., Patterson, N., Loh, P.R., *et al.* (2013a). Reconstructing Roma history from genome-wide data. *PLoS ONE*, **8**(3), e58633.

Moorjani, P., Thangaraj, K., Patterson, N., *et al.* (2013b). Genetic evidence for recent population mixture in India. *Am. J. Hum. Genet.*, **93**(3), 422–438.

Morgenstierne, G. (1973). *Irano-Dardica*. Dr. Ludwig Reichert Verlag, Wiesbaden.

Nakano, M., Ikeda, Y., Taniguchi, T., *et al.* (2009). Three susceptible loci associated with primary open-angle glaucoma identified by genome-wide association study in a Japanese population. *Proc. Natl Acad. Sci. USA*, **106**(31), 12838–12842.

Nalini, A., Thennarasu, K., Yamini, B.K., Shivashankar, D. and Krishna, N. (2008). Madras motor neuron disease (MMND): clinical description and survival pattern of 116 patients from Southern India seen over 36 years (1971–2007). *J. Neurol. Sci.*, **269**(1–2), 65–73.

Nalini, A., Yamini, B.K., Gayatri, N., Thennarasu, K. and Gope, R. (2006). Familial Madras motor neuron disease (FMMND): study of 15 families from southern India. *J. Neurol. Sci.*, **250**(1–2), 140–146.

Narang, A., Jha, P., Rawat, V., *et al.* (2011). Recent admixture in an Indian population of African ancestry. *Am. J. Hum. Genet.*, **89**(1), 111–120.

Nei, M. (1982). Evolution of human races at the gene level. *Progr. Clin. Biol. Res.*, **103**(Pt A), 167–181.

Palanichamy, M.G., Agrawal, S., Yao, Y.G., *et al.* (2006). Comment on "Reconstructing the origin of Andaman islanders". *Science*, **311**(5760), 470; author reply 470.

Palanichamy, M.G., Zhang, C.L., Mitra, B., *et al.* (2010). Mitochondrial haplogroup N1a phylogeography, with implication to the origin of European farmers. *BMC Evol. Biol.*, **10**, 304.

Pandit, J.J., Gopa, S. and Arora, J. (2011). A hypothesis to explain the high prevalence of pseudo-cholinesterase deficiency in specific population groups. *Eur. J. Anaesthesiol.*, **28**(8), 550–552.

Papiha, S.S. (1996). Genetic variation in India. *Hum. Biol.*, **68**(5), 607–628.

Patterson, N., Price, A.L. and Reich, D. (2006). Population structure and eigenanalysis. *PLoS Genet.*, **2**(12), e190.

Peterlin, B., Zorn, B., Volk, M. and Kunej, T. (2006). Association between the apolipoprotein B signal peptide gene insertion/deletion polymorphism and male infertility. *Molec. Hum. Reprod.*, **12**(12), 777–779.

Pott, A.F. (1844). *Die Zigeuner in Europa und Asien : Ethnographischlinguistische untersuchung, vornehmlich ihrer herkunft und sprache*. E. Heynemann, Halle.

Rafiq, S., Venkata, K.K., Gupta, V., *et al.* (2012). Evaluation of seven common lipid associated loci in a large Indian sib pair study. *Lipids Health Dis.*, **11**, 155.

Rai, N., Chaubey, G., Tamang, R., *et al.* (2012). The phylogeography of Y-chromosome haplogroup h1a1a-m82 reveals the likely Indian origin of the European Romani populations. *PLoS ONE*, **7**(11), e48477.

Rai, R., Sharma, K.L., Tiwari, S., *et al.* (2013). DCC (deleted in colorectal carcinoma) gene variants confer increased susceptibility to gallbladder cancer (Ref. No.: Gene-D-12-01446). *Gene*, **518**(2), 303–309.

Rao, K.N., Kaur, I. and Chakrabarti, S. (2009). Lack of association of three primary open-angle glaucoma-susceptible loci with primary glaucomas in an Indian population. *Proc. Natl Acad. Sci. USA*, **106** (44), E125–126; author reply E127.

Rajender, S., Vijayalakshmi, K., Pooja, S., *et al.* (2009). Longer (TA)n repeat but not A49T and V89L polymorphisms in *SRD5A2* gene may confer prostate cancer risk in South Indian men. *J. Androl.*, **30**(6), 703–710.

Ramana, G.V., Su, B., Jin, L., *et al.* (2001). Y-chromosome SNP haplotypes suggest evidence of gene flow among caste, tribe, and the migrant Siddi populations of Andhra Pradesh, South India. *Eur. J. Hum. Genet.*, **9**(9), 695–700.

Rani, D.S., Dhandapany, P.S., Nallari, P., *et al.* (2010). Mitochondrial DNA haplogroup 'R' is associated with Noonan syndrome of south India. *Mitochondrion*, **10**(2), 166–173.

Rao, P.R. and Gopalam, K.B. (1979). High incidence of the silent allele at cholinesterase locus I in Vysyas of Andhra Pradesh (S. India). *Hum. Genet.*, **52**(1), 139–141.

Rao, R.P., Yadav, N., Vahia, M.N., *et al.* (2009). Entropic evidence for linguistic structure in the Indus script. *Science*, **324**(5931), 1165.

Reich, D., Thangaraj, K., Patterson, N., Price, A.L. and Singh, L. (2009). Reconstructing Indian population history. *Nature*, **461**(7263), 489–494.

Roychoudhury, A.K. and Nei, M. (1985). Genetic relationships between Indians and their neighboring populations. *Hum. Hered.*, **35**(4), 201–206.

Sadhukhan, T., Vishal, M., Das, G., *et al.* (2012). Evaluation of the role of *LRRK2* gene in Parkinson's disease in an East Indian cohort. *Dis. Mark.*, **32**(6), 355–362.

Sahoo, S. and Kashyap, V.K. (2006). Phylogeography of mitochondrial DNA and Y-chromosome haplogroups reveal asymmetric gene flow in populations of Eastern India. *Am. J. Phys. Anthropol.*, **131**(1), 84–97.

Sahoo, S., Singh, A., Himabindu, G., *et al.* (2006). A prehistory of Indian Y chromosomes: evaluating demic diffusion scenarios. *Proc. Natl Acad. Sci. USA*, **103**(4), 843–848.

Satake, W., Nakabayashi, Y., Mizuta, I., *et al.* (2009). Genome-wide association study identifies common variants at four loci as genetic risk factors for Parkinson's disease. *Nature Genet.*, **41**(12), 1303–1307.

Saxena, R., Saleheen, D., Been, L.F., *et al.* (2013). Genome-wide association study identifies a novel locus contributing to type 2 diabetes susceptibility in Sikhs of Punjabi origin from India. *Diabetes*, **62**(5), 1746–1755.

SEARCH Collaborative Group, Link, E., Parish, S., *et al.* (2008). *SLCO1B1* variants and statin-induced myopathy – a genomewide study. *New Engl. J. Med.*, **359**(8), 789–799.

Sengupta, S., Zhivotovsky, L.A., King, R., *et al.* (2006). Polarity and temporality of high-resolution y-chromosome distributions in India identify both indigenous and exogenous expansions and reveal minor genetic influence of Central Asian pastoralists. *Am. J. Hum. Genet.*, **78**(2), 202–221.

Shah, A.M., Tamang, R., Moorjani, P., *et al.* (2011). Indian Siddis: African descendants with Indian admixture. *Am. J. Hum. Genet.*, **89**(1), 154–161.

Sharma, G., Tamang, R., Chaudhary, R., *et al.* (2012). Genetic affinities of the central Indian tribal populations. *PLoS ONE*, **7**(2), e32546.

Tanwar, V.S., Chand, M.P., Kumar, J., *et al.* (2013). Common variant in *FUT2* gene is associated with levels of vitamin B(12) in Indian population. *Gene*, **515**(1), 224–228.

Teng, Y.N., Lin, Y.M., Lin, Y.H., *et al.* (2002). Association of a single-nucleotide polymorphism of the deleted-in-azoospermia-like gene with susceptibility to spermatogenic failure. *J. Clin. Endocrinol. Metab.*, **87**(11), 5258–5264.

Thangaraj, K., Ramana, G.V. and Singh, L. (1999). Y-chromosome and mitochondrial DNA polymorphisms in Indian populations. *Electrophoresis*, **20**(8), 1743–1747.

Thangaraj, K., Joshi, M.B., Reddy, A.G., *et al.* (2002). CAG repeat expansion in the androgen receptor gene is not associated with male infertility in Indian populations. *J. Androl.*, **23**(6), 815–818.

Thangaraj, K., Chaubey, G., Kivisild, T., *et al.* (2005). Reconstructing the origin of Andaman Islanders. *Science*, **308**(5724), 996.

Thangaraj, K., Chaubey, G., Singh, V.K., *et al.* (2006a). *In situ* origin of deep rooting lineages of mitochondrial Macrohaplogroup 'M' in India. *BMC Genom.*, **7**, 151.

Thangaraj, K., Deepa, S.R., Pavani, K., *et al.* (2006b). A to G transitions at 260, 386 and 437 in *DAZL* gene are not associated with spermatogenic failure in Indian population. *Int. J. Androl.*, **29**(5), 510–514.

Traill, G.W. (1851). *Statistical Sketch of Kumaon*. John Murray, London.

Turner, R.L. (1927). *The Position of Romani in Indo-Aryan*. B. Quaritch, London.

Underhill, P.A., Myres, N.M., Rootsi, S., *et al.* (2010). Separating the post-Glacial coancestry of European and Asian Y chromosomes within haplogroup R1a. *Eur. J. Hum. Genet.*, **18** (4), 479–484.

Vasan, S.K., Neville, M.J., Antonisamy, B., *et al.* (2011). Absence of birth-weight lowering effect of *ADCY5* and near CCNL, but association of impaired glucose-insulin homeostasis with *ADCY5* in Asian Indians. *PLoS ONE*, **6**(6), e21331.

Voight, B.F., Scott, L.J., Steinthorsdottir, V., *et al.* (2010). Twelve type 2 diabetes susceptibility loci identified through large-scale association analysis. *Nature Genet.*, **42**(7), 579–589.

Willer, C.J., Sanna, S., Jackson, A.U., et al. (2008). Newly identified loci that influence lipid concentrations and risk of coronary artery disease. *Nature Genet.*, **40**(2), 161–169.

Yasuda, S.U., Zhang, L. and Huang, S.M. (2008). The role of ethnicity in variability in response to drugs: focus on clinical pharmacology studies. *Clin. Pharmacol. Therapeut.*, **84**(3), 417–423.

22 How to solve genetic disease on a population scale

Barry Merriman

22.1. Introduction: the future of personalized medicine

The future of personalized medicine will be built on a foundation of knowing all disease-causing genes and variants, and the ability to diagnose disease with precision informed by this knowledge. This "solving of disease," in terms of gene discovery and resulting diagnostics, is made possible by advances in genome sequencing technology, and progress towards the "$1000 genome." Here, we give an overview of the status of this, and the major global projects that are deploying genome sequencing to solve and diagnose disease on the population-scale. We further outline the key design considerations and ideal architecture for such projects, as a model for countries to follow as they fully personalize medicine for their unique genetic backgrounds. A critical point is the need for global enterprise service providers that can provide end-to-end support for countries building out their capacity for genomic medicine. This is essential for delivering the benefits of genomic medicine – lower healthcare costs and improved patient care – globally, as quickly as possible, by removing the risks, gaps, and bottlenecks that will otherwise stall or derail such large, complex projects.

Personalized medicine is an emerging field that holds promise for major advances in prevention and care at the patient level, and major reductions in the cost of healthcare at the societal level. Of course, the trend towards personalization is not new: medicine always takes action at the level of treating individual patients, and tools such as medical imaging and blood-type matching incorporate a great deal of knowledge about the individual into medical practice. However, as powerful as such advances have been, they are still relatively crude compared to the ideal of complete information and action at the cellular and molecular level of the patient, put in context by the knowledge that comes from accumulating and analyzing such information across millions of patients. The new era of personalizing medicine will come from empowering the practice of medicine with this

Genome-Wide Association Studies: From Polymorphism to Personalized Medicine, ed. K. Appasani. Published by Cambridge University Press.

much deeper way of seeing and acting, and this much broader integration of knowledge. The ideal is precise and comprehensive assessment of each individual's biological state, and the use of this, and the knowledge of millions of other such patients, to chart the course of action for the individual, including the use of therapeutics tailored to their personal biology (Auffray *et al.*, 2010).

22.1.1. The role of genomics

The two most important foundational elements of the new era of personalization are genome sequencing and stem cells, which provide a new basis for diagnosis and therapy, respectively. These define two major categories of personalized medicine activity: genomic medicine, and cellular medicine. The latter broadly refers to relying on patient-derived cells as the basis for selecting or developing a therapy, with stem cell-based tissue regeneration providing the most powerful potential for revolutionary therapy. Using patient cells for "disease-in-a-dish" modeling and drug screening is another example of cellular medicine activity.

In personalized medicine, genomics generally plays the role of the ultimate molecular diagnostic. Genomic medicine in the narrow sense would be the use of the individual's genomic DNA sequence across all diagnostic activities; that is, to inform risk assessment, prevention, diagnosis, prognosis, therapy selection, and monitoring. More broadly interpreted, "genomic medicine" includes the similar use of related "-omic" comprehensive molecular assays, such as assaying the epigenome, microbiome, transcriptome, proteome, and metabolome, as well as such assays done in patient-related cells, such as tumors, the microbes that comprise the microbiome, or infectious disease agents.

22.1.2. Solving versus curing disease

Confronted with the general phenomenon of human disease, we may draw a fundamental distinction between "solving" versus "curing" a disease. Curing a disease is the traditional goal of medicine, of course. However, the new era of genomic molecular assessment opens up the potential to "solve" a disease; that is, figure out what causes the disease at the fundamental molecular level. In particular, for genetic disease, the goal is to figure out what genes and gene variants cause the disease, or, more generally, contribute to disease risk. Thus, in particular, by solving a disease, we mean solving for the genetic basis of the disease. At the translational research level, this means gene hunting efforts to find the causal genes and variants. At the clinical practice level, this means precision diagnosis, based on the patient DNA sequence for such genes. While curing disease is the ultimate goal, solving disease is important, because it provides the ultimate rational basis for diagnosis, subsequent actions, and the search for therapies. Moreover, accurate diagnosis is of great value in its own right, in terms of providing resolution for the patient, and for taking preventative action, such as carrier screening in the context of preventing severe inherited genetic diseases.

Solving disease is also a much more tractable problem than curing disease, because it can be achieved through systematic, population-scale efforts enabled and empowered by the advances of genome-sequencing technology. Such efforts, in current practice and, moreover, in their ideal form, are the subject of this chapter.

22.1.3. Importance of population-scale and global perspective

By far the greatest promise for near-term genomic medicine lies through solving disease, as the current tools of genomics – DNA sequencing and statistical genetics – are ideal for finding disease-causing genes. In order to be maximally powerful, the effort to solve disease must be done at a population scale – 100,000 to millions of subjects – to get the large numbers of disease patients needed to represent the full complexity of genetic disease. Specifically the 5000+ severe genetic diseases, the modifiers of these diseases that control severity, the many genetic factors in complex, common diseases such as diabetes, heart disease, and mental illness, and the genes that control the response to drugs used in treating disease. This requires very large projects done at a national scale, well beyond the scale of traditional academic studies.

Also, it is very important to maintain a global perspective: ethnic differences in gross genetic background can play a large role in determining the specific genetic risk factors for disease, and disease-causing genes and especially gene variants can vary dramatically between groups. For example, this is illustrated in the extreme by the different variants of the beta globin gene that cause anemia (thalassemia) in the different Arab ethnicities (Zahed, 2001). Thus, in principle, every ethnic group needs its own separate genetic disease studies to solve for their unique disease genetics. In practice, this means that countries need to undertake their own population-scale studies, to assess their own people, to augment, build on, and optimize for their people the findings of traditional major academic research studies performed primarily on US or European populations. This in turn means the 195 countries of the world need to ultimately take individual responsibility for fully solving the diseases relevant to their people, and for creating the optimized genomic foundation for personalized medicine in their country.

22.2. Practical genomic medicine for national healthcare systems

While there is still a great deal of discovery to be made in solving genetic diseases, for large-scale healthcare providers, it is most important to understand what genomic medicine could do *now* to impact patient care and reduce healthcare costs. This can be broken down to three major areas of immediate, high-impact clinical utility, summarized below. All of these areas will benefit in the future from more discovery, done at scale, and proper personalization for the different countries – but nonetheless, there is still substantial utility to be had from what is possible now. Notably absent below is risk assessment for common complex disease, as the genetics of these are still poorly understood, and the ability to identify those of high risk based on genetics is of limited value at present; instead, the value of genomics for these disorders lies more in selection of medication, as described below.

22.2.1. Diagnosis of rare disease

Current knowledge and sequencing technology can very efficiently diagnose thousands of severe, rare, genetic disorders, in a way that is a major advance over the current standards of care, in terms of being comprehensive, precise, accurate, fast, low-cost, and minimally invasive. This includes, in particular: diagnosis of all the solved single-gene disorders without undergoing a diagnostic odyssey, as well as non-invasive prenatal testing (NIPT) for Down syndrome (trisomy 21) and other major syndromes from circulating fetal DNA, and pre-conception carrier screening to avoid at-risk pregnancies. Collectively, such disorders impact roughly 3–8% of all pregnancies depending on the country (Christianson *et al.*, 2005), and thus genomic diagnostics can have substantial impact on both patients and healthcare costs.

22.2.2. Safe and effective medication selection

Existing knowledge of drug–gene interactions, or pharmaco-genomics, provides hundreds of cases where a genomic profile could be used to avoid adverse drug reaction (ADR), or to select the most effective drug from a set of options, or to moderate drug dose based on individual metabolic efficiency. Because the majority of the population receives a prescription drug every year (e.g. 70% in the US), and because ADRs are a major source of suffering and cost, this is an area where genomic medicine can touch the largest fraction of the population, and make major impact.

22.2.3. Cancer therapy guidance

There are over 100 genes that, when mutated in a tumor, indicate a preferred clinical action according to current established practices of oncology. Thus, a proper genomic profile of patient tumors can provide the maximum benefit from genomic medicine to cancer treatment, which again could impact the 10% or more of the population newly diagnosed or living with cancer. This is widely perceived as a major area of value of genomic medicine, as reflected by numerous commercial offerings for such diagnostic gene panels (e.g. OncoMine from Thermo Fisher, OncoPlex from KEW, and FoundationOne from Foundation Medicine).

22.3. A global survey of population-scale genomics projects

At present, many countries or large healthcare systems around the world are working to place genomic medicine on their planning roadmaps. However, despite a great deal of interest and discussion, very few have taken action in actually funding, planning, and executing population-scale projects in translational genomic medicine research or deployment of clinical genomic medicine.

In particular, the following are the only true large-scale projects properly organized, funded, and in execution as of August 2014.

22.3.1. The US VA Million Veteran Program (MVP)

This effort initiated in 2011 is funded by the Veterans Healthcare Association (VHA), which is the largest healthcare system in the US, serving over 23 million US military veterans, with over 8 million patients treated each year, in a national network of 152 Veterans Administration (VA) hospitals, and a total of over 1700 care sites. In 2006 the VHA established a program in genomic medicine, to develop their roadmap, and in 2011 they formally launched the MVP as their flagship effort. This translational research project is intended to establish the foundation for genomic medicine in the VHA system. This is the largest translational research project in genomic medicine in the world. The simple architecture of this program is to biobank blood and DNA from one million patients within the VHA system, sequence their genomes using exome and whole-genome sequencing, establish a database of their genomic and clinical data, and thereby support large-scale discovery efforts by VHA-sponsored researchers to solve genetic conditions especially relevant to the VHA mission. Beyond the common disease concerns, they must solve disorders uniquely enriched in the veteran population, such as post-traumatic stress disorder and exposure to battlefield chemicals. The biobank presently has over 400,000 samples (August 2014), after 3 years of active collection, and is growing at the rate of 2000 samples per week. In 2013 the MVP bid out major sequencing service contracts to existing service providers, to serve their genome-sequencing needs. Claritas Genomics, using the Thermo Fisher Ion Torrent Sequencing platform, and Personalis, using the Illumina HiSeq platform, are presently the approved service providers. In particular, the task order for 2014 included the sequencing of 19,000 exomes at Claritas, which is the largest service contract ever awarded in terms of genomes analyzed and dollar amount. The project has the capacity to sequence over 200,000 exomes over the 5-year period of the service contract (see the Million Veteran Program website, http://www.research.va.gov/MVP/).

22.3.2. The Saudi Genome Project

This project, formally initiated in 2012 and publicly announced in December 2013, is funded by the Saudi national science-funding agency, KACST, and is the most comprehensive national translational genomics project in the world, funded at the $300 million level for its primary 5-year phase. The project goals are to build national competency and infrastructure in genomics, and to create the foundation for personalized medicine in the Kingdom, in the form of a knowledge base of all disease-causing genes for the Saudi population. This in turn will be generated by discovery and gene-hunting effort based on sequencing over 100,000 individuals over 5 years, through a combination of gene panel, exome and whole-genome sequencing. This effort includes both single-gene and common complex genetic disease. The discovery effort is organized into major stages, with incoming patients being diagnosed against panels of known disease genes, followed by exome and genome sequencing to support discovery work. The sequencing work is being done at a national network of 15 laboratories being set up to support the project, using primarily the Thermo Fisher Ion Torrent platform for the initial phases of gene panel and exome sequencing. Thermo Fisher Scientific and their local distributor, Integrated Gulf Biosytems, provide general contractor-style, end-to-end support for the planning and execution of the project (see The Saudi Genome Project website, http://rc.kfshrc.edu.sa/sgp/index.asp).

22.3.3. Genome England

This project, initiated in 2013, is funded by the government of England and the English National Health Service (NHS), and is the largest effort ever to employ genome sequencing in the practice of medicine. The project intends to sequence 100,000 whole genomes from NHS patients over 5 years, and – importantly – to return data to patients to impact their care. The major focus of this effort will be tumor sequencing for cancer patients, with particular emphasis on lung cancer. The project will also focus to a lesser extent on diagnosis for rare inherited disease and infectious disease. The project as of June 2014 has over $600 million in allocated funding. The sequencing and informatics aspects of the project are being bid out to service providers, although the sequencing solution was ultimately architected as a joint venture between the project office and sequencer vendor Illumina, Inc. (see the Genome England project website, http://www.genomicsengland.co.uk/).

22.3.4. Genome Qatar

This project, initiated in December 2013, and funded principally by the Qatar Foundation, intends to sequence the genome of every Qatari citizen, of the total population of approximately 300,000, and couple these data into their Cerner-based national Electronic Medical Records system, to establish the foundation for genomic medicine in Qatar. This is the most comprehensive national effort to deploy genomic medicine at the whole-population scale. The sequencing will be done at a centralized facility being set up and operated by the project, located at the Sidra Medical Center, using a HiSeqX10 system from vendor Illumina, Inc. (see the Sidra Genome Qatar project website, http://www.sidra.org/sidra-to-help-qatar-grow-as-a-hub-for-genetic-medicine/).

22.3.5. Taiwan Biobank Project

This project, formally initiated in 2012, but in planning and piloting stages for nearly a decade, is sponsored by the Academica Sinica of Taiwan. This national translational research project, with the deep involvement of over 10 major medical centers, is recruiting nearly 200,000 patients, and will sequence them in a combination of exome and whole-genome sequencing, in order to establish a database that will support solving genetic diseases in the Taiwanese people. This in turn will provide the foundation for future genomic medicine in the country. The sequencing is being bid out to in-country service providers, with contracts to do the first 1000 normal reference whole genomes in 2014. Sequencing platforms from both vendors Thermo Fisher Scientific and Illumina are supporting the initial work at the service providers (see the Taiwan Biobank Project website, http://newsletter.sinica.edu.tw/en/file/file/1/125.pdf).

22.3.6. Human Longevity, Inc.

This is a unique, privately funded venture to develop comprehensive genomic and cellular medicine products and services. This commercial venture was launched in 2013 by Dr. Craig Venter, and publicly announced in 2014, as it took on $85 million in A-round funding. A cornerstone of the effort will be genome sequencing of 100,000+ disease patients within 3 years, with longer-term goals reaching the 1 million patient scale. This will be used to solve for the genetics of the common diseases of aging, and as a foundation for developing diagnostics and therapeutics for personalized medicine. Sequencing will be done at the private laboratories established at the company, using vendor Illumina's HiSeqX10 platform. This effort is unique, globally, as the only privately funded effort operating on the population scale (see the Human Longevity Inc. website, http://www.humanlongevity.com/). As a historical note, this has a precedent in the pioneering effort of deCODE Genetics, Inc., a company set up in 1996 to mine genomic and clinical data from the 200,000+ Icelandic population in support of developing new drugs with Pharma. This was the very first population-scale effort to solve disease with modern genomic tools, and had numerous scientific successes using the linkage approaches available at the time. Their effort was ultimately not a sustainable business, and ended in 2012, when deCode was acquired for $415 million by AmGen, with their genomics platform subsequently going into a 2014 spin-out company, NextCode Health (see the Nextcode Health website, https://www.nextcode.com/).

22.4. Other notable large-scale efforts and emerging trends

The following are not highly organized and focused population-scale genomics projects as above, but represent related, more diffuse large-scale efforts to solve disease, as well as near-term potential or trends towards establishing new population-scale initiatives.

22.4.1. Major pediatric genomic medicine initiatives

Pediatrics is an ideal setting for making an impact through genomic medicine, because of the thousands of rare pediatric disorders that are difficult to diagnose or are not yet solved for their causal genes. The power of genomics to diagnose or discover in such cases is well established. Because of this, there is a trend toward large-scale genomic initiatives within pediatric medicine. The rare nature of the disorders and the cost and complexity of establishing the diagnostic and discovery sequencing capabilities encourage hospitals to join together in shared efforts, to increase patient numbers for statistical power and independent validation of gene discoveries, and to make efficient use of resources. Two notable recent efforts are as follows.

22.4.1.1. Boston Children's Hospital (BCH)/Claritas Genomics

In 2013, BCH spun out its well-established molecular diagnostics laboratory into a joint venture with Thermo Fisher Scientific to establish a for-profit diagnostics laboratory, Claritas Genomics, Inc. This company performs all the classical molecular diagnostics, diagnosing thousands of diseases, as well as a rapidly growing menu of genomic diagnostics based on exome sequencing and gene panels, for comprehensive, precision, cost-effective diagnosis. Moreover, Claritas is developing a franchise business model, to allow other US and international pediatric hospitals to join into a meaningful diagnostic network and grow towards the population-scale desirable in pediatric genomics. The notable elements of this approach – forming an external venture, and creating a coalition of hospitals – are both emerging trends to overcome the genomic medicine cost and complexity constraints and satisfy the need for scale (see the Claritas Genomics website, http://claritasgenomics.com/).

22.4.1.2. Rady Children's Hospital

In 2014, billionaire philanthropist Ernest Rady pledged a $120 million donation to the Rady Children's Hospital of San Diego, to support development of genomic medicine at the hospital. This illustrates the level of funding required for a single institution to undertake its own genomic medicine effort, in contrast to the BCH–Claritas coalition–spin-out approach (see the Rady Children's Hospital website, http://www.rchsd.org/).

22.4.2. Major biobanking initiatives

Establishing population-scale patient biobanks is a natural and powerful first step towards population-scale efforts to solve disease. Two notable examples of such national-scale biobanking that are also in transition towards large scale genomic sequencing, are the following.

22.4.2.1. Finland

In aggregate, Finland has disease biobanks comprising over 200,000 samples covering a wide range of disease, which reflects their historical global leadership in genetics research. While there is substantial funding for isolated sequencing projects involving these resources, there is as yet no organized funding on the $200+ million dollar scale that would be required to sequence all banked samples and create the unified, parallel data resource to support solving disease.

22.4.2.2. Estonia

The Estonian Biobank houses 50,000 samples, collected over the past 15 years, and representing a cross-section of nearly 5% of the Estonian population. This sophisticated biobank has been the basis for a substantial amount of disease research, and the experience from this project is currently translating into plans for a national personalized medicine initiative funded by EU development funds, to impact the healthcare for their population of one million. Estonia is especially well positioned, globally, for such a national effort, because of their long experience with nation-scale genetic research, their small and socially progressive population, and, foremost, their comprehensive national IT infrastructure, X-Roads, that in particular has full integration of whole population electronic medical records and electronic pharmacy records. This would provide an ideal setting for the integration of genomic diagnostics into their healthcare system, including support for future data mining and discovery efforts at the whole-population scale (see the Estonian Biobank website, http://www.geenivaramu.ee/en).

22.4.3. Major translational research initiatives

There are several notable, large-scale translation research programs that have emerged globally. These are not as highly focused and centrally organized as the population-scale projects above, but they share a similar aim of advancing the deployment of genomic medicine and the solving for genetic basis of disease.

22.4.3.1. US NIH Programs: NCATS and NHGRI

In the US, in 2011 the NIH established a new major national institute, the National Center for Advancing Translational Science (NCATS), with the broad mission of supporting translation of research into medical practice, by removing the major bottlenecks in the process, and a $665 million dollar annual budget. Much of this work supports translation in the area of genomic medicine. In addition, in the specific area of solving disease, the National Human Genome Research Institute (NHGRI), also with a $500 million dollar annual budget, is putting nearly $50 million in total funding towards the creation and 4 years of subsidized operation of three major academic sequencing centers that comprise the Centers for Mendelian Genomics, with the mission to solve rare genetic disease, located at Yale, University of Washington, and Baylor/Johns Hopkins. Individual researchers can make use of these centers to solve rare genetic diseases, at reduced prices for sequencing and services. This represents the latest incarnation of the NHGRI funding campaigns in genomic medicine that began with the NIH support for the Human Genome Project in the 1990s, and later the International HapMap and 1000 Genomes Projects in the 2000s (see the Centers for Mendelian Genomics website, http://www.mendelian.org/, the National Center for Advancing Translational Science website, http://www.ncats.nih.gov/, and the NHGRI website, http://www.genome.gov/).

22.4.3.2. China BGI Programs

In China, The Beijing Genome Institute, now BGI, underwent a major expansion in 2010 as the recipient of a $1.5 billion dollar loan from the Chinese central development bank, with the intent of establishing China as the global leader in genomics. BGI undertakes diverse sequencing-related projects and activities. In the area of solving human genetic disease, in 2012 they launched their "1000 Mendelian Disease" initiative in collaboration with Children's Hospital of Philadelphia, one of the leading US pediatric hospitals, to solve thousands of rare Mendelian disorders through a combination of service and collaborative projects globally (see the BGI website, http://www.genomics.cn/en/index).

22.4.4. Major charitable foundation initiatives

A natural area for the early introduction of population-scale genomics to solve disease would be to support the research activities of major national or global charities for genetic diseases. While this is indeed a very promising direction, and there are large, international charities for highly genetic diseases such as cleft palate, autism, and birth defects, this idea has yet to be realized. Nonetheless, initial efforts are underway, one notable example being the CCFA.

22.4.4.1. CCFA

The Crohns and Colitis Foundation of America (CCFA) is creating a centralized biobank and IT resource to support its research towards a cure for the one million people affected in the US alone. This effort, entering a pilot stage in 2014, will include support for genomic big data and sequencing of biobanked samples, in conjunction with creating the biorepository, IT infrastructure, and patient clinical database (see the Crohns and Colitis Foundation of America website, http://www.ccfa.org/).

22.5. The $1000 genome: status and relevance to solving disease

Dating back to 2001, the "$1000 genome" has been put forth as the critical price point at which genome sequencing could enter widespread medical practice. Since that time, the NHGRI has maintained a roadmap to advance sequencing technology towards this price point in 2014, and championed this through their "$100,000 Genome" and later "$1000 Genome" technology development funding programs.

While it is undoubtedly true that, relative to the $1 billion dollars it cost to sequence a human genome circa the year 2001, it was critical to move the cost point much closer to $1000 for practical purposes, and to be in line with the costs of other advanced diagnostics in clinical use, there is much ambiguity as to the precise status of this goal, and its precise relevance to genomic medicine, now that it is at hand. In particular:

22.5.1. Whose $1000 genome?

Sequencing a genome, considered as the result of running an instrument, has a cost that could be measured at many points along the supply chain: the cost of the raw commodity materials, the cost of the refined components and chemicals involved, the cost to the system integrator that makes the instrument and reagent kit, the cost to the primary service lab that runs the assay, or the cost to an end user, such as a major project contracting a service, or a doctor that orders a "genome test." For example, for several years, the cost of sequencing a genome internally at the major sequencer manufacturers (cost of goods and services, or "COGS") has been less than $1000, so in that weak sense a $1000 genome was probably reached circa 2012. The next major milestone, realizing this cost at a large-scale service laboratory, was just achieved in 2014 (oddly in accord with the NHGRI roadmap) by instrument vendor Illumina, with the release of the HiSeqX10 system. This provides a service laboratory with an all-in cost of $1000 per genome, for very large-scale projects (80,000 genomes amortized over 4 years, with a $10 million instrumentation capital expense for the X10 system). Not yet reached is the point at which this could be delivered to a small-scale service laboratory. The Ion Torrent Proton platform from vendor Thermo Fisher Scientific has promised a $1000 genome for small-scale service laboratories with the release of their P2 chip, also slated for the end of 2014. Also not yet reached is this price point for an end consumer, such as a large project bidding-out sequencing services, or a doctor ordering a test. At present in 2014, large-scale service contracts can deliver genomes to very large projects (1000+ genomes) for $2000, such as for the VA MVP. The price of a whole-genome diagnostic-grade test that a physician might order remains in the range of $5000–9500 (e.g., Knome, and Illumina TruSeq services, respectively). Thus, while the needle has moved to the point where the "$1000" genome is available in 2014, the actual costs associated with genome sequencing lie in the range of $500–10,000 depending on the context of who is paying, and at what scale, as outlined above.

22.5.2. What is a genome?

Also not explicitly noted in the "$1000" genome is what actually constitutes the genome output. Current convention in research would be the reads and variant calls from a 30× coverage (100 billion bases) human genome sequence data set. However, coverage standards are sometimes considered to be 40× or higher to achieve clinical grade, and 80× or more for tumor whole genomes. All of these raise the cost in proportion to coverage. Beyond coverage, there are the issues of whether the data are phased into maternal and paternal chromosomes, and whether copy number and rearrangement variants are also provided, all of which may require a combination of special analyses and assays, which are available, but which would further raise the cost several fold beyond that of the $1000 base genome. Finally, beyond variants, is the issue of what form of annotation and interpretation come with the genome, and in what electronic format it is stored and maintained. The costs associated with that are largely unaddressed, but as noted again result in 5- to 10-fold higher costs at present.

22.5.3. Which assay: (whole) genome, exome, gene panel or marker panel?

A more relevant question for large-scale genomic medicine is whether the whole genome is the most important assay, versus exomes or gene panels, or even panels of markers typed by other modalities, such as array or qPCR. For the purpose of establishing population references, such as 1000 Genome Project-style references for specific ethnic populations, whole-genome sequencing is definitely to be preferred. However, in practice, for large-scale translational research projects, exomes can provide a cost-effective intermediate, with prices fivefold lower than whole genomes. This approach has been pursued by the VA MVP, which is doing the great majority of its present sequencing service orders for exomes. Gene panels provide a very cost-effective way to deploy assays at a population scale for healthcare systems, and to triage incoming samples for whether they need genome- or exome-based discovery-oriented sequencing versus existing diagnosis. Panels are also uniquely important for tumor sequencing, where extreme depth of coverage (1000× to 10,000×) is needed to look for mutations in minority cell populations in the tumor, extending down to 1% or less of the tumor cells. Finally, maximal economy of deployment is had with marker panels, especially typed via qPCR for the scale of 10–100 markers. This approach does not enable much future discovery, but minimizes costs, which is often a major concern of payers for population-scale deployment. The summation is that while whole-genome sequencing represents an ideal, the cost of this assay is often not the most important factor for the economics of genomic medicine. The costs of exome sequencing or gene panels are far more critical in many projects and settings, and this can be 5- to 10-fold lower assay. These reduced assays are also substantially easier to interpret.

22.5.4. The role of other -omics

As noted, the whole-genome sequence is the most prominent and fundamental form of omic data, but comprehensive genomic medicine projects will often involve other -omics, and their economics may or may not be closely correlated with that of whole genomes, depending on whether the assay is based on sequencing, versus other measurement modalities such as mass spectrometry.

22.5.5. Practical genome costs and relevance for population-scale projects

In highly practical terms, large-scale projects in 2014 – meaning those performing more than $10 M in sequencing, or thousands of samples, for example, such as the VA MVP – can definitely achieve an all-in price of $2000 for a 30× whole genome, delivered to the project as a service, for example. This price is likely to drop only slowly at this point, with a long trend towards $1000. Exome prices are likely to be substantially lower, around $500 for an 80× coverage, and dropping slowly towards $200. Gene panels are lower still, around $200 (e.g. for a 500 gene panel, at 200× coverage) and dropping towards $50. On the other hand, gold standard whole genomes, such as for population references, which include phasing, copy number and rearrangement information, will be several-fold higher, as will any of these assays applied to cancer genomes due to the requirements for greater depth of coverage to see rare mutations. The important point is that, as of 2014, none of these costs represents any substantial barrier to population-scale genomic medicine or translational research, and large projects face larger problems in terms of data interpretation, integration, and logistics. Thus, given nearly any project budget above the $10 million dollar scale, a high-impact program can be devised that relies on a well-chosen combination of genomes, exomes, gene panel, and marker panels to make a major impact in national-scale genomic medicine, and within that budget allocation.

22.6. Key considerations for population-scale projects to solve disease

The above major projects provide a comprehensive picture of what *is* occurring in the world of population-scale genomic medicine and translational research. However, each of these major projects has notable idiosyncrasies in terms of their formulation, often stemming from their history and the complexities of getting the stakeholder buy-in required. This in turn can make them far from ideal in various respects, and therefore not the best models for other countries to emulate.

In order to guide future country-scale projects closer to the ideal of delivering the full benefit and impact of genomic medicine to their people, we here consider abstractly the major design criteria that go into architecting a successful large-scale project. At a high level, this breaks down to four project design choices, which we characterize as choosing:

- the right technology,
- the right disease,
- the right country, and
- the right approach

as considered in detail below.

22.6.1. The right technology

Population-scale genome sequencing – meaning ultimately hundreds of millions of people per year, globally – requires careful choice of a technology that can scale up to this level, and is also suitable for global clinical deployment.

In the field of genome sequencing, for the past decade there has been a general perception that advanced nanotechnologies will produce the ultimate sequencer, directly reading human genomes, in minutes, for dollars, with clinical precision. This has most recently been embodied by expectations that Nanopore sequencing would deliver a new scale of speed and low cost. In practice, these nanotechnology solutions are exceedingly difficult to develop, are still years from entering the market, and will require years more to fully surpass the scaling abilities of the current technologies already in the market. As a result, these nanotechnology solutions actually have no relevance to planning practical population-scale projects, and the right technologies are actually as of 2014 already available.

The ideal sequencing technology for the future of genomic medicine would be one which is mass-manufacturable with well-established, precision, low-cost manufacturing supply chains, and resulting in a compact, robust, low-cost instrument suitable for broad global clinical deployment. Indeed, it would be a computer-like device, because the manufacturing of computers and computer chips is the highest level of precision mass manufacturing yet achieved by civilization, and is the only field of production (perhaps rivaled by injection molded plastic) that supplies billions of low-cost, precision devices each year – specifically (CMOS) chip CPUs, DRAMs, and digital camera imagers. In this regard, the Ion Torrent sequencer from vendor Thermo Fisher Scientific is positioned as an ideal technology for the future of global clinical genomic sequencing, because the instrument is simple and computer-like, and the key consumable is a sequencing chip, which is actually a disposable CMOS sensor array chip, very similar to a digital camera chip, but at a 10× higher "pixel" count, 100 million to 1 billion pixels. This leverages the exact same fabs that are used to produce logic and memory and digital camera chips. This makes the Ion Torrent platform a good choice for many aspects of population-scale genomic medicine, such as gene panel and exome sequencing done at scale (Merriman *et al.*, 2012). For example, in the VA MVP project, a 32-instrument Ion Torrent installation is delivering data at the rate of approximately 50,000 exomes per year. Collectively, over 30,000 gene panels are run on Ion Torrent platforms globally every *month*, in clinical or translational settings. However, the technology on the market in 2014 has not yet achieved the performance necessary to properly support large-scale whole-genome sequencing projects, which require similar numbers of whole genomes. While rapid progress is being made towards that, it may in practice require 2–3 years to fully mature for this application – a factor that is very important for near-term project planning.

An alternative and valid technology strategy to enable population genome sequencing would be to take a relatively mature existing technology and optimize it for the creation of large-scale sequencing factories – specifically in the form of instruments and chemistries optimized for high-throughput sequence data generation per instrument. In this approach, large centralized factories would support major projects, or enable large-scale service providers. The company Complete Genomics was based on this principle, but their effort was hampered by the limitations of their relatively immature technology. They have since been acquired by BGI, and it remains to be seen if their sequencing factory aspirations will become viable. In contrast, the HiSeqX10 introduced in 2014 by vendor Illumina best embodies this approach. This 10-instrument system is capable of sequencing 20,000 genomes per year, and delivering an all-in price point of $1000 per genome to the service laboratory. However, there is a high up-front capital expense of $10 million dollars just for the instruments, which is a significant risk factor and barrier in planning and executing large projects. Nonetheless, this provides an enabling technology foundation for those projects that require large-scale sequencing of whole genomes. Such is the case for the Genome England project, which is partnering with Illumina to create an X10-based sequencing factory in England to support their initial project goal of sequencing 100,000 genomes over 5 years.

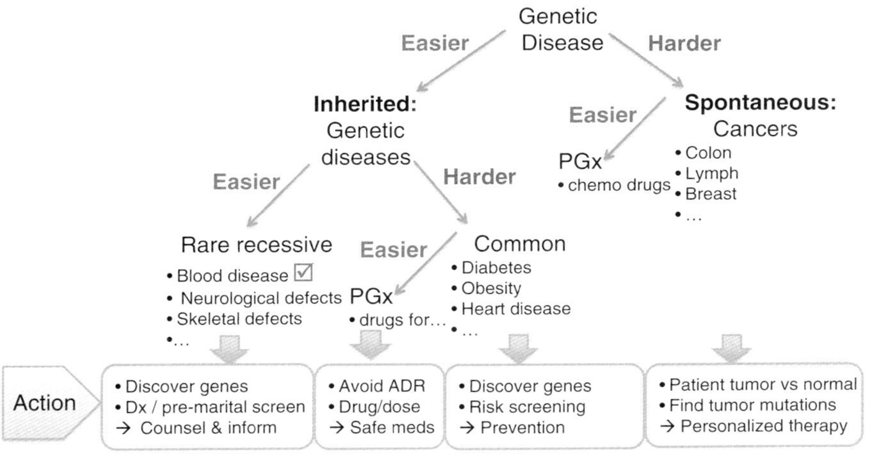

Figure 22.1. Major applications of genomics in healthcare. Shown are the major disease areas from a genetic perspective, with an indication of which are easier or harder to address in population-scale programs to solve disease. Also shown are the major actions that can be taken in each case. PGx, pharmaco-genomics. The most tractable areas of genomic medicine and in diagnosis or carrier screening for rare disease, and PGx for proper drug selection relevant to common diseases and cancer. A black and white version of this figure will appear in some formats. For the color version, please refer to the plate section.

Thus, the current Thermo Fisher Ion Torrent and Illumina HiSeqX10 platforms offer strong foundations for architecting population-scale genomic projects. There is no need for projects to wait further for technology improvements, significant cost declines, or greater capability to sequence at scale.

22.6.2. The right disease

The key to making a major impact and having near-term project success – which is very important for stable support from stakeholders – is to properly focus on the diseases where genomics can be most impactful. The overall impact of applying genomics to a particular disease – either as a clinical diagnostic or as a translational gene discovery tool – depends on the number of people with the disease, the costs of the disease, the ability of genomic tools to yield answers for the disease, and the ability to take impactful action based on those answers. If we consider the general realm of genetic disease in a population from this perspective, the classes of disorder can be prioritized from "easier" to "harder" in terms of the ability of genomics to make impact (see Figure 22.1).

22.6.2.1. Within genetic disease: cancer versus inherited disease

Cancer is harder, because we have limited ability to turn the spectrum of mutations in a tumor into clinical actions, and because of the great complexity of using genome sequencing to survey the mutations in a tumor, especially in regards to the depth of coverage required to see rare mutations. Moreover, there is the major issue that the genome view is only one view of a tumor, and views of the tumor from other -omics, such as methylation state and protein expression, as well as histological and cytological views, are also important in assessing the tumor. In contrast, it is much easier to catalog the germline variants in a person, and correlate those to disease, which is the easier problem encountered in inherited disease.

22.6.2.2. Within inherited disease: common versus Mendelian disorders

Common disorders – such as type II diabetes and heart disease – are harder, because there are many genes to be discovered, with multiple risk factors acting in each person, many are of which are of weak effect, so that very large studies ($N = 10{,}000$–$100{,}000$) must be undertaken to discover such factors. The validation of such discoveries presents many challenges as well. Moreover, there are substantial environmental effects to be understood for common disease, greatly complicating both study design and clinical actions. In addition, translating these findings into meaningful actions is challenging. In contrast, Mendelian (single gene) disorders are much more readily diagnosed or solved, and genome-sequencing approaches are extremely powerful and effective for this. Thus, both clinical diagnosis and disease gene discovery are "quick win" activities in which projects can excel early. Even though each individual Mendelian disease is quite rare, typically in the 1:1000 to 1:1,000,000 range, collectively (considered as severe genetic birth defects) they impact 3–8% of births based on the March of Dimes global assessment, depending on country (Christianson *et al.*, 2005). Thus, collectively, Mendelian disease impacts a notable fraction of the population, and such disorders are often quite burdensome, costly, and difficult to diagnose. In the area of actionability, early diagnosis can substantially improve outcomes, such as in newborn screening for metabolic disorders, where diet modification and drugs given early can avoid devastating consequences. Also, the ability to perform pre-conception carrier screening for such disorders ultimately enables parents to choose not to have affected children, which is an extremely high impact intervention.

22.6.2.3. The exception: pharmaco-genomics of common disease

The one great exception to common disease being harder is that the pharmaco-genomics (PGx) of drug therapy for such diseases can be relatively easy and high-impact, being rather more Mendelian in nature. Thus, in the general area of common disease, such as diabetes, the high-impact, easier application of genomic medicine and translational work is in the area of using it to understand which drugs and doses to use among the therapeutic options, and particularly to avoid ADRs, which are a major cost in healthcare and a major cause of death.

Thus, in terms of genomic project impact and short-term success, addressing Mendelian disease and PGx are the two ideal focus areas. Addressing common disease and cancer are important, but can be done on a longer timescale, with proper balance of project resources – aside from delivering specific actionable information for cancer via gene panels, as noted in the section above on practical genomic medicine.

22.6.3. The right country

There are 195 countries in the world, and while all will eventually need genomic medicine, they are not equal when it comes to planning successful, high-impact projects to solve diseases or establish clinical deployments. There are three major factors which go into identifying which countries provide the best opportunities for high-impact programs. Assessing all these factors, below, shows that the Arab Gulf Cooperation Council (GCC) countries in particular are globally ideal countries to most aggressively pursue population-scale projects to solve disease and deploy genomic medicine. The major factors for focusing on the right countries are:

- money,
- disease burden, and
- social readiness

as described below.

22.6.3.1. Money

Lead countries must be relatively wealthy, because these leading-edge genomic medicine projects are necessarily relatively expensive. More specifically, countries need ready access to funding on the $100 million dollar scale. Where this money could come from may vary: there could be a general strong economy, as in England, there could be extreme concentrations of oil wealth, as in the Arabian GCC countries or Norway, or there could be large-scale economic stimulus funding, such as the EU development funds provided to smaller EU countries, such as Estonia.

22.6.3.2. Swords into plowshares

A novel, untapped source of national genomics funding is the offset obligations owed to countries by the defense contractors, as a result of major weapons purchases. A defense contractor – such as Lockheed Martin selling F-35 fighter jets at $200 million each – typically has obligations to invest hundreds of millions of dollars back as in-country ventures that improve the economy and social well-being of the purchasing country. Genomic medicine certainly would ideally satisfy these criteria. Offset commitments total on the order of $10 billion dollars per year, globally, and there are commitments to over 45 countries. If such could be satisfied through major genomic medicine projects, it could bring genomic medicine to many countries. This could be a modern version of “beating swords into plowshares” – in this case, “beating missiles into sequencers.”

22.6.3.3. Disease burden

Lead countries should have a high genetic disease burden, so that there is real need and real impact to be made through genomic medicine, both in improving quality of life for the individuals and in reducing healthcare costs for governments. While money is necessary, a high disease burden is the true motivation for undertaking genomic medicine projects now, versus waiting. If one looks at the global burdens of genetic disease – both Mendelian and common – the highest concentrations are found in the Arab countries – roughly double the rates in the West, both in terms of Mendelian disease (8% of births verus 4% in the US or UK), and in the common disease such as diabetes (20% versus 10% in the US or UK; see Global diabetes prevalence rates, http://www.allcountries.org/ranks/diabetes_prevalence_country_ranks.html). The high Mendelian disease burden comes from the cultural practice of marrying close relatives, which is common in the Arab cultures. The much higher incidence of common disease is of less clear origin, perhaps recent changes in lifestyle combined with intrinsic genetic factors. Thus, in terms of sheer "need," in terms of genetic disease burden, the Arab countries are especially ideal places for near-term population-scale efforts to solve genetic disease.

22.6.3.4. Social readiness

Lead countries should have a social structure that is ready to make use of this new genomic information to improve healthcare. This is actually the greatest challenge, because the scientific discoveries, no matter how profound, cannot penetrate and make impact in the society if it is not ready to make use of information about the genes that cause disease. In this regard, a unique global situation is that most countries in the "malaria belt" have well-organized national testing programs for carrier status of the recessive inherited anemias – thalassemia and sickle cell anemia – that are carried in the population at a high frequency as a result of the protective effect of such mutations against malaria. In such countries, the carrier rate can approach 20% of the population, resulting in high incidence of these severe recessive diseases. As a result, carrier screening and appropriate actions have become the accepted norms of behavior. Moreover, such countries have a well-developed system and process for using knowledge of disease genes in public health. This provides an ideal existing framework for simply inserting much more content into the current carrier screening programs.

22.6.3.5. Saudi example

For example, in Saudi Arabia, in order to get a marriage license there is mandatory carrier screening for thalassemia and sickle cell anemia, and at-risk couples are simply counseled but are otherwise free to marry and plan families as they see fit. Yet, the net result after just 5 years of this mandatory screen being in place is that half of at-risk couples choose not to marry – thus in just 5 years the incidence of these severe genetic diseases has been cut in half, solely through a genetic test and information. This is an ideal model for inserting much more content for all Mendelian disease, and in principle this could achieve similar results of cutting the overall incidence of Mendelian disease in half in just a few years.

This level of immediate, high-impact genomic medicine via carrier screening for Mendelian disease is simply not achievable by any other means. Thus, countries with existing pre-marital or pre-conception screening for anemias would be ideally ready for much more comprehensive genetic screens.

Similar existing testing infrastructure is present in countries that have well-developed newborn screening programs, and this could also provide a channel for much broader content deployment versus the 31 metabolic conditions typically tested. However, this route to deploying content is in fact limited, in that such testing needs to be done at a very low price point, probably below $20 per sample, and moreover there is much more limited actionability at the newborn stage, versus pre-conception, so the opportunity to truly reduce disease burdens is not as great.

22.6.3.6. The uniquely ideal Arabian situation

As noted, the above factors of money, need, and readiness all favor the Arab GCC countries in particular as ideal places to lead in population-scale efforts to solve disease and deploy genomic medicine. This is indeed reflected by the Saudi Genome Program and Genome Qatar projects launched in the past two years, and being two of the few global initiatives currently underway. Yet, there remains great opportunity for the rest of the six GCC countries to create national genomic medicine initiatives, and to assist in establishing such efforts throughout the greater Arab world, which comprises 22 countries and over 400 million people. Of special and critical importance, the median age in these countries is approximately 25, due to a baby boom in the region over past decades, and from the disease perspective, this will lead to two great waves of genetic disease burden: most critically, as the boom of young adults has children, there will be a boom in Mendelian disease with 8% of births affected; then, as the boom approaches 50, in two decades, there will be a boom in diabetes, due to the 20%+ incidence. These two waves represent a looming crisis in healthcare costs, and personal disease burden. This also creates great national motivation to develop genomic personalized medicine as quickly as possible, as the only viable way to avoid these major waves of disease.

22.6.4. The right approach

The great majority of countries do not have the organic capability to efficiently plan and execute a population-scale project in solving genetic disease or deploying genomic medicine. Such work requires not only deep domain expertise, but also the ability to handle the considerable logistical complexity of integrating all the elements of the solution, in the context of a large, schedule-driven project (see Figure 22.2).

22.6.4.1. The enterprise solutions approach

For such projects to achieve success in most countries, there must be a global "enterprise solutions" provider that can act as a general contractor and systems integrator to supply an end-to-end solution in support of the project goals (see Figure 22.3). This is indeed how it is done for all other cases of high-value, high-complexity national infrastructure projects: for example, if a country needs nuclear power, they could send a generation of students to MIT to study nuclear

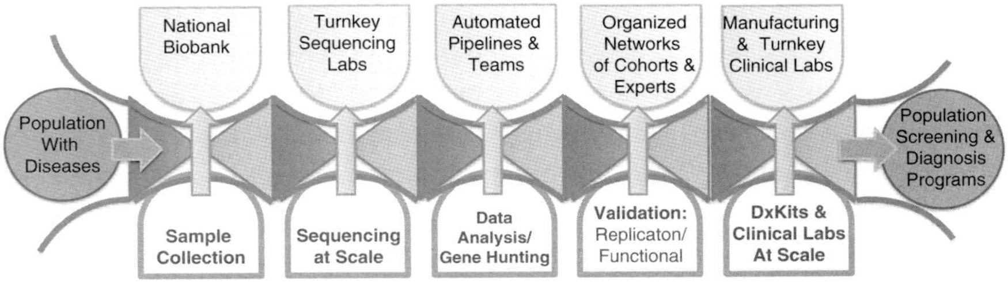

Figure 22.2. How to solve disease on a whole-population scale: process and critical bottlenecks. The major steps of such programs are shown, starting from a population with disease, and ending with that population having personalized screening and diagnosis programs. The lower track shows the functional activities, and the upper track shows the solutions that support these. At each stage there is a major bottleneck of obtaining the needed functionality in a coherent way to support the goals of the project. Both discovery and clinical tracks share similar bottlenecks at this level of description. A black and white version of this figure will appear in some formats. For the color version, please refer to the plate section.

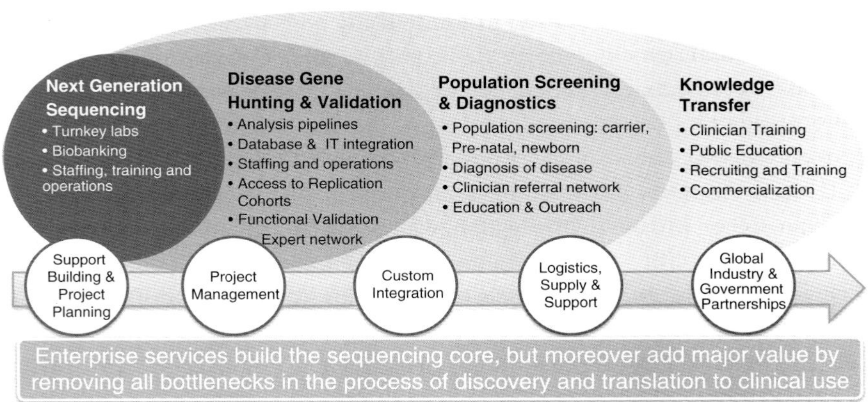

Figure 22.3. Enterprise genomics solutions for solving disease at the population scale. The range of services and solutions that are ideally provided by an enterprise end-to-end support provider. Establishing biobanking and next-generation sequencing facilities is the core, at left, but much more must be done, as indicated by the expanding layers of support activities, to fill all gaps and remove all bottlenecks in the process. At the two extremes of the end-to-end process, at the start there is help building support among sponsors, and at the end there is assistance commercializing the results of the program. A black and white version of this figure will appear in some formats. For the color version, please refer to the plate section.

engineering, and proceed to mine uranium, build plants, etc. – but a much more effective way is to call GE, and GE will design, build, commission, and operate nuclear power plants in the country, and integrate them into their power grid. Similarly, if a country needs a cell telephone system, they can call on IBM to design, install, and operationalize a cell telephone network for the country, as an end-to-end, turnkey solution. These global enterprise solutions providers greatly de-risk such large-scale national projects, and greatly accelerate delivering benefits to the country.

22.6.4.2. The DNA fingerprinting example

A precedent from the genomics industry is the manner in which the use of DNA fingerprinting has spread rapidly around the world in the past 20 years, with at present 49 countries having national DNA fingerprinting systems, containing 50 million samples, and another 20 countries in process. This remarkably rapid global adoption of national genomics infrastructure was achieved so quickly and efficiently because the company that developed the technology, ABI, also established enterprise end-to-end solutions support to set up such systems for entire countries. This is precisely the right approach to delivering genomic medicine, globally, as well, in order to drive global adoption and deliver the real benefits to the countries and their people.

More specifically for genomic medicine, when one considers the totality of elements of a population-scale effort, either in translation research to solve diseases or clinical application of genomic diagnostics (see Figure 22.2), there are major steps of collecting and biobanking samples, sequencing at scale, data management, analysis, and interpretation, validation, and reporting of findings, and the production of custom assay kits at population scale to ultimately deliver genomic tests to the greater population. If there is a gap in capability at any stage, the entire process of delivering the benefits of genomic medicine to the population is delayed or halted entirely.

A proper enterprise genomics solutions group is capable of developing projects from the stage of conception, shaping and stakeholder management, through detailed project architecting, planning, execution and operations, and filling any and all gaps along the way (see Figure 22.3). A notable example of this is the Thermo Fisher Enterprise Genomics Solutions group, which adheres to this model, and has played such a critical role both within the Saudi Genome Project, and the VA MVP sequencing services project, and the BCH/Clarita Global pediatric genomics franchise, as the three largest examples. Lockheed Martin also has a genomic solutions group within their Life Sciences Division, for example, which is also capable of playing the role of end-to-end systems integrator for such efforts. In particular, this group also supports elements of the MVP and the Saudi Genome Program.

Thus, the "right approach" to population-scale efforts to solve disease and deploy genomic medicine is the enterprise services end-to-end support model, which provides the most efficient way to deliver genomic medicine to the majority of countries in the world. As there are now such enterprise service providers with proven experience, hopefully this can spur many governments into action, with the knowledge that they do not need to rely on inefficient, high-risk, low-impact organic efforts.

22.7. The ideal project: how to solve disease at the population scale

We describe here an ideal high-level architecture of a project to solve disease at a population scale for a country. The goal is to deliver the ultimate benefit of genomic medicine to a population, as quickly and impactfully as possible, including near term benefits, and properly establishing genomic medicine optimized for their population, and all within the constraints of responsible spending.

22.7.1. Stable support

It is not often discussed explicitly, but in order to be successful, these large, visible projects need to carefully build stable support and alignment among sponsors and major stakeholders. Ideally, there would be a champion sponsor who is well informed on the value of personalized medicine, and has the available budget required to at least establish (if not fully fund) the project, in order to catalyze quick action without relying on a committee-based decision process. The project must also have a way to deliver prestige, visible credit and quick wins to all major stakeholders, especially those who would otherwise oppose the project, or attempt to compete with it. Creating stable support can benefit tremendously from having a major global Enterprise group promising end-to-end support, as the project will otherwise lack a credible execution path. Building support also requires having a clear, strong value proposition.

22.7.2. Project value proposition

The value proposition to national stakeholders and sponsors is to deliver the tangible twin benefits of this major new field of genomic personalized medicine to the country:

- improved patient care,
- lower cost of healthcare,

as well as the broader national benefits of:

- laying the foundation for future personalized medicine for the country,
- stimulating economic growth in this important future sector, and
- establishing a place among elite global leaders in personalized medicine.

22.7.3. Project architecture

At the high level, the project would have two parallel tracks that interact: a clinical track, which delivers the practical benefits of the most impactful, currently available genomic medicine to the population now, and a discovery track, which will define new genetic factors unique to the population. The major interplay is that the clinical track will route informative patients into the discovery track, and the discovery track will provide new content and interpretation for future updates of the clinical diagnostics, thereby personalizing them to the population. The clinical effort broadly encompasses population screening programs for genetic disease or risk, such as pre-marital, pre-conception, pre-natal and newborn screening. Ideally, the clinical track should be organized to acquire – via genomic diagnostics – data suitable for mining on a much larger number of patients than what would be ascertained for discovery studies, so that this can be used to achieve the ultimate detailed interpretation and personalization for the population.

22.7.4. Population reference genomes

The initial phase of effort in the project should include a campaign to establish a high-quality population reference, typically on the order of 1000 whole genomes that are sequenced to very high, gold standard levels of quality. Ideally this would include phasing of variants, and cataloging copy number and rearrangement variants in addition to local variants. A major output of this campaign is the master table of allele frequencies for the population, which are critical for interpreting observed variants in both the clinical and discovery processes. Similarly, with phasing, this provides a way to stratify the genetic background of the population based on observed haplotypes, which further refines the ability to interpret variants, now in the context of the their distinct ethnic origins.

22.7.5. Clinical

The clinical track would focus on deploying existing rare disease diagnosis, PGx profiling, and assays for actionable mutations in cancer patient tumors. This would progress through a pilot phase to educate clinicians, and then on to broader deployment, and the content of these diagnostics will undergo periodic refinement and personalization for the population. These diagnostics will ideally be based on gene sequencing, not just typing markers, with this used to create a large, unified clinical database that can drive further clinical research and discovery, leveraging the large numbers of patients. The clinical effort also encompasses population screening programs for genetic disease or risk, such as pre-marital, pre-conception, pre-natal and newborn screening – especially for comprehensive assessment of Mendelian diseases, which involves screening on panels comprising approximately 4000 genes.

22.7.6. Discovery

The discovery track would focus on gene-hunting efforts, and the validation of these findings, especially through comparison with independent patient cohorts in other global population-scale efforts. This sort of validation is much more scalable than functional studies in model organisms and cells, although such work must occur as well. In terms of disease, the discovery efforts would focus first on solving Mendelian disease, and then transition towards campaigns focused on the most impactful complex common disorders, which for many countries would be diabetes. Such efforts should particularly include the PGx of drugs for such diseases, as well as hunting for the primary risk factors. Discovery in cancer is extremely research-intensive and complex, and therefore difficult and costly to industrialize, and is at this point best left to the international cancer genomics research efforts, such as the TCGA and ICGC projects. The discovery efforts for Mendelian disease will require assessing tens of thousands of patients, and efforts for complex disease such as diabetes may require on the order of 100,000 patients. The potential high cost of this can be mitigated by proper triage and subject selection for the discovery efforts, good study design, and judicious use of gene panel, exome and whole-genome sequencing. Validated discoveries will be released into future content upgrades of the clinical diagnostics and population screens.

22.7.7. Roadmap from solving to curing

Such projects, which are geared to solving/diagnosing disease, should also present a vision of how to enable and couple to follow-on programs to cure disease, especially with the techniques of cellular medicine, such as stem cell therapy to regenerate tissue, and disease-in-a-dish modeling for personalized drug screening.

22.7.8. Timelines and costs

Projects done in this way will typically cost on the scale of $100 million to several times this amount, and involve the scale of 100,000 patients on the discovery side, with a 5-year timeline. On the clinical side, it requires an initial $10 million scale expenditure to establish the required core infrastructure, and then per-patient cost on the order of $100 when deployed at scale, with judicious design of diagnostic assays. The clinical infrastructure requires 2 years to build out and pilot, and 3 more years to achieve full population-scale deployment. The aggressive 5-year timelines for clinical and discovery phases are only realistically achievable with enterprise support.

22.7.9. Enterprise support

Lastly, but far from least, such projects should be undertaken by a country with the support of a proven, global enterprise genomics solutions provider, to provide end-to-end project support, fill all gaps, eliminate all bottlenecks, and de-risk execution. This includes support from the earliest stages of stakeholder alignment and project conception. Such a supporting global system integrator and domain expert can stabilize the project politically, by giving stakeholders confidence that an experienced, capable guide will substantially de-risk the project, and provide discipline in adhering to the budget and timelines.

22.7.10. The best place to start

Such projects are particularly ideal for the Arab GCC countries, which are ideal in terms of access to funding, very high disease burden for Mendelian disease and diabetes, and social readiness based on their experience with using population carrier screening to manage inherited anemias.

22.8. Conclusion: dancing with elephants

Solving disease at a population scale necessarily involves aligning very powerful stakeholders, as these projects can range from the $100 million to $1 billion dollar scale for medium-sized countries or major healthcare systems. This often involves obtaining the support of kings, princes, presidents, ministers or billionaires. This is a "dance with elephants," as Craig Venter once described his relationship with

US President Bill Clinton, alluding to the fact that Clinton originally gave a huge lift to his company Celera's effort to sequence the first human genome, but then turned around and crushed their value by establishing the Clinton–Blair agreement that genes could not be patented.

Nonetheless, even though it has required dancing with elephants, the first efforts to solve disease at the population scale through the power of genomics are fully underway, and it is a valuable and worthwhile effort that all countries of the world must address to get the full benefits of genomic medicine and personalized medicine. Moreover, this dance will go more smoothly, and with far fewer injuries, if careful thought is given to architecting such projects rationally – with the right technology, right disease targets, and for the right countries – and, most of all, if the planning and execution of such projects are undertaken with the support of global enterprise solutions providers. This will allow countries to de-risk and accelerate execution, remove barriers of competency and capacity, and focus on integrating the benefits of genomics into their national healthcare systems for the benefit of their people.

REFERENCES

Auffray, C., Charron, D. and Hood, L. (2010). Predictive, preventive, personalized and participatory medicine: back to the future. *Genome Med.*, **2**(8), 57.

Christianson, A., Howson, C. and Modell, B. (2005). *March of Dimes Global Report on Birth Defects: The Hidden Toll of Dying and Disabled Children*. March of Dimes Birth Defects Foundation, White Plains, NY.

Merriman, B., Ion Torrent R&D Team and Rothberg, J. (2012). Progress in Ion Torrent semiconductor chip based sequencing. *Electrophoresis*, 33(23), 3397–3417.

Zahed, L. (2001). The spectrum of β-thalassemia mutations in the Arab populations. *BioMed. Res. Int.*, **1**(3), 129–132.

BGI website: http://www.genomics.cn/en/index

Centers for Mendelian Genomics website: http://www.mendelian.org/

Claritas Genomics website: http://claritasgenomics.com/

Crohns and Colitis Foundation of America website: http://www.ccfa.org/

Estonian Biobank website: http://www.geenivaramu.ee/en

Genome England project website: http://www.genomicsengland.co.uk/

Global diabetes prevalence rates: http://www.allcountries.org/ranks/diabetes_prevalence_country_ranks.html

Human Longevity Inc. website: http://www.humanlongevity.com/

Million Veteran Program website: http://www.research.va.gov/MVP/

National Center for Advancing Translational Science website: http://www.ncats.nih.gov/

Nextcode Health website: https://www.nextcode.com/

NHGRI website: http://www.genome.gov/

Rady Children's Hospital website: http://www.rchsd.org/

Sidra Genome Qatar project website: http://www.sidra.org/sidra-to-help-qatar-grow-as-a-hub-for-genetic-medicine/

Taiwan Biobank Project: http://newsletter.sinica.edu.tw/en/file/file/1/125.pdf

The Saudi Genome Project website: http://rc.kfshrc.edu.sa/sgp/index.asp

23 Economics of personalized medicine

Katherine Payne and Martin Eden

23.1. Introduction

Use of the term "personalized medicine" is now ubiquitous. However, there is little consistency in how personalized medicine is either defined or used in clinical practice. Various authors have attempted to bring together the different definitions of personalized medicine (for example: Redekop and Mladsi, 2013; Schleidgen *et al.*, 2013). A framework has been put forward, which can be used to understand the implications of personalized medicine in terms of: the questions (e.g., what is the diagnosis?); the methods used to answer them (e.g., a test); and the available actions (e.g., to give or not give a particular drug). Alongside this proposed framework, Redekop and Mladsi offer a useful working definition of personalized medicine as: "the use of combined knowledge (genetic or otherwise) about a person to predict disease susceptibility, disease prognosis, or treatment response and thereby improve that person's health." This is the definition adopted in this chapter, because it facilitates discussion of the broad implications when considering the economics of personalized medicine, rather than adopting the narrower focus on using genetics to target medicines.

Irrespective of how personalized medicine is defined, decision-makers charged with allocating finite healthcare budgets require robust and timely evidence to support the cost-effectiveness of technologies used to "personalize" the diagnosis, prevention, or treatment of diseases. Information on the value for money of technologies used to personalize medicine is needed before they can be recommended for funding and used in clinical practice. This chapter outlines the underlying concepts that inform the framework of economic evaluation which can be applied to provide decision-makers with the requisite information. The different types of economic evaluation are described and a summary of the current level of evidence supporting the use of a specific type of personalized medicine, using pharmaco-genetic and pharmaco-genomic technologies, is provided. An

Genome-Wide Association Studies: From Polymorphism to Personalized Medicine, ed. K. Appasani. Published by Cambridge University Press.

overview of the key stages in the design and conduct of an economic evaluation is presented. The chapter concludes by presenting an overview of the issues to be considered when evaluating the economic impact of technologies to personalize medicine.

23.2. Informing decision making

The existence of finite healthcare budgets and the scarcity of healthcare resources mean that there is an opportunity cost associated with every decision made to provide an intervention. The decision to allocate resources to a particular service excludes those resources from alternative possible uses within a healthcare system. It follows that there is a need to demonstrate that technologies used to personalize medicine offer added value, in terms of the relative costs and benefits, compared with current practice.

The discipline of economics provides theories and methods to study the allocation of resources and aims to generate information on how to achieve the most efficient use of scarce resources. A subdiscipline of economics has emerged which involves the application of economic theory and practice within the context of healthcare provision. Economic evaluation is one method used by health economists to generate evidence to inform decision making. It has been defined as "the comparative analysis of alternative courses of action in terms of both their costs and consequences" (Drummond, 2005). The underlying premise of economic evaluation is the notion of opportunity cost in relation to budgets specifically allocated to provide healthcare services, treatments, and interventions. With this in mind, the economic evaluation framework enables assessment of an intervention's incremental benefits and costs over existing treatments or the alternative of doing nothing.

Good-quality economic evidence is now routinely demanded for decision making in healthcare systems throughout the world. Consequently, a series of guidelines (country-specific) for the design and conduct of economic evaluations have been published (ISPOR, 2014a). Some countries, such as Australia, use evidence from economic evaluations as a requirement for the reimbursement of medicines (PBS, 2014). Countries, such as the United Kingdom, include evidence from economic evaluations as part of health technology assessment reports to inform the production of national guidance on technologies, public health interventions, diagnostics and clinical guidelines. In the United States, the legislative focus is on comparative effectiveness rather than cost-effectiveness *per se*, but there is still considerable production of economic evidence to inform decision making (Weinstein and Skinner, 2010). Some countries, such as Greece, to date have not embraced the use of economic evidence at a national level (ISPOR, 2014b). Even though there are inter-country differences in the design and application of economic evaluation, it has clearly become a key requirement for the informed introduction of new technologies into clinical practice.

23.3. Types of economic evaluation

The need for applying the principles and framework of economic evaluation to personalized medicine is well recognized (Phillips and Van Bebber, 2005; Conti *et al.*, 2010). The two main approaches to economic evaluation are cost-effectiveness analysis (CEA) and cost–benefit analysis (CBA). Within healthcare, a specific type of CEA termed cost–utility analysis has been developed and represents the predominant approach to economic evaluation in a number of healthcare systems. In addition to CEA and CBA, there are two types of study which have received attention in the literature on healthcare decision making: cost–consequence analysis and cost-minimization analyses. Cost–consequence analysis has been suggested as a potentially useful method for economic evaluation, particularly in the context of complex interventions (Coast, 2004), but has been criticized because it places the onus on the decision-maker to decide which outcome is of primary interest (Brazier *et al.*, 2005; Claxton *et al.*, 2005). Cost-minimization analysis has been used in the literature, but is no longer considered useful for decision making due to assumptions made in the analyses and limitations in the methods used to handle uncertainty (Briggs and O'Brien, 2001).

Table 23.1 summarizes the types of economic evaluation and illustrates that resources and costs are identified and quantified in the same way, no matter what type of economic evaluation is specified. The difference between the types of economic evaluation is based on how outcomes, defined as the benefit to patients, are defined and valued. In the literature, the terms cost-effectiveness analysis and cost–utility are sometimes interchanged because the theoretical underpinning of the two methods is the same. The difference in how outcomes are defined and valued in cost-effectiveness (utility) analysis and CBA is underpinned by theories and "views of the world," extra-welfarist and welfarist, respectively (see Brouwer *et al.*, 2008, for a good explanation of these theories).

CBA fits within the welfarist view of the world. Welfarism, derived from utilitarian ethics, assumes that individuals are the best judge of their own welfare (or utility) and that welfare is taken to be the satisfaction of an individual's preferences. A CBA approach to economic evaluation is aimed at determining the net societal benefit of projects through the measurement and collation of individuals' preferences for a proposed intervention.

To gauge the strength of preference for an intervention when market provision is absent, which is typical in the healthcare sector, contingent valuation methods (most commonly applied using willingness-to-pay (WTP) techniques). Contingent valuation methods provide a measure of benefit on the same monetary scale as costs. The use of contingent valuation methods is described in the health economics literature, but is not currently advocated, or being used, in health policy decisions. The paucity of examples of full CBA in healthcare (Smith and Sach, 2009, 2010) may be the result of ongoing methodological challenges and decision-makers recognizing ethical challenges in applying contingent valuation methods to measure the benefits of healthcare interventions.

Table 23.1. Types of economic evaluation

Economic evaluation	Costs	Outcome
Cost-minimization analysis (CMA)	Costs may be identified and measured as: (1) direct costs. These may be medical (costs to the hospital or health service which are directly associated with treatment, eg., costs of drugs, equipment, and staff), or non-medical costs (costs to social services); and (2) indirect costs (costs to society, e.g., lost productivity through morbidity and mortality)	The outcomes of the service/treatments being compared are taken to be the same. This assumption must be based on robust evidence
Cost-effectiveness analysis (CEA)		The outcomes of the service/treatments being compared are measured using a single, natural unit. Examples include: response to treatment measured using a clinical indicator or life-years gained. Treatments that are aimed at achieving the same effect can be compared
Cost–utility analysis (CUA)		The outcomes of the service/treatments being compared are measured using utility values, which are then converted into a Quality Adjusted Life Year (QALY). The QALY takes into account the effect of treatment on both the quality (utility) and quantity of life. It is possible to compare treatments for different conditions
Cost–benefit analysis (CBA)		The outcomes of the service/treatments being compared are measured using monetary values. Contingent valuation methods are used to measure WTP that represents the monetary value. An example would be: how much are you willing to pay for the information generated from having a genetic-based diagnostic test?
Cost–consequences analysis (CCA)		The outcomes of the service/treatments being compared are not aggregated into a single measure. All relevant outcomes that were quantified are presented. Examples could be: (i) diagnostic test accuracy, (ii) impact on health status, (iii) impact on subsequent uptake on treatment

The extra-welfarist perspective has been developed in response to the failure of traditional welfarist approaches to economic evaluation in healthcare. Extra-welfarism provides the theoretical foundation for the use of CEA and allows for additional factors alongside the preferences of individuals to be considered in decision making, which strict utilitarian-based approaches do not. For healthcare policies and interventions, it is generally argued that the appropriate measure of benefit is "health" or "health status" because a specific budget has been allocated to provide healthcare. The methods of CEA and CUA, underpinned by extra-welfarism, have become the dominant type of economic evaluation in healthcare.

23.3.1. Design and conduct of CEA and CUA

To conduct a CEA (CUA), two main vehicles to generate evidence exist: trial-based or model-based evaluations (Sculpher *et al.*, 2006). Trial-based evaluations (also called prospective economic evaluations) result in an analysis based on data collected for an individual-level patient sample recruited to a randomized controlled trial. The collection of data necessary for the economic evaluation, such as resource use, effectiveness, and health status, should be an integral component of the clinical trial protocol. Trial-based studies can be time- and resource-intensive, requiring considerable funding, the results are often not sufficiently timely for decision making in practice, and will clearly only generate results for the specified technology and comparator. This may limit the generalizability of the trial findings to other settings. There are also specific challenges to be considered when designing trial-based evaluations of technologies to personalize medicine.

Thompson *et al.* (2014) illustrate some of the general and specific challenges in a published prospective trial-based CEA in the context of personalized medicine. In the study, a pharmaco-genetic test was used to identify patients at risk of profound neutropenia before starting azathiopine in the management of inflammatory diseases. The study suggested that genotyping potentially offered a less-expensive alternative than current practice, but it may also have a small negative impact on health status. Considerable uncertainty surrounded the findings. Therefore, these results pose difficult challenges to decision-makers regarding whether or not to introduce the pharmaco-genetic test in this context.

Key challenges, relating to the measurement of an intervention's effectiveness, had to be addressed during the design phase of Thompson *et al.*'s study. A fundamental challenge relevant to designing trials generating both clinical and economic evidence is how to select and define the relevant primary outcome for the analysis. This issue is common to all clinical trials, but is further compounded when designing a trial that aims to detect a change in a poorly defined endpoint: reduction of an adverse drug reaction (ADR). Questions immediately present themselves. Which ADRs should be included? If a specific set of ADRs is to be used, such as neutropenia, then how will this be defined? Should the focus be on all grades of neutropenia or specific ones, and if specific ones, then how should these be chosen? There is no published guidance on how this specific issue was resolved. The selected primary endpoint chosen for the economic evaluation by Thompson *et al.* was any instance of azathioprine being discontinued due to ADR in the first four months of treatment. This allowed the design of a pragmatic study, but did not facilitate an analysis of that which *a priori* may have been considered to be the key outcome in question: reduction in cases of profound neutropenia. The selection of primary outcome is, in part, influenced by the statistical power of the study and choosing a "rare" outcome such as profound neutropenia would generate a prohibitively large sample size to be able to detect a statistically significant difference in effect through the introduction of this example of personalized medicine.

A second related challenge concerned the proposed setting, the English National Health Service (NHS), for this form of genotyping in the management of inflammatory diseases. From the perspective of the NHS, the preferred outcome for use in an economic evaluation would be health status that allows estimation of the change in Quality Adjusted Life Years (QALYS) thereby facilitating cost–utility analysis and, consequently, better-informed resource allocation decision making.

Again, using health status (measured for example by the EQ-5D measure: see www.euroqol.org) poses a challenge in two ways. First, the difference in health status with (intervention arm) and without (control arm) the intervention is likely to be very small and hence a large sample size would be needed to show a difference in effect between the two study arms.

Also, a specific challenge to linking the impact of experiencing an ADR on health status is introduced as the trial design will need to specify key data collection points. ADRs may occur between these key data collection points and may also be transient, which makes the timing of collecting health status data another challenge to address to enable the impact of the ADR on health status to be quantified. Thompson *et al.* (2014) suggest that the issue of timing of collecting health status data and experiencing ADRs is a particular issue when generating evidence of clinical and cost-effectiveness for pharmaco-genetic tests that aim to predict the risk of an ADR.

Although trial-based evaluations are generally still viewed as the gold standard when generating evidence of clinical effectiveness, the use of model-based economic evaluations is preferred by some decision-making bodies. Model-based evaluations have the advantage that they can: incorporate all currently available evidence; use an appropriate time horizon to consider long-term costs and outcomes if needed; and be used to simultaneously compare all relevant interventions (Sculpher *et al.*, 2006). A model-based economic evaluation allows data from many different sources to be systematically assimilated and used. There are many potential sources of data such as systematic reviews of trial-based effectiveness data, resource use data from observational studies, micro-costing studies, reviews of utility data and expert opinion (Philips *et al.*, 2006; Leal *et al.*, 2007; Sullivan and Payne, 2011). The advantage of model-type studies is that they are generally less time- and resource-intensive, and allow for extrapolation of data for the lifetime horizon and exploration of the sources and impact of uncertainty in the data. The identification of uncertainty in a model-based study is a key component of the analysis (Philips *et al.*, 2006). Sensitivity analysis is therefore used in all model-based economic evaluation and methods. Techniques such as probabilistic sensitivity analysis should be used to explore the joint impact of parameter uncertainty (Claxton, 2008).

Both trial- and model-based economic evaluations have important roles within different stages of the iterative process of evidence gathering. Model-type studies are particularly useful when assessing the costs and benefits of an intervention in the early stages of its development and use, and by including value of information analysis (Claxton *et al.*, 2004) could be used to ascertain the value of further research. Value of information analysis involves quantifying the extent of uncertainty in the parameters used to populate the model-based economic evaluation and assign a cost to indicate the value of conducting further research to address the existent uncertainty in estimates of cost-effectiveness. The next step would be to generate further estimates for key parameters such as clinical effectiveness, which are then used to repopulate a more complex model. This iterative approach to economic evaluation, as recommended by Sculpher *et al.* (1997), is an extremely useful approach to evaluate the economic impact of technologies to personalize medicine as the early clinical evidence to support them is generally quite weak and limited, as illustrated by Fleeman *et al.* (2013). Using the iterative model-based approach as and when new evidence becomes available is a practical solution to this challenge.

23.4. Applied examples of economic evaluations of personalized medicine

There is an extensive number of published systematic reviews that have identified and summarized published economic evaluations of personalized medicines (Giacomini *et al.*, 2003; Phillips and Van Bebber, 2004; Carlson *et al.*, 2005; Jarrett and Mugford, 2006; Vegter *et al.*, 2008, 2010; Wong *et al.*, 2010; Hatz *et al.*, 2014; Phillips *et al.*, 2014). The findings from these reviews cannot be compared directly because different definitions of personalized medicine, different search strategies, and inconsistent study inclusion criteria have been used. Some of the reviews focussed on identifying economic evaluations of genomic-based approaches to personalize medicine and others more broadly considered genomic-based diagnostic tests. The definition of what constituted economic evidence also differed between the reviews, with some identifying all economic evaluations, including CBA, and some restricting their focus to CUAs. The first of these reviews was published in 2004 and aimed to identify all published cost-effectiveness and cost–utility analyses of pharmaco-genomics interventions, defined as the use of genetics to guide drug development and prescribing, finding just 11 studies. More recent examples include reviews by Hatz *et al.* (2014) and Phillips *et al.* (2014) which focussed on identifying cost–utility analyses of genomic-based strategies to diagnose and target treatments. A common finding from all of the published systematic reviews is that the quantity and quality of the economic evidence base is improving, but some key methodological challenges remain for the design and conduct of CEAs of personalized medicine.

23.5. Issues for the design of cost-effectiveness analyses of personalized medicine

A number of commentators have published papers on the specific issues for the design of CEAs of personalized medicine (for example, see: Conti *et al.*, 2010; Faulkner *et al.*, 2012; Annemans *et al.*, 2013; Buchanan *et al.*, 2013). Generally, these papers have focussed on issues for the economic evaluation of technologies concerned with identifying a single gene to stratify patient populations into subgroups more likely to respond safely and effectively to a targeted treatment. Danzon and Towse (2002) explored some specific issues related to the economics of gene therapy, concluding that such extreme examples of the "personalization of medicines," down to the individual patient level, are likely to be prohibitively expensive given the relative size of the benefits to patient populations (Danzon and Towse, 2002). Buchanan *et al.* (2013) described some key methodological challenges for conducting economic evaluations of genomic-based diagnostic or prognostic tests, including those used to target treatment strategies. The methodological challenges described related to how to define the relevant comparator, study perspective, and time horizon for the analysis. They also described challenges associated with identifying and quantifying resource use and effectiveness data in an environment in which there are limited trials of the interventions being evaluated and there is limited evidence on patient and clinician behavior in terms of if/how they use genomic information in practice. Finally, the authors outlined the challenge of how to quantify the benefits to patients from genomic technologies because sometimes the use of health status, as valued in a CEA or CUA, may not capture all the relevant outcomes of interest. Annemans *et al.* (2013) used a more focussed definition of personalized medicine and described 10 methodological challenges for generating estimates of cost-effectiveness using model-based economic evaluations of strategies used to stratify patient populations into groups of safe and effective responders to screening or treatment options. Some key issues described mirrored points raised by Buchanan *et al.* (2013), including the correct framing of the research question, data collection challenges, and how to measure the value of tests. Additional points described were specific to designing model-based evaluations of complex interventions that require a link to be made between if/how tests are used, the process of obtaining a diagnosis, and interpretation of test results and informing a subsequent treatment decision to which patients adhere. Furthermore, these authors described how the analysis can become very complex if different tests are combined or used sequentially, as there is a need to calculate the combined sensitivities and specificities of the tests. Each of these issues leads to a higher degree of uncertainty in economic models designed to identify and quantify the incremental costs and benefits of personalized medicine compared with their simple pharmaceutical counterparts.

In the following sections, a summary of some key issues for the design and conduct of CEAs of personalized medicine are presented that have been informed by the existing literature. Here, a broad definition of personalized medicine is used to include genomic-based strategies to target screening or treatment options together with genomic-based diagnostic and predictive tests. Importantly, these section describes some of the emerging issues that will be necessary to consider when evaluating the impact of strategies now made possible by new sequencing techniques, such as the use of array tests and whole exome/genome sequencing. These issues are presented using key headings from a checklist designed to guide authors how to report the key details about methods and results to enhance the robust reporting of published economic evaluations (see Table 23.2). Readers who are interested in a general introduction to methods of economic evaluation are directed to some examples of introductory (for example, Elliott and Payne, 2005) and more advanced textbooks (for example, Drummond, 2005; Morris *et al.*, 2007).

Table 23.2. Key components in the design of a cost-effectiveness analysis

Method	Description
Define the research question	Present the study question and its relevance for health policy or practice decisions
Target population and subgroups	Describe characteristics of the base case population and subgroups analyzed, including why they were chosen
Setting and location	State relevant aspects of the system(s) in which the decision(s) need(s) to be made
Study perspective	Describe the perspective of the study and relate this to the costs being evaluated
Comparators	Describe the interventions or strategies being compared and state why they were chosen
Time horizon	State the time horizon(s) over which costs and consequences are being evaluated and say why appropriate
Discount rate	Report the choice of discount rate(s) used for costs and outcomes and say why appropriate
Choice of health outcomes	Describe what outcomes were used as the measure(s) of benefit in the evaluation and their relevance for the type of analysis performed
Measurement of effectiveness	Single study-based estimates: Describe fully the design features of the single effectiveness study and why the single study was a sufficient source of clinical effectiveness data Synthesis-based estimates: Describe fully the methods used for identification of included studies and synthesis of clinical effectiveness data
Measurement and valuation of preference-based outcomes	If applicable, describe the population and methods used to elicit preferences for outcomes
Estimating resources and costs	Single study-based economic evaluation: Describe approaches used to estimate resource use associated with the alternative interventions Model-based economic evaluation: Describe approaches and data sources used to estimate resource use associated with model health states
Currency and price date	Report the dates of the estimated resource quantities and unit costs
Choice of model	Describe and give reasons for the specific type of decision-analytic model used
Analytical methods	Describe all analytical methods supporting the evaluation. This could include methods for dealing with skewed, missing, or censored data; extrapolation methods; methods for pooling data; approaches to validate or make adjustments (such as half cycle corrections) to a model; and methods for handling population heterogeneity and uncertainty

Adapted from Husereau *et al.* (2013).

23.5.1. Define the research question

This is clearly a vital key step in the design of any economic evaluation. Defining a research question involves being explicit about the intervention to be evaluated and in what context. Furthermore, the reason for generating the economic evidence also needs to be clearly understood. The challenge for personalized medicine has been previously described by a number of authors, including for example Fleeman *et al.* (2013), Buchanan *et al.* (2013) and Annemans *et al.* (2013). The common issue that these authors raise is that interventions to personalize medicine can be poorly defined in the literature, or heterogeneous in the type of potential genomic technologies, for example marketed tests or bespoke laboratory tests, which are likely to use laboratory-specific protocols. This issue is further complicated when the intervention involves multiple gene testing, such as in the case of array testing or whole exome/genome sequencing. Furthermore, most interventions used to personalize medicine are a complex intervention potentially involving a number of discrete but interlinked component parts and associated behaviors. Therefore, it is essential that the analyst designing the CEA is clear what intervention is to be evaluated, which will require input from individuals with the appropriate expertise such as clinical scientists and clinicians who will use the intervention in practice.

23.5.2. Target population and subgroups

For simple interventions, such as pharmaceuticals, the target population is generally defined by the eligible patient population listed in the product licence provided by the manufacturer. Subgroups of this total population may be identified by RCTs that have been conducted as part of the drug licensing process supported by subsequent clinical experience with treating the condition of interest. For personalized medicine involving single-gene tests, again the target population of interest is likely to be readily defined by either the relevant family history of the genetic condition of interest or the specific condition that is being targeted for treatment. However, as we move to using panels of genes in tests then the target population which stands to benefit from personalized medicine extends in line with the number of genes being tested up to the extreme situation in which the whole population could be eligible for testing using whole-genome sequencing. Buchanan *et al.* (2013) and Annemans *et al.* (2013) have also touched on this issue, which is perhaps one of the major challenges for analysts designing robust CEA of strategies to personalized medicine because the results produced will only be relevant to the population of interest defined in the analysis. Subgroup analysis can aid understanding of the implications of heterogeneity in the results of the CEA, but to reflect this heterogeneity it is necessary to be as specific as possible about the characteristics of the recipients of new interventions, which requires an associated evidence base (Sculpher, 2008). This evidence base may not be available for genetically diverse populations within and across countries. A simple solution to this challenge may be to describe the target population in terms of presenting phenotype, but even this strategy may result in a diverse group of individuals who may be eligible for an intervention. In addition, it is also necessary to recognize that for genomic-based tests that the persons being tested are not the only relevant study population, which may extend to other family members.

23.5.3. Setting and location

Ideally, the evidence base used to populate an economic model should reflect the target population and setting of the evaluation. Healthcare systems are country-specific and the cost and outcomes of interest introducing new technologies are specific to the system in which they are put forward for use. The need to clearly specify the setting and location of the CEA links in with the challenge of defining relevant subgroups and using an evidence base that relates to the genetic characteristics of the population of interest. As the interventions of interest for personalization of medicine are genomic-based, then it is likely to be necessary to be clear whether the evidence base used to populate the model matches the setting and location chosen for the CEA.

23.5.4. Study perspective

The study perspective defines the relevant costs and outcomes to be included in the CEA. The study perspective of interest can range from societal, healthcare system to the patient perspective. If a healthcare system perspective is chosen, then the objective of the analysis becomes how to spend healthcare resources to achieve maximum health benefits to patient populations. Alternative perspectives may incorporate different objectives and concern different budgets. The chosen perspective relevant for the analysis is guided by the research question of interest and the location for the CEA (ISPOR, 2014a). Therefore, the issues associated with defining the relevant perspective for an evaluation of a personalized medicine strategy are closely related to those described for the selection of the relevant study population, setting and location of the CEA.

23.5.5. Comparators

The need to define all relevant comparators for the intervention of interest is underpinned by the concept of opportunity cost. The aim of the CEA is to quantify the relative costs and outcomes of an intervention. Consequently, it is necessary to define and describe a relevant set of comparator interventions that could conceivably be used to address the clinical issue in question. For a personalized medicine or screening strategy, then the comparator can be relatively easily defined by describing the process of current treatment or screening without a test to stratify the patient population. However, defining the relevant set of comparator interventions when conducting a CEA of a genomic-based test, and particularly one involving multiple genes, then the task becomes more complex because multiple options may be available. Using a new genomic technology to diagnose a condition will bring the test to the start of the diagnostic care pathway and/or introduce testing in a new population. In both instances this makes it very challenging to decide what the relevant comparator should be. For example, consider designing an economic evaluation of NGS for inherited eye conditions, such as retinitis pigmentosa, for which over 100 causative genes have been identified. The relevant comparators could be specified as each gene as a single test, conducted using Sanger sequencing, but it would also be necessary to recognize that for some genes there is no current test available. This creates a complex problem for the analyst trying to define a relevant comparator.

23.5.6. Time horizon and discount rate

In a similar manner to the choice of the study perspective, the time horizon selected for the CEA will dictate the relevant costs and outcomes to be included in the CEA. The time horizon can be relatively short to just cover, for example, the diagnostic process, but more usefully it should reflect the period of time in which the technology is likely to affect costs and outcomes of interest which could imply a lifetime horizon of the patient population. Using a lifetime horizon generally necessitates the use of model-based CEA. A discount rate is used in a CEA to calculate the value of costs and benefits that extend beyond the first year of the analysis. This is done to reflect the assumption derived from economic theory that people prefer to receive payouts (outcomes) now but defer costs, and they have a different time preference for costs and outcomes.

The chosen time horizon for the analysis is also guided by the research question of interest and the location for the CEA (ISPOR, 2014a). A potential additional challenge when evaluating genomic-based diagnostic tests is whether it is necessary to take into account inter-generational issues. The lifetime horizon assumes the relevant patient population is the person being tested, but for inherited conditions this may need extending, in theory, to cover future generations. Leaving aside arguments for and against extended time horizons to include future generations in economic evaluation, the challenge of how to incorporate requisite costs and outcomes is likely to preclude any attempts at a CEA with such an extensive scope.

23.5.7. Measurement of effectiveness

Generally, in a CEA the measurement of effectiveness will be informed by the clinical area of interest. In the context of genomic tests the first two components of the ACCE (analytic validity, clinical validity, clinical utility and associated ethical, legal and social implications) framework (CDC, 2014) are useful starting points to think about the different relevant measures of effectiveness. It is then necessary to be clear how to link these measures of clinical effectiveness with the final patient outcomes to be used in the CEA. Two vehicles may be used to collect and calculate the measurement of effectiveness: trial-based or model-based economic evaluations. Each approach has strengths and weaknesses that are generally similar no matter what type of intervention is being evaluated. However, in the context of personalized medicines, some key characteristics of an intervention may limit the value of using a trial-based evaluation. One possible example includes evaluating a pharmaco-genetic test to predict the risk of a rare ADR, which would preclude a trial on the basis that the required sample size to achieve adequate power makes the design unfeasible. Another example would be that in which the delivery of a genetics service might preclude the practical implementation of a randomization process necessary for an RCT comparing NGS with current diagnostic strategies, which would be further complicated by the prohibitively large research budget needed to follow-up patient outcomes and resource use over a sufficient time horizon to produce meaningful results for decision-makers.

23.5.8. Choice and measurement of outcomes

The choice of outcomes should reflect the patient benefits the intervention is designed, or anticipated, to influence and also consider the study perspective and location of the CEA. Ideally, a final outcome relevant to the patient, rather than an intermediate or proxy outcome, should be used. If the objective of the analysis is to understand how to spend a healthcare budget, then the relevant patient outcome could be health gain, as is currently the case for decision making in the NHS. In the UK, this is generally measured using a generic health status measure such as the EQ-5D (EuroQol, 2014). Published preference weights are then used to adjust the remaining years of life to take into account quality of life. This allows the calculation of quality-adjusted life years (QALYs), which are the product of the remaining years and quality of life. The use of QALYs is appropriate and relevant in the context of stratified approaches to personalized medicine (Payne and Thompson, 2013). However, the use of QALYs becomes potentially problematic when measuring the outcomes for genomic-based diagnostics that may be used to diagnose conditions for which there are no obvious health gains. Buchanan *et al.* (2013) together with Payne *et al.* (2013) and Grosse *et al.* (2008), in the context of clinical genetic testing, describe in detail the challenge associated with the valuing outcomes using QALYs. There is no current solution to this challenge. Grosse *et al.* (2008) suggest moving beyond CEA and conducting CBAs of genomic tests. One practical example of trying to elicit monetary outcomes has been published in the context of genetic testing for inherited forms of retinitis pigmentosa (Eden *et al.*, 2013). However, for both theoretical and practical reasons, this approach is not likely to replace CEA as the preferred method for generating economic evidence to inform national decision making (Smith and Sach, 2009). This is why there is some early research that has explored if and how non-health status measures of outcome, such as being able to make an informed decision, could be used in the context of complex interventions, such as clinical genetic testing services (Payne *et al.*, 2013).

2.5.9. Estimating resources and costs

In a similar manner to generating estimates of effectiveness, two vehicles may be used to collect and calculate the resource use: trial-based or model-based economic evaluations. The challenges associated with collecting resource use in the context of personalized medicine have been described in detail by Buchanan *et al.* (2013) and Annemans *et al.* (2013). Two general issues emerge relating to (1) a lack of published studies to identify resource use and associated unit costs, and (2) methodological challenges of how to collect a broad range of costs potentially relevant given the chosen study perspective, time horizon and patient population. Furthermore, even completing the perhaps seemingly simple task of attaching a unit cost of the test can be complicated by the lack of national price lists of genomic-based diagnostics.

23.5.10. Choice of model and analytical methods

Design and statistical issues similar to those relevant for clinical trials, such as consideration of clustering, sample size calculations, and appropriate methods of data analysis, must be also addressed in the planning of a prospective economic study along with specific challenges to the analysis of economic data (Mihaylova *et al.*, 2011). There are useful published guidelines that describe the key criteria to use when selecting the appropriate model type for the conduct of a model-based CEA (see Brennan *et al.*, 2006). In general, there are no specific issues that make the choice of model and analytical methods different when designing CEA of a personalized medicine intervention compared with any other intervention. Published systematic reviews of personalized medicine support this premise and report that the most commonly used model-based approaches use decision-tree or state-transition models. However, one example of a particular challenge that means different modeling approaches may sometimes be appropriate can be illustrated when considering how to quantify the costs and benefits of a genomic-based diagnostic, such as for example, next-generation sequencing (NGS) for *BRCA1/2* (Jacob and Payne, 2012). In general, model-based CEAs aim to capture the impact on healthcare resource use and diagnostic benefits that led to improved test accuracy. In the context of NGS, additional benefits of diagnostics may include: reduced laboratory time; reduced time to results; and capacity to increase the number of tests performed. Current *BRCA1/2* testing using Sanger sequencing technologies are limited by long (one-year) turnaround times, which together with limited resources to increase the volume of tests and associated genetic counseling has led to the use in some countries of a "risk threshold" to target women eligible for *BRCA1/2* testing. In the medium-term, the cost of providing testing using NGS is likely to be equivalent to that of current technologies. Furthermore, a new diagnostic test using NGS will only be provided for use in a clinical setting if the technology meets current levels of sensitivity, specificity and predictive values. Using NGS offers the opportunity of decreased turnaround time and increased volume of *BRCA1/2* tests, which will impact on the benefits and costs associated with the diagnostic service. Using a decision tree or state transition model will not allow the evaluation of the impact of capacity constraints and increased turnaround time on the costs and benefits of this new diagnostic technology. For this reason, patient-level modeling using discrete event simulation may be preferred. Such models have a further advantage in that the approach allows the assessment of the impact on costs and benefits by considering changes in patient pathways through the introduction of new genomic-based diagnostic technologies.

23.6. Concluding remarks

Producing economic evidence within the emerging field of personalized medicine is vital to support the timely and evidence-based introduction of technologies which have a potentially beneficial role in healthcare systems worldwide. Methods of economic evaluation, and specifically CEA, have developed a crucial role in resource allocation decision-making. Examples of good practice in the application of CEAs of personalized medicine exist, but there is a need for health economists to continue to develop new economic evidence while being cognizant of the methodological challenges as the complexity of the interventions, level of personalization, and the remit of personalized medicine continue to increase.

REFERENCES

Annemans, L., Redekop, K. and Payne, K. (2013). Current methodological issues in the economic assessment of personalized medicine. *Val. Health*, **16**, S20–26.

Brazier, J., McCabe, C. and Edlin, R. (2005). Health economics and cost consequences analysis: a step back in time. BMJ Rapid Response to Coast, J., 2004. Is economic evaluation in touch with society's health values? *Br. Med. J.*, **329**, 1233–1236.

Brennan, A., Chick, S.E. and Davies, R. (2006). A taxonomy of model structures for economic evaluation of health technologies. *Health Econ.*, **15**, 1295–1310.

Briggs, A.H. and O'Brien, B.J. (2001). The death of cost-minimization analysis? *Health Econ.*, **10**, 179–184.

Brouwer, W.B.F., Culyer, A.J., van Exel, N.J.A. and Rutten, F.F.H. (2008). Welfarism vs. extra-welfarism. *J. Health Econ.*, **27**, 325–338.

Buchanan, J., Wordsworth, S. and Schuh, A. (2013). Issues surrounding the health economic evaluation of genomic technologies. *Pharmacogenomics*, **14**, 1833–1847.

Carlson, J.J., Henrikson, N.B., Veenstra, D.L. and Ramsey, S.D. (2005). Economic analyses of human genetics services: a systematic review. *Genet. Med.*, **7**, 519–523.

CDC.gov. (2014). Genomics|Genetic Testing|ACCE. http://www.cdc.gov/genomics/gtesting/ACCE/ (accessed April 30, 2014).

Claxton, K. (2008). Exploring uncertainty in cost-effectiveness analysis. *PharmacoEconomics*, **26**, 781–798.

Claxton, K., Ginnelly, L., Sculpher, M., Philips, Z. and Palmer, S. (2004). A pilot study on the use of decision theory and value of information analysis as part of the NHS Health Technology Assessment programme. *Health Technol. Assess.*, **8**, 1–103, iii.

Claxton, K.P., Sculpher, M.J. and Ades, T. (2005). Cost consequences: implicit, opaque and anti scientific. BMJ Rapid Response to Coast, J., 2004. Is economic evaluation in touch with society's health values? *Br. Med. J.*, **329**, 1233–1236.

Coast, J., 2004. Is economic evaluation in touch with society's health values? *Br. Med. J.*, **329**, 1233–1236.

Conti, R., Veenstra, D.L., Armstrong, K., Lesko, L.J. and Grosse, S.D. (2010). Personalized medicine and genomics: challenges and opportunities in assessing effectiveness, cost-effectiveness, and future research priorities. *Med. Decis. Making*, **30**, 328–340.

Danzon, P. and Towse, A. (2002). The economics of gene therapy and of pharmacogenetics. *Value Health J. Int. Soc. Pharmacoecon. Outcomes Res.*, **5**, 5–13.

Drummond, M.F. (2005). *Methods for the Economic Evaluation of Health Care Programmes*. Oxford University Press, Oxford.

Eden, M., Payne, K., Combs, R.M., *et al.* (2013). Valuing the benefits of genetic testing for retinitis pigmentosa: a pilot application of the contingent valuation method. *Br. J. Ophthalmol.*, **97**, 1051–1056.

Elliott, R. and Payne, K. (2005). *Essentials of Economic Evaluation in Healthcare*. Pharmaceutical Press, London.

EuroQol. (2014). EuroQol – Home. http://www.euroqol.org (accessed April 30, 2014).

Faulkner, E., Annemans, L., Garrison, L., *et al.* (2012). Challenges in the development and reimbursement of personalized medicine-payer and manufacturer perspectives and implications for health economics and outcomes research: a report of the ISPOR personalized medicine special interest group. *Val. Health*, **15**, 1162–1171.

Fleeman, N., Payne, K., Newman, W.G., *et al.* (2013). Are health technology assessments of pharmacogenetic tests feasible? A case study of CYP2D6 testing in the treatment od breast cancer with tamoxifen. *Person. Med.*, **10**, 601–611.

Giacomini, M., Miller, F. and O'Brien, B.J. (2003). Economic considerations for health insurance coverage of emerging genetic tests. *Commun. Genet.*, **6**, 61–73.

Grosse, S.D., Wordsworth, S. and Payne, K. (2008). Economic methods for valuing the outcomes of genetic testing: beyond cost-effectiveness analysis. *Genet. Med.*, **10**(9), 648–655.

Hatz, M.H.M., Schremser, K. and Rogowski, W.H. (2014). Is individualized medicine more cost-effective? A systematic review. *PharmacoEconomics*, **32**(5), 443–455.

Husereau, D., Drummond, M., Petrou, S., *et al.* (2013). Consolidated Health Economic Evaluation Reporting Standards (CHEERS) – explanation and elaboration: a report of the ISPOR health economic evaluations publication guidelines good reporting practices task force. *Val. Health*, **16**, 231–250.

ISPOR. (2014a). Pharmacoeconomic Guidelines Around The World. http://www.ispor.org/PEguidelines/index.asp (accessed April 20, 2014).

ISPOR. (2014b). Pharmaceutical HTA and Reimbursement Processes – Greece. http://www.ispor.org/HTARoadMaps/Greece.asp (accessed April 30, 2014).

Jacob, I. and Payne, K. (2012). Evaluating the economic impact of technological advances in diagnostics: the case of high throughput sequencing for hereditary breast cancer. Health Economists' Study Group. Discussion Paper, June 2012.

Jarrett, J. and Mugford, M. (2006). Genetic health technology and economic evaluation: a critical review. *Appl. Health Econ. Health Pol.*, **5**, 27–35.

Leal, J., Wordsworth, S., Legood, R. and Blair, E. (2007). Eliciting expert opinion for economic models: an applied example. *Val. Health*, **10**, 195–203.

Mihaylova, B., Briggs, A., O'Hagan, A. and Thompson, S.G. (2011). Review of statistical methods for analysing healthcare resources and costs. *Health Econ.*, **20**, 897–916.

Morris, S., Devlin, N.J. and Parkin, D. (2007). *Economic Analysis in Health Care*. John Wiley & Sons, Chichester.

Payne, K. and Thompson, A.R. (2013). Economics of pharmacogenomics: rethinking beyond QALYs? *Pharmacogenom. Pers. Med.*, **11**, 187–195.

Payne, K., McAllister, M. and Davies, L.M. (2013). Valuing the economic benefits of complex interventions: when maximising health is not sufficient. *Health Econ.*, **22**, 258–271.

PBS. (2014). Pharmaceutical Benefits Scheme (PBS) – Economics Sub Committee (ESC). http://www.pbs.gov.au/info/industry/listing/participants/economics-subcommittee-esc (accessed April 30, 2014).

Philips, Z., Bojke, L., Sculpher, M., Claxton, K. and Golder, S. (2006). Good practice guidelines for decision-analytic modelling in health technology assessment: a review and consolidation of quality assessment. *Pharmacoeconomics*, **24**, 355–371.

Phillips, K.A. and Van Bebber, S.L. (2004). A systematic review of cost-effectiveness analyses of pharmacogenomic interventions. *Pharmacogenomics*, **5**, 1139–1149.

Phillips, K.A. and Van Bebber, S.L. (2005). Measuring the value of pharmacogenomics. *Nature Rev. Drug Discov.*, **4**, 500–509.

Phillips, K.A., Ann Sakowski, J., Trosman, J., *et al.* (2014). The economic value of personalized medicine tests: what we know and what we need to know. *Genet. Med.*, **16**, 251–257.

Redekop, W.K. and Mladsi, D. (2013). The faces of personalized medicine: a framework for understanding its meaning and scope. *Val. Health*, **16**, S4–9.

Schleidgen, S., Klingler, C., Bertram, T., Rogowski, W.H. and Marckmann, G. (2013). What is personalized medicine: sharpening a vague term based on a systematic literature review. *BMC Med. Ethics*, **14**, 55.

Sculpher, M. (2008). Subgroups and heterogeneity in cost-effectiveness analysis. *PharmacoEconomics*, **26**, 799–806.

Sculpher, M., Drummond, M. and Buxton, M. (1997). The iterative use of economic evaluation as part of the process of health technology assessment. *J. Health Serv. Res. Pol.*, **2**, 26–30.

Sculpher, M.J., Claxton, K., Drummond, M. and McCabe, C. (2006). Whither trial-based economic evaluation for health care decision making? *Health Econ.*, **15**, 677–687.

Smith, R.D. and Sach, T.H. (2009). Contingent valuation: (still) on the road to nowhere? *Health Econ.*, **18**, 863–866.

Smith, R.D. and Sach, T.H. (2010). Contingent valuation: what needs to be done? *Health Econ. Policy Law*, **5**, 91–111.

Sullivan, W. and Payne, K. (2011). The appropriate elicitation of expert opinion in economic models: making expert data fit for purpose. *PharmacoEconomics*, **29**, 455–459.

Thompson, A.R., Newman, W.G., Elliott, R.A., *et al.* (2014). The cost-effectiveness of a pharmacogenetic test: a trial-based evaluation of TPMT testing for azathioprine. *Val. Health*, **17**, 22–33.

Vegter, S., Jansen, E., Postma, M.J. and Boersma, C. (2010). Economic evaluations of pharmacogenetic and genomic screening programs: update of the literature. *Drug Develop. Res.*, **71**(8), 492–501.

Vegter, S., Boersma, C., Rozenbaum, M., *et al.* (2008). Pharmacoeconomic evaluations of pharmacogenetic and genomic screening programmes: a systematic review on content and adherence to guidelines. *PharmacoEconomics*, **26**, 569–587.

Weinstein, M.C. and Skinner, J.A. (2010). Comparative effectiveness and health care spending implications for reform. *New Engl. J. Med.*, **362**, 460–465.

Wong, W.B., Carlson, J.J., Thariani, R. and Veenstra, D.L. (2010). Cost effectiveness of pharmacogenomics: a critical and systematic review. *PharmacoEconomics*, **28**, 1001–1013.

Index